Jets, Wakes and Separated Flows

Toshihiko Shakouchi

Jets, Wakes and Separated Flows

Fundamentals and Some Topics

 Springer

Toshihiko Shakouchi
School of Engineering (Mechanical
Engineering)
Mie University
Tsu, Japan

ISBN 978-981-96-9438-9   ISBN 978-981-96-9439-6 (eBook)
https://doi.org/10.1007/978-981-96-9439-6

This Springer imprint is published by the registered company Springer Nature Singapore Pte Ltd.
The registered company address is: 152 Beach Road, #21-01/04 Gateway East, Singapore 189721,
Singapore

If disposing of this product, please recycle the paper.

Preface

The so-called jet flow, a phenomenon in which a fluid with velocity issues into a space from a small hole or slit, is an interesting and important flow phenomenon in the fields of fluids, thermodynamics, and engineering. It includes the major fundamentals of fluid dynamics, namely boundary layer flow, turbulent flow, high velocity and supersonic flow, high temperature flow, multiphase flow, and so on. In fact, various jet flows can be used, for example, high-velocity and high-temperature jets from jet engines and rocket engines for thrust, and high-velocity water jets from jet ships or jet skis, to hot or cold air jets with low velocity from domestic air conditioners. It is used in a wide variety of ways from daily life to many industrial fields. Also, the flow after passing through a body, the so-called wake flows, are also an interesting and important flow phenomenon in the field of fluids, thermodynamics and engineering. In other words, because the size of the wake is related to the magnitude of the object's flow resistance (form resistance), efforts have been made to reduce it. In addition, under certain conditions, the wake can exhibit an oscillation phenomenon (Karman vortex), and this phenomenon has been studied in detail. In this way, jets, wakes and separated flows are not only of academic interest in the fields of fluid dynamics, thermodynamics and engineering, but are also important in practical applications, i.e. industrially.

In this book, first, the basics of theoretical treatment of jets, wakes and separated flows are examined, and then some interesting topics are discussed. I have also presented details on related some topics in the form of research papers. However, the phenomena of jets, wakes and separated flows range from simple to extremely complex, and the author must point out that what is covered in this book is a small part of them. In addition, this book is written as a book for learning and providing topics for graduate school and university students specializing in fluids, thermodynamics, and engineering, as well as up-and-coming researchers and engineers.

In writing this book, the author received considerable cooperation from many research colleagues and Doctoral and Master's students, as well as Undergraduate students, at the Flow Phenomenology, Fluid- and Thermo-dynamics and Engineering, and Flow Control Laboratories in the Graduate School of Engineering at my university, Mie University, Japan. The author also referred to many research papers and books and would like to express my sincere gratitude to them.

April 2025

Toshihiko Shakouchi, Dr. Eng.
Professor Emeritus
Mie University
Tsu, Japan
shako@mach.mie-u.ac.jp

Competing Interests The author has no competing interests to declare that are relevant to the content of this manuscript.

Contents

Chapter 1
Introduction

The flow that ejects into space from a small slit or nozzle, the so-called jet flow, and the flow behind an object (wake flow) are free and wall shear layer (boundary layer) flows. It is not only important and interesting from the viewpoint of fluid- and thermo-dynamics, but also has practical applications in many fields, such as industry, everyday life around us and nature. Here, the fluid dynamic properties of jets, wakes and separated flows are described, and some their practical applications are shown.

Up until now, fluid mechanics and engineering have focused on individual phenomena, such as inviscid flow, viscous flow, laminar flow/turbulent flow, free boundary layer, wall boundary layer, flow stability, compressible supersonic flow, multiphase flow, etc. Although it feels like we have treated each of these phenomena individually theoretically and experimentally, jet flow, wake flow, and separation flow encompass all of these phenomena, and they are different from conventional fluid mechanics and engineering.

It is thought that it is possible to look at it from a different perspective and summarize it.

Figure 1 shows the relation between jet and wake flows and the above-mentioned individual phenomenon. For example, in order to elucidate the flow and heat transfer characteristics of impinging jets, these phenomena can be studied: two- and three-dimensional jets, free and wall jets, free and wall turbulence, free and wall shear flows (free and wall boundary layers)., oscillating flow, heat transfer, impingement heat transfer, heat transfer enhancement, cooling of various heating surfaces, etc., and are intricately related to various flow and heat transfer phenomena, so it is important to clarify their relationships. It is believed that such a perspective can improve and deepen conventional thinking in fluid, thermodynamics, and engineering.

T. Shakouchi, *Jets, Wakes and Separated Flows,*
https://doi.org/10.1007/978-981-96-9439-6_1

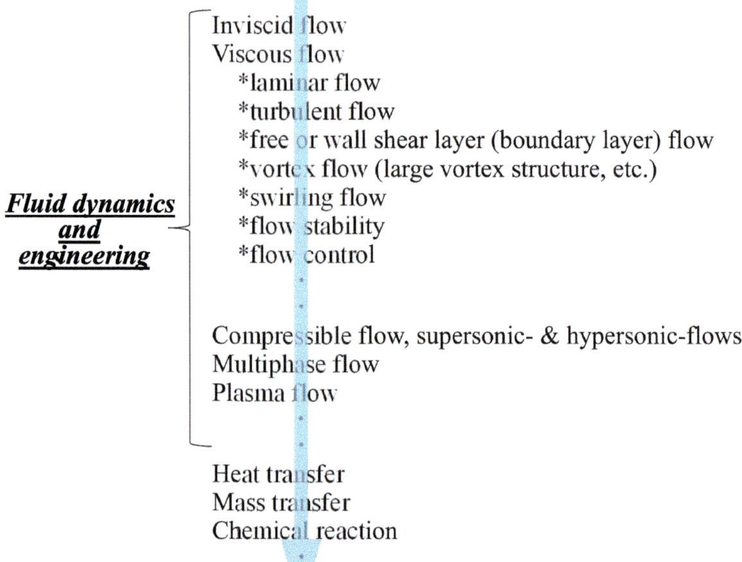

Fig. 1 Relation between jets, wakes and separated flows and individual phenomena of fluid- and thermo-dynamics and engineering

1   Jet Flows

One of the most common types of je flows is the high-temperature, high-pressure, and high-speed jets of combustion gases from rocket engines and jet engines.

Figure 2 shows the launch of the H3 rocket No. 3, Japan. The rocket is equipped with liquid hydrogen and liquid oxygen, and uses hydrogen as fuel to eject high-temperature, high-pressure, and high-speed combustion gas from the rocket nozzle as a supersonic jet to obtain thrust. You can see how high-temperature, high-speed combustion gas (the bright part) erupts from the nozzle and diffuses laterally.

Figures 3 shows one example of a submerged water free jet issued from a circular nozzle of $d_0 = 10.0$ mm into still water visualized by dye of fluorescein sodium aqueous solution. Figure 3a is the jet flow from a long pipe nozzle with $L/d_0 = 50$ and nozzle Reynolds number $Re = 3000$. In this case, the jet takes on the appearance of a laminar jet flow and maintains a nearly constant diameter without spreading up to the flow direction $x/d_0 \doteqdot 8$. On the other hand, the jet from the orifice nozzle with the contraction area ratio $CR = 0.11$ shown in Fig. 3b does not diffuse in the

Fig. 2 Launch of the H3 rocker No. 3, on March 07, 2023, Japan (by *Yomiuri Shimbun, Japan*)

direction perpendicular to the flow for an extremely short distance from the nozzle exit to $x/d_0 \fallingdotseq 1.3$, and then spreads approximately in proportion to the distance in the flow direction. At the outer edge of the jet a large-scale vortex structure that fluctuates irregularly and three-dimensionally is seen. At this time, the jet mixes and diffuses (momentum exchange) with the surrounding fluid, and the flow resistance of the nozzle increases than the Pipe nozzle.

In this way, even with the same nozzle diameter, same Reynolds number, and flow rate, the flow characteristics vary greatly depending on the nozzle shape. In other words, it is possible to control the flow characteristics of the jet stream to some extent by changing the nozzle shape.

***Classification of jet flow phenomena**

Jet flow phenomena are various types, and their flow patterns vary greatly depending on parameters such as the types of flowing fluids (whether the jet and surrounding fluid are gas or liquid, or single-phase or multiphase), the jet velocity (whether the jet is compressible or incompressible), the temperature (whether it is high or low), the dimension (whether it is two-dimensional or three-dimensional), the time (whether it is steady or unsteady), and the shape of the nozzle and flow path.

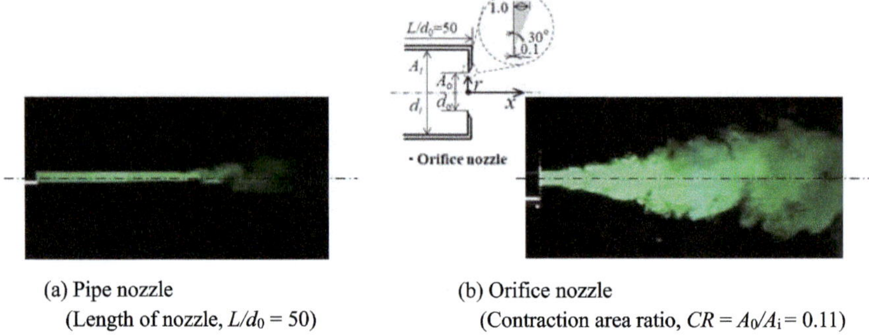

(a) Pipe nozzle   (b) Orifice nozzle
(Length of nozzle, L/d_0 = 50)   (Contraction area ratio, $CR = A_0/A_i$ = 0.11)

Fig. 3 Visualized circular submerged water free jet (Shakouchi et al. 2008) ($Re = 3000$, Nozzle exit diameter: $d_0 = 10.0$ mm)

Table 1 shows these in summary, and it is clear that jet flow phenomena are extremely diverse and wide-ranging.

Table 1 Classification of jet flow phenomena

Fluid	Flow form of jet	
1. Working fluid	2. Flow form I (Dimension)	3. Flow form II (Time)

1.1 Incompressible jet
  (a) Laminar jet
  (b) Turbulent jet
    (velocity)
	2.1. Two-dimensional jet
	2.2. Three-dimensional jet
		3.1. Steady jet
		3.2. Unsteady jet

1.2 Compressible jet
  (a) Supersonic jet
  (b) Hypersonic jet
1.3 High temperature, flame jet
1.4 Low temperature jet
1.5 Gas jet into gas
  (Submerged jet)
1.6 Gas jet into liquid
  (Gas–liquid, two-phase
    flow)
1.7 Liquid jet into gas
  (Gas–solid, two-phase
    flow)
1.8 Liquid jet into liquid
  (Submerged jet)
1.9 Spray jet
1.10 Particle laden gas jet
1.11 Particle laden liquid jet
    etc.
	4. Flow form III (**Jet into infinite or confined space**)
	
	4.1. Free jet
	4.2. Reattached jet
	4.3. Wall jet
	4.4. Impinging jet
	4.5 Opposing jet
	4.6 Accompanying jet etc.
		5. Flow form IV (Nozzle or jet shape)
		
		5.1. Circular jet
		5.2. Non-circular jet
		  (a) Rectangular jet
		  (b) Triangle jet
		  (c) Lobed jet etc.

*** Jet flow phenomena and their applications**

Table 2 shows examples of various jet flow phenomena and their applications, such as their purpose of use and type. As mentioned earlier, it is understood that they are in very wide-regions.

Table 2 Jet flow phenomena and their applications

Phenomena purpose	Jet flows, types and applications
* thrust	supersonic, compressible jet (rocket-, jet-engine) water jet (jet ship, jet ski)
* mixing, diffusion, absorption	various jets (free-, wall-, attached-, impinging, swirling-, curved-jet, etc.) diffusion, bubble, plunging-jet, micro-bubble jet (aeration, carbonation)
* transportation	entrainment jet (jet pump, ejector, aspirator)
* heating, cooling,	blowing jet (air conditioner, heater), impinging jet, wall jet (film-, mist-cooling)
* insulation	air curtain, wall jet
* combustion, flame	combustion-, flame-jet
* spray, atomization	atomization by spray jet (atomization of droplet, injection jet, fuel injection, spray painting)
* ink jet	ink jet printer, 3-D printer (manufacture of parts of resin or metal)
* melting snow	water jet (sprinkler)
* digestion	fire water jet, spray jet (sprinkler)
* washing	high pressure water jet (jet cleaning), high pressure gas jet, gas–solid jet (sand blast, ejector)
* moisturizing	mist jet
* processing	super high pressure water jet (jet cutting), gas–solid, gas–water jet (micro blast, jet peening), globular formation of fine solid particles by high temperature gas jet
* welding, fusing	high temperature gas jet
* pulverization	gas–solid impinging jet (jet pulverization)
* milling	gas-, water-atomization (metal powder)
* spinning	interlace jet
* classification	gas–solid swirling jet, attached jet (Coanda attached jet)
* control	attached-, impinging-, swirling-, oscillating-jet (fluidics) boundary layer control by blowing or suction jet
* display	various jets, bubble jet, smoke jet

2 Wakes and Separated Flows

The flow behind an object, the so-called wake, is an important phenomenon in relation to the flow resistance, heat transfer characteristics, etc. of objects moving in the flow, such as aircraft, rockets, ships, etc.

Table 3 shows the examples of classification of various jet phenomena, such as their purpose of use and type. It is understood that they are in very wide-regions.

Figure 4 shows one example of a flow (wake) behind a circular cylinder set in a uniform oil flow visualized by dye. It can be seen that behind the cylinder, the flow separated from the cylinder wallforms a vortex region and the separation vortices may be shed regularly and alternately from the upper and lower sides of the cylinder, Karman vortex streets.

According to Karman's theoretical calculations in a perfect fluid, a stable Karman vortex street is formed where the ratio of the interval h between two vortex streets

Table 3 Wakes and separated flows and their applications

Phenomena purpose	Wakes and separated flows, type and their applications

*wake of cylinder	Karman vortex;
	    *Karman vortex in nature,
	      Huge Karman vortex from the Hallasan Mountain, Jeju Island, South Korea, observed from a meteorological satellite
	    *Harm: vibration and collapse of structures, bridges, noise, etc.
	    *Use: Karman vortex flowmeter, eolian sound, edge tone

*wake of forward facing step	flow separation, generation of separation bubble. increase of flow resistance, flow control, reduction of flow resistance, heat transfer
*wake of backward facing step	flow separation, generation of reattached bubble, heat transfer
*wake of wing	flow separation, flow control, boundary layer control, blowing or suction of boundary layer, enhancement of wing lift, STOL, VTOL Coanda effect
*wake of wing tip	wing tip vortex, increase of flow resistance, flow control, reduction of resistance, winglet
*wake of vehicle	cars, racing cars, trucks, trains, flow control by wing

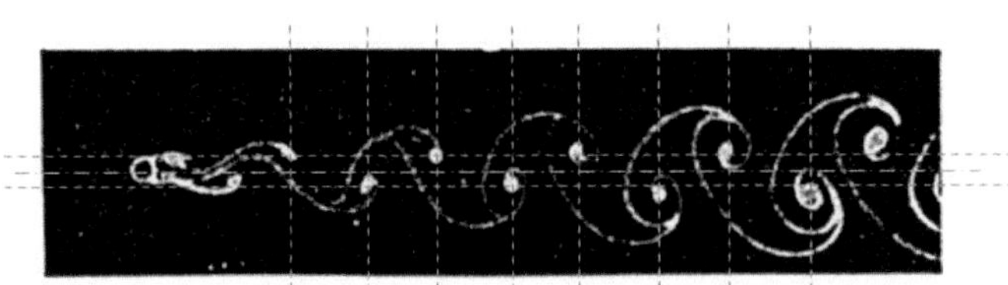

Fig. 4 Visualized wake flow around and behind a single circular cylinder (Re = 71, by Homann 1936)

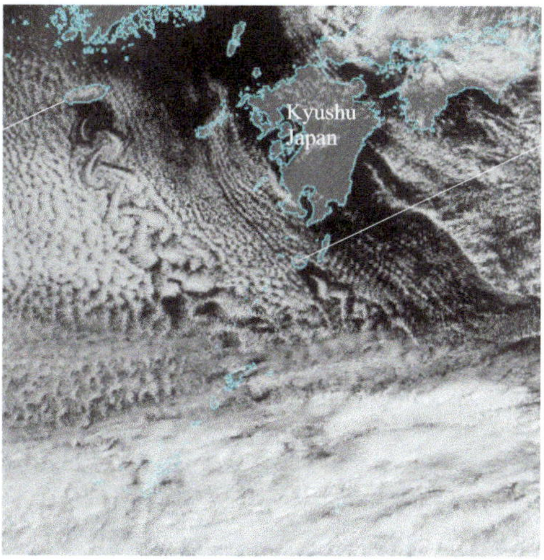

Fig. 5 Karman vortex street in nature. Photo took from the Weather satellite of Japan named Himawari (Sunflower) on February 21st, 2000, In the upper left of the photo, the Karman vortex street can be seen southeast of Jeju Island in South Korea and southeast of Yakushima Island in Japan (Kochi University, Weather Home, Japan, https://weather.is.kochi-.ac.jp/wiki/inwxhome).

to the interval, pitch, a between one vortex street is $h/a = 0.281$. However, in actual flows, the pitch a of the vortices is often not constant.

Karman vortices occur under certain flow conditions, but when Karman vortices occur in the wake of a structure, the structure may be excited and vibrate depending on the shedding frequency, leading to a grave. Note that the vortex shedding frequency f of the Karman vortex is proportional to the velocity u_i flowing into the object, so f is measured from a columnar object installed in a pipe, u_i is calculated, and the volumetric flow rate is determined. This is the Karman vortex flowmeter, which is widely commercially available and used in many industrial fields. Additionally, Karman vortices occur widely in nature depending on the conditions. For example, high-voltage power lines emit a whistling sound when exposed to the winter northwest monsoon, or the Karman vortex visible in the flow of clouds from an earth observation satellite, for example, the Karman vortex from the hillside of Hallasan Mountain on Jeju Island, South Korea, is visualized by the clouds hanging over it due to the constant northwestern seasonal wind in winter (Fig. 5). The wind speed is $u \fallingdotseq 10$ m/s and the size of the reference length is $d \fallingdotseq 30$ km, then the Reynolds number is $Re \fallingdotseq 2 \times 10^{10}$.

The importance of jets, wakes and separated flows has been outlined using a few examples, and the details of each will be described below.

References

Homann F (1936) Einfluß großer Zähigkeit bei Strömung um Zylinder. Forschg Ing-Wes 7:1–10

Schlichting H (1968) Boundary Layer Theory, 6th edn. McGraw-Hill

Shakouchi T (2004) Jet flow engineering (in Japanese). Morikita Pub. Co. Ltd., Japan. ISBN 4-627-67201-2

Shakouchi T, Kito M, Sakamoto T, Tsujimoto K, Ando T (2008) Flow analysis of orifice free jet (Effects of contraction ratio). Trans. of JSME 74(737 B):42–48. https://doi.org/10.1299/kikaib.74.42

Chapter 2
Fluid Mechanics of Jets, Wakes and Separated Flows (Basic Equations and Numerical Analysis)

1 Equation of Motion of Fluid

In order to theoretically analyze the behavior of a fluid flow, such as a jet or wake flow, it is necessary to describe its motion mathematically.

1.1 Navie-Stokes Equation

The equations describing the motion of incompressible, viscous fluids, the so-called Navier–Stokes equations of motion, are given as follows when expressed in Cartesian coordinates:

$$\frac{\mathrm{D}u}{\mathrm{D}t} = F_x - \frac{1}{\rho}\frac{\partial p}{\partial x} + \frac{\mu}{\rho}\left(\frac{\partial^2 u}{\partial x^2} + \frac{\partial^2 u}{\partial y^2} + \frac{\partial^2 u}{\partial z^2}\right) \tag{1}$$

$$\frac{\mathrm{D}v}{\mathrm{D}t} = F_y - \frac{1}{\rho}\frac{\partial p}{\partial y} + \frac{\mu}{\rho}\left(\frac{\partial^2 v}{\partial x^2} + \frac{\partial^2 v}{\partial y^2} + \frac{\partial^2 v}{\partial z^2}\right) \tag{2}$$

$$\frac{\mathrm{D}w}{\mathrm{D}t} = F_z - \frac{1}{\rho}\frac{\partial p}{\partial z} + \frac{\mu}{\rho}\left(\frac{\partial^2 w}{\partial x^2} + \frac{\partial^2 w}{\partial y^2} + \frac{\partial^2 w}{\partial z^2}\right) \tag{3}$$

where,

$$\frac{\mathrm{D}}{\mathrm{D}t} = \frac{\partial}{\partial t} + u\frac{\partial}{\partial x} + v\frac{\partial}{\partial y} + w\frac{\partial}{\partial z} \tag{4}$$

$$\left\{ \begin{array}{l} x,\ y,\ z : \text{Cartesian coordinate system} \\ u,\ v,\ w : \text{velocity components in the } x,\ y,\ z \text{ direction, respectively} \\ F_x, F_y, F_w : \text{external forces in the } x,\ y,\ z \text{ directions, } t;\ \text{time, } p;\ \text{pressure} \\ \mu,\ v,\ \text{and } \rho : \text{viscosity, kinematic viscosity and density of fluid, respectively} \end{array} \right.$$

Also, the equation of continuity for an incompressible fluid flow is

$$\frac{\partial u}{\partial x} + \frac{\partial v}{\partial y} + \frac{\partial w}{\partial z} = 0 \tag{5}$$

To theoretically clarify the state of motion of an incompressible fluid, it is sufficient to simultaneously solve the equation of motion in Eqs. (1)–(3) and the continuity equation in Eq. (5) under certain initial conditions and boundary conditions, but in general, it is extremely difficult to analytically solve nonlinear equations of motion except in certain special cases.

Therefore, various theoretical and numerical analysis solutions have been proposed. Here, we will briefly describe the main ones.

1.2   Turbulent Flow, Reynolds Stress and Reynolds Equation

The behavior of a flow when it exceeds a certain critical Reynolds number and becomes turbulent is expressed, for example, by the Reynolds equation, which expresses the flow as a time-averaged characteristic. Almost all flow phenomena handled in engineering involve turbulent flow. The vortex motion in turbulent flow is unsteady and its size is small compared to the overall flow field, so direct calculation would be extremely costly.

However, in reality, it is rarely necessary to know even the minute eddy movements; in most cases, it is sufficient to predict macroscopic quantities such as flow resistance. Therefore, we will consider a time-averaged flow field, that is Reynolds Averaged Navie-Stokes Equations, Reynolds equation. It can be obtained as follows. Now, if we express the flow velocity and pressure at a certain moment as the sum of their time average value ($^-$) and fluctuation component ('), we get

$$u = \bar{u} + u\prime,\ v = \bar{v} + v\prime,\ w = \bar{w} + w\prime,\ p = \bar{p} + p\prime \tag{6}$$

If we take the time average of these values over a sufficiently long period of time, for example, for u,

$$\bar{u} = \frac{1}{T} \int_{t-T/2}^{t+T/2} u\,dt,\quad \bar{u'} = \frac{1}{T} \int_{t-T/2}^{t+T/2} u'\,dt = 0 \tag{7}$$

By substituting equation Eq. (6) into the equation consisting of Eqs. (1)–(3) and (5) and taking the time average, we obtain the following Reynolds equation.

$$\frac{\partial \bar{u}}{\partial t} + \bar{u}\frac{\partial \bar{u}}{\partial x} + \bar{v}\frac{\partial \bar{u}}{\partial y} + \bar{w}\frac{\partial \bar{u}}{\partial z} = F_x - \frac{1}{\rho}\frac{\partial \bar{p}}{\partial x} + \frac{\mu}{\rho}\nabla^2 \bar{u} - \left(\frac{\partial \overline{u'^2}}{\partial x} + \frac{\partial \overline{u'v'}}{\partial y} + \frac{\partial \overline{u'w'}}{\partial z}\right)$$

(8)

$$\frac{\partial \bar{v}}{\partial t} + \bar{u}\frac{\partial \bar{v}}{\partial x} + \bar{v}\frac{\partial \bar{v}}{\partial y} + \bar{w}\frac{\partial \bar{v}}{\partial z} = F_y - \frac{1}{\rho}\frac{\partial \bar{p}}{\partial y} + \frac{\mu}{\rho}\nabla^2 \bar{v} - \left(\frac{\partial \overline{u'v'}}{\partial x} + \frac{\partial \overline{v'^2}}{\partial y} + \frac{\partial \overline{v'w'}}{\partial z}\right)$$

(9)

$$\frac{\partial \bar{w}}{\partial t} + \bar{u}\frac{\partial \bar{w}}{\partial x} + \bar{v}\frac{\partial \bar{w}}{\partial y} + \bar{w}\frac{\partial \bar{w}}{\partial z} = F_z - \frac{1}{\rho}\frac{\partial \bar{p}}{\partial z} + \frac{\mu}{\rho}\nabla^2 \bar{w} - \left(\frac{\partial \overline{w'u'}}{\partial x} + \frac{\partial \overline{w'v'}}{\partial y} + \frac{\partial \overline{w'^2}}{\partial z}\right)$$

(10)

where, ∇^2 is a vector operator

$$\nabla^2 \equiv \frac{\partial^2}{\partial x^2} + \frac{\partial^2}{\partial y^2} + \frac{\partial^2}{\partial z^2}$$

(11)

In other words, the Reynolds equation expresses the flow velocity and pressure in the Navier–Stokes equation as time-averaged values, and adds the stress increment (Reynolds stress) caused by turbulent velocity fluctuations. By solving this, the flow characteristics of the field, i.e., the average velocity distribution and pressure distribution, can be determined. Note that in order to solve the Reynolds equation, it is necessary to give Reynolds stress, and various evaluation methods have been proposed for this.

2 Energy Equation

Energy equation is given by the following equation.

$$\rho c_p \frac{DT}{Dt} = \frac{\partial}{\partial x}\left(k\frac{\partial T}{\partial x}\right) + \frac{\partial}{\partial y}\left(k\frac{\partial T}{\partial y}\right) + \frac{\partial}{\partial z}\left(k\frac{\partial T}{\partial z}\right) + \dot{q} - \frac{T}{\rho}\left(\frac{\partial \rho}{\partial T}\right)_p \frac{DT}{Dt} + \mu\Phi$$

(12)

where, c_p: specific heat, k: thermal conductivity, $\dot{q}$: heat generation rate per unit volume [J/(m^2 s)],

T: temperature, $\mu\Phi$: viscous dissipation (usually ignored as small).

In the case of incompressible flow,

$$\rho c_p \frac{DT}{Dt} = \frac{\partial}{\partial x}\left(k\frac{\partial T}{\partial x}\right) + \frac{\partial}{\partial y}\left(k\frac{\partial T}{\partial y}\right) + \frac{\partial}{\partial z}\left(k\frac{\partial T}{\partial z}\right) + \dot{q} \tag{13}$$

If k is constant,

$$\rho c_p \frac{DT}{Dt} = k\left(\frac{\partial^2 T}{\partial x^2} + \frac{\partial^2 T}{\partial y^2} + \frac{\partial^2 T}{\partial z^2}\right) + \dot{q} \tag{14}$$

3 Two-Dimensional Free Jet

3.1 Equation of Motion

Now, consider the equation of motion of a two-dimensional (x–y plane), steady free jet. Since the spread of the jet in the y direction (Fig. 1) is small, the changes in velocity and stress in the y direction are much larger than those in the x direction. Taking these things into consideration, and assuming that u and v are average velocity components, the Reynolds equation of Eqs. (8)–(10) and the continuity equation of Eq. (11) are as follows.

$$u\frac{\partial u}{\partial x} + v\frac{\partial u}{\partial y} = -\frac{1}{\rho}\frac{\partial p}{\partial x} + \frac{\mu}{\rho}\frac{\partial^2 u}{\partial y^2} - \frac{\partial \overline{u'v'}}{\partial y} - \frac{\partial \overline{u'^2}}{\partial x} \tag{15}$$

$$-\frac{1}{\rho}\frac{\partial p}{\partial y} - \frac{\partial \overline{v'^2}}{\partial y^2} = 0 \tag{16}$$

$$\frac{\partial u}{\partial x} + \frac{\partial v}{\partial y} = 0 \tag{17}$$

Integrating Eq. (16) regarding pressure from y to ∞, substituting it into Eq. (15) and ignoring minute terms, we obtain the following equation of motion.

$$u\frac{\partial u}{\partial x} + v\frac{\partial u}{\partial y} = -\frac{1}{\rho}\frac{\partial p_\infty}{\partial x} + \frac{\mu}{\rho}\frac{\partial^2 u}{\partial y^2} - \frac{\partial \overline{u'v'}}{\partial y} \tag{18}$$

By the way, the second and third terms on the right side of Eq. (18) can also be written as follows using the shear stress of laminar flow and turbulent flow, τ_l and τ_t.

$$\frac{\mu}{\rho}\frac{\partial^2 u}{\partial y^2} - \frac{\partial \overline{u'v'}}{\partial y} = \frac{1}{\rho}\frac{\partial}{\partial y}\left(\mu\frac{\partial u}{\partial y}\right) + \frac{1}{\rho}\frac{\partial}{\partial y}\left(-\rho\overline{u'v'}\right) = \frac{1}{\rho}\frac{\partial}{\partial y}(\tau_l + \tau_t) \tag{19}$$

Since there is no solid boundary in a free jet, if $\tau_t \gg \tau_l$, ignoring τ_l and ignoring the pressure gradient in the jet axis direction (x-direction), the equation of motion of a two-dimensional free jet is obtained as follows.

$$u\frac{\partial u}{\partial x} + v\frac{\partial u}{\partial y} = \frac{1}{\rho}\frac{\partial \tau_\mathrm{t}}{\partial y}. \tag{20}$$

The equation of continuity is in Eq. (17).

The flow characteristics, velocity field, of a two-dimensional free jet can be obtained by giving τ_t and solving Eqs. (20) and (17) simultaneously.

3.2 Momentum Integral Equation and Conservation of Momentum

Since no external forces act on a two-dimensional free jet that is ejected into an infinitely large, stationary space with a constant pressure, the axial momentum of the jet is conserved. This can be explained as follows.

Multiplying Eq. (20) by ρ and integrating from $y = -\infty$ to $y = \infty$ (Fig. 1) gives us the following equation.

$$\rho \int_{-\infty}^{\infty} u\frac{\partial u}{\partial x}dy + \rho \int_{-\infty}^{\infty} v\frac{\partial u}{\partial y}dy = \int_{-\infty}^{\infty} \frac{\partial \tau}{\partial y}dy \tag{21}$$

The first term on the left side of Eq. (21) can be obtained as follows using Leipnitz's law.

$$\rho \int_{-\infty}^{\infty} u\frac{\partial u}{\partial x}dy = \frac{1}{2} \int_{-\infty}^{\infty} \frac{\partial}{\partial y}(\rho u^2)dy = \frac{1}{2}\frac{d}{dx} \int_{-\infty}^{\infty} \rho u^2 dy \tag{22}$$

Since $u = u_\mathrm{m}$, $v = 0$ at $y = 0$, and $u = 0$, $v = v_\mathrm{e}$ (v_e: entrainment velocity) at $y = \infty$, the second term on the left side is,

$$\rho \int_{-\infty}^{\infty} v\frac{\partial u}{\partial y}dy = \rho[uv]_{-\infty}^{\infty} - \rho \int_{-\infty}^{\infty} u\frac{\partial v}{\partial y}dy = \rho \int_{-\infty}^{\infty} u\frac{\partial u}{\partial x}dy = \frac{1}{2}\frac{d}{dx} \int_{-\infty}^{\infty} \rho u^2 dy \tag{23}$$

Therefore, the left side of Eq. (21) is

$$\frac{d}{dx} \int_{-\infty}^{\infty} \rho u^2 dy \tag{24}$$

The right side of Eq. (21) can be interpreted as follows:

Fig. 1 Shear stress in the jet

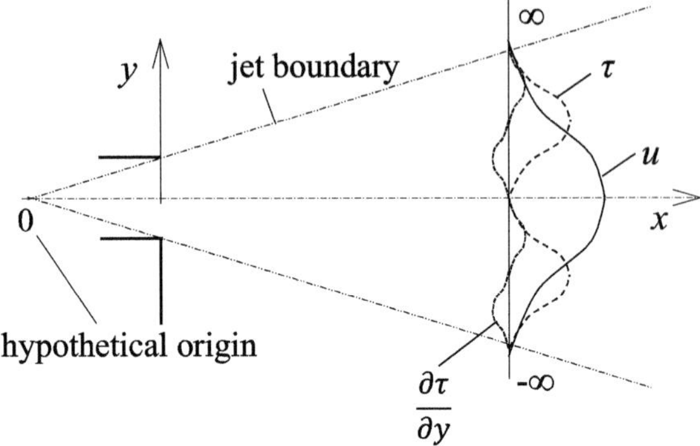

* Shear stress in the jet

The distribution of the shear stress τ with respect to the distribution of the axial velocity u of the free jet is shown by the dashed line in Fig. 1. Therefore, $\partial\tau/\partial y$ appears as the short-dashed line and then,

$$\int_{-\infty}^{\infty} \frac{\partial\tau}{\partial y}dy = 0 \tag{25}$$

Therefore, the shear force τ within the jet is balanced by the viscous force of the fluid, and the external force is zero.

In the end, Eq. (21) is

$$\frac{d}{dx}\int_{0}^{\infty} \rho u^2 dy = 0 \tag{26}$$

This shows that momentum is conserved in the axial direction.

4 Three-Dimensional Circular Free Jet

* Equation of Motion

In this case, since the phenomenon is axisymmetric, it is more convenient to use cylindrical coordinates $(r, \varphi. z)$.

The general equation of motion (Reynolds equation) that describes the behavior of a steady jet is

$$v_r\frac{\partial v_r}{\partial r} + v_z\frac{\partial v_r}{\partial z} - \frac{v_\phi^2}{r} = -\frac{1}{\rho}\frac{\partial p}{\partial r} + \frac{\mu}{\rho}\left(\frac{\partial^2 v_r}{\partial r^2} + \frac{1}{r}\frac{\partial v_r}{\partial r} - \frac{v_r}{r^2} + \frac{\partial^2 v_r}{\partial z^2}\right)$$
$$- \left(\frac{\partial \overline{v_r'^2}}{\partial r} + \frac{\partial}{\partial z}\overline{v_r'v_z'} + \frac{\overline{v_r'^2}}{r} - \frac{\overline{v_\phi'^2}}{r}\right) \tag{27}$$

$$v_r\frac{\partial v_\phi}{\partial r} + v_z\frac{\partial v_\phi}{\partial z} + \frac{v_r v_\phi}{r} = \frac{\mu}{\rho}\left(\frac{\partial^2 v_\phi}{\partial r^2} + \frac{1}{r}\frac{\partial v_\phi}{\partial r} - \frac{v_\phi}{r^2} + \frac{\partial^2 v_\phi}{\partial z^2}\right)$$
$$- \left(\frac{\partial}{\partial r}\overline{v_r'v_\phi'} + \frac{\partial}{\partial z}\overline{v_\phi'v_z'} + 2\frac{\overline{v_r'v_\phi'}}{r}\right) \tag{28}$$

$$v_r\frac{\partial v_z}{\partial r} + v_z\frac{\partial v_z}{\partial z} = -\frac{1}{\rho}\frac{\partial p}{\partial z} + \frac{\mu}{\rho}\left(\frac{\partial^2 v_z}{\partial r^2} + \frac{1}{r}\frac{\partial v_z}{\partial r} + \frac{\partial^2 v_z}{\partial z^2}\right)$$
$$- \left(\frac{\partial}{\partial r}\overline{v_r'v_z'} + \frac{\partial \overline{v_z'^2}}{\partial z} + \frac{\overline{v_r'v_z'}}{r}\right) \tag{29}$$

where, r, ϕ, z; coordinates of radius-, peripheral- and jet axis-directions of circular jet, respectively

v_r, v_ϕ, v_ϕ, v_z: mean velocity components in r, ϕ, z directions
v_r', v_ϕ', v_z': fluctuating velocity, turbulence, components in r, ϕ, z directions
Also, the continuity equation is

$$\frac{\partial}{\partial r}(rv_r) + \frac{\partial}{\partial z}(rv_z) = 0 \tag{30}$$

For an axisymmetric jet with no swirl component, Eq. (27) becomes

$$v_r\frac{\partial v_r}{\partial r} + v_z\frac{\partial v_z}{\partial z} = -\frac{1}{\rho}\frac{\partial p}{\partial r} + \frac{\mu}{\rho}\left(\frac{\partial^2 v_r}{\partial r^2} + \frac{1}{r}\frac{\partial v_r}{\partial r} - \frac{v_z}{r^2} + \frac{\partial^2 v_r}{\partial z^2}\right)$$
$$- \left(\frac{\partial \overline{v_r'^2}}{\partial r} + \frac{\partial}{\partial z}\overline{v_r'v_z'} + \frac{\overline{v_r'^2}}{r}\right) \tag{31}$$

Also, since it is $v_\phi = 0$, Eq. (28) disappears and Eqs. (29), (30) remains unchanged.

Now, since $v_z \gg v_r$ and the flow is turbulent, viscous stress is extremely small compared to turbulent friction stress, and turbulent normal stress has almost equal radius and circumferential components.

Equations (31) and (29) become as follows, respectively.

$$\frac{1}{\rho}\frac{\partial p}{\partial r} = -\frac{\partial \overline{v_r'^2}}{\partial r} \tag{32}$$

$$v_r\frac{\partial v_z}{\partial r} + v_z\frac{\partial v_z}{\partial z} = -\frac{1}{\rho}\frac{\partial p}{\partial z} - \left(\frac{\partial}{\partial r}\overline{v_r'v_z'} + \frac{\partial \overline{v_z'^2}}{\partial z} + \frac{\overline{v_r'v_z'}}{r}\right) \tag{33}$$

The continuity Eq. (30) remains unchanged.

Integrating Eq. (31) to the outer edge of jet, we get

$$p = p_\infty - \rho \overline{v_r'^2} \tag{34}$$

Substituting this into Eq. (33) and simplifying it as in the case of two-dimensional free jet, we get

$$v_r \frac{\partial v_z}{\partial r} + v_z \frac{\partial v_z}{\partial z} = -\frac{1}{\rho} \frac{\partial p}{\partial z} - \frac{1}{r} \frac{\partial}{\partial r} \left(r \overline{v_r' v_z'} \right) \tag{35}$$

Now, since $dp/dz \doteq 0$ we get

$$v_r \frac{\partial v_z}{\partial r} + v_z \frac{\partial v_z}{\partial z} = \frac{1}{\rho r} \frac{\partial (r\tau)}{\partial z} \tag{36}$$

where, $\tau = -\rho \overline{v_r' v_z'}$.

Here, we have discussed the equations of motion, Navier–Stokes equations, etc. that describe the behavior of two-dimensional and three-dimensional circular free jets, but as mentioned earlier, these are nonlinear differential equations that cannot be solved analytically except in special cases. It is very difficult to obtain. Therefore, in practice, approximate theoretical analysis and numerical analysis are performed.

In the next, we will discuss numerical analysis methods and turbulent models.

5 Numerical Analysis of Flow

To clarify flow characteristics, theoretical, experimental and numerical analysis are generally performed. Theoretical analysis has progressed with advances in mathematics, experimental analysis with advances in sensors, measuring instruments and computers for data processing, and numerical analysis with the development of mathematical models of phenomena and the remarkable development of computers.

Here, only the essentials of numerical analysis of flows will be presented.

5.1 Numerical Analysis of Jet and Wake Flows

To investigate the behavior of incompressible jet and wake flows theoretically and analytically, it is sufficient to find analytical solutions to the governing equations of the flow, namely the Navier–Stokes equations of motion and the equation of continuity, but in general these equations are nonlinear and it is very difficult to solve them. In such a case, a solution is obtained numerically.

In the case of compressible fluids, it is necessary to simultaneously solve the equation of state and the energy equation relating to thermodynamic quantities.

Numerical flow analysis methods can be broadly divided into the finite difference method, the finite element method, and the boundary element method.

This article provides an overview of the numerical analysis of incompressible fluid flows using the finite difference method and the boundary element method (discrete vortex method).

5.2 Finite Difference Method

In order to numerically analyze the governing equations of flow, the Navier Stokes equations, it is necessary to discretize the derivatives. One method for doing this is the finite difference method. The finite difference method is a method for expressing the Navier Stokes equations and other equations as simultaneous equations of variables at discretely given points, and then solving these. In doing so, it is necessary to express the derivatives of the variables using the variables of the discretely given points (for example, x_{i-1}, x_i, x_{i+1}, the intervals between each point are Δx_{i-1} and Δx_i, respectively) (Fig. 2).

If $\Delta x_{i-1} = \Delta x_i = \Delta x$, the first and second derivatives of a continuous function $f(x)$ of a variable x at x_i can be approximately expressed as a difference equation (algebraic equation) using the following Taylor expansion,

$$f_{i+1} = f_i + \left(\frac{\partial f}{\partial x}\right)_i \Delta x + \frac{1}{2!}\left(\frac{\partial^2 f}{\partial x^2}\right)_i \Delta x^2 + \frac{1}{3!}\left(\frac{\partial^3 f}{\partial x^3}\right)_i \Delta x^3 + \ldots \tag{37}$$

$$f_{i-1} = f_i - \left(\frac{\partial f}{\partial x}\right)_i \Delta x + \frac{1}{2!}\left(\frac{\partial^2 f}{\partial x^2}\right)_i \Delta x^2 - \frac{1}{3!}\left(\frac{\partial^3 f}{\partial x^3}\right)_i \Delta x^3 + \ldots \tag{38}$$

Fig. 2 Differential lattice

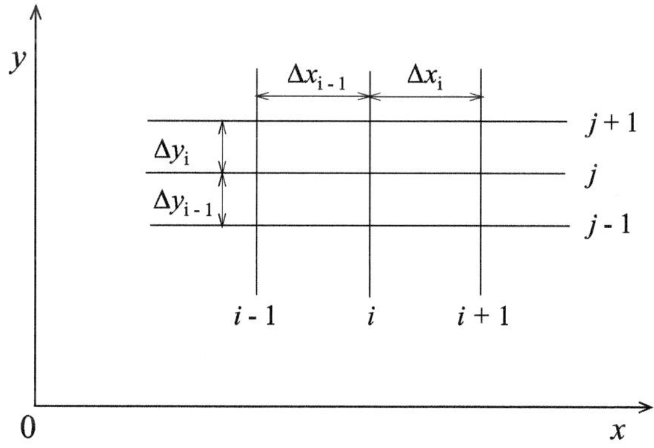

In other words, if we take the difference between equation Q and equation F, we get

$$\left(\frac{\partial f}{\partial x}\right)_i = \frac{f_{i+1} - f_{i-1}}{2\Delta x} - \frac{1}{6}\left(\frac{\partial^3 f}{\partial x^3}\right)_i \Delta x^2 + \cdots = \frac{f_{i+1} - f_{i-1}}{2\Delta x} + O(\Delta x^2) \tag{39}$$

$O(\Delta x^2)$ the equation represents the sum of the terms above Δx^2. This term is ignored, so it is called the truncation error, and Eq. (39) is called a second-order difference equation. Note that since $(\partial f/\partial x)_i$ is approximated by f_{i+1} and f_{i-1}, it is called the central difference. If $(\partial f/\partial x)_i$ is expressed as a forward difference and a backward difference, they are

$$\left(\frac{\partial f}{\partial x}\right)_i = \frac{f_{i+1} - f_i}{\Delta x} + O(\Delta x), \quad \left(\frac{\partial f}{\partial x}\right)_i = \frac{f_i - f_{i-1}}{\Delta x} + O(\Delta x) \tag{40}$$

Next, if we add up Eqs. (37) and (38), $(\partial^2 f/\partial x^2)$ is obtained as follows.

$$\left(\frac{\partial^2 f}{\partial x^2}\right)_i = \frac{f_{i+1} - 2f_i + f_{i-1}}{\Delta x^2} - \frac{1}{12}\left(\frac{\partial^4 f}{\partial x^4}\right)_i \Delta x^2 + \cdots = \frac{f_{i+1} - 2f_i + f_{i-1}}{\Delta x^2} + O(\Delta x^2)$$

$$\tag{41}$$

5.3   Basic Equation

For simplicity, let us assume a two-dimensional, steady, incompressible flow. From Eqs. (1)–(5) [in II, 1.1], the equation describing the fluid motion is as follows:

$$u\frac{\partial u}{\partial x} + v\frac{\partial u}{\partial y} = \frac{\partial p}{\partial x} + \frac{1}{Re}\left(\frac{\partial^2 u}{\partial x^2} + \frac{\partial^2 u}{\partial y^2}\right) \tag{42}$$

$$u\frac{\partial v}{\partial x} + v\frac{\partial v}{\partial y} = -\frac{\partial p}{\partial x} + \frac{1}{Re}\left(\frac{\partial^2 u}{\partial x^2} + \frac{\partial^2 u}{\partial y^2}\right) \tag{43}$$

$$\frac{\partial u}{\partial x} + \frac{\partial v}{\partial y} = 0 \tag{44}$$

where, $Re = u_\infty L/v$, u_∞: reference velocity, L: reference length.

It is difficult to solve the above Equations of motion (42) and (43), and the continuity Eq. (44) in a linked manner. Here, the solution methods using the stream function/vorticity method and the velocity/pressure method will be briefly explained.

5.4 Solution Based on Stream Function and Vorticity Method

Now, if we differentiate Eqs. (43) and (42) with respect to x and y respectively and subtract them, we get the following equations.

$$u\frac{\partial \zeta}{\partial x} + v\frac{\partial \zeta}{\partial y} = \frac{1}{Re}\left(\frac{\partial^2 \zeta}{\partial x^2} + \frac{\partial^2 \zeta}{\partial y^2}\right) = \frac{1}{Re}\nabla^2 \zeta \qquad (45)$$

Here, ζ is the vorticity, and by introducing the stream function ψ ($u = \partial\psi/\partial y$, $v = -\partial\psi/\partial x$), the following Poisson's equation is obtained.

$$\zeta = \frac{\partial v}{\partial x} - \frac{\partial u}{\partial y} = -\left(\frac{\partial^2 \psi}{\partial x^2} + \frac{\partial^2 \psi}{\partial y^2}\right) = -\nabla^2 \psi \qquad (46)$$

In actual numerical calculations, Eqs. (45) and (46) are expressed as differential equations and solved simultaneously to obtain ψ and ζ, and then the velocities u and v are calculated based on these.

5.5 Solution Based on Velocity and Pressure Method

In the stream function-vorticity method, the velocity and pressure are expressed as φ and ζ to reduce the number of dependent variables, but in the velocity–pressure method, velocity and pressure are used as dependent variables. Pressure is calculated by differentiating Eqs. (42) and (43) with respect to x and y, adding them together, and then rearranging them to obtain the following Poisson's equation.

$$\left(\frac{\partial^2 p}{\partial x^2} + \frac{\partial^2 p}{\partial y^2}\right) = -\frac{\partial^2 u}{\partial x^2} - 2\frac{\partial^2 uv}{\partial x \partial y} - \frac{\partial^2 v}{\partial y^2} - \frac{\partial V_g}{\partial t} + \frac{1}{Re}\left(\frac{\partial^2 V_g}{\partial x^2} + \frac{\partial^2 V_g}{\partial y^2}\right) \qquad (47)$$

where, $V_g \equiv \partial u/\partial x + \partial v/\partial y$.

In the final solution, $V_g = 0$, and the continuity equation is satisfied.

5.6 Turbulent Flow

As mentioned before, the vortex motion in turbulent flow is unsteady and its size is small compared to the overall flow field, so direct calculation would be extremely costly. However, in reality, it is rarely necessary to know even the minute eddy movements; in most cases, it is sufficient to predict macroscopic quantities such as flow resistance. Therefore, we will consider a time-averaged flow field, Reynolds averaged Navie-Stokes equations (RANS), Reynolds equation. There are two methods for

RANS: eddy viscosity model and Reynolds stress equation model (RSM). Velocity differences in the flow disrupt the flow and increase flow resistance. The resistance force that acts on the flow due to this turbulence is called "Reynolds stress."

Here, we consider modeling of the effects of turbulence on flow characteristics using "eddy viscosity".

5.6.1 Reynolds Averaging Navier–Stokes Model, RANS Model

A model for calculating the Reynolds stress τ_t in the equation of motion will be described.

(a) 0-equation model: This model expresses the Reynolds stress as $\tau_t = \rho\varepsilon(du/dy)$ or $\tau_t = \rho\overline{u'v'} = \rho l^2 (du/dy)^2$, where l is mixing length.

(b) 1-equation model: This model finds the turbulent kinetic energy k from the transport equation, and determines the characteristic length l of the turbulence using an algebraic formula.

(c) 2-equation model: The k–ε turbulence model is a model in which both k and l are determined from the transport equation, but uses the dissipation rate of turbulent energy ε instead of l.

In general, the k-ε turbulence model assumes isotropic turbulence, so it is not suitable for calculating separated flows or swirling flows that are highly anisotropic. However, since there is no relatively simple method to replace it, the so-called general-purpose k-ε turbulence model is often used industrially. Non-isotropic k-ε turbulence models have also been developed, but they are not sufficient.

5.6.2 Eddy Viscosity Model

Similar to the case of viscous force, a model is used in which Reynolds stress is proportional to the velocity difference, and in this case, the proportionality coefficient of Reynolds stress is called the "eddy viscosity". There are three methods for eddy viscosity model: 0-equation model (mixing length model), 1-equation model and 2-equation model (k-ε model, etc.)

*** k–ε model**

The eddy viscosity is a physical quantity that changes depending on the flow conditions. We will explain the k-ε model, which is the most commonly used method for calculating the eddy viscosity. In the k-ε model, the eddy viscosity is calculated from the turbulent energy k and the turbulent dissipation rate ε as follows.

$$\mu_t = c_\mu \rho \frac{k^2}{\varepsilon} \tag{48}$$

This formula means that the larger the turbulence, the larger the eddy viscosity, and the faster the turbulence disappears, the smaller the eddy viscosity.

On the other hand, the part that is inversely proportional to ε means that short-lived turbulence (large ε) has little effect on the flow. It is said that small vortex tubes have a short lifespan, and when small vortex tubes are clustered together, their contribution to the increase in flow resistance is small even under the same degree of turbulence.

5.6.3 Large Eddy Simulation, LES

Although RANS only looks at the mean flow field, there are many situations in which it is desired to calculate unsteady phenomena, and there are limits to the analytical accuracy of RANS, so expectations for LES are increasing. In a turbulent flow field, eddies of various sizes are generated. In LES, large vortices that can be captured by the mesh are calculated directly, and small vortices that cannot be captured by the mesh are calculated by the model, taking into account their effects. A model that simulates the effects of small eddies is called a "Sub-grid scale (SGS) model" and is modeled using a virtual stress called SGS stress. Since LES directly calculates all large eddies, it relies less on models than RANS, making it a more versatile calculation method.

*** Smagorinsky model**

A typical example of the SGS model is the Smagorinsky model. In RANS, the eddy viscosity coefficient ε is given by the turbulence energy and turbulence dissipation rate obtained by solving an equation separately from the flow velocity, but in the Smagorinsky model, the equation is not solved and it is proportional to the mesh size M_s and the magnitude of the velocity gradient V_g. It is given as follows.

$$\varepsilon = C_s M_s^2 V_g \tag{49}$$

As shown in the above equation, the equation for the eddy viscosity of the Smagorinsky model includes a model constant called C_s. It is known that C_s takes a value between 0.1 and 0.2 when tuned in a typical flow field, but there is no clear guideline for determining the value, causing ambiguity in analysis. The Smagorinsky model is easy to use and is the mainstream SGS model, but it has the drawback that it cannot express the backflow of energy from small eddies to large eddies. The Smagorinsky model is easy to use and is the mainstream SGS model, but it has the drawback that it cannot express the backflow of energy from small eddies to large eddies. There are models to solve this problem, such as the Scale-Similarity Model, but I will omit the details here and leave them to others.

5.6.4 Direct Numerical Simulation, DNS

Direct Numerical Simulation (DNS) is a method of turbulent flow analysis that performs high-precision discretization without modeling turbulence for the basic equations consisting of the Navier–Stokes equations of motion and the continuity equation. DNS requires very fine mesh division to resolve small-scale vortices, which results in an extremely large computational load. When the Reynolds number in a flow increases, fine-scale turbulence occurs and the flow becomes more complex. In that case, the ratio of the maximum scale to the minimum scale of turbulence is estimated to be proportional to $Re^{3/4}$ (Re: Reynolds number). The number of meshes N required for 3D calculations is known to be $N = Re^{9/4}$ as a guideline.

In other words, even for flows with a relatively low Reynolds number, several hundred billion computational meshes are required. With the calculation speed and memory capacity of current computers, it is difficult to perform DNS from the perspective of the scale of calculations. Now, most DNS is used for research purposes, and in most practical calculations, turbulent flow modeling using the RANS or LES methods mentioned above is used.

References

Abramovich GN (1963) The theory of turbulent jets. MIT Press, Cambridge, Mass., Chap. 1. Fundamentals of Turbulent Jets, Microsoft Word—CH1-Fundamentals of Turbulent Jet-EV

Behrouzi P, Feng T, McGuirk JJ (2008) Active flow control of jet mixing using steady and pulsed fluid tabs. In: Proceedings of IMechE 2008, p. 222 (Part I); J Syst Control Eng 381–392. https://doi.org/10.1243/09596518JSCE543

Blevins RD (1984) Applied fluid dynamics handbook. Van Nostrand Reinhold Co. ISBN 0-442-21296-8

Brown GL, Roshko A (1974) On density effects and large structure in turbulent mixing layer. J Fluid Mech 64(4):775–816. https://doi.org/10.1017/S002211207400190X

Chanpagne FH, Wygnanski IJ (1971) An experimental investigation of coaxial turbulent jets. Int J Heat Mass Transfer 14:1445–1464. https://doi.org/10.1016/0017-9310(71)90191-8

Foerthman E (1936) Turbulent jet expansion, NACA TM-789. (original paper in Germany, 1934: Ing.Archiv., 5)

Goertler H (1942) Berechnung von Auhgaben der freien Turbulenz auf Grundeines neunen Naeherungsatzes. ZAMM 22:224–254

Heskestadt G (1965) Hot-wire measurements in a plane turbulent jet. Trans ASME J Appl Mech 32(4):721–734. https://doi.org/10.1115/1.3627309

Ishihara T, Tsuchida A (n.d.) Jet flow. J JSME 66(537):1333–1340

Nakayama Y (1989) Fluid mechanics (in Japanese). Yokendo Ltd. Pub. Japan. ISBN4-8425-8906-x

Ninomiya N, Kasagi N (1993) Turbulence statistics in the self-preserving region of an axisymmetric free jet. Trans JSME 59(561):1532–1538. https://doi.org/10.1299/kikaib.59.1532

Rajaratnam N (1976) Turbulent Jets. Elsevier

Reichardt H (1942) Gasetzmaessigkeiten der freien Turbulenz. VDI-Fotshungsheft, p 414

Rembold B, AdamsNA KL (2002) Direct numerical simulation of transitional rectangular jet. Int J Heat Fluid Flow 23:547–553

Schlichting H (1933) Laminare Strhlenausbreitung. ZAMM 13:260–263

Schlichting H (1968) Boundary layer theory, 6th edn. McGraw-Hill

Schlichiting H, Gersten K (2000) Boundary-layer theory, 8th edn. Springer

Seo IL W, Lyu S, Park YS (2002) Experimental investigation of the velocity measurements and turbulent structures of turbulent jets using PIV. In: Advances in fluid modeling and turbulence measurements, pp 185–192. https://doi.org/10.1142/9789812777591_0022

Shakouchi T (2004) Jet flow engineering (in Japanese). Morikita Pub. Co. Ltd., Japan. ISBN 4-627-67201-2

Shakouchi T, Tsujimoto K, Maeda T (2008) Fluid mechanics (in Japanese). Yokendo Ltd. Pub., Japan. ISBN: 9784842504377

Tollmien W (1926) Berechnung Turbulenter Ausbreitungsvorgaenge. ZAMM 6:468–478

Trentacoste N, Sforza PM (1967) Further experimental results for three-dimensional free jets. AIAA J 5(5):885–891

Chapter 3
Jet Flows

1 Nozzle

1.1 Nozzle Shape

Generally, a jet flow is often ejected into a large space, which is called free jet, or a confined space, called confined jet, from a nozzle with a circular or rectangular exit cross-section as shown in Fig. 1.

The nozzle consists of a straight pipe, a slowly contracting or diverging tube and a rapidly contracting tube from a high-pressure reserve tank. In addition, there are nozzles with special shapes such as quadrant-, orifice-, Laval- (which is convergent-divergent nozzle used to efficiently generates supersonic jet flow), lobed- (which is petal shaped nozzle with a large wet perimeter), chevron-nozzle (which has notches at the nozzle exit and is used in jet engines to reduce noise), rocket engine nozzle, etc. depending on the purpose of use.

1.2 Effects of Nozzle Shape

The flow characteristics of the jet, such as velocity distribution and spread of the jet, vary greatly depending on the nozzle shape. For example, as shown in Fig. 2, we show a visualization of the flow of a submerged water free jet at $Re = 3000$ from a long pipe (circular), orifice with contraction area ratio $CR = 0.11$, and 6 petal- (lobed-) shaped nozzle. The diameter of circular and orifice nozzle $d_e = 10.0$ mm and hydraulic diameter of 6 petal-shaped nozzle is $d_e = 12.5$ mm. In contrast to circular jet, which hardly diffuses in the downstream direction and is not disturbed, in orifice and petal-shaped jets, they diffuse downstream while entraining the surrounding fluid. In addition, the diffusion behavior differs greatly between orifice and petal-shaped jets. In petal-shaped jet, cross-sectional visualization images are also shown.

© The Author(s), under exclusive license to Springer Nature Singapore Pte Ltd. 2025

T. Shakouchi, *Jets, Wakes and Separated Flows,*

https://doi.org/10.1007/978-981-96-9439-6_3

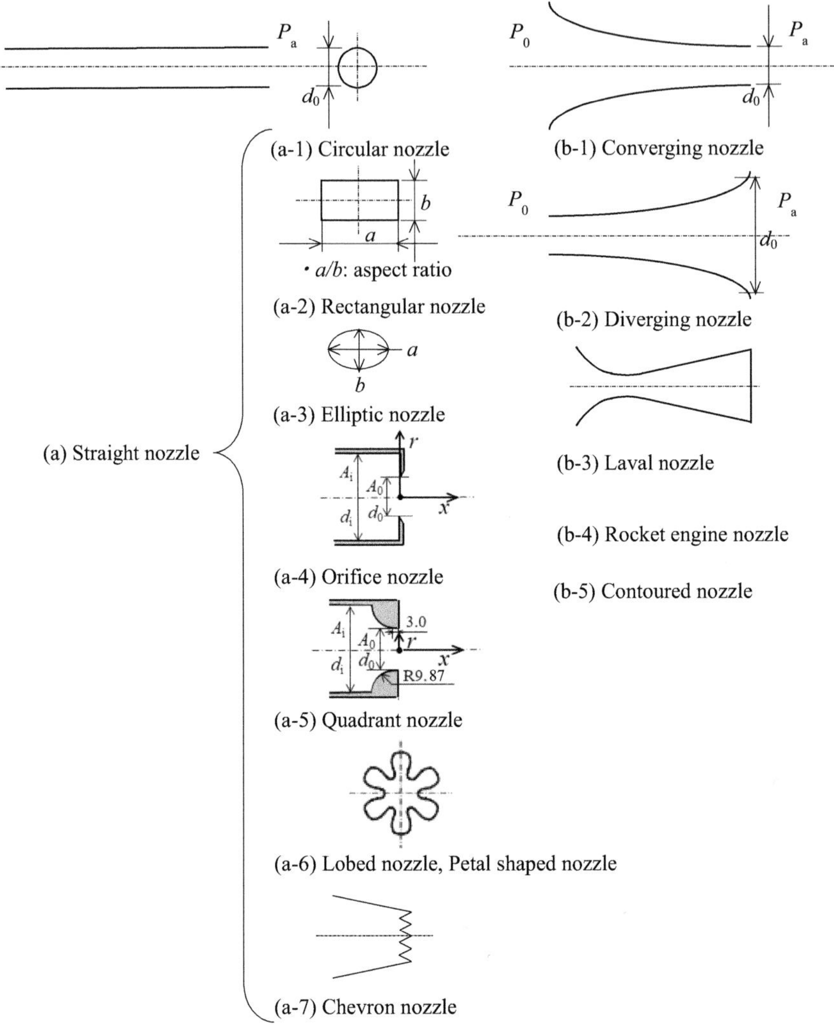

Fig. 1 Basic nozzle

Figure 3 shows a sketch by Fiedler (1998) of how the vortex rings generated in jets from elliptical, rectangular, and triangular nozzles change in the downstream direction. The vortex rings generated at each corner split downstream, then coalesce and deform.

In this way, even with the same Reynold number, the flow characteristics vary greatly depending on the nozzle shape. In other words, the flow characteristics of the jet flow can be controlled or passively controlled to some extent by changing the nozzle shape.

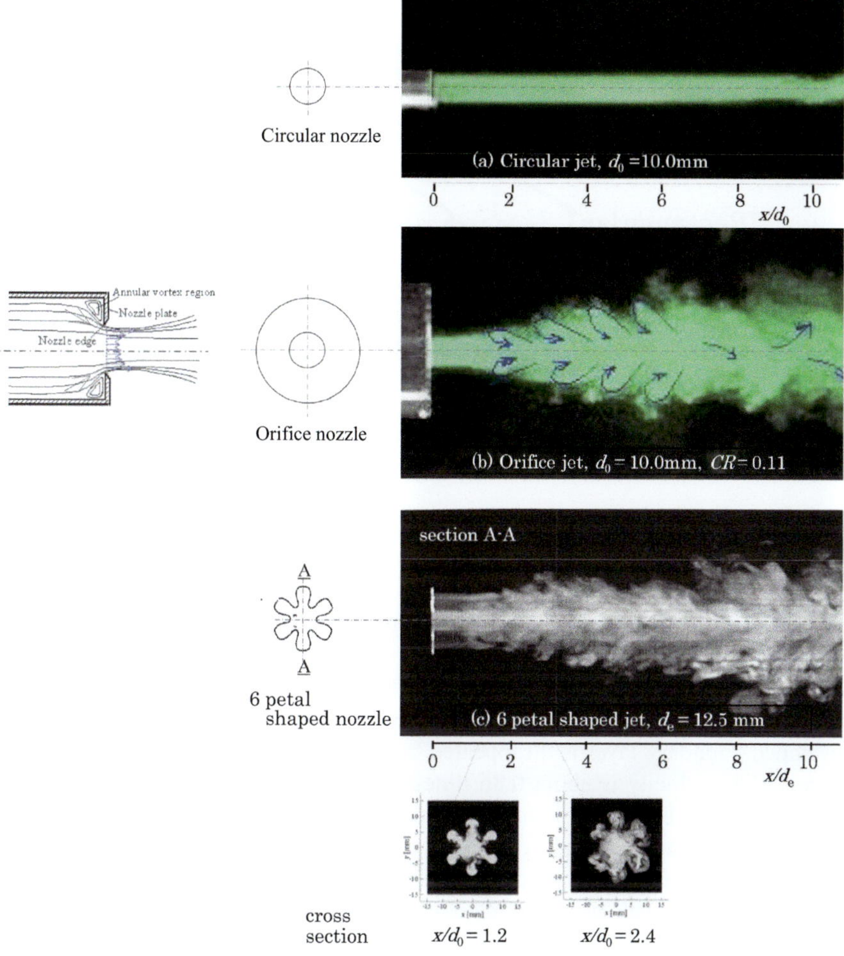

Fig. 2 Submerged water jet flow from a long pipe-, orifice-, and petal-shaped nozzle (Re = 3,000)

Next, we will show the effect of contraction area ratio CR of orifice nozzle on the flow characteristics of the orifice jet, that is, the velocity distribution immediately after the nozzle exit. The effect of nozzle geometry on the mean and fluctuating velocity distributions at just after the nozzle exit of $x/d_0 = 0.2$ is shown in Fig. 4 using a circular, orifice and quadrant nozzles. The results are those measured using a constant temperature hot wire anemometry. The contraction area ratio of orifice nozzle is $CR = A_0/A_i = 0.69$ to 0.11, and $CR = 1.0$ means the circular nozzle of long straight pipe with the length of $L/d_0 = 50$, where d_0 is inner diameter and 10.0 mm. The nozzle Reynolds number is $Re = u_m d_0 /v = 1.5 \times 10^4$, where u_m is mean velocity at the nozzle exit and v is kinematic viscosity. Since the distribution shape was axisymmetric, the left half of the figure shows mean velocity u/u_m in

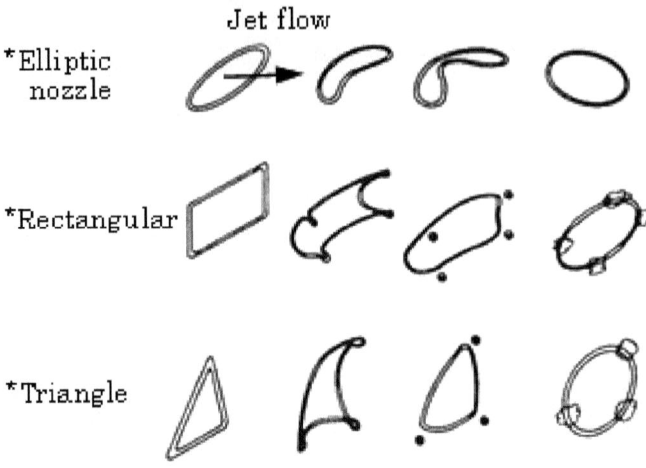

Fig. 3 Vortex ring from oval-, rectangular-. and triangular-nozzles (by Fiedler 1998) *Passive jet flow control by nozzle shape

the flow direction of x, and the right half shows turbulence intensity u'/u_{m}, where u' is the rms value. The velocity profile of the pipe nozzle has the 1/7th power law of a fully developed pipe flow because the flow passes a long enough distance of $L/d_0 = 50$. The distribution shape of the quadrant nozzle is almost flat because a developed boundary layer is disappeared by a sudden smooth contraction, and the jet centerline velocity is about 0.86 times of the pipe nozzle. On the other hand, the velocity distribution shape of the jet from orifice nozzle largely depends on the contraction area ratio CR, and as CR becomes smaller, the jet width becomes smaller and the velocity gradient at the outer edge of the jet becomes larger. At $CR = 0.11$, the maximum flow velocity at the outer edge of the jet is approximately 1.25 times that at the center.

The turbulence intensity u'/u_{m} is large at the outer edge of the jet where the velocity gradient is large, and its maximum value increases as CR becomes smaller, and at $CR = 0.11$, it is about 2.5 times that of a pipe nozzle.

As mentioned above, the flow characteristics of the jet greatly depend on the nozzle shape. In other words, by changing the nozzle shape, the flow characteristics of the jet flow can be controlled to some extent (passive control).

2   Free Jet Flows

A lot of research has been done for a long time on the flow characteristics of two-dimensional free jet, which is the most fundamental jet phenomena, when a jet from a two-dimensional nozzle (slit with a large aspect ratio) is issued into

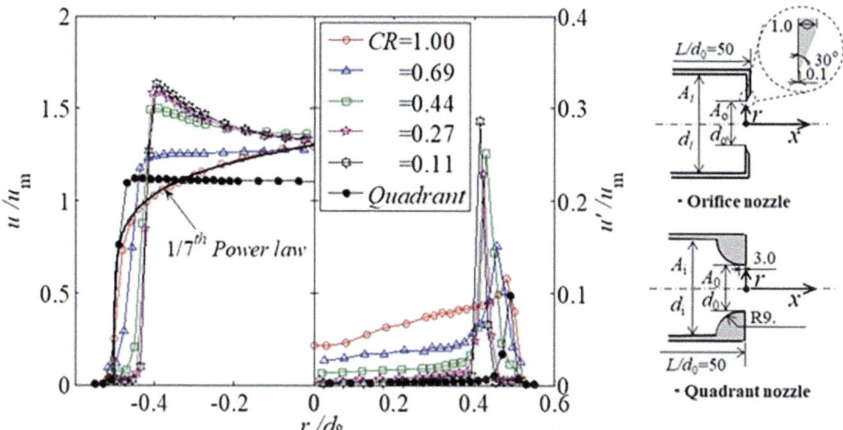

Fig. 4 Mean and fluctuating velocities at nozzle exit of $x/d_0 = 0.2$ ($d_0 = 10.0$ mm, $Re = 1.5 \times 10^4$)

the same static fluid of infinite size, the so-called submerged two-dimensional jet, and the velocity distribution and large-scale vortex structure have been clarified by theoretical, experimental and numerical analyses.

Figure 5a shows the schematic of a two-dimensional, turbulent submerged free jet and coordinate system, and the jet centerline velocity is shown in Fig. 5b. The same applies to a three-dimensional, turbulent submerged free jet issued from a circular nozzle. Figure 5a shows how the jet with a uniform velocity u_0 is ejected into a large still space and spreads while entraining in the surrounding fluid. However, a uniform velocity distribution of the jet at the nozzle exit does not actually exist because a boundary layer develops on the inner wall surface of the nozzle. In addition, there is a velocity difference between the outer edge of the jet and the surrounding fluid, and since the fluid has viscosity, the jet entrains the surrounding fluid, and momentum exchange occurs at that time. As shown in the figure, the shape of velocity distribution is divided into two regions: a region near the nozzle and a region downstream from it. The region near the nozzle is called the initial region, and there is a potential core region in the center where the velocity is not attenuated, and a mixing region where the jet and surrounding fluid mix around it. The mixing region expands in the y direction downstream from the nozzle exit, and the core region shrinks and eventually disappears. The region downstream from this is called the developed region, and if the velocity distribution u of each cross section is expressed as dimensionless by the maximum flow velocity u_{max} and half width $b_{1/2}$ at that cross section, the distribution shape at each cross section becomes a similar distribution shape (self-preservation law). From the experimental results, the length xc of the initial region is $x_c \fallingdotseq 6b$ in a two-dimensional free jet and $x_c \fallingdotseq 5d_0$ in a three-dimensional circular free jet. Furthermore, the jet width b is actually difficult to determine because the boundary of the outer edge of the jet (the $\pm y$ position where the velocity u becomes zero) is a free boundary, unlike the case of a clear solid boundary. Therefore, for example,

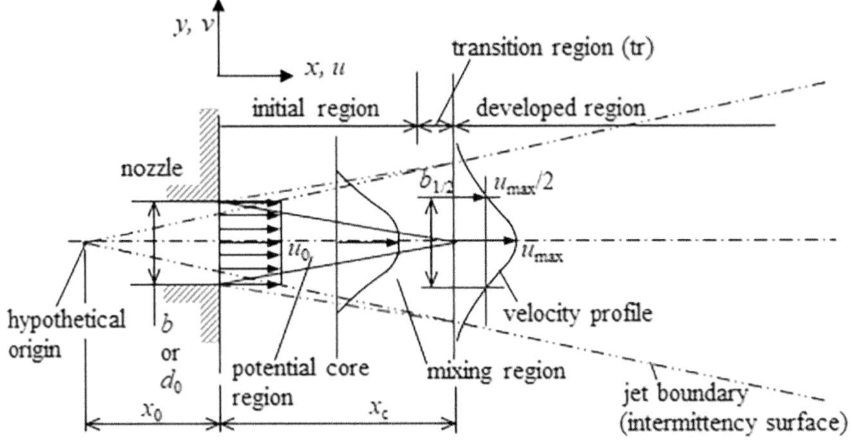

(a) Flow model of two- or three-dimensional circular free jet

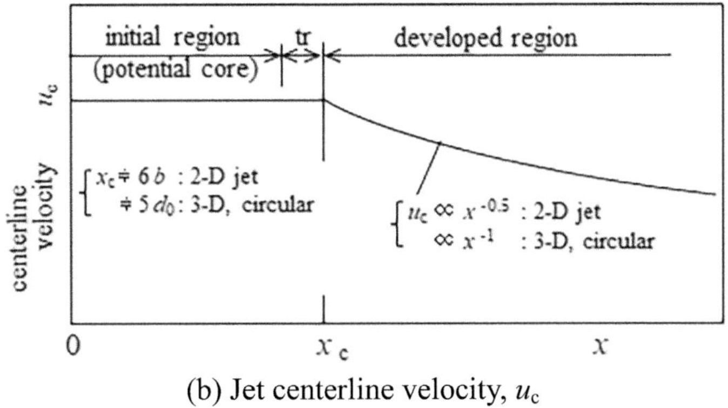

(b) Jet centerline velocity, u_c

Fig. 5 Flow model of two- or three-dimensional circular turbulent submerged free jet

the $\pm y$ position of $u/u_{max} = 0.1$ is defined as the jet outer edge, and the jet width is determined.

By the way, if the outer edges of the jet in the developed region are extended upstream, they intersect at one point, as shown in the figure. This point is called the hypothetical origin, and the distance from there to the nozzle exit is called the hypothetical origin distance x_0.

If we think about it this way, we can treat the jet as if it were ejected from a single point, which may be useful when theoretically dealing with the behavior of the jet.

The velocity distributions of two-dimensional laminar and turbulent jets are shown below.

2.1 Two-Dimensional Free Jet

A lot of research has been done for a long time on the flow characteristics of two-dimensional free jet, which is the most fundamental jet flow phenomena, when a jet from a two-dimensional nozzle (slit with a large aspect ratio) is issued into the same static fluid of infinite size, the so-called submerged two-dimensional jet, and the velocity distribution and large-scale vortex structure have been clarified by theoretical, experimental and numerical analyses. The velocity distributions of two-dimensional laminar and turbulent jets are shown below.

2.1.1 Two-Dimensional Laminar Free Jet

Figure 6 shows a flow model of a two-dimensional laminar jet. In the figure, ψ indicates the streamline. The jet is assumed to originate from one point of $x = -x_0$, that is, the virtual origin. The jet flows down while entraining the surrounding fluid, widening its width, and decreasing its initial velocity.

The virtual origin position x_0 is obtained with the following assumptions.

(1) matching or equalizing the flow rate Q at the nozzle outlet $x = 0$ with the actual one
(2) matching the kinetic energy of the jet at $x = 0$ with the actual one
(3) the nozzle exit is tangential to the appropriate streamline.

A theoretical solution for the velocity distribution of a two-dimensional laminar jet has been obtained by Schlichting (1933).

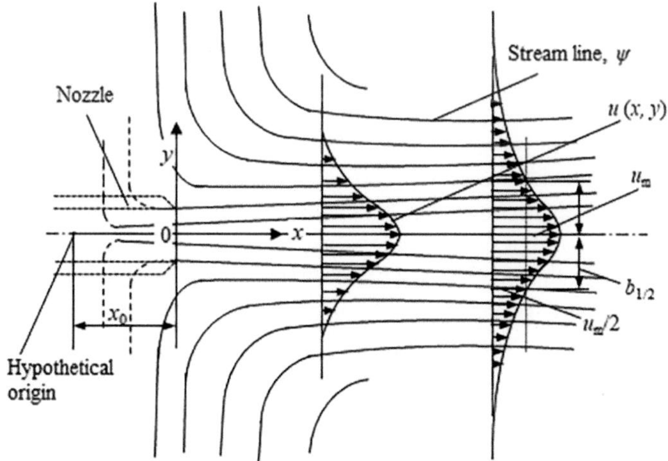

Fig. 6 Flow model of two-dimensional laminar free jet

*** Velocity distribution of two-dimensional laminar free jet**

$$u = 0.4543\left(\frac{K^2}{vx}\right)^{1/3}\left(1 - \tanh^2 \xi\right) \tag{1}$$

$$v = 0.5503\left(\frac{v\,K}{x^2}\right)^{1/3}\left\{2\xi\left(1 - \tanh^2 \xi\right) - \tanh \xi\right\} \tag{2}$$

where, $\xi = 0.2752\,(K/v^2)^{1/3}\,(y/x^{2/3})$

$$K = \frac{J}{\rho} = \int_{-\infty}^{\infty} u^2 dy \tag{3}$$

2.1.2 Two-Dimensional Turbulent Free Jet

(a) **Velocity distribution by Goertler** (1942)

Figure 5 shows a flow model of a two-dimensional turbulent jet. The velocity distribution of a two-dimensional turbulent jet was first described by Tollmien (1926), who expressed the turbulent shear stress as $\tau_1 = \rho\,l^2|\partial u/\partial y|\,(\partial u/\partial y)$, ($l$: mixing length). It is calculated using his Prandtl's mixing length theory.

In addition, Goertler (1942) calculated the velocity distribution of a two-dimensional free jet by changing the eddy viscosity coefficient ε in the shear stress equation $\tau_t = \rho\varepsilon(\partial u/\partial y)$ to the following relationship assumed by Prandtl (Prandtl's second assumption),

$$\varepsilon = k_1 b_{uc} \tag{4}$$

where, k_1: constant, b: jet width ($\sim x$), u_c: jet centerline velocity ($\sim x^{-1/2}$) is used to represent and calculate. The details are shown below.

⌈By the way, Eq. (4) were given by Prandtl as follows. Since the velocity gradient in the y direction at the center of the jet is du/dy 3 0, if the mixing length l is set to $l = $ const., then $\varepsilon = l^2|du/dy| = 0$ and the shape of the velocity distribution at the center of the jet becomes a sharp shape different from the actual one. Prandtl proposed that the concept of mixing length cannot be applied when the size of the fluid parcel is larger than the characteristic length of the flow field (e.g., the width of the jet), and that ε is expressed by Eq. (4). The actual velocity distribution of the jet can be fairly well described by introducing Eq. (4).⌋

In addition, the following assumptions were made for the analysis.

(1) Nozzle width is an infinitesimal size.
(2) Jets ejected from infinitesimally narrow nozzles have similar velocity profiles.
(3) The momentum passing through the cross section perpendicular to the jet axis is constant.

(4) No pressure gradient.

From the equation of motion for two-dimensional flow and Eq. (4) for the two-dimensional flow obtained earlier,

$$u\frac{\partial u}{\partial x} + v\frac{\partial u}{\partial y} = \varepsilon\frac{\partial^2 u}{\partial^2 y} \tag{5}$$

Let u_c and b at $x = x_s$ be u_s and b_s, respectively, then

$$u_c = u_s\left(\frac{x}{x_s}\right)^{-1/2}, b = b_s\left(\frac{x}{x_s}\right) \tag{6}$$

and obtain the following equation.

$$\varepsilon = \varepsilon_s(x/x_s)^{1/2} \tag{7}$$

However, $\varepsilon_s = k_1 b_{sus}$, and also, if we put $\eta = \sigma\,(y/x)$, σ: spreading parameter ($=$ const.) and the stream function ψ

$$\psi = \sigma^{-1}u_s x_s^{1/2}x^{1/2}F(\eta) \tag{8}$$

Then, we obtain the following relationship.

$$u = \partial\psi/\partial y = u_s(x/x_s)^{-(1/2)}F'_s \tag{9}$$

$$v = -\partial\psi/\partial x = \sigma^{-1}u_s x_s^{1/2}x^{-1/2}\left(\eta F' - F/2\right) \tag{10}$$

⌈By the way, the above stream function ψ (Eq. 8) is given as follows. The velocity distribution of the jet becomes a similar form when u and y are non-dimensionalized by the centerline velocity u_c and the half width of the jet (the value of y at the position where the velocity u is $u_c/2$) $b_{1/2}$, respectively. is done.

$$u/u_c = f(\xi) \tag{11}$$

where, $\xi = y/b_{1/2}$.

In fact, the dimensionless velocity distribution in the development region of the jet is a so-called self-preserving flow, which is determined by one reference velocity um and one reference length ξ.

Now, if we put $u_c = C_1 u_{m0}(b_0/x)^{1/2}$, $\eta = \sigma\,(y/x)$ and use $F'(\eta)$ instead of $f(\xi)$,

$$u = C_1 u_0(b_0/x)^{1/2}F'(\eta) \tag{12}$$

where, $\eta = \sigma C_2\xi$, u_{m0}: maximum flow velocity at the nozzle exit.

Since $u = \partial\psi/\partial y$, the stream function ψ is given by the following equation.

$$\psi = \int_0^y u\,dy = \int_0^y C_1 u_{m0}\left(\frac{b_0}{x}\right)^{\frac{1}{2}} F'(\eta)\frac{x}{\sigma}d\eta = C_1 u_{m0}\left(\frac{b_0}{x}\right)^{1/2}\frac{x}{\sigma}F(\eta) \qquad (13)$$

Now,

$$\frac{u_c}{u_{m0}} = C_1\left(\frac{x}{b_0}\right)^{-1/2}, \quad \frac{u_c}{u_s} = \left(\frac{x}{x_s}\right)^{-1/2} \qquad (14)$$

then the steam function ψ is obtained by Eq. (8).

Also, now

$$\frac{u_c}{u_{m0}} = C_1\left(\frac{x}{b_0}\right)^{-1/2}, \quad \frac{u_c}{u_s} = \left(\frac{x}{x_s}\right)^{-1/2} \qquad (15)$$

Therefore, the stream function ψ is expressed by Eq. (13). $\rfloor$
From Eqs. (9), (10) and (8),

$$\frac{F'^2}{2} + \frac{FF''}{2} + \left(\frac{\varepsilon_s}{u_s x_s}\right)\sigma^2 F''' = 0 \qquad (16)$$

where, boundary conditions are $F = 0$, $F' = 1$ at $\eta = 0$, $F' = 0$ at $\eta = \infty$.
Also, ε_s is can be expressed as

$$\sigma = \frac{1}{2}\left(\frac{u_s x_s}{\varepsilon_s}\right)^{1/2} \qquad (17)$$

Substituting this into Eq. (15) and integrating twice, we get the following equation.

$$F^2 + F' = 1 \qquad (18)$$

The solution to the above equation is $F(\eta) = \tanh\eta = \frac{1-e^{-2\eta}}{1+e^{-2\eta}}$, so

$$u = u_s\left(\frac{x}{x_s}\right)^{-1/2}\left(1 - \tanh^2\eta\right) \qquad (19)$$

The momentum per unit depth of the jet is

$$J = \rho\int_{-\infty}^{\infty} u^2 dy \qquad (20)$$

Therefore,

$$J = \frac{4u_s^2 x_s}{3\sigma} \qquad (21)$$

Now, if we put $J/\rho \equiv K$, the following velocity distribution equation of a two-dimensional turbulent free jet can be obtained.

$$u = \left(\frac{3K\sigma}{4x}\right)^{1/2}\left(1 - \tanh^2 \eta\right) \tag{22}$$

$$v = \left(\frac{3K}{16\,\sigma\,x}\right)^{1/2}\left\{2\eta\left(1 - \tanh^2 \eta\right) - \tanh \eta\right\} \tag{23}$$

The spreading parameter σ represents the extent of jet spread, and in the case of a two-dimensional turbulent free jet, it was experimentally determined to be $\sigma = 7.67$ by Reichardt (1942).

Figure 7 shows an example of experimental results of the velocity distribution u of a two-dimensional turbulent free jet (Foerthmann 1934). A jet with a nearly rectangular velocity profile ($u_0 = 35\,\text{m/s}$) at the nozzle exit of $x = 0$ has the same maximum velocity u_m at $x = 10$ cm, but the maximum velocity u_m attenuates as it goes downstream, and the velocity increases in the y direction.

Figure 8 shows the relation between the velocity u and the coordinate y, which are made dimensionless by u_m and the jet half-width $b_{1/2}$, respectively. Curve ② in the figure is the result of theoretical calculation using Eq. (22). It can be seen that the velocity distribution u in each x-section of the developed region is similar and represented by a single curve, and the calculated results represent the experimental results well except for the outer edge of the jet.

Figure 9 shows the experimental results of the velocity distribution v in the y direction and the theoretical calculation results by Eq. (23), rendered dimensionless. The profile is symmetrical about the $y/b_{1/2}$ axis. The velocity profile of v at each

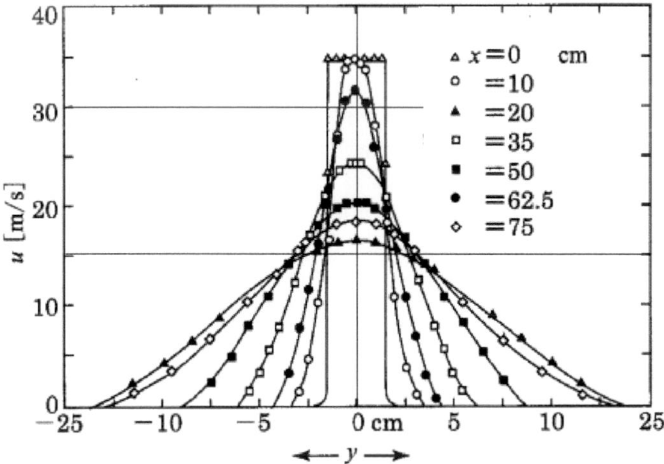

Fig. 7 Velocity distribution of two-dimensional turbulent free jet (Foerthman 1942)

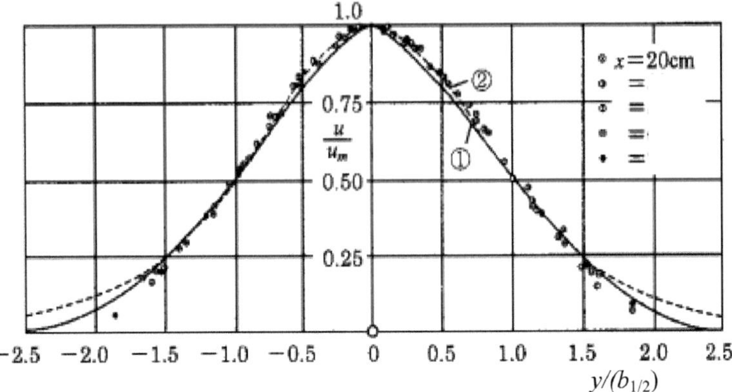

Fig. 8 Non-dimensional velocity distribution of two-dimensional turbulent free jet (Measurements: by Foerthman 1942, curve ①: Theory by Tollmien, curve ②: Eq. 22)

Fig. 9 $v/u_\mathrm{m} - y/b_{1/2}$

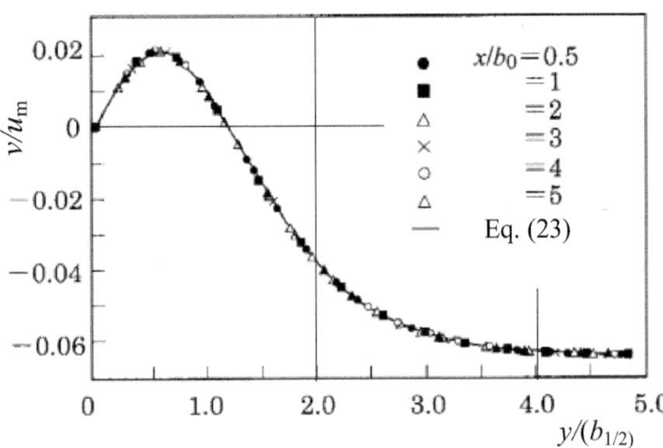

x-section is expressed by a single curved line, when $y/b_{1/2} > 1.5$, v becomes a negative value and the jet flows down while entraining in the surrounding fluid.

Figure 10 shows how the centerline (maximum) flow velocity u_c of a two-dimensional turbulent free jet changes in the downstream direction. As mentioned above, the jet spreads downstream while drawing in the surrounding fluid. The entrainment velocity V_e is defined by the following equation.

$$V_\mathrm{e} = \frac{d}{dx} \int_{-\infty}^{\infty} u\,dy = \frac{dQ_e}{dx} \tag{24}$$

where, Q_e is the volumetric entrainment flow rate.

(b) **Velocity distribution by Tollmien** (1926)

In order to theoretically obtain the velocity distribution of the two-dimensional free jet, Tollmien solved the equation of motion using Prandtl's mixing length equation.

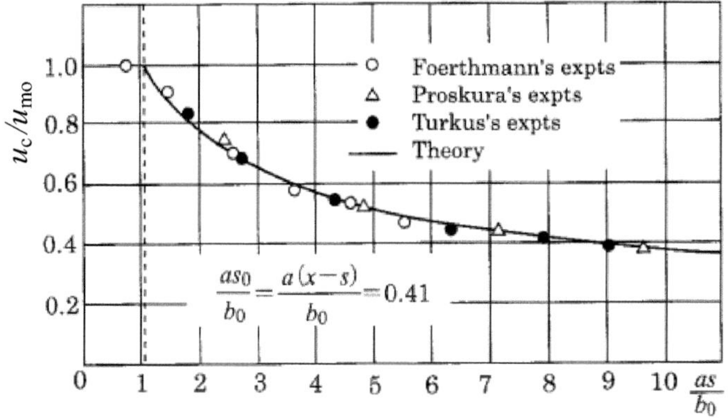

Fig. 10 Jet centerline velocity of two-dimensional turbulent free jet (Abramovich 1963)

$$\tau = -\varrho l^2 \left(\frac{\partial u}{\partial y}\right)^2 \tag{25}$$

where, l: mixing length

In addition, if we consider the dimensions of an arbitrary cross section, we obtain the following relationship. Where, β is constant.

$$l \propto b \quad or \quad l = \beta b \quad or \quad l = \beta C_2 x$$

Under these assumptions, the equation of motion was solved to obtain the velocity distribution. The results are shown in Fig. 8.

(c) **Turbulent properties**

Turbulence intensity, $\sqrt{\overline{u'^2}}/u_m$, in x the direction at the jet centerline is shown in Fig. 11 by Heskestadt measured with a hot-wire anemometry. $\sqrt{\overline{u'^2}}/u_m$ increases rapidly up to $x/2b_0 < 25$ and then increases gradually as $\propto (2b_0)^{0.0018}$.

Figure 12a–c shot the turbulence intensities, $\overline{u'^2}/u_m^2$, $\overline{v'^2}/u_m^2$, and $\overline{w'^2}/u_m^2$ in the x, y, and z directions, at $x/2b_0 = 101$, the downstream region where the flow is fully

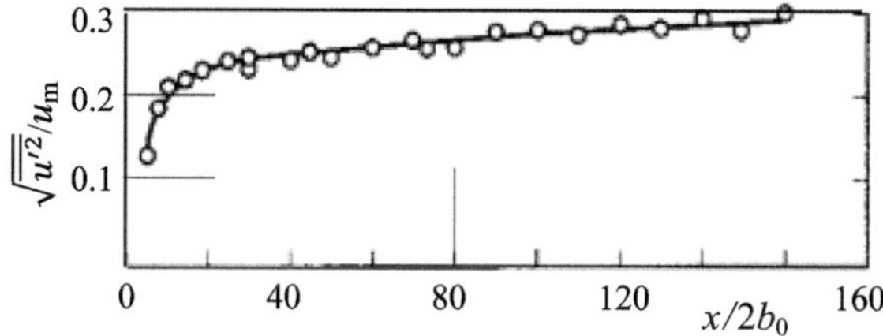

Fig. 11 Turbulent intensity, $\sqrt{\overline{u'^2}}/u_m$, in the x direction at the jet centerline

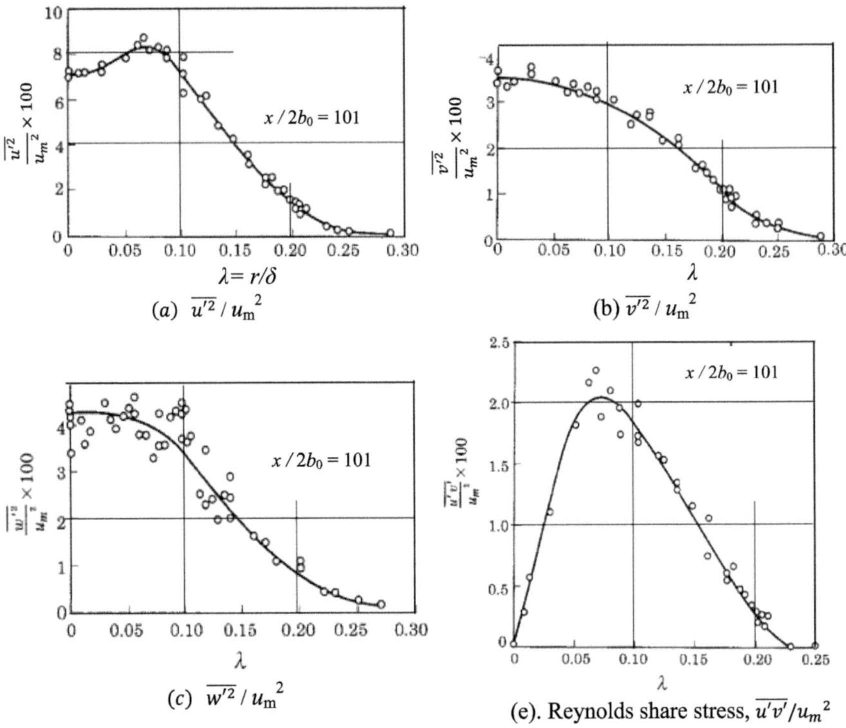

Fig. 12   Turbulent intensity in the $x-\lambda$ plane at $x/2b_0 = 101$

developed, are shown. $\overline{u'^2}/u_\mathrm{m}^2$ is a large value in the shear layer with a large velocity gradient, and $\overline{v'^2}/u_\mathrm{m}^2$ and $\overline{w'^2}/u_\mathrm{m}^2$ have almost the same distribution profile.

Figure 12e indicates the Reynolds shear stress, $\overline{u'v'}/u_m^2$. The Reynolds shear stress takes zero at jet center and the maximum in the shear layer.

2.2   Three-Dimensional Axisymmetric Circular Free Jet

The velocity distribution of a three-dimensional axisymmetric circular free jet has been also determined by the following theoretical formula, as in the case of the two-dimensional free jet described above.

2.2.1   Three-Dimensional Laminar Free Jet

The theoretical analysis for a circular jet was given by Schlichting and the equation of velocity distribution has been shown as follows.

$$u = \frac{3}{8\pi}\left(\frac{K}{\nu x}\right)\frac{1}{\left(1 + \xi^2/4\right)^2} \tag{26}$$

$$v = \frac{1}{4}\sqrt{\frac{3}{\pi}}\frac{\sqrt{K}}{x}\frac{\xi - \xi^2/4}{\left(1 + \xi^2/4\right)^2} \tag{27}$$

where,

$$\xi = \frac{1}{4}\sqrt{\frac{3}{\pi}}\frac{\sqrt{K}}{\nu}\frac{y}{x},\ K = \frac{J}{\rho} = 2\pi\int_0^\infty u^2 y\,dy \tag{28}$$

2.2.2 Three-Dimensional Turbulent Free Jet

(a) Velocity distribution by Goertler

The flow model of three-dimensional circular jet is shown in Fig. 5 in Sect. 2, in this chapter. Goertler analyzed the three-dimensional, axisymmetric, circular, turbulent free jet, and obtained the velocity distribution using Prandtl's eddy viscosity model.

The equation of motion and stream function ψ in the axial direction (x-direction) of the axisymmetric flow are

$$\frac{\partial u}{\partial t} + u\frac{\partial u}{\partial x} + v\frac{\partial u}{\partial r} = -\frac{1}{\rho}\frac{\partial p}{\partial x} + \frac{\partial}{\partial r}\left\{(\varepsilon + \nu)\frac{\partial u}{\partial r}\right\} + \frac{(\varepsilon + \nu)}{r}\frac{\partial u}{\partial r} \tag{29}$$

where, ε: eddy viscosity

$$u = \frac{1}{r}\frac{\partial \psi}{\partial r} \tag{30}$$

$$v = -\frac{1}{r}\frac{\partial \psi}{\partial x} \tag{31}$$

We are considering a steady jet with no pressure gradient in the flow direction, so the time and pressure terms are zero. Also, $\varepsilon \gg \nu$.

Now, let $\eta = \frac{r}{x^\alpha}$ and $f(\eta)\frac{\psi}{x^\beta}$ and then $\partial\eta/\partial r = x^{-\alpha}$, $\partial\eta/\partial x = -\alpha x^{-\alpha-1}r = -\alpha x^{-1}\eta$.

Calculating each term of Eq. (29), we get

$$u = \frac{1}{r}\frac{\partial \psi}{\partial r} = x^{\beta-2\alpha}\eta^{-1}f' \tag{32}$$

$$v = -\frac{1}{r}\frac{\partial \psi}{\partial x} = -\beta x^{\beta-\alpha-1}\eta^{-1}f + \alpha x^{\beta-\alpha-1}f' \tag{33}$$

$$\frac{\partial u}{\partial x} = (\beta - \alpha)x^{\beta-2\alpha-1}\eta^{-1}f' - \alpha x^{\beta-2\alpha-1}f'' \tag{34}$$

$$\frac{\partial u}{r\partial r} = -x^{\beta-4\alpha}\eta^{-3}f' + x^{\beta-4\alpha}\eta^{-2}f'' \tag{35}$$

$$\varepsilon = ku_{1/2}(x)b_{1/2}(x) = \alpha x^{\beta-\alpha} \tag{36}$$

where, $u_{1/2}$: half the maximum velocity at $\eta_{1/2}$, $b_{1/2}$: half width of jet ($=$ jet width at $\eta_{1/2}$)

$$\frac{\partial}{\partial r}\left(\varepsilon\frac{\partial u}{\partial r}\right) = \alpha x^{2\beta-5\alpha}\left(2\eta^{-3}f' - 2\eta^{-2}f'' + \eta^{-1}f'''\right) \tag{37}$$

Now, the momentum of the jet is

$$M = \varrho \int\limits_0^\infty \int\limits_0^{2\pi} u^2 r\,d\theta\,dr = 2\pi\rho \int\limits_0^\infty u^2 r\,dr = 2\pi\rho \int\limits_0^\infty x^{2\beta-2\alpha}\eta^{-2}f'^2\,dr \tag{38}$$

Since the momentum is conserved in the x direction,

$$\alpha = \beta \tag{39}$$

As a result, Eq. (36) becomes a constant.
Substituting Eqs. (32)–(35) into (29), taking Eq. (39) into account, we get

$$\alpha x^{\alpha-1}\eta^{-2}ff' - \alpha x^{\alpha-1}\eta^{-1}ff'' - \alpha x^{\alpha-1}\eta^{-1}f'^2 = \alpha\left(\eta^{-2}f' - \eta^{-1}f'' + f'''\right) \tag{40}$$

Since we are looking for a similar solution, this equation is not a function of x. Therefore,

$$\alpha = 1 \tag{41}$$

Then, Eq. (40) is changed to

$$ff' = \alpha\left(f' - \eta f''\right) + C \tag{42}$$

From Eq. (32), we get

$$f'(\eta) = x\eta u \tag{43}$$

Then, the boundary condition is

$$f(0) = 0, f' = 0 \tag{44}$$

Then, $C = 0$, and if we set $f(\eta) = aF(\eta)$ again, we get the following equation.

$$FF' = F' - \eta F''$$
(45)

The boundary condition is

$$
\begin{cases}
u = u_{\mathrm{m}} \text{ at } r = 0, & \text{that is, } F'(0) = 0 \text{ at } \eta = 0 \\
v = 0 \text{ at } r = 0, & \text{that is, } F(0) = 0 \text{ at } \eta = 0 \\
r = \infty, u = 0, & \text{that is, } F''(\infty) = 0 \text{ at } \eta = \infty
\end{cases}
$$

Then, the solution of Eq. (45) is

$$F(\eta) = \frac{0.5\eta^2}{1 + 0.125\eta^2}$$
(46)

The velocity u is expressed as follows from Eq. (31).

$$u = x^{-1}\frac{f'(\eta)}{\eta}$$
(47)

Next, the velocity distribution is expressed by the maximum velocity u_{m} at the center ($r = 0$) of the jet and the radial position $r_{1/2}$ where u_{m} is half. u_{m} is expressed as follows from Eq. (47).

$$u_m = x^{-1} \lim_{\eta \to 0} \frac{aF'(\eta)}{\eta} = x^{-1} \lim_{\eta \to 0} aF''(\eta) = ax^{-1}$$
(48)

where,

$$F'(\eta) = \frac{\eta}{\left(1 + 0.125\eta^2\right)^2}$$
(49)

$$F''(\eta) = \frac{1 - 0.375\eta^2}{\left(1 + 0.125\eta^2\right)^3}$$
(50)

Similarly, the radial position $\eta_{1/2} = r_{1/2}/x$ where $u = u_{\mathrm{m}}/2$ is

$$\frac{u_m}{2} \equiv u_{1/2} = x^{-1}\frac{aF'(\eta_{1/2})}{\eta_{1/2}} = u_m\frac{F'(\eta_{1/2})}{\eta_{1/2}} = u_m\frac{1}{\left(1 + 0.125\eta_{1/2}^2\right)^2}$$
(51)

Therefore, the velocity distribution of the circular jet is expressed by the following equation from Eq. (47):

$$u = x^{-1}\frac{aF'(\eta)}{\eta} = \frac{u_m}{\left(1 + 0.125\eta^2\right)^2} = \frac{u_m}{\left[1 + \left(\sqrt{2} - 1\right)\frac{r^2}{r_{1/2}^2}\right]^2} \tag{52}$$

And, also,

$$v = \frac{u_m\left(\eta - 0.125\eta^2\right)}{2\sigma\left(1 + 0.125\eta^2\right)^2} \tag{53}$$

where, $\sigma = 1.81\left(x/b_{1/2}\right)$.

Figure 13 shows an example of the experimental results for the velocity distribution u of a three-dimensional axisymmetric circular free jet. The profile is axially symmetrical about the y-axis. As in the case of the two-dimensional jet shown in Fig. 6, the jet spreads in the y direction while decreasing its maximum velocity as it goes downstream. The results were measured using a pitot tube and are old, but they are basic results.

Figure 14 shows the results in Fig. 13, with the velocity u and coordinate y made dimensionless by the maximum velocity u_m and half-width $b_{1/2}$ of each cross section. The velocity profile of each x-section is similar and can be well represented by a single curve. Also, it agrees well with the theoretical calculation result by Tollmien (1926, solid line in the figure).

Figure 15 shows the Reichardt's experiments and Tollmien and Goertler-type solutions for the tree-dimensional circular free jet. The Goertler solution does not match the experimental results at the outer edge of the jet.

Figure 16 shows the results of other researchers. In the figure, the Gaussian profile is the result of the following formula (Ninomiya and Kasagi 1993).

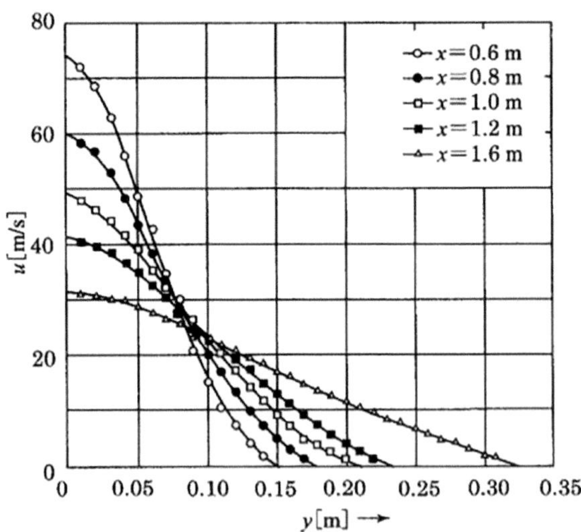

Fig. 13 Velocity distribution u of three-dimensional axisymmetric circular free jet, u–y (Truepel 1915)

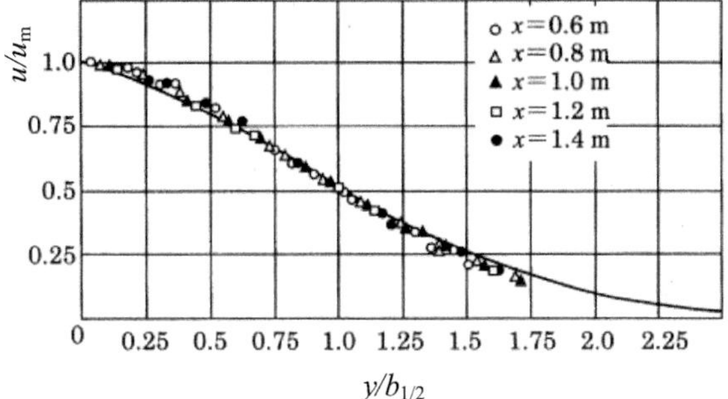

Fig. 14 Nondimensional velocity distribution of three-dimensional axisymmetric circular free jet, $u/u_\mathrm{m}-y/b_{1/2}$ (Exps.: Truepel 1915, Theory: Tollmien 1926)

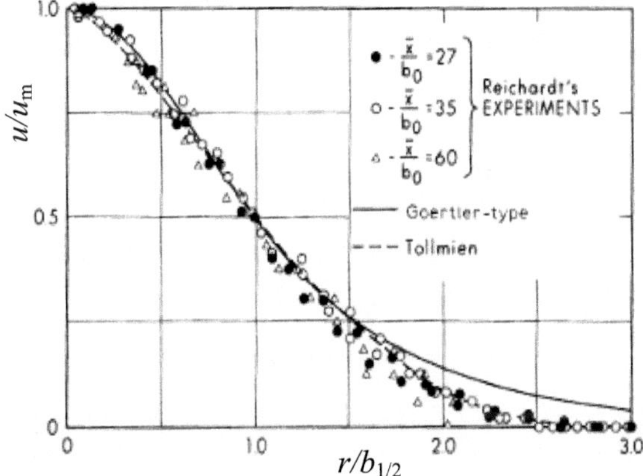

Fig. 15 Reichardt's experiments and Tollmien and Goertler-type solutions (Abramovich 1963. The Theory of Turbulent Jets, MIT Press.)

$$\frac{u}{u_m} = \left[1 + 0.225\left(\frac{r}{b_{1/2}}\right)^2 + 0.164\left(\frac{r}{b_{1/2}}\right)^4\right]\exp\left[-1.022\left(\frac{r}{b_{1/2}}\right)^2\right] \quad (54)$$

The Gaussian profile, Eq. (54), can be expressed well the experiments.

Figure 17 shows the experimental results of jet centerline velocity u_c by Truepel (1915) and the theoretical calculation result in the developed region of $u_\mathrm{c} \propto x^{-1}$ by Tollmien (1926). The core region is in the region of $x/d_0 < 5.0$ and u_c decreases as $u_\mathrm{c} \propto x^{-1}$ and can be expressed by $u_\mathrm{c}/u_\mathrm{m} = 12(r_0/x)$. The theoretical calculation result can be expressed well the experimental results.

Figure 18 the entrainment flow rate Q/Q_0 obtained by the velocity distribution u. Q/Q_0 increases almost linearly to the downstream and the turbulent jet has a much larger than that of the laminar jet. For example, Q/Q_0 at $x/d_0 = 24$ for the turbulent jet is about 5.2 times that for laminar jet.

(b) **Turbulent properties**

Figure 19a–c show the turbulent intensities in x, y and z directions, respectively, and (d) is Reynold shear stress. In the figure, "present measurement" is by Ninomiya and

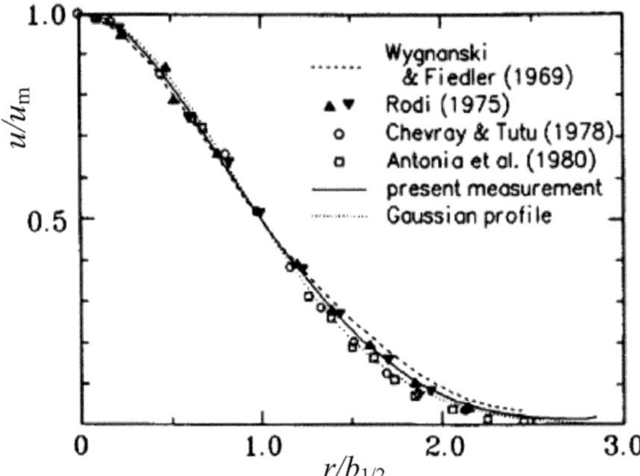

Fig. 16 Nondimensional velocity distribution of three-dimensional axisymmetric circular free jet, $u/u_\mathrm{m}-r/b_{1/2}$ ("present measurement" in the figure is by Reichardt 1942)

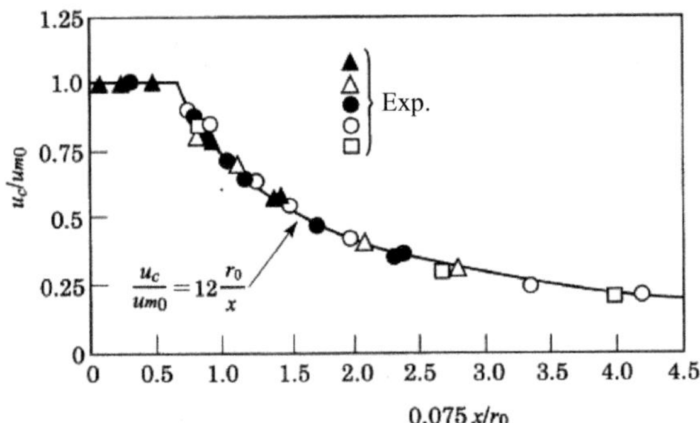

Fig. 17 Nondimensional centerline velocity of three-dimensional axisymmetric circular free jet, u_c-x/r_0 (Exps: Truepel 1915, Theory: Tollmien 1926)

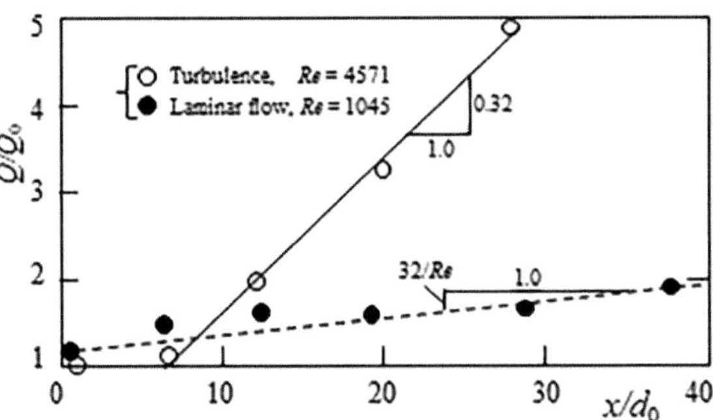

Fig. 18 Entrainment flow rate of three-dimensional axisymmetric circular free jet

Kasagi (1993) and δ is half width of circular jet. Although the measurement results differ in the measurement equipment and methods, they generally match well.

Figure 20 shows the flow characteristics of submerged round water jet measured by a particle image velocimetry, PIV (Seo et al. 2002). From the velocity distribution in Fig. 20b, it can be seen that there is a core region just after the nozzle exit, and

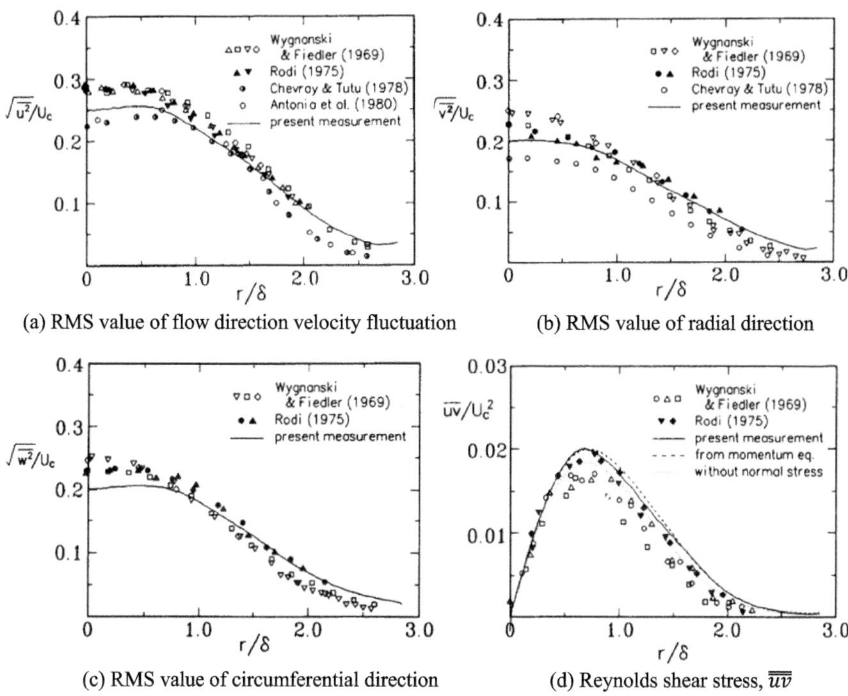

(a) RMS value of flow direction velocity fluctuation

(b) RMS value of radial direction

(c) RMS value of circumferential direction

(d) Reynolds shear stress, $\overline{uv}$

Fig. 19 Turbulent of three-dimensional axisymmetric circular free jet (by Ninomiya and Kasagi 1993)

from the turbulent intensity in Fig. 20c, there is a region of large turbulent intensity at the outer edge of the jet where the velocity gradient is large.

Regarding the flow characteristics of the free jet, such as velocity distribution, other examples of results are summarized in Tables 1 and 2 (Blevins 1984).

In Tables 1 and 2 the flow characteristics such as velocity distribution, jet width etc. of two- and three-dimensional submerged free jet flow are summarized.

2.2.3 Three-Dimensional Turbulent Free Jet from Long Pipe Nozzle

Hear, we show the flow characteristics of turbulent free jet from a long pipe nozzle with the diameter $d_0 = 10.0$ mm and length $L/d_0 = 50$ of pipe nozzle. The jet at the nozzle exit has a fully developed velocity profile of the internal flow that follows the so-called 1/7th power law since the pipe nozzle has an enough length.

Figure 21 shows the centerline velocity distribution u_c/u_m and turbulence intensity distribution u_c'/u_m of the air jet from the pipe nozzle with length $L/d_0 = 50$. A potential core region where the velocity does not change is formed from the nozzle exit, and its length decreases with increasing Reynolds number at $Re = 3000, 5000$, and 7000, and $x/d_0 \approx 9, 7$, and 3, respectively. After that, it passes through a transition

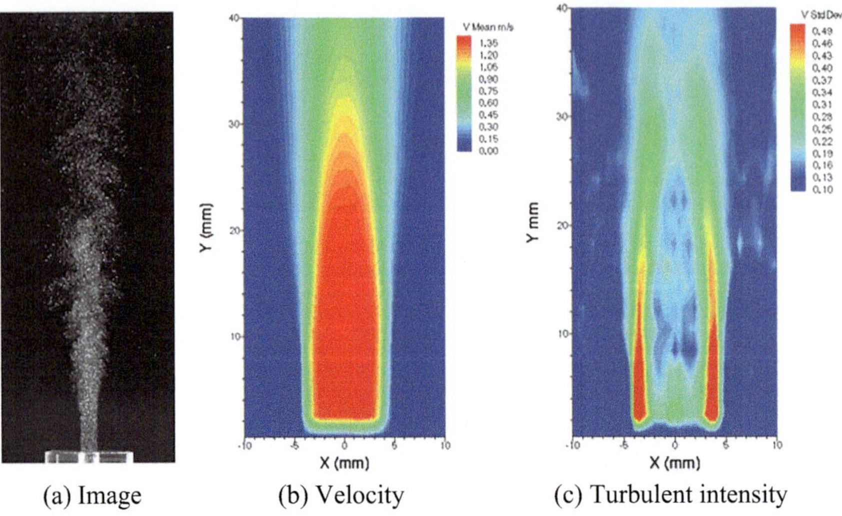

(a) Image   (b) Velocity   (c) Turbulent intensity

Fig. 20 Jet flow characteristics measured by PIV (Seo et al. 2002)

region and reaches a fully developed region. The velocity distribution in the fully developed region is expressed by the following equation.

$$\frac{u_c}{u_m} \propto \left(\frac{x}{d_0}\right)^a \tag{55}$$

Here, the coefficient a is -1.96, -1.38, -1.16 for $Re = 3000$, 5000, and 7000, respectively, and decreases as the number of Re increases because the turbulence component increases. At $Re = 1000$ and 3000, the transition region is hardly seen, but at $Re \geqq 7000$, where the Reynolds number is large, it exists in the range of $4 < x/d_0 < 8$. The centerline turbulence intensity u_c'/u_m has a distribution shape with a maximum value regardless of the Reynolds number, but as the Reynolds number increases, the maximum value moves toward the nozzle exit side in response to the core region length. Furthermore, when the Reynolds number is large ($Re \geqq 7000$), u_c'/u_m increases rapidly in the core region compared to low Reynolds numbers, but its maximum value decreases as the Reynolds number increases.

2.3 Large-Scale Vortex Structure of Circular Free Jet

2.3.1 Generation and Collapse of Large-Scale Vortex Structures

A regularly occurring large-scale vortex structure exists in the shear layer at the outer edge of the jet after exiting the nozzle. It was pointed out for the first time by Brown

Table 1 Flow characteristics of submerged laminar jet flow

Flow characteristics	Two-dimensional jet (Plane jet)	Three-dimensional, axisymmetric, circular jet
a. Stream function, ψ	$\psi = 1.651(Jvx)^{1/3}\tan\xi$ where, $\xi = 0.2752(J/v^2)^{1/3}(y/x^{2/3})$ $J = \int u^2 dy = \text{const}$	$\psi = vx\xi^2/(1+\xi^2/4)$ where, $\xi = 0.2443(J^{1//2}/v)(r/x)$ $J = 2\pi \int u^2 rdr = \text{const}$
b. Centerline (maximum) velocity, u_m	$u_\mathrm{m} = 0.4543\,[J^2/(vx)]^{1/3}$	$u_\mathrm{m} = 0.1194[J/(vx)]$
c. Inline velocity, u	$u = 0.4543[J^2/(vx)]^{1/3} \times (1-\tanh^2\xi)$	$u = 0.1194[J/(vx)]/(1+\xi^2/4)^2$
d. Transverse velocity, v	$v = 0.5503(Jv/x^2)^{1/2}[2\xi \times (1-\tanh^2\xi)-\tanh\xi]$	$v = 0.2443(J^{1/2}/x)(\xi-\xi^3/4)/(1+\xi^2/4)^2$
e. Half width, $b_{1/2}$ or $y_{1/2}$	$b_{1/2} = 3.203(v^2/J)^{1/3}x^{2/3}$	$b_{1/2} = 5.269(v^2/J)^{1/2}x$
f. Volumetric flow rate, Q	$Q = 3.3019(Jvx)^{1/3}$	$Q = 25.13vx$
g. Reynolds number $Re = u_\mathrm{m}b_0/v$	$Re = 1.455\,(Jx/v)^{1/3}$	$Re = 0.6289\,(J^{1/2}/v)$
h. Critical Reynolds number, Rec	$Rec = u_0 b_0/v \doteq 30$	$Rec = u_0\,(2r_0)/v \doteq 1000$

where, r_0: radius of circular nozzle, u_0: mean velocity at nozzle exit, J: Momentum flux of jet

Table 2 Flow characteristics of submerged turbulent jet flow

Flow characteristics	Two-dimensional jet (Plane jet)	Three-dimensional, axisymmetric, circular jet
a. Length of initial region, x_c	$x_c = 6b_0$	$x_c = 10\,r_0$
b. Centerline (maximum) velocity, u_m	$u_m = 3.4[b_0/(2x)]^{1/2}\,u_0$	$u_m = 12r_0u_0/x$
c. Velocity distribution, u/u_m or u	$u/u_m = \exp\left[-57(y/x)^2\right]$ $u =$ $[3K\sigma/(4x)]^{1/2}\left(1-\tanh^2\eta\right)$ where, $\eta = \sigma\,y/x$	$u/u_m = \exp\left[-94(y/x)^2\right]$ $u =$ $3K/\left[(8\pi\varepsilon_0 x)\left(1+\eta^2/4\right)\right]$ where, $\eta = [3K/(16\pi\varepsilon_0)]^{1/2}(y/x)$ $\varepsilon_0 = 0.0161K^{1/2}$
f. Half width, $b_{1/2}$ or $y_{1/2}$	$b_{1/2} = 0.11\,x$	$b_{1/2} = 0.086\,x$
g. Volumetric flow rate, Q	$Q = 0.44\,(2x/b_0)^{1/2}$	$Q = 0.16\,(x/r_0)$
h. Entrainment velocity, v_e	$v_e = 0.053\,u_m$	$v_e = 0.031\,u_m$
i. Kinetic energy, J	$J = 2.6[b_0/(2x)]^{1/2}J_0$	$J = 8.2(r_0/2x)J_0$

where, $J_0{:} = \rho u_0{}^2/2$, $K = J/\rho$

and Roshko (1974) that an organized vortex structure exists even in turbulent flow. For example, a jet from an orifice nozzle with sharp edge generates a small-scale vortex ring streets by amplifying the instability of a thin shear layer separated from the nozzle edge (Kelvin–Helmholtz instability). They coalesce to form a large-scale vortex street, which then spreads and collapses.

Figure 22 shows a visualized photograph of the flow of a water jet ejected into still water from an orifice nozzle with a diameter $d_0 = 10.0$ mm and a contraction area ratio $CR = 0.21$. The jet was visualized with a dye (Fluorescein sodium aqueous solution) in its central cross section. You can see how the above vortex ring streets is generated, coalesced, diffused, and collapsed.

In the case of an orifice jet, when the generation frequency of the vortex ring is f, the Strouhal number St is experimentally expressed by $St = f\,d_0/u_0 = 0.68 \sim 0.8$ (u_0: nozzle exit mean velocity).

When the boundary layer is laminar in a contour nozzle with a gently narrowed cross section, a vortex street is formed due to the same instability, and if f is the frequency at which the vortex street occurs at $x/d_0 = 3 \sim 4$, then $St = 0.24 \sim 0.51$ (Toyoda and Hussain 1989; Toyoda et al. 1992). When the jet boundary layer at the nozzle exit is turbulent, regular vortex street formations due to shear layer instability are not observed near the nozzle exit as in the case of laminar flow, but at $x/d_0 = 2 \sim 3$, due to column instability, a regular large-scale vortex structure is generated. In the case of pipe jet with a fully developed turbulent velocity distribution, a distinguished large-scale vortex structure appears due to column instability with $St = 0.38$ (Toyoda and Hussain 1989; Toyoda et al. 1992).

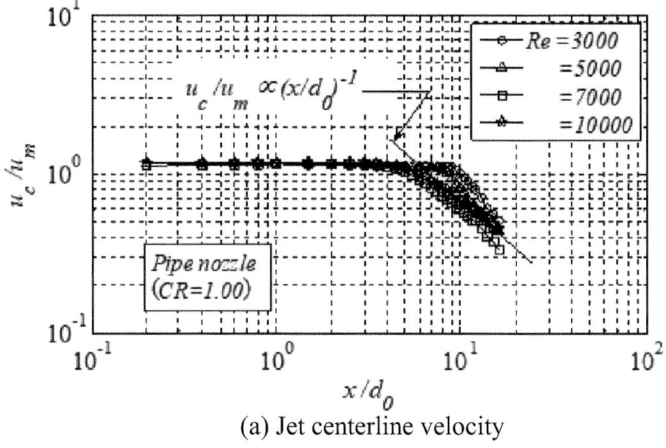

(a) Jet centerline velocity

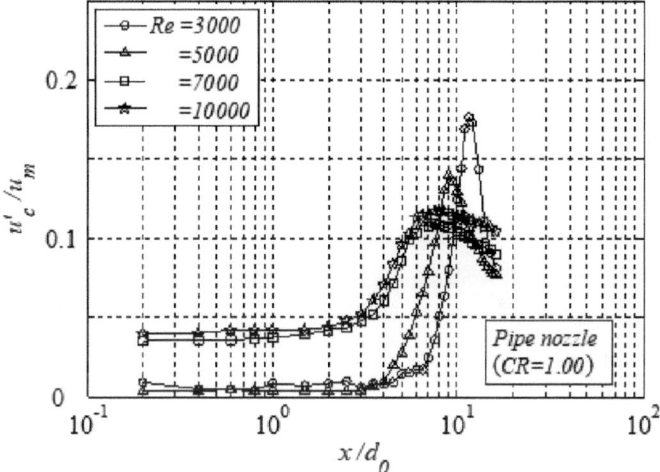

(b) Turbulent intensity on jet centerline Jet

Fig. 21 Pipe jet ($CR = 1.0$)

In Fig. 23, the results for a pipe jet with a diameter of pipe nozzle $d_0 = 10.0$ mm, length $L/d_0 = 50$, and Reynolds number $Re = 5.0 \times 10^3$ are shown. In the figure, visualized photographs of jet cross sections at $x/d_0 = 2, 4, 8$ are also shown.

The behavior of vortices in circular jets, vortex interaction, coalescence, etc., have been studied in detail by Hussain and Zaman (1981), and those of non-circular jets, such as ellipsoids and rectangles, have been studied in detail by Husain and Hussain (1983), Toyoda and Hussain (1989), Toyoda et al. (1992).

Large-scale vortices generated in the jet flow approach each other and coalesce due to their induced velocity. This coalescence phenomenon has been investigated

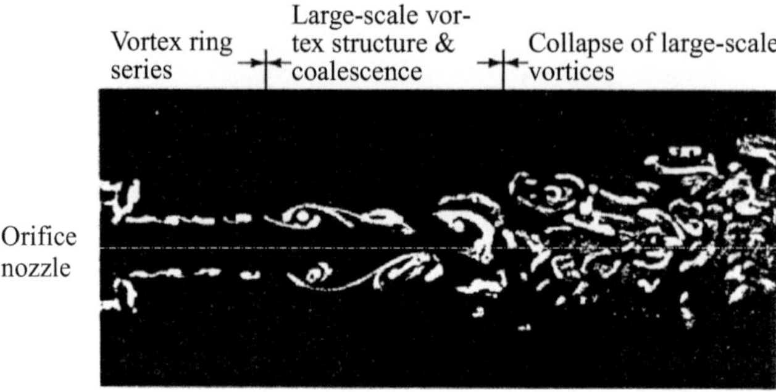

Fig. 22   Visualized flow pattern of orifice free jet ($d_0 = 10.0$ mm, $CR = 0.21$, $Re = 1 \times 10^3$)

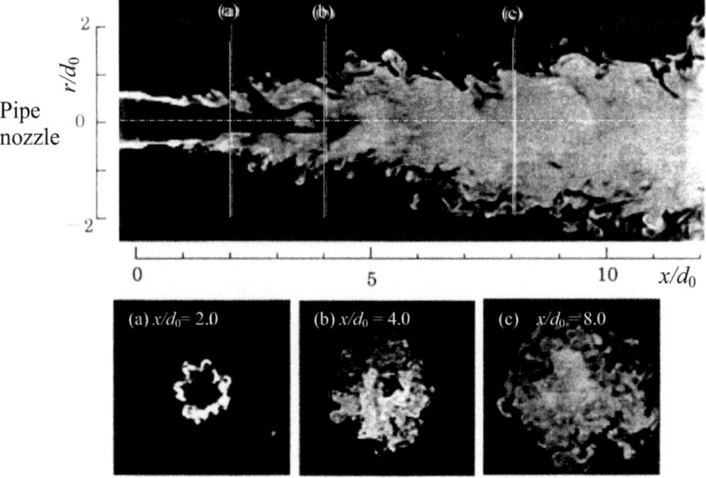

Fig. 23   Visualized flow pattern of pipe free jet ($d_0 = 10.0$ mm, $L/d_0 = 50$, $Re = 5 \times 10^3$)

in detail by Hussain and Zaman (1980). When a circular jet is excited by $St = 0.85$, stable coalescence of vortex rings occurs, and the vorticity distribution at that time becomes, for example, as shown in Fig. 23b. The numbers in the figure are the values obtained by dividing the vorticity by the excitation frequency.

The large-scale vortex street structure mentioned above collapses at the end of the potential region of $x/d_0 = 4 \sim 5$. Although there are many unknowns about the details of the collapse mechanism of large-scale vortex rings, for details of this collapse process, a model as shown in Fig. 25 has been proposed. The circular vortex ring structure deforms into waves in the circumferential direction and splits into small-scale vortices. It has been revealed that the circumferential deformation of the vortex ring is associated with a longitudinal vortex with an axis in the flow direction.

Regarding the details of the collapse mechanism of large-scale vortex rings, there are researches by Toyoda (2005) and Matsuda and Sakakibara (2005).

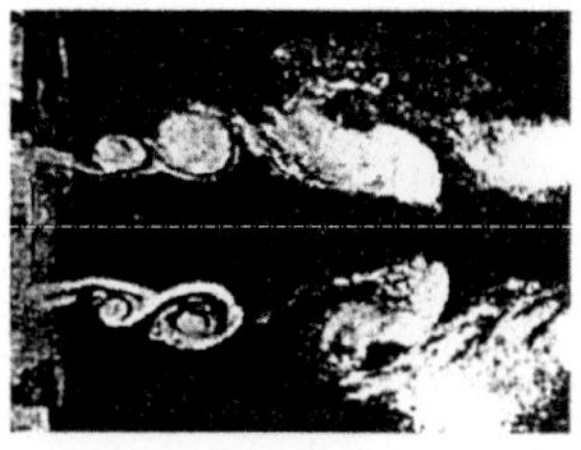

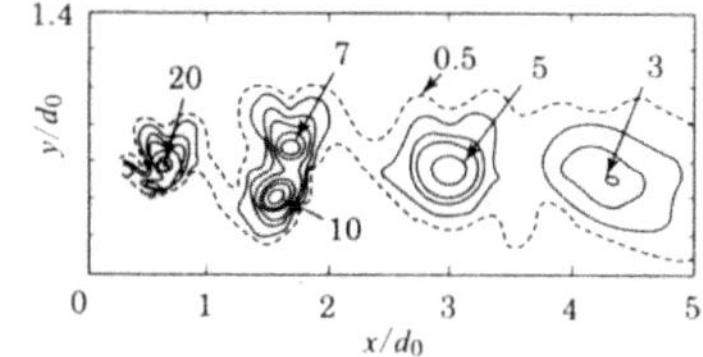

(a) Interaction and coalescence of vortex

(b) Coalescence of vortex in shear layer, vorticity distribution

Fig. 24 Vortex interaction and coalescence (Husssain and Zaman 1980)

Fig. 25 Collapse mechanism of large-scale vortex ring (Husssain 1986)

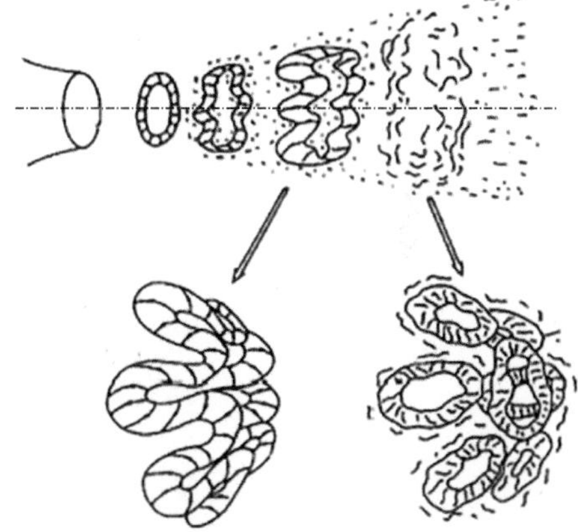

2.3.2 Control of Large-Scale Vortex Structure

When a jet is excited at a certain frequency f_e, a corresponding unstable wave begins to grow and a concentration of vorticity occurs, resulting in a significant increase in Reynolds shear stress as well as an increase in entrainment and jet width.

Figure 26 shows the state when the jet near the nozzle exit is excited in the zero-order mode, shear layer mode.

Figure 27 shows the situation when a two-dimensional jet is excited in the first mode at x_s in the development region. At that time, when excited at the excitation frequency f_E, the jet spreads greatly in the developed region, and the Strouhal number becomes $St_{xs} = f_E\, x_s/u_0 \fallingdotseq 1.0$. The amplitude of the disturbance during excitation is $(0.01 \sim 1.0)u_0$.

In addition, Hussain and Zamann (1980) revealed that in a circular jet, stable vortex coalescence can be obtained when the jet is excited with $St = 0.85$.

Table 3 shows control of large-scale vortex structure by jet excitation.

Lee and Reynolds (1985) showed that by simultaneously applying nozzle precession and axial disturbance of the jet, double mode excitation, the blooming jet shown in Fig. 28 and the branching jet of bi-/tri-/n–furcation can be realized.

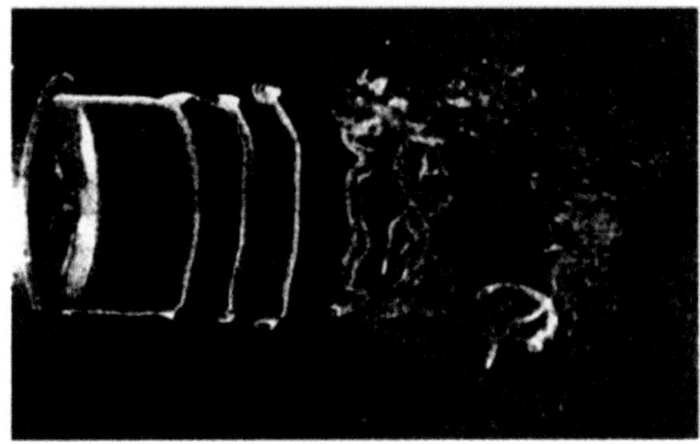

Fig. 26 Excitation of a circular jet in the zero-order mode (shear layer mode) (Michalke and Wahrmann 1964)

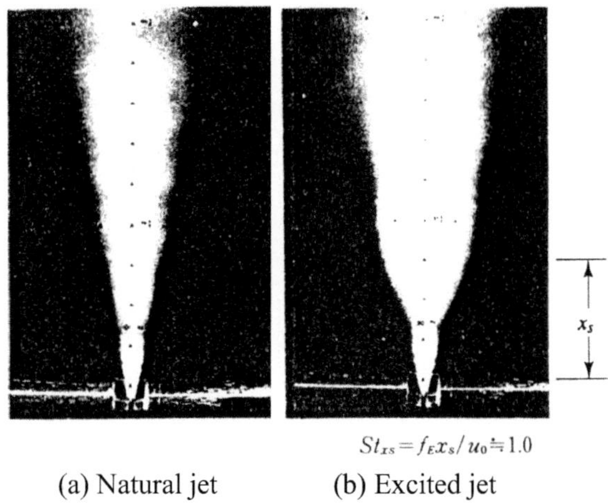

Fig. 27 Excitation of a two-dimensional jet in the first mode (Korschelt 1980; Hilberg 1996)

$$St_{xs} = f_E x_s / u_0 \doteqdot 1.0$$

(a) Natural jet　　　　(b) Excited jet

Table 3 Control of large-scale vortex structure by jet excitation

Control of core region;	
· Shear layer mode (Zero-order mode)	$0.8 \leqq St_{x,\max}(= f_E x/u_0) \leqq 1.5$
· Optimal mode	$0.2 \leqq f_E d_0/u_0 \leqq 1.5$
Control of developed region;	
· Two-dimensional jet	$St_{x,\max} = f_E\, x/u_0 \doteqdot 1.5$ (primary mode)
· Axisymmetric jet	$St_{x,\max} \doteqdot 0.5$ (axisymmetric mode)

Gohil et al. (2015) and Tyliszczak (2015) has been studied blooming jet by Computational Fluid Dynamics, CFD, Direct Numerical Simulation, DNS (Echigo et al. 2020) and examined the mean characteristic of vortical structure and the detailed frequency characteristics.

Thus, manipulating the vortex structure in some way (for example, by exciting it at a certain frequency) in order to attenuate, collapse, or amplify the large-scale

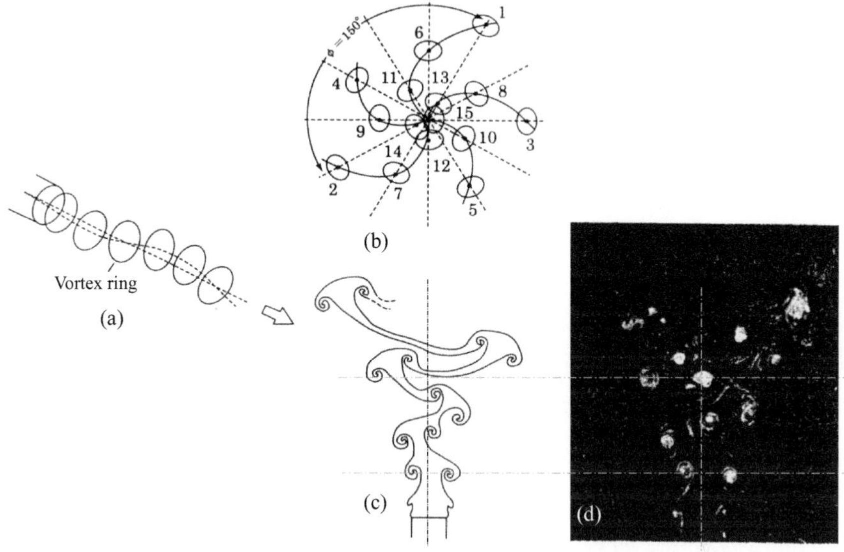

(a): Continuous vortex ring due to dual mode excitation
(b): Location of the vortices of the blooming jet
(c): Location of vortices in branched jets
(d): Visualization photo of branched jets using smoke

Fig. 28 Excitation of axisymmetric jet (Lee and Reynolds 1985)

vortex structure can, to some extent, control and manipulate the unique properties of the jet.

Regarding operations such as suppression and amplification of the large-scale vortex structure of the jet flow, there are methods other than simply applying an oscillating flow, such as the use of non-circular nozzles, sound waves, and micro-actuators as shown below. Other methods include the use of tabs, ribs, and vortex generators, the use of resonant jets, synthetic jets, and plasma jets.

2.3.3 Jet Flow Control by Non-circular Nozzle

Conventionally, in addition to jets from circular nozzles, the behavior and flow characteristics of jets from nozzles of various shapes such as square (Pani 1972), rectangle (Yevdjevich 1966; Trentacoste and Sforza 1967; Sfeir 1979; Tsuchiya and Horikoshi 1986; Shakouchi and Onohara 1990; Rembold and Adams 2002), ellipse (Husain and Hussain 1983), triangle, and cross (Fujita et al. 1984) have been investigated. There are also studies that use these non-circular nozzles to control the behavior of jets, such as controlling the mixing and diffusion of jets (Toyoda and Hussain 1989; Hussain and Husain 1989; Toyoda et al. 1992).

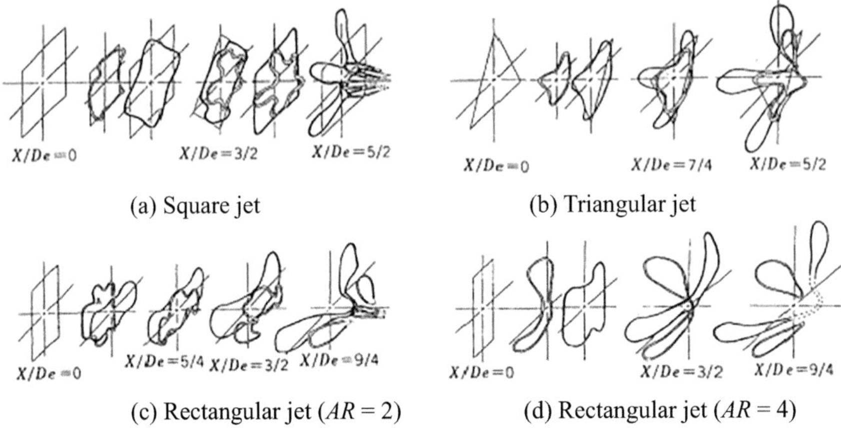

Fig. 29 Interference model of vortex structure of jet from non-circular vortex ring (Toyoda 2005)

Figure 29a shows an example of vortex street pairs and interference in square jet (Toyoda 2005). The thick line in the figure shows the leading vortex, and the double line shows the tracking vortex. In a square jet, the leading vortex is deformed so that its vertical and diagonal axes swap, and the tracking vortex enters into the leading vortex. Thereafter, the two vortex rings partially merge, and the overtaken portion of the tracking vortex is transported by the high-speed flow near the jet center and stretched downstream. As a result of this interference, the two vortex rings are significantly deformed and a vortex structure with an axis in the flow direction is formed. The velocity induced by the vortex structure in the flow direction promotes fluid transport inside and outside the jet, thereby contributing to increased diffusion and mixing. In Fig. 29b–d, the triangular jet, rectangular jets with $AR = 2, 4$ are shown, respectively. The behavior of the interaction between tracking and leading vortices can be seen.

2.3.4 Jet Flow Control by Micro-Actuator

(a) Diffusion control of two-dimensional jet flow by wake of small object in nozzle

Miyakoshi and Haniu (2003) showed a two-dimensional jet flow control by a small object, circular cylinder or semi-circular cylinder, set in the nozzle contraction as shown in Fig. 30. Where, Reynolds number is $Re = u_m h/v = 2508$ [h: nozzle width $= 15.0$ mm (nozzle aspect ratio: 20), u_m: mean velocity at the nozzle exit, v: kinematic viscosity of the fluid] and x_c is the distance of cylinder from nozzle exit.

Figure 30 shows the visualized flow pattern by a dye of submerged water jet. The shedding vortex, Karman vortex street, from the semi-circular cylinder set in the nozzle (Fig. 30b) excites the jet fluctuation after the nozzle exit and consequently the

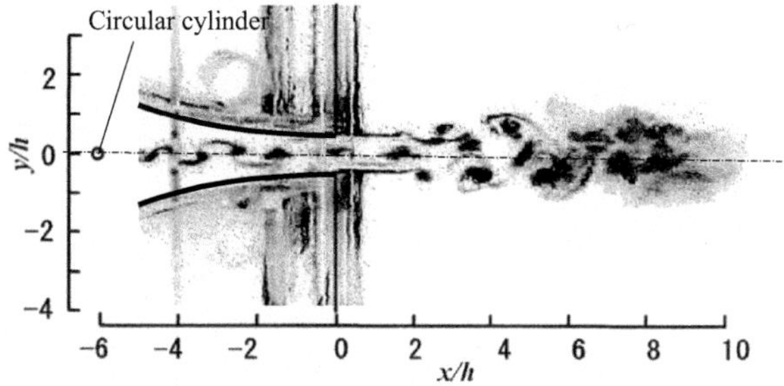

(a) Wake of circular cylinder in nozzle contraction and shear layer of the jet
($d/h = 0.266$, $x_c/h = 6$, $Re = 3200$)

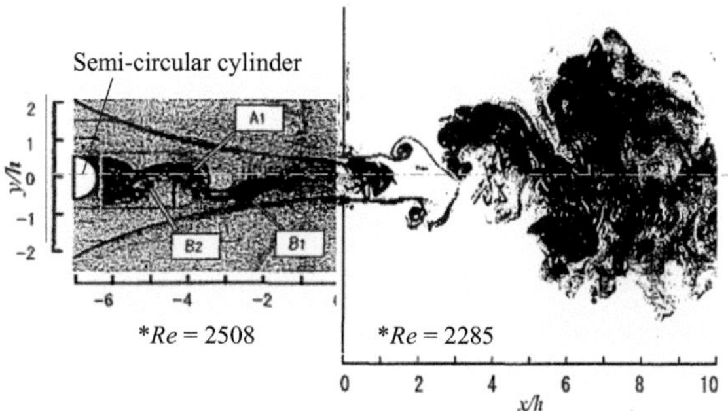

(b) Wake of semi-circular cylinder in nozzle contraction and shear layer of the jet
($d/h = 1.20$, $x_c/h = 7$)

Fig. 30 Visualized flow pattern of the wake of cylinder in nozzle contraction and shear layer of
the jet (Miyakoshi and Haniu 2003)

jet spreads widely in the perpendicular direction to the nozzle axis. This enhances
the jet entrainment. The visualized photographs of the flow in and after the nozzle
exit were taken separately, and the Reynolds numbers are $Re = 2508$ and 2285,
respectively.

(b) **Active jet flow control by intelligent nozzle**

As shown in Fig. 31, Suzuki et al. (2004), Kurimoto et al. (2004) and Kasagi (2006)
demonstrated the active jet flow control by an intelligent nozzle with 18 electro-
magnetic flap actuators. A thin flap-type electromagnetic actuator with a size of 9×3 mm is evenly placed on the inner surface or the end face of a circular nozzle with an
outlet diameter of 20 mm. These flap actuators are cantilever and their one end can
be moved inward of the nozzle, and their movements can be controlled arbitrarily.
By controlling the movement of each flap arbitrary, various kinds of flow patterns
can be realized. One example can be seen in Fig. 31b. This shows the flow pattern of
jet bifurcation induced by the actuation of alternate modes at the Strouhal number,

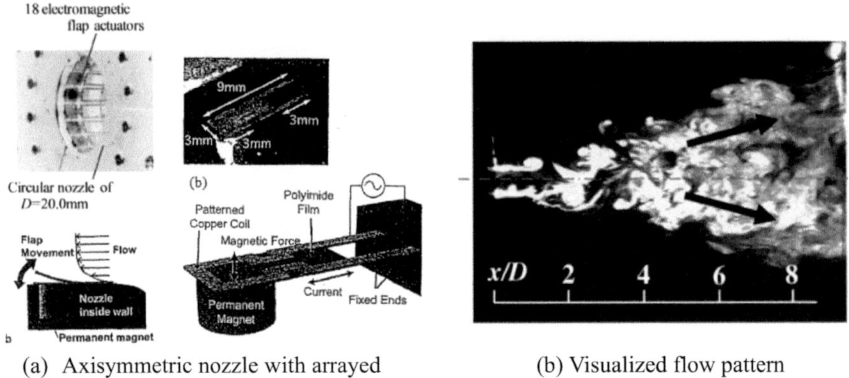

(a) Axisymmetric nozzle with arrayed   (b) Visualized flow pattern
     18 electromagnetic flap actuators

Fig. 31 Jet bifurcation induced by the actuation of alternate mode at St = 0.25 (by Ninomiya and Kasagi 1993, Suzuki et al. 2004, Kasagi 2006)

$St = 0.25$. As a result, it was shown that significant mixing promotion is possible by selectively controlling the large-scale vortex structure of the jet flow.

2.4 Orifice Free Jet (Effects of Contraction Area Ratio of Nozzle)

The submerged free jet from the orifice nozzle (hereinafter referred to as the orifice jet) has a contraction flow immediately downstream of the nozzle exit, and its flow characteristics (velocity distribution, mixing/diffusion characteristics, etc.) are similar to those from the pipe nozzle (hereinafter referred to as the pipe jet) is very different. Here, we will experimentally analyze and discuss the effects of the contraction aera ratio $CR = A_0/A_i$ (A_0: nozzle exit cross-sectional area, A_i: original pipe cross-sectional area) of the orifice nozzle on the flow characteristics of the orifice free jet (Shakouchi et al. 2008a, b).

* Experimental Apparatus and Procedure

Figure 32 shows the nozzle shape and coordinate system. In the experiment, a pipe nozzle with a nozzle diameter $d_0 = 10.0$ mm ($CR = 1.00$, straight pipe length $L = 500$ mm, $L/d_0 = 50$) and a pipe nozzle with a nozzle diameter of $d_0 = 10.0$ mm were used. Orifice nozzles with $d_i = 12.00, 15.00, 19.23, 29.75$ mm ($CR = 0.69, 0.44, 0.27, 0.11$, respectively) were used. The length of the straight pipe section is $L = 500$ mm, and the wall thickness is 1.0 mm. In this experiment, since $L = 500$ mm = const., the comparison between the length of the straight pipe section and the diameter d_i varies from $L/d_i = 50$ to 16.8, but in the case of $L/d_i = 16.8$ ($d_i = 29.75$ mm) The velocity distribution at the outlet of the straight pipe (with the orifice plate removed)

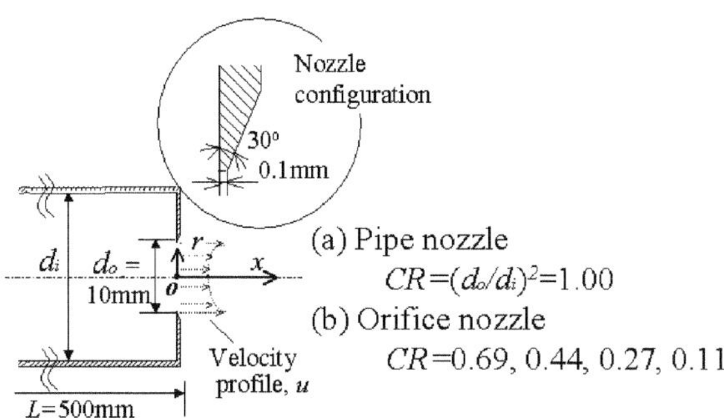

Fig. 32 Orifice nozzle and coordinate system

was a well-developed velocity distribution inside the pipe that matched that of L/d_i = 50 (results for CR = 1.00 in Fig. 39). The coordinate system has the nozzle exit center as the origin 0, the nozzle center downstream direction as the x coordinate, and the nozzle radial direction as the r coordinate.

To visualize and observe the flow, a fluorescein sodium aqueous solution is spouted from a nozzle at a relatively low velocity (Reynolds number $Re = u_m d_0/v = 1000$ to 3000) in a sufficiently large water tank filled with water, and the flow at the center cross section is visualized and observed. The state was photographed and recorded with a digital video camera, and the large-scale vortex structure and diffusion of the jet were examined.

The average and fluctuating velocity (turbulence strength) distribution of the air jet was measured using a constant temperature hot wire anemometer/Single probe (hot wire: 5 μm diameter tungsten wire, prong spacing: 2.0 mm, sensor length: 0.8 mm). The range of experiments using air jets is $Re = 3000$ to 15,000.

2.4.1 Orifice Free Jet in Low Reynolds Number Region (Mainly in *Re* = 1000 ~ 3000)

*** Flow pattern, large-scale vortex structure**

The influence of the orifice area ratio CR of the orifice nozzle on the flow field of the orifice jet was investigated using visualized images of water jets with a low Reynolds number (Re = 1000 to 3000).

Figure 33a–c show visualization images of the flow in the jet center axis cross section at Re = 1000 and CR = 1.00, 0.69, and 0.11, respectively. In a pipe jet (CR = 1.00), since the straight pipe part is long with L/d_i = 50, the jet is ejected from the nozzle with a fully-developed velocity distribution of laminar flow inside the pipe, and then flows down in a band shape with almost no mixing with the surroundings. On the other hand, in the orifice jet (CR = 0.69, 0.11), an axially symmetrical vortex ring (large-scale vortex structure) is released periodically in both cases, and its size increases as the orifice area ratio decreases. Furthermore, at CR = 0.11, the large-scale vortex structure collapses downstream and the flow diffuses in the radial direction. Therefore, the orifice area ratio CR is greatly involved in the generation of large-scale vortex structures in the jet, and the smaller its value, the more unstable the jet becomes.

Similar visualization and observation of the flow was also performed at $Re = 3000$, and as in the case of $Re = 1000$, the generation of large-scale vortices that was not seen in the pipe jet was confirmed in the orifice jet. We also found that CR is significantly involved in the generation of a large-scale vortex structure in the jet, and the smaller its value, the more unstable the jet becomes, and the result is that the jet width expands significantly by involving the surrounding fluid.

Figure 33 shows the results for $CR = 0.11$ as an example, but the situation is very different compared to the case of $Re = 1000$ (Fig. 33c), and the jet spread is large.

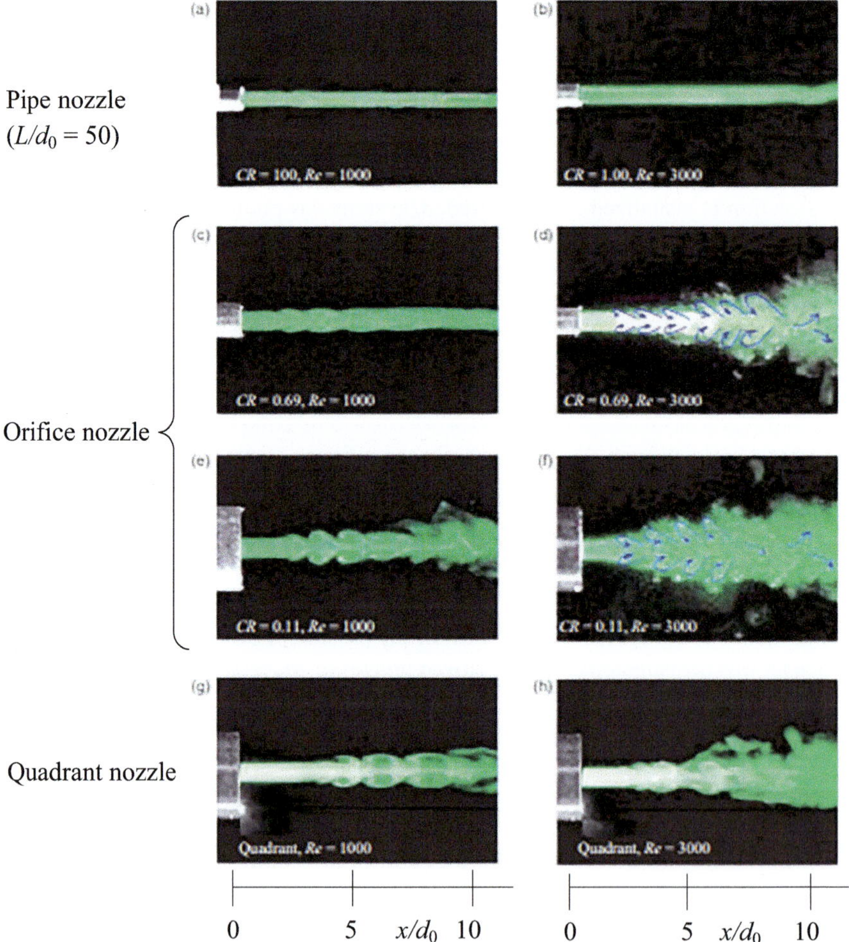

Fig. 33 Visualized flow pattern of pipe, orifice and quadrant jets ($Re = 1000$ and 3000)

2.4.2 Jet Centerline Velocity

Figure 34 shows, as an example, the distributions of centerline velocity uc/um and turbulence intensity uc'/um of an orifice jet with $CR = 0.11$. In an orifice jet, the velocity vector at the nozzle exit points in the direction of the center axis and the flow is accelerated, resulting in contracted flow and the centerline velocity increases from the nozzle exit. After that, a potential core region where the velocity does not change is formed, and a fully developed region is reached through a transition region. The velocity distribution in the fully developed region is expressed by Eq. (55) similarly to the pipe jet when $Re > 7000$, and $a = -1$, but in the case of low Re numbers, $a = -1$ at $Re = 3000$ and 5000, respectively. 1.53, –0.85, and a increases as the Reynolds number increases.

The turbulent intensity is different from that of a pipe jet, and at low Reynolds numbers, the distribution shape has two peaks. The first peak is seen at $x/d_0 = 2$ to 4, which is related to the generation of large-scale vortices confirmed by the flow visualization images, and it is thought that the turbulence strength increases due to the generation of vortices.

Fig. 34 Orifice nozzle ($CR = 0.11$)

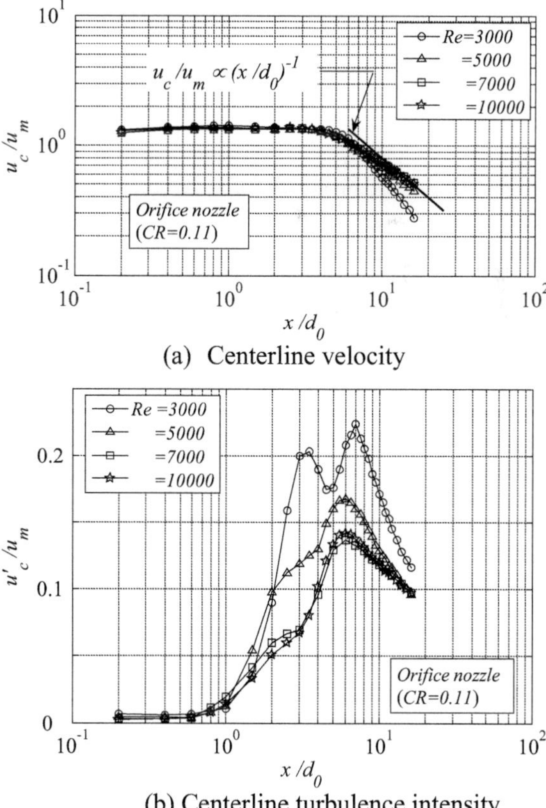

(a) Centerline velocity

(b) Centerline turbulence intensity

In addition, when $CR = 0.69$ and $Re = 3000$, the distribution shape was similar to that of the pipe jet. Therefore, it is thought that the influence of large-scale eddies seen in the visualized images does not affect the center line at low Reynolds numbers. However, in the case of $CR = 0.11$, the first peak is observed even at $Re = 3000$, and the influence of large-scale eddies appears significantly even on the center line. However, at high Reynolds numbers, the collapse of the large-scale eddies occurs, and the first peak disappears, resulting in a distribution similar to that of a pipe jet. In addition, in the orifice jet, a rapid increase in turbulence is seen closer to the exit at $x/d_0 \fallingdotseq 1$ than in the case of the pipe jet, and the effect of contractile flow does not appear on the centerline until $x/d_0 \fallingdotseq 1$.

2.4.3 Fluctuation Characteristics and Power Spectrum Distribution

Figure 35 shows the power spectrum distribution of the centerline velocity u_c of the pipe jet ($CR = 1.00$) and orifice jet ($CR = 0.69, 0.11$) at $Re = 3000$. A sharp peak in the spectrum appeared at $x/d_0 \fallingdotseq 3$ for pipe jets and at $x/d_0 \fallingdotseq 2$ for orifice jets, and the frequency remained constant until $x/d_0 \fallingdotseq 8$ for pipe jets, and was not observed downstream. This agrees with the result of the visualization image in which the pipe jet maintains a laminar flow state for a while after ejecting from the nozzle at $Re = 3000$. On the other hand, in the orifice jet, the spectrum peak gradually attenuates and disappears when $x/d_0 \fallingdotseq 6$. This is because the potential core area is smaller in the orifice jet than in the pipe jet. In addition, the orifice jet entrains the surrounding fluid and forms a large-scale vortex, but at the same time, the turbulence strength increases rapidly, causing the large-scale vortex to collapse downstream. The collapse of this large-scale eddy is thought to attenuate the amplitude of the peak frequency. In addition, in the case of a pipe nozzle, the turbulence intensity is smaller than in the case of an orifice jet, so it is thought that the peak will be in a narrow band.

In order to obtain the dominant frequency of jet fluctuation where the jet reached its maximum velocity at $x/d_0 = 2.0$, the power spectrum was calculated from the velocity fluctuations at the jet center, as plotted in Fig. 36. The power spectrum was normalized integrating the function Φ_u^* as follows: $\int \Phi_u^*(f*)df^* = 1$ as proposed by Mi et al. (2001) for a comparison. This equation means that when the peak frequency Φ_u^* decreases, the predominant frequency of the flow, including other frequency bands, becomes unclear.

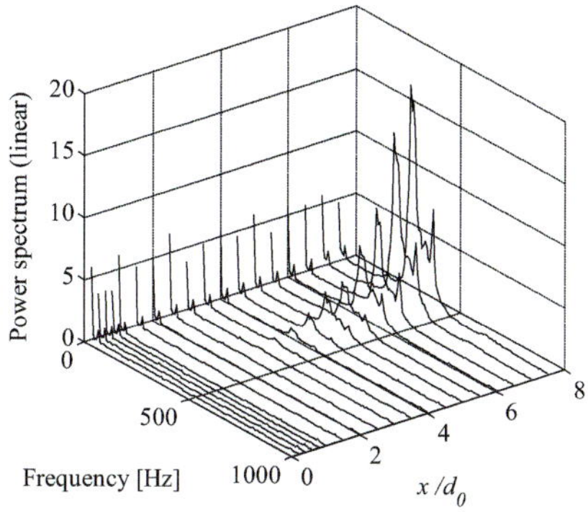

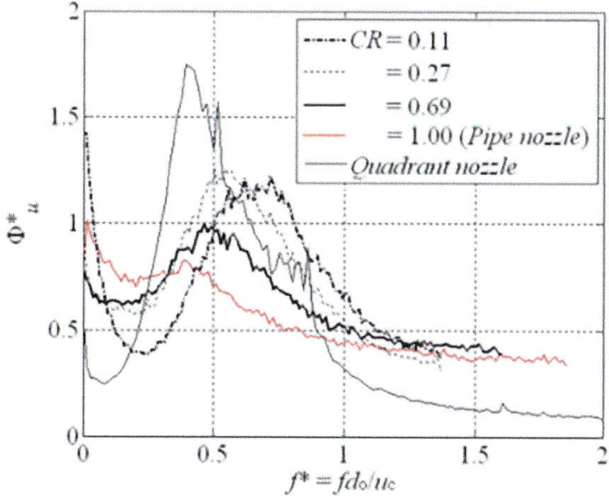

(a) Pipe nozzle ($CR = 1.0$)

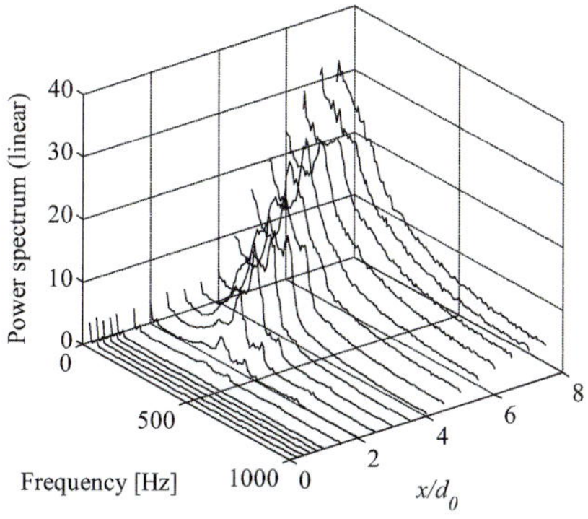

(b) Orifice nozzle ($CR = 0.69$)

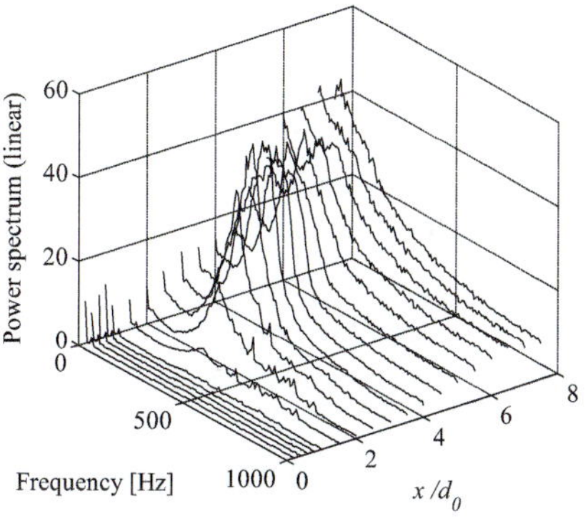

(c) Orifice nozzle ($CR = 0.11$)

Fig. 35 Power spectrum of u_c ($Re = 3000$)

Fig. 36 Dimensionless power spectrum distribution at $x/d_0 = 2.0$

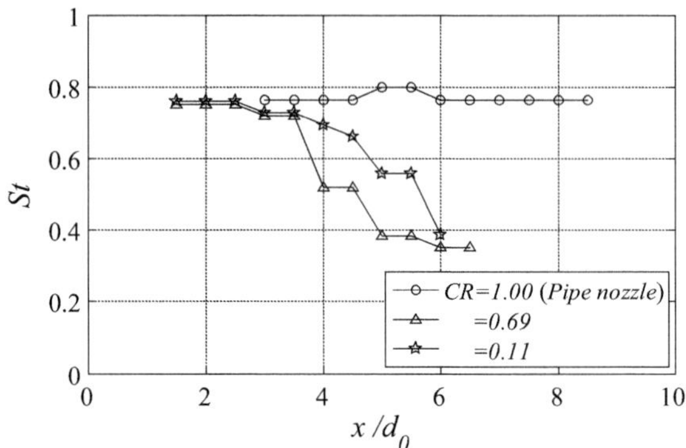

Fig. 37 Strouhal number St ($Re = 3000$)

The peak frequencies obtained for orifice jet at $CR = 0.11$, 0.69, and 1.0 (pipe jet) and a quadrant nozzle were $f^* = 0.71$, 0.48, and 0.39, respectively, although the peak was not clear for $CR = 1.0$. The values for the quadrant nozzle and $CR = 0.11$ are consistent with those reported by Mi et al. (2001). They presented values of $f^*{=}0.40$ for a contoured nozzle and $f^*{=}0.70$ for an orifice. Clearly, the peak frequency decreases with increasing CR. In addition, the magnitude of Φ_u^* at the peak also decreases and the peak becomes obscure, reflecting the large coherent vortex disintegrations. These results suggest that the same coherent vortex structure seen in Fig. 33 at low Reynolds number might exist at high Reynolds number as well.

Next, Fig. 37 shows the Strouhal number St determined from the peak frequency of the spectrum. In the case of a pipe jet, $St{=}0.76 =$ const. In the case of an orifice jet, the Strouhal number increases downstream from $St \approx 0.76$ because the centerline velocity downstream increases compared to the nozzle exit average velocity due to contraction, and the peak frequency attenuates due to the collapse of large-scale vortices. It decreases to $St{=}0.4$ when $x/d_0{=}6$, but the rate of attenuation is larger when CR is larger. Note that when $Re > 7000$, the jet contains various frequency components due to the collapse of the large-scale eddy, and no dominant frequency was observed.

2.4.4 Behavior of Free Jet in High Reynolds Number Region

Here, we will experimentally examine the effects of CR on the distributions of air jet velocity and turbulent intensity in the case of fully developed turbulence flow at $Re = 15,000$ without considering compressibility.

(a) Flow characteristics of nozzle and velocity distribution

Figure 4, already shown, shows the time-averaged velocity distribution u/u_m at the nozzle exit $x/d_0 = 0.2$ of the pipe jet ($CR = 1.00$) and orifice jet ($CR = 0.69$, 0.44, 0.11) (left half of the figure) and the turbulence intensity distribution u'/u_m (right half of the figure). Since the distribution shape was axially symmetric, only half the area in the radial direction was shown. The shape of velocity distribution at the nozzle exit differs greatly depending on the CR. In other words, since the pipe jet has a straight pipe length ($L/d_0 = 50$) necessary for the flow to fully develop, the jet ejects with a fully-developed velocity distribution shape in the pipe with a thick boundary layer. In the orifice jet, the distribution shape is close to a top hat at $CR = 0.69$, and the influence of contraction flow becomes noticeable at $CR = 0.11$, and the distribution shape has a maximum velocity at the outer edge of the jet $r/d_0 = \pm0.39$ (the maximum value is at the nozzle exit). The average speed is 1.6 times u_m, $u_\mathrm{c}/u_\mathrm{m} = 1.6$). In all cases of CR, the center line velocity of the jet is $u_\mathrm{c}/u_\mathrm{m} \fallingdotseq 1.3$, and CR affects the velocity on the outer side of the jet, and as CR decreases, the jet width decreases (contraction becomes stronger). In all cases, the turbulence intensity is minimum at the center of the jet where there is no velocity gradient, and maximum at the outer edge of the jet where the velocity gradient is greatest. Note that the smaller the CR, the smaller the value at the center, and the larger the maximum value at the outer edge. For example, when $CR = 0.11$, $u'/u_\mathrm{m} = 1.8 \times 10^{-3}$ at the center and the maximum value at the outer edge is $u'/u_\mathrm{m} = 0.29$. The error of velocity measurement is approximately 1.5% of its maximum value.

Figure 38 shows the velocity distribution u/u_m and turbulence intensity distribution u'/u_m at each downstream cross section. At the nozzle exit, the velocity gradient at the outer edge of the jet becomes larger as CR becomes smaller, resulting in stronger shear force and increased turbulent intensity. In addition, the velocity gradient in the shear layer increases as the CR decreases, and the flow rate increases downstream due to its strong entrainment. As CR decreases, the velocity near the centerline increases downstream.

Next, Fig. 39 shows the velocity distribution in each cross section, rendered dimensionless by the center velocity uc and the half-width $b_{1/2}$. For comparison, the figure also shows Tollmien's theoretical analysis results for an axisymmetric jet. Both the pipe jet ($CR = 1.00$) and the orifice jet with $CR = 0.11$ almost match Tollmien's theoretical solution after $x/d_0 = 5$, and the shape of the velocity distribution at the nozzle exit does not affect the distance to reach the fully developed region. Note that similar results were obtained when $CR = 0.69$.

(b) Centerline velocity and turbulent intensity distributions

Figure 40 shows the centerline velocity distribution $u_\mathrm{c}/u_\mathrm{m}$ and turbulent intensity distribution $u_\mathrm{c}'/u_\mathrm{m}$ at the contraction area ratio $CR = 1.00, 0.69, 0.27$, and 0.11. In the pipe jet, a potential core region is formed from the nozzle exit, and its length is $x_\mathrm{c} \fallingdotseq 4d_0$. In addition, in the downstream development region, $u_\mathrm{c}/u_\mathrm{m} \propto (x/d_0)^{-1.0}$. In the orifice jet, the centerline velocity increases from the nozzle exit due to contracted flow, and this increases as the orifice contraction area ratio decreases, and both reach the maximum velocity at $x/d_0 \fallingdotseq 2.0$. After that a potential core region is formed, and as in the case of a pipe jet, a fully developed region is reached through a transition region. The turbulent intensity at the nozzle exit decreases as CR decreases, and in pipe jets, it increases greatly from near the end of the potential core to the fully developed region (Fig. 40b). On the other hand, in the orifice jet, the turbulent intensity begins to increase upstream due to contracted flow, and the position shifts to the upstream

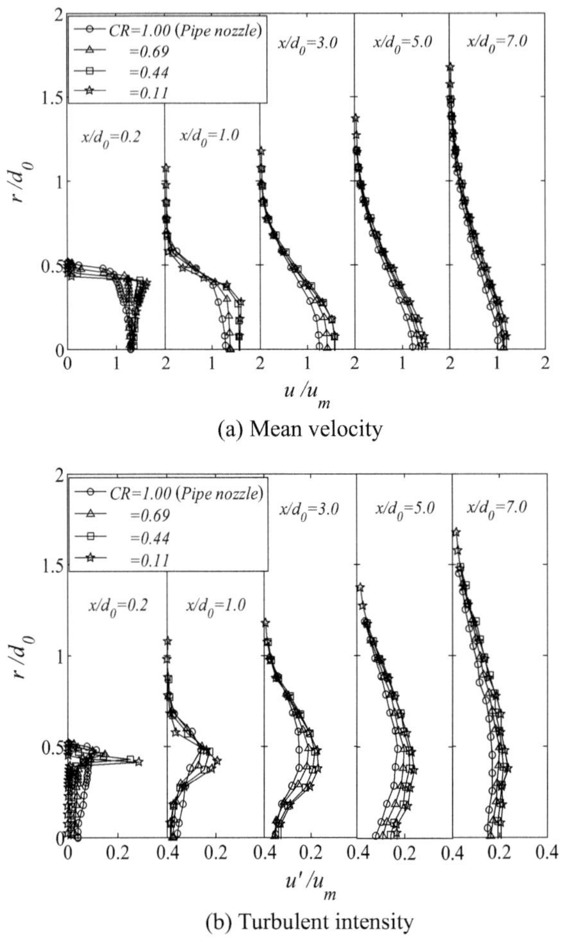

(a) Mean velocity

(b) Turbulent intensity

Fig. 38 Mean and fluctuating velocities

side as the *CR* becomes smaller. For example, when $CR = 0.11$, it starts to increase rapidly at $x/d_0 \doteq 0.4$, reaches the maximum at $x/d_0 \doteq 7$, and becomes larger than for other nozzles.

Figure 41 shows the relationship between center line velocity u_c/u_m and contraction area ratio *CR* at $x/d_0 = 2.0$. u_c/u_m increases as *CR* increases, reaching a maximum of $u_c/u_m = 1.7$ at $CR = 0.27$. This is because in the orifice jet, the exit velocity vector points in the direction of the center axis, which causes contraction and accelerates the center line velocity toward the downstream. However, as *CR* becomes smaller (pipe diameter increases), u_c/u_m decreases. This is thought to be due to the fact that the vortex region formed at the end of the constriction section in the nozzle, which had acted to accelerate the centerline velocity uc, conversely acts to decelerate u_c. That is, the velocity vector pointing toward the central axis increases until $CR =$

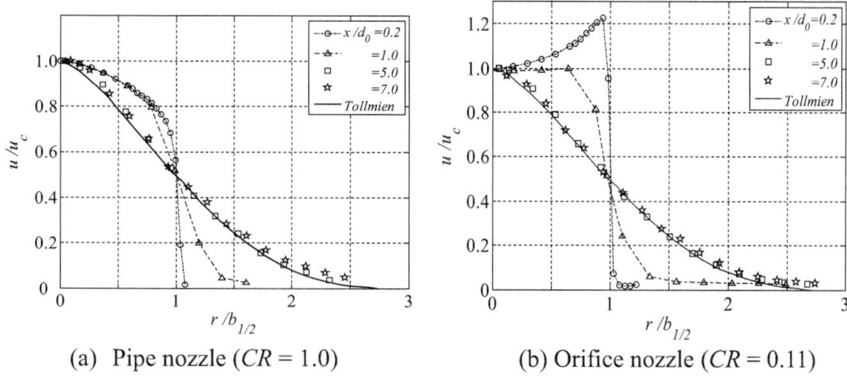

(a) Pipe nozzle ($CR = 1.0$)   (b) Orifice nozzle ($CR = 0.11$)

Fig. 39 Velocity profile

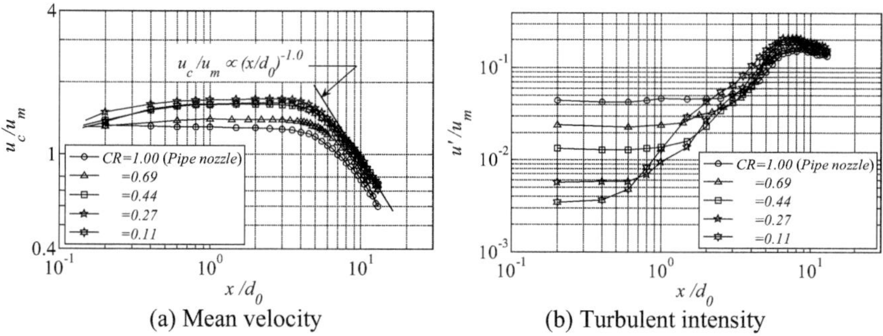

(a) Mean velocity   (b) Turbulent intensity

Fig. 40 Mean and fluctuating velocities on jet centerline

0.27, reaches a maximum of $u_c/u_m = 1.7$ at $CR = 0.27$, and then decreases. The relationship between u_c/u_m and CR is given by the following formula. Note that this experiment is considered to be valid in a turbulent state where the Reynolds number is sufficiently large (in a range where compressibility does not need to be considered).

$$\frac{u_c}{u_m} = 1.9CR^3 - 3.55CR^2 + 1.38CR + 1.53 \tag{56}$$

(c) **Spreading of jet flow**

Figure 42 shows how the jet spreads in terms of the half-width $b_{1/2}/d_0$ and the jet width b/d_0. The jet width was determined by defining the jet edge at the position where the maximum velocity was 10% in each cross section. In either case, it increases linearly from the nozzle exit, and the slope changes from $x/d_0 = 5.0$. When $x/d_0 > 5.0$, the half width and jet width are the same regardless of CR.

Note that the half width is expressed by the following formula (Table 4).

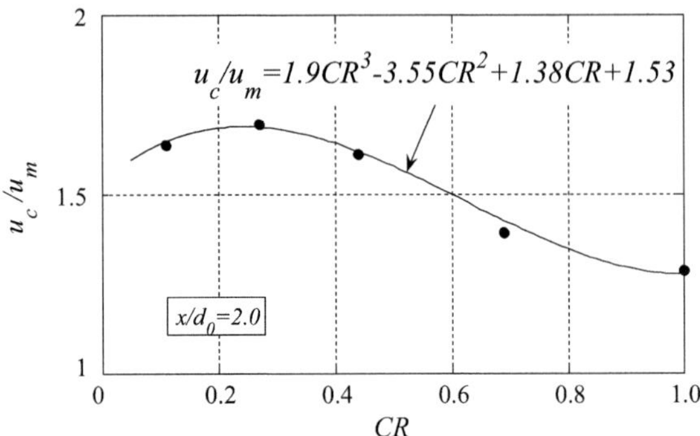

Fig. 41 Center line maximum velocity ($x/d_0 = 2.0$)

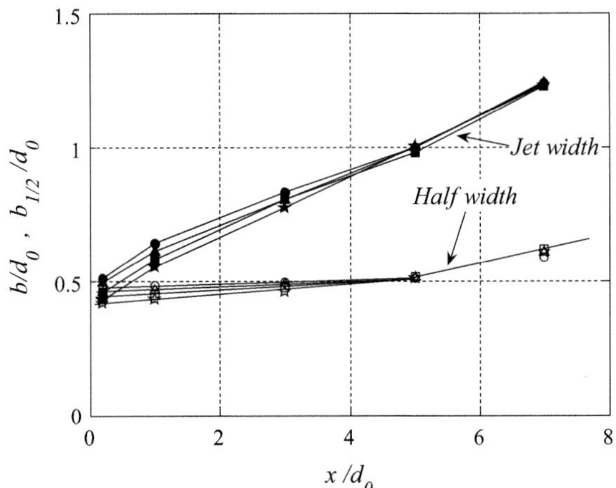

Fig. 42 Jet width and half width

Table 4 Half width

* $2 < x/d_0 < 5$	* Half width, $b_{1/2}$	
$CR = 1.00$	$b_{1/2}/d_0 = 0.006\,(x/d_0) + 0.481$	
$= 0.69$	$= 0.012\,(x/d_0) + 0.458$	
$= 0.44$	$= 0.017\,(x/d_0) + 0.431$	
$= 0.11$	$= 0.019\,(x/d_0) + 0.414$	(57)
* $5 < x/d_0$		
$CR = 1.00 \sim 0.11$	$= 0.048\,(x/d_0) + 0.273$	(58)

Here, we experimentally clarified the influence of the nozzle contraction area ratio *CR* on the flow characteristics of an orifice free jet, which has not been systematically clarified in the past.

To summarize, the following is the case.

(1) Visualized images of the flow of the orifice free jet showed that the orifice aperture area ratio has a large effect on the large-scale vortex structure.
(2) Contracted flow occurs in orifice jets, so the centerline velocity and turbulence strength distribution are significantly different from those in pipe jets. Also, the relationship between the centerline velocity uc/um of the orifice jet and CR is given by the following equation, and the maximum uc/um = 1.7 when $CR = 0.27$.

Note that this empirical formula is considered to hold true in turbulent flow conditions where the Reynolds number is sufficiently large (in a range where compressibility does not need to be taken into account).

$$u_c/u_m = 1.9CR^3 - 55CR^2 + 1.38CR + 1.53$$

(3) The velocity distribution shape $u/u_c - r/b_{1/2}$ has a similar distribution shape when $x/d_o \geqq 5$ for all CRs, which agrees well with Tollmien's analytical solution.

The centerline velocity in the fully developed region is expressed by the following equation, as in the case of pipe nozzles.

$$u_c/u_m \propto (x/d_o)^{-1.0}$$

(4) The centerline velocity fluctuation at $Re = 3000$ was $St \fallingdotseq 0.76 = $ const. in the case of the pipe jet, but in the case of the orifice jet, it changed from $St \fallingdotseq 0.76$ to $x/d_o \fallingdotseq 2$ with $x/d_o \fallingdotseq 2.6$, it decreases to $St \fallingdotseq 0.4$, and its appearance depends on CR.

2.5 Cone Orifice Free Jet

Because the flow path is constricted inside the orifice nozzle, the streamline passing through the nozzle edge looks like A in Fig. 43 (Kito et al. 2008a, b).

In addition, since the flow at the nozzle outlet has a velocity component toward the center of the nozzle, a contraction occurs in the flow near the outlet (the jet diameter d_j becomes smaller than the outlet diameter d_0). As a result, the centerline flow velocity increases and the nozzle exit velocity distribution takes on a distribution shape with the maximum velocity near the outer edge of the jet. However, if the flow path is suddenly narrowed to a large extent, an annular vortex region is formed at the corner immediately before the exit, increasing the flow resistance of the nozzle. Therefore, for example, a smooth nozzle shape along streamline A in Fig. 43 or a quadrant nozzle shape (see Fig. 44) is adopted. Note that in these cases, the centerline flow velocity of the jet from the orifice increases as a result of contraction immediately after the nozzle exit, whereas the velocity distribution shape at the nozzle exit is uniform and flat (due to a thin boundary layer). Therefore, it is possible to create a nozzle in which contracted flow occurs and the flow resistance is not so large, by making the

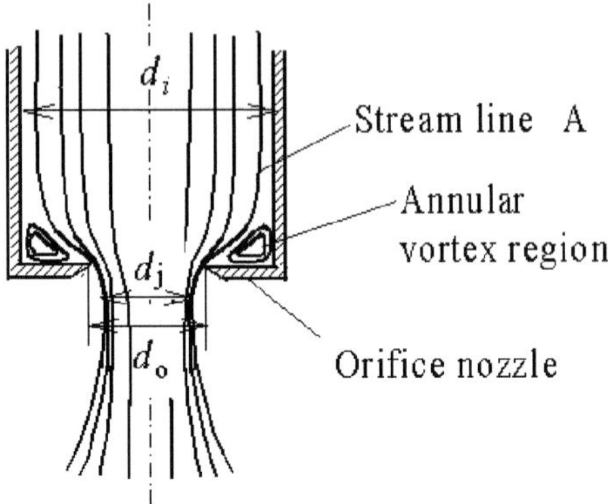

Fig. 43 Flow model of orifice jet

inner shape of the orifice nozzle conical. Furthermore, the contraction flow increases the centerline flow velocity of the jet, which is considered to be advantageous for improving the heat transfer characteristics near the stagnation point of the impinging jet. Note that this will be referred to as a cone orifice nozzle from now on (Fig. 44).

Here, we experimentally clarify the influence of the cone angle α of a cone orifice nozzle (see Fig. 44) on the flow characteristics of a conc orifice free jet and the heat transfer characteristics of the impinging jet at the impingement plate. For comparison, we investigated the flow characteristics of a jet from a quadrant nozzle (hereinafter referred to as a quadrant jet) and a jet from a pipe nozzle (hereinafter referred to as a pipe jet).

Figure 45 shows the sketch of stream line in the nozzle. The flow in the cone orifice nozzle with $\alpha = 15°$ seems to be smoother and has less flow resistance than the others.

Fig. 44 Nozzle configuration and coordinate system

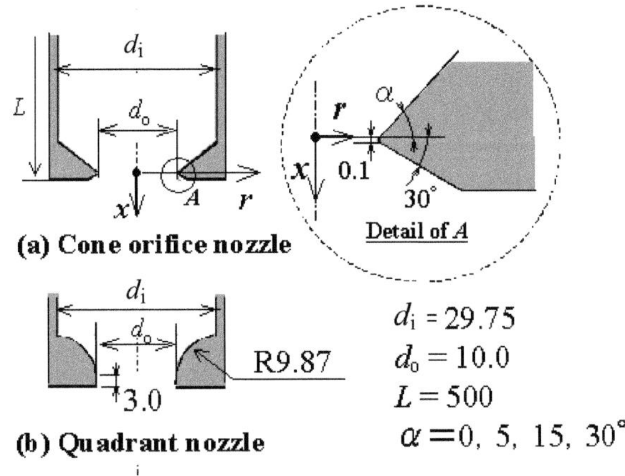

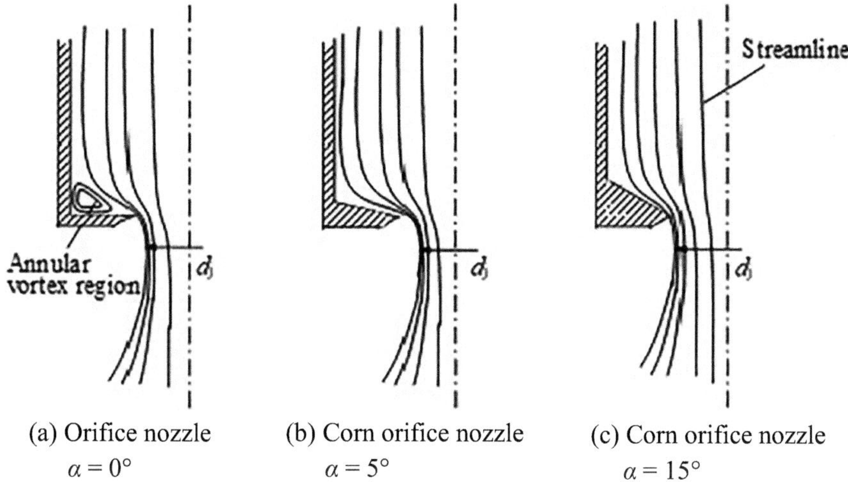

(a) Orifice nozzle   (b) Corn orifice nozzle   (c) Corn orifice nozzle
      $\alpha = 0°$ $\alpha = 5°$ $\alpha = 15°$

Fig. 45 Image of flow pattern of orifice and corn orifice nozzle

2.5.1 Velocity Distribution at Nozzle Exit

Figure 46 shows the velocity distribution u/u_m (left half in the figure) and turbulence intensity distribution u'/u_m (right half in the figure). Since the distribution shape was axially symmetric, only half the area in the radial direction is shown. The velocity distribution of the orifice jet at the nozzle exit becomes one with the maximum velocity at the outer edge of the jet due to the effect of contraction. At its maximum value, as cone angle α (see, Fig. 44) decreases, the effect of contraction becomes more pronounced, its position shifts toward the center of the jet, and the velocity gradient at the outer edge of the jet becomes larger. At $\alpha = 5°$, stronger contraction occurs than at $\alpha = 0°$, and the jet diameter becomes the smallest. This is thought to be because the flow inside the cone orifice nozzle has the maximum velocity component toward the center at $\alpha = 5°$. In other words, as α increases, the annular vortex generated at the corner immediately before the exit shown in Fig. 44a attenuated and reduced, causing the flow to flow along the inner wall of the nozzle, and the velocity component toward the center becomes smaller. On the other hand, when $\alpha = 0°$, it is thought that some velocity component toward the center is lost due to the annular vortex. In all cases, the turbulent intensity is maximum at the outer edge of the jet where the velocity gradient is greatest, and it increases as α becomes smaller. On the other hand, since the flow path of the quadrant jet is smoothly constricted, no contraction occurs and the velocity distribution is almost uniform throughout the nozzle exit. In addition, the pipe jet has a well-developed turbulent velocity distribution within the pipe, and the turbulent intensity near the center of the jet is larger than that of other nozzles.

The results shown here are for the case where the nozzle Reynolds number is $Re = u_m d_0/\nu = 1.5 \times 10^4$.

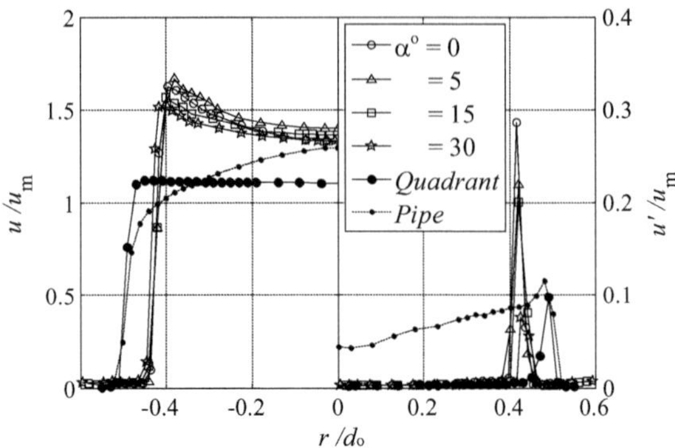

Fig. 46 Mean and fluctuating velocity distributions at nozzle exit (at $x/d_0 = 0.2$)

2.5.2 Jet Centerline Velocity

Figure 47a, b show the centerline flow velocity distribution u_c/u_m and turbulence intensity distribution u_c'/u_m, respectively. In either case, the velocity vector of a cone orifice jet is directed toward the center of the jet due to contraction, so the centerline flow velocity increases from the nozzle exit. After that, a potential core region is formed, and a fully developed region is reached through a transition region ($5 < x/d_0 < 6.5$), whose distribution shape is expressed as $u_c/u_m \propto (x/d_0)^{-1.0}$. The increase in centerline flow velocity is larger as α is smaller, but depending on the degree of contraction confirmed in the velocity distribution at the nozzle exit, the distribution shapes at $\alpha = 0°$ and $15°$ are almost the same, and $\alpha = 5°$ has a rather large distribution shape. Also, both reach their maximum velocity when $x/d_0 \fallingdotseq 2$ to In the case of a quadrant jet, the centerline flow velocity passes from the nozzle exit to the potential core region ($0 \leq x/d_0 < 5$), the transition region ($5 < x/d_0 < 8$), and then to the development region [$u_c/u_m \propto (x/d_0)^{-1.0}$]. The potential core region was defined as the point where the centerline flow velocity reaches its maximum velocity and the point after which the velocity attenuates and reaches 97% of the maximum velocity. In contrast, the conical orifice jet has a short x/d_0 of 4, which indicates that mixing is promoted further upstream. In the pipe jet, the potential core region ($0 \leq x/d_0 < 3$), the transition region ($3 < x/d_0 < 8$), and then the development region [$u_c/u_m \propto (x/d_0)^{-1.0}$] are reached. The turbulence intensity is small at the nozzle exit for all nozzles, and then increases rapidly until it reaches the fully developed region. The larger α is, the smaller the turbulence strength at the nozzle exit is. In the quadrant jet, a sharp increase in turbulence is seen downstream than in other nozzles, and even downstream it is smaller than in the conical orifice jet. The turbulence intensity of the pipe jet is much larger near the nozzle exit than elsewhere, but it becomes smaller than that of the cone orifice jet downstream.

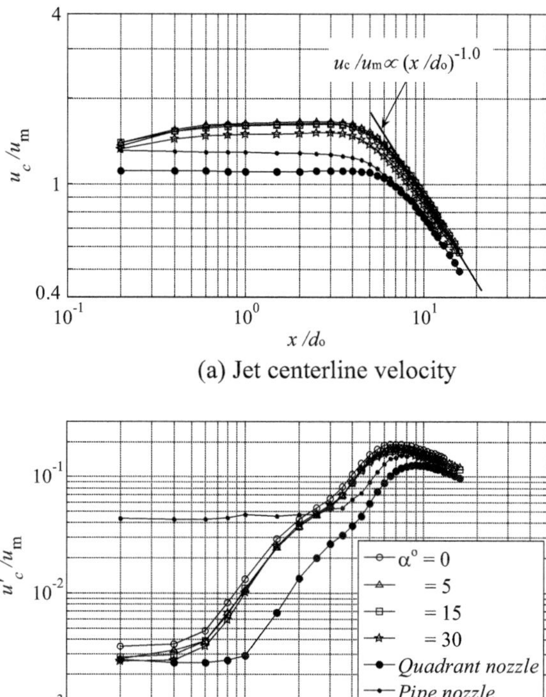

Fig. 47 Mean and fluctuating velocity distributions on jet centerline

(a) Jet centerline velocity

(b) Turbulent intensity

2.5.3 Fluctuation Characteristics and Power Spectrum Distribution

Figure 48 shows the dimensionless power spectrum distribution at $x/d_0 = 0$ (see, Sect. 2.4.3, in this chapter). In the pipe jet, no clear dominant frequency is observed, i.e., there is no periodic vortex emission. In the quadrant jet, $St=0.4$, and in the cone jets, an antecedent frequency of $St=0.5$, which is somewhat larger than that of the quadrant jet, is observed, and regular vortex emission corresponding to them is inferred.

2.5.4 Effects of Cone Angle

Figure 49 shows the influence of the cone angle α on the centerline maximum flow velocity u_{cm}/u_m (at $x/d_0=2 \sim 3$). The relationship between u_{cm}/u_m and α° is expressed by the following formula.

$$u_{cm}/u_m = 1.92 \times 10^{-6}\alpha^3 - 3.31 \times 10^{-4}\alpha^2 + 3.9 \times 10^{-3}\alpha + 1.64 \qquad (59)$$

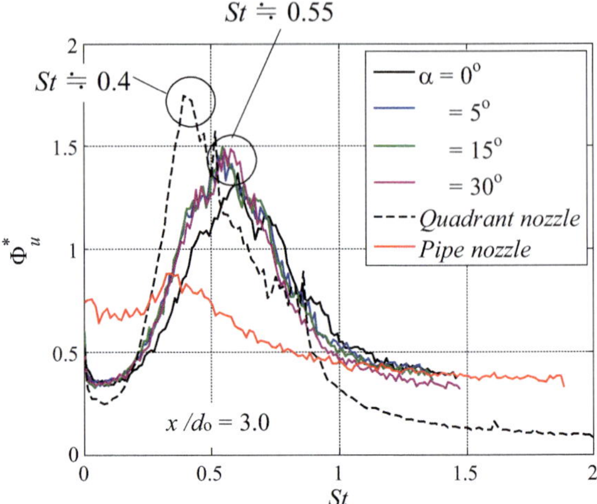

Fig. 48 Dimensionless power spectrum distribution at $x/d_0 = 0$

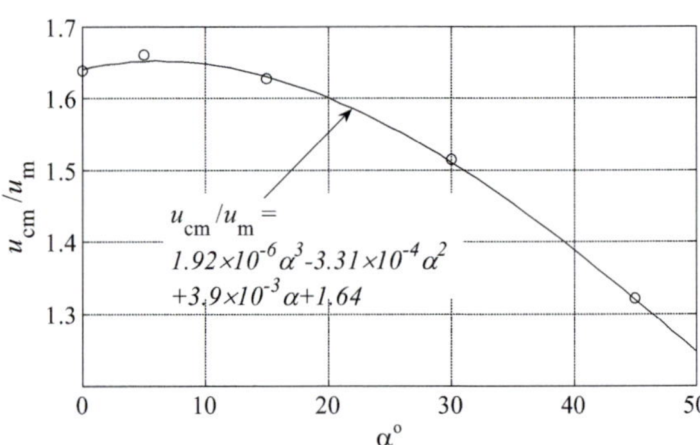

Fig. 49 Maximum centerline velocity at $x/d_0 ≒ 2 \sim 3$(Effects of cone angle, α)

The maximum centerline velocity reaches a maximum of $u_{cm}/u_m \approx 1.66$ at $\alpha ≒ 5°$, and this is thought to be due to the largest effect of contracted flow, as seen in the nozzle exit velocity distribution (see, Fig. 46).

2.5.5 Spreading of Jet and Flow Rate

Figure 50 shows the effect of the cone angle of the nozzle on the spread of the jet in terms of the jet half-width $b_{1/2}/d_0$ and the jet width b/d_0. The jet width was set at 10% of the maximum flow velocity in each cross section. Both the half-width and the jet width of the pipe jet without contraction and the quadrant jet are larger near the nozzle exit than in the case of the cone orifice jet with contraction.

Since the potential core length of the quadrant jet was $x/d_0 ≒ 5$, the jet almost maintained its momentum at the outlet until downstream, and its half-width was

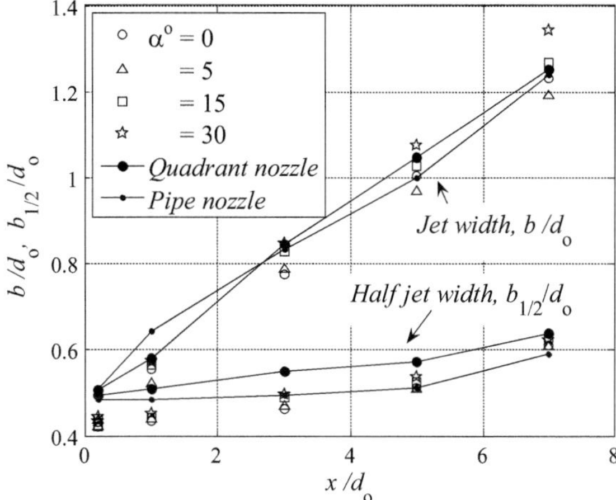

Fig. 50 Jet and Half jet widths

larger than the others. The jet width increases as the centerline flow velocity attenuates faster, and decreases as the contraction flow becomes stronger in each cross section.

The effect of contraction flow is also seen in the entrainment volumetric flow rate. Figure 51 shows the entrainment flow rate ratio $(Q_x - Q_0)/Q_0$ determined from the velocity distribution. Q_x and Q_0 are the flow rates at arbitrary position x and nozzle exit ($x/d_0 = 0.2$), respectively, and were calculated from the following equation.

$$Q = 2\pi \int_0^\delta u r dr \tag{60}$$

where, δ is the jet width, which was set at the radial position of 1% of the center flow velocity at the nozzle exit, as in Boguslawski et al.

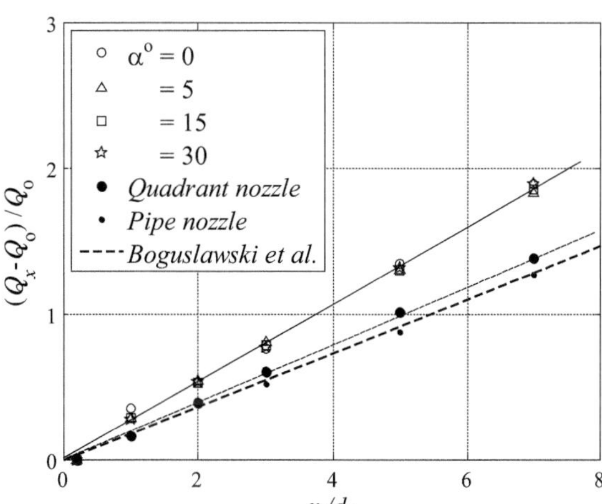

Fig. 51 Entrainment flow rate ratio

Figure 51 also shows the results of a pipe jet with a fully-developed turbulent velocity distribution in the pipe $[(Q_x-Q_0)/Q_0 = 0.183(x/d_0)]$ by Boguslawski et al. The pipe jet results of this experiment were in good agreement with the results of Boguslawski et al.

The entrainment flow rate is as follows for a quadrant nozzle:

$$\frac{Q_x - Q_0}{Q_0} = 0.20\left(\frac{x}{d_0}\right) \tag{61}$$

This is close to the result for a pipe jet.

In addition, in the case of a conical orifice nozzle,

$$\frac{Q_x - Q_0}{Q_0} = 0.26\left(\frac{x}{d_0}\right) \tag{62}$$

The effect of the cone angle α is not observed, but compared to a quadrant jet or a pipe jet, for example, when $x/d_0 = 7.0$, it is approximately 1.35 times and 1.5 times, respectively, and the mixing effect with the surrounding fluid is large.

Figure 52 shows the flow resistance ΔP of the nozzle. For reference, the case of $\alpha = 45°$ is also shown. ΔP is almost the same when $\alpha = 0°$ and $5°$, but decreases as α increases. For example, for the same Reynolds number ($= 1.5 \times 10^4$), ΔP at $\alpha = 15°$ and $45°$ are approximately 4.8 and 34.5% lower than when $\alpha = 0°$, respectively.

By using the newly proposed conical orifice nozzle described above, the flow resistance of the nozzle can be reduced compared to that of an orifice nozzle, and furthermore, by utilizing the contraction flow effect (increased centerline flow velocity), the heat transfer characteristics of the impinging jet can be improved compared to that of other nozzles.

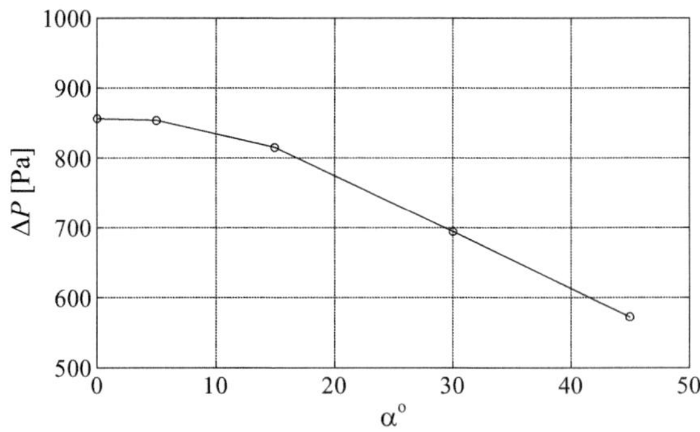

Fig. 52 Flow resistance (pressure loss, ΔP) of cone nozzle

2.6 Notched Orifice Free Jet

The jet flow from the orifice nozzle (hereinafter referred to as the orifice jet flow) has a contraction phenomenon (the jet diameter d_j downstream of the nozzle exit becomes smaller than the orifice diameter d_0 (Fig. 53). As a result, the maximum flow velocity also increases downstream of the nozzle exit), and a large velocity gradient occurs at the outer edge of the jet, resulting in an increased entrainment flow rate from the surroundings and excellent mixing and diffusion characteristics. The authors previously clarified the effect of the orifice contraction area ratio ($d_0/d_i)^2$ on the flow characteristics of the orifice jet (Shakouchi et al. 2008a, b, 2011, 2019). In addition, with an orifice jet, the flow path at the nozzle rapidly contracts, which increases the flow resistance of the nozzle. Therefore, from the perspective of increasing the efficiency of equipment using orifice jets, it is necessary to reduce the flow resistance while maintaining the contracted flow effect is desired. Regarding this, we proposed the use of a cone orifice nozzle and clarified its flow characteristics (Kito et al. 2008a, b).

By the way, in order to further improve the flow characteristics of the orifice jet, it may be possible to create a notch in the orifice or conical orifice nozzle to increase the turbulence of the jet while maintaining the contraction effect (Shakouchi et al. 2008a, b, 2011, 2019). Here, a triangular cutout or notch is provided in the orifice nozzle to reduce the nozzle's flow resistance, increase turbulence, and increase entrainment while maintaining the flow contraction effect. Furthermore, the above points are considered to be significant in improving the heat transfer characteristics of impinging jets. We also examine the flow switching phenomenon caused by providing a triangular notch.

Fig. 53 Flow model of orifice jet

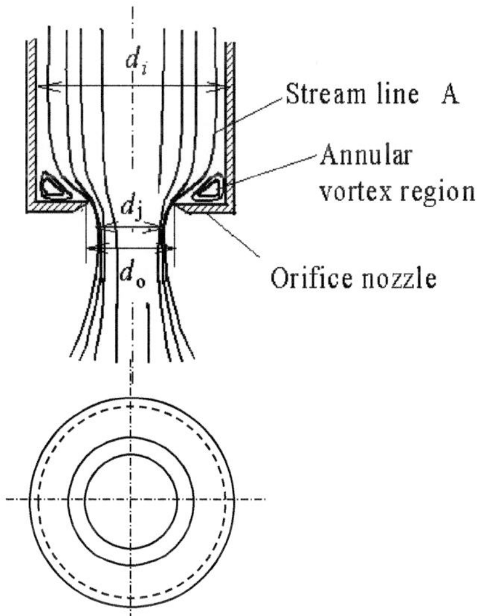

2.6.1 Notched Orifice Nozzle

Figure 54a shows the orifice nozzle (contraction area ratio: 0.27) with nozzle exit diameter $d_0 = 10.0$ mm and pipe diameter $d_i = 19.23$ mm. Figure 54b, c respectively show a notch nozzle with four or eight triangular notches placed at equal intervals around the circumference of the orifice. The shape of the notch is triangular, with an opening angle of 60° and a depth of 1.0 mm. In addition, a pipe nozzle ($d_0 = 10.0$ mm) was used for comparison. The length and wall thickness of the straight nozzle pipe section are $L = 500$ mm and wall thickness 1.0 mm in all cases. The coordinate system has the nozzle exit center as the origin 0, the radial direction of the nozzle as the r coordinate, and the downstream direction of the nozzle center axis as the x coordinate (Fig. 54).

The mean and turbulent velocity of the jet flow were measured using an single, I-type hot-wire probe, a constant-temperature hot-wire anemometer, and a traverse device. The flow resistance and operating power W in the nozzle section are determined by providing a pressure hole with a diameter of 0.8 mm in the nozzle pipe wall 20° upstream from the nozzle outlet, and measuring the pressure difference ΔP between the static pressure there and the atmospheric pressure using a manometer. and the flow rate Q using the following formula.

$$W = Q\left(\Delta P + \frac{\rho u_{pm}^2}{2}\right) \tag{63}$$

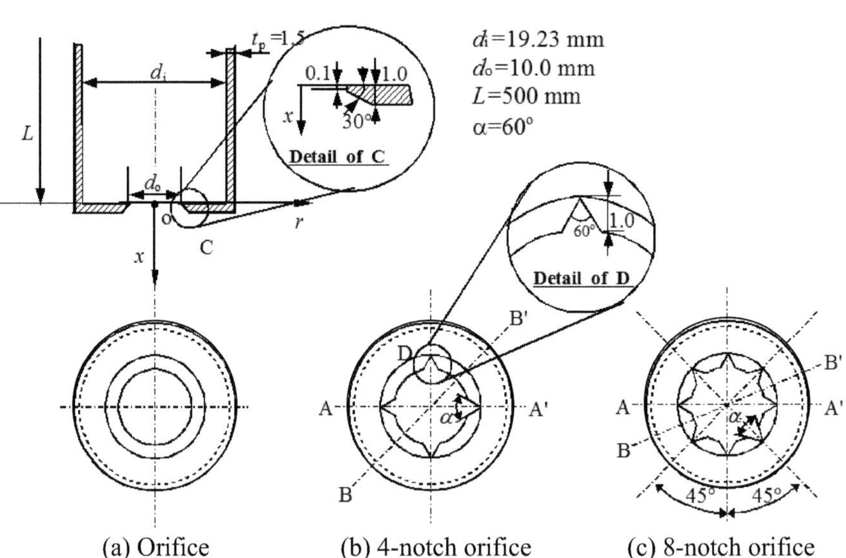

Fig. 54 Nozzle geometry

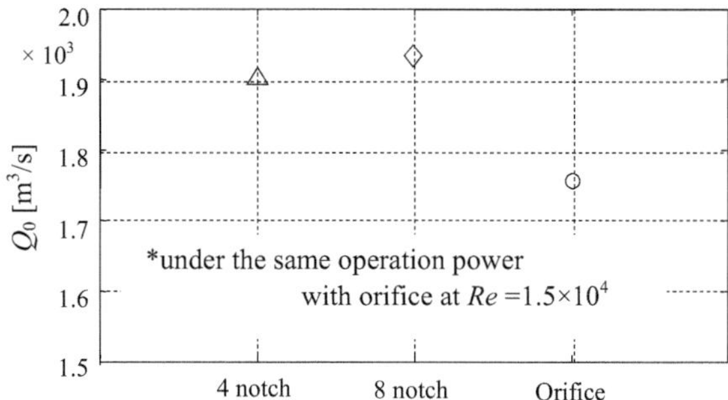

Fig. 55 Volumetric flow rate at nozzle exit (at $x/d_0 = 0.2$)

Note that the nozzle Reynolds number is $Re = u_m d_0/v = 1.5 \times 10^4 = $ const. for orifice nozzles unless otherwise noted.

2.6.2 Reference Length

Normally, the nozzle diameter is used as the reference length to generally express the jet phenomenon. However, in this orifice nozzle with a notch, even if we use the wetted edge or area-based equivalent diameter, the turbulence characteristics etc. are greatly different (flow patterns and characteristics are not similar), so it seems that it is difficult to express the phenomenon generally. Therefore, here we summarized the flow characteristics by assuming the same operating power for each nozzle (at $Re = 1.5 \times 10^4$ for the orifice nozzle), which was calculated based on the flow resistance.

Figure 55 shows the nozzle outlet flow rate Q_0 of each nozzle when the operating power is the same as above. In this case, the nozzle exit cross-sectional area of the notched orifice jet is larger than that of a general orifice jet, so its flow resistance decreases and Q_0 increases. In other words, for 4-notch and 8-notch jets, Q_0 increases by approximately 8.5 and 10.2%, respectively, compared to orifice jets.

2.6.3 Flow Characteristics of Nozzle

Figure 56a shows the mean and turbulent velocity distributions u/u_{me} (u_{me}: mean flow velocity at nozzle exit) and u'/u_{me} at the A-A' cross section of the orifice jet, notched orifice jet, and pipe jet at the nozzle exit $x/d_0 = 0.2$ are shown. Since the distribution shape was axially symmetric, u/u_{me} and u'/u_{me} are shown in the left and right halves of the figure, respectively. Similarly, Fig. 56b shows the results for the B-B' cross section. In an orifice jet, the flow just before the nozzle exit is accelerated toward the center of the nozzle, resulting in contracted flow and the jet diameter just after the nozzle exit becomes smaller than the nozzle exit diameter, resulting in a concave velocity distribution shape with the maximum velocity near the outer edge of the jet.

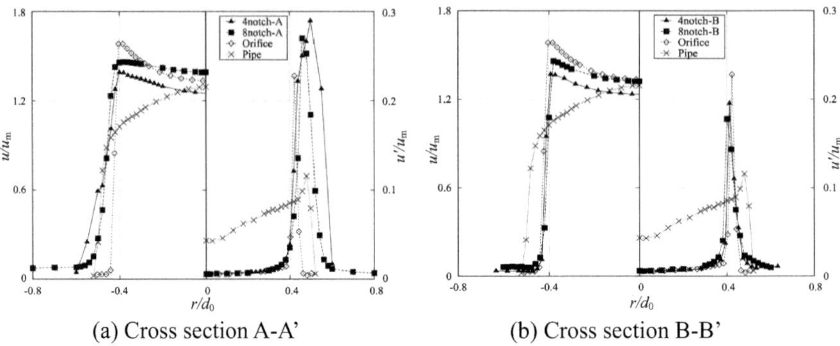

Fig. 56 Mean and fluctuating velocity distributions at nozzle exit, $x/d_0 = 0.2$)

The width of the notched orifice jet is large on the A-A' cross section, corresponding to the notch, and on the B-B' cross section, the distribution shape is similar to that of the orifice jet due to the effect of contraction. In the case of a 4-notch orifice jet, the jet width is large at the A-A' cross section, and the velocity distribution shape is significantly different between the A-A' and B-B' cross sections.

On the other hand, in the case of an 8-notch orifice jet, the vortices generated from adjacent notches sections. The maximum turbulence intensity occurs at the position of the maximum velocity gradient in both cases, and is larger in the notched orifice jet than in the general orifice jet, especially at the A-A' cross section. In addition, since the pipe jet has a long pipe length of $L/d_o = 50$, it has a fully-developed turbulent velocity distribution shape in the pipe, which is well expressed by the so-called 1/7th power law. The turbulence intensity near the center of the jet is larger than in other nozzles where the flow is contracted and accelerated toward the center.

2.6.4 Jet Centerline Velocity

Figure 57 shows the centerline velocity distribution u_c when the operation power is the same. In the case of an orifice jet, the centerline velocity increases from the nozzle exit as a result of contraction. After that, a potential core region is formed, and a development region is reached via a transition region. The same applies to the case of a notched orifice jet. In the case of a 4-notch, as shown in Fig. 54b, the nozzle exit velocity distribution shapes in the A-A' and B-B' cross sections are significantly different, so the mixing of the jet flow in the radial direction is large and its influence extends along the center line. As a result, the potential core region is short compared to other jets. In addition, the turbulent intensity u'/u_m on the center line shown in Fig. 47b increases by up to 47.2% at $x/d_0 = 3.5$ with respect to the orifice jet in the case of 4-notch, and in the case of 8-notch, when $x/d_0 \geq 8$, it becomes larger than other jets. Additionally, a rapid increase in turbulence is seen downstream compared to the case of orifice jets. This is thought to be due to the interference of the vortices generated by the notch, which increases the turbulence component.

Fig. 57 Mean and fluctuating velocity distributions along jet centerline

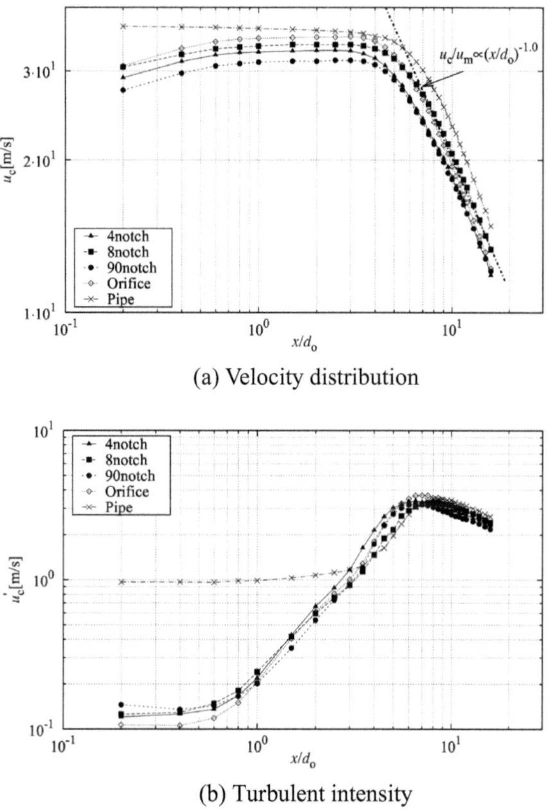

(a) Velocity distribution

(b) Turbulent intensity

In the case of a pipe jet, the turbulence is larger near the nozzle exit than in other jets, but the increase in turbulence is gradual and becomes smaller than in other jets when $x/d_0 \geqq 3$.

2.6.5 Velocity Distribution at Cross Section

Figure 58a shows the velocity distribution u/u_m on the x–r plane of the 4-notch when the operation power is constant. Since the distribution shape was symmetrical about the central axis, only the results for half the area in the radial direction were shown. As mentioned above, the velocity distribution shape at the nozzle exit is greatly influenced by the nozzle shape, and this influence remains considerably until $x/d_0 = 7.0$. The spread of the jet is quite different in the A-A' and B-B' cross sections of $x/d_0 = 7$, 4-notch, and the velocity attenuation in the center region is large in the pipe nozzle.

The velocity distribution of the B-B' cross section becomes almost the same. In this case as well, the pipe nozzle has a large velocity attenuation in the center region.

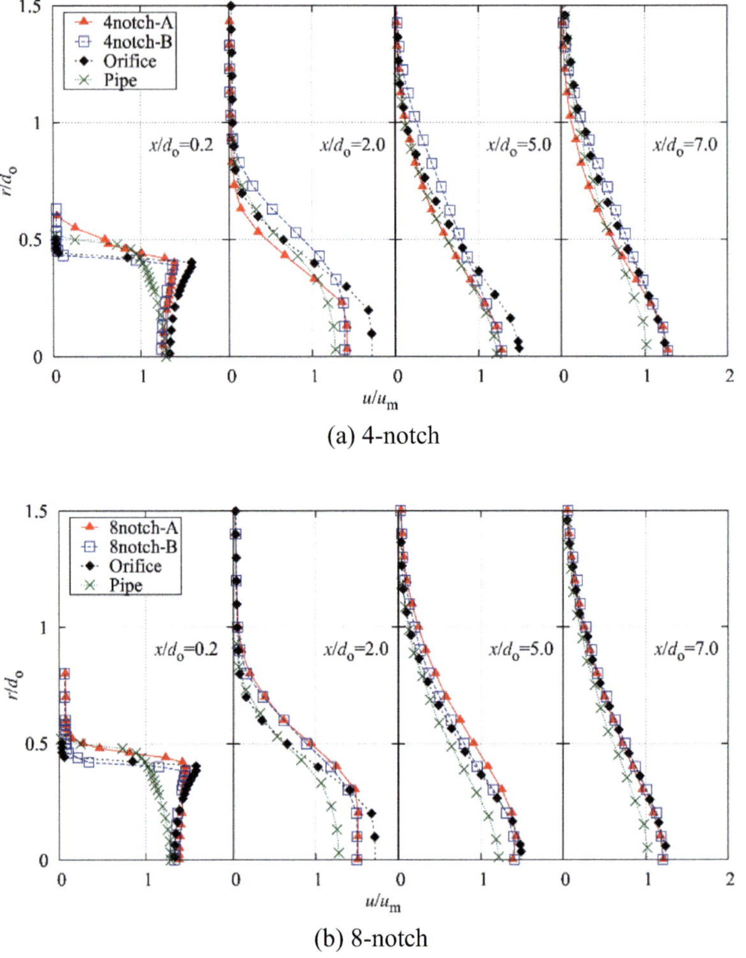

(a) 4-notch

(b) 8-notch

Fig. 58   Velocity distribution in x–r plane

Figure 59 shows the results for 8-notch as an example of the distribution of the turbulent intensity u'/u_{m} in the downstream direction. It is maximum at A-A' at the nozzle exit, but the orifice becomes larger downstream. This is because the orifice has a larger contraction effect, and for example, the velocity gradient at the outer edge of the jet at $x/d_0 = 2$ is larger than that of the 8-notch (Fig. 58b).

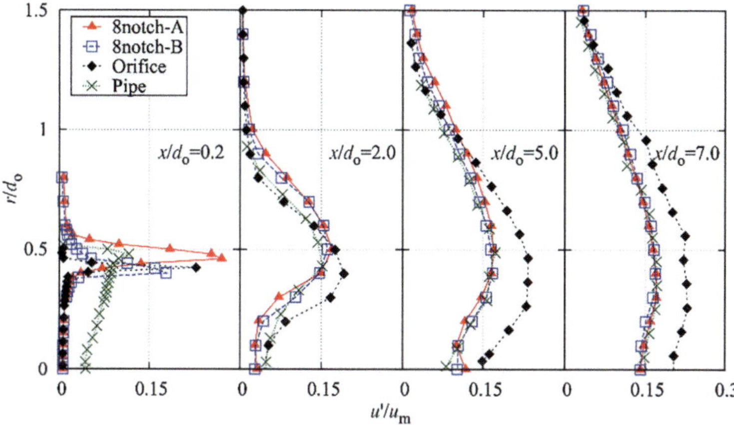

Fig. 59 Turbulence intensity for 8-notch

2.6.6 Spreading of Jet Flow and Switching Phenomenon

From the measurement results of the velocity distribution of the jet described above, the spread of the jet, that is, the jet width b/d_0 and the half-width $b_{1/2}/d_0$ were determined.

Figure 60 shows, as an example, the jet width b/d_0 and half-width $b_{1/2}/d_0$ of an 8-notch when the operating power is constant. In either case, the 8-notch has a larger radial spread than the orifice or pipe nozzle. This is because the nozzle is originally large in the radial direction at the notch, and the following switching phenomenon occurs, which increases mixing. Furthermore, for 8-notch, the jet width and half-width are reversed between $0.2 < x/d_0 < 2.0$ in the A-A' and B-B' sections.

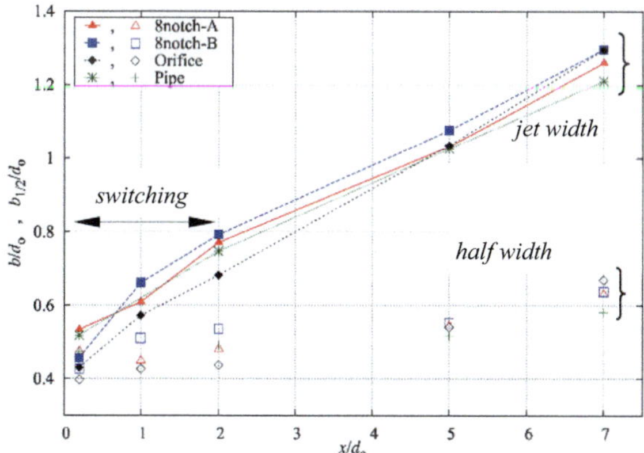

Fig. 60 Jet spreading for 8-notch, jet width b/d_0 and half width $b_{1/2}/d_0$

In addition, Fig. 61 shows a photo of the 8-notch jet stream visualized using the Schlieren method. The white line in the figure indicates the approximate half-width position of the jet. In this experiment, a heated jet with a temperature approximately 20 °C higher than the ambient temperature was used, and $Re = 1.5 \times 10^4$. In this case as well, due to the switching phenomenon, the magnitude of the half-width is reversed between the A-A' and B-B' sections between $0.2 < x/d_0 < 2.0$. That is, Fig. 62a, b show isolines of half-width in the case of 4-notch and 8-notch, respectively. As shown by the inward arrow in the figure, the velocity of the convex part at the nozzle exit becomes a concave part at $x = d_0$, whereas the velocity of the concave part at the nozzle exit becomes a convex part at $x = d_0$ (in the figure, the diagonal outward arrow). The same applies to other convex portions and concave portions. A so-called switching phenomenon occurs regarding the jet width and half-width. This causes a vortex flow in the direction shown in Fig. 63 at the end of the flow that exits the triangular notch, causing the tip (A-A' plane) to become concave. Also, as a result, the area between the tips (B-B' surface) protrudes. In general, it has been reported that a so-called switching phenomenon occurs in elliptical jets, cross-shaped jets with different nozzle lengths and widths, and the like, in which the long and short axes are swapped downstream of the nozzle exit. However, as in this study, it was shown that switching phenomena occur even in orifice jets with small notches.

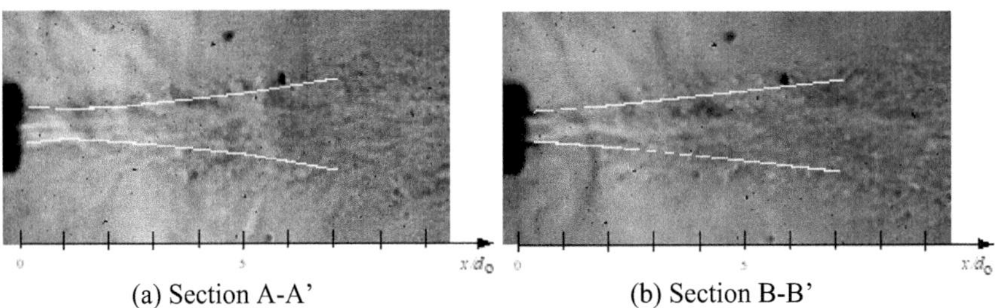

(a) Section A-A'   (b) Section B-B'

Fig. 61 Flow pattern (by Schlieren photograph)

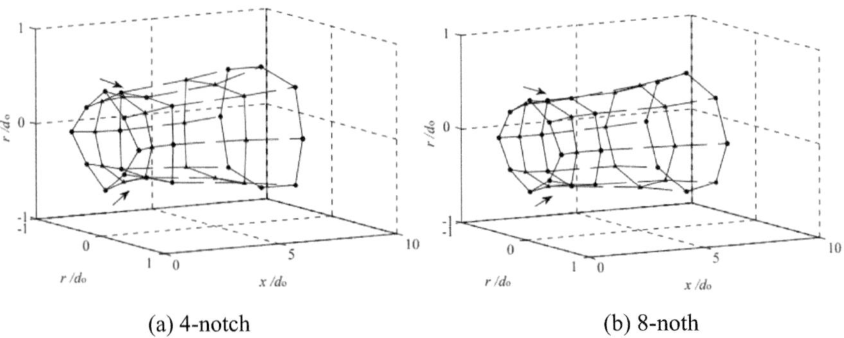

(a) 4-notch   (b) 8-noth

Fig. 62 Equi velocity profile of half-width

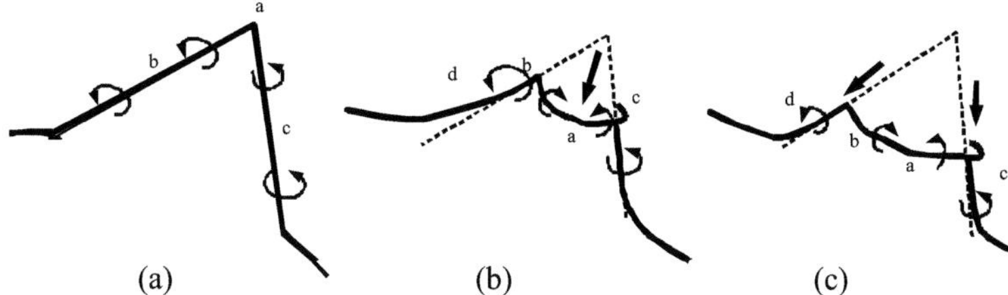

Fig. 63 Switching phenomena of notched orifice jet

Small notches greatly affect the flow structure of jets and reduce the nozzle resistance or operating power as well as increase turbulence at the nozzle exit. The turbulence intensity of the four-notched orifice jet increases along the centerline by ~47.2% at $x/d_o = 5$ compared with the other jets under the same operating power.

Axis-switching phenomena occur in the notched-orifice jets, although the notches are relatively small.

2.7 *Effects of Nozzle Plate Thickness on Orifice Free Jet*

The jet from the orifice nozzle separates at the outer edge of the nozzle exit (A in Fig. 64), and the flow is directed toward the center of the nozzle axis by the vortex area (B in Fig. 64) that is generated at the inner corner of the nozzle exit. Therefore, the jet width is smaller than the nozzle exit diameter. In other words, contracted flow occurs, and the velocity distribution at the nozzle exit becomes concave, with a large velocity at the outer edge (Fig. 67).

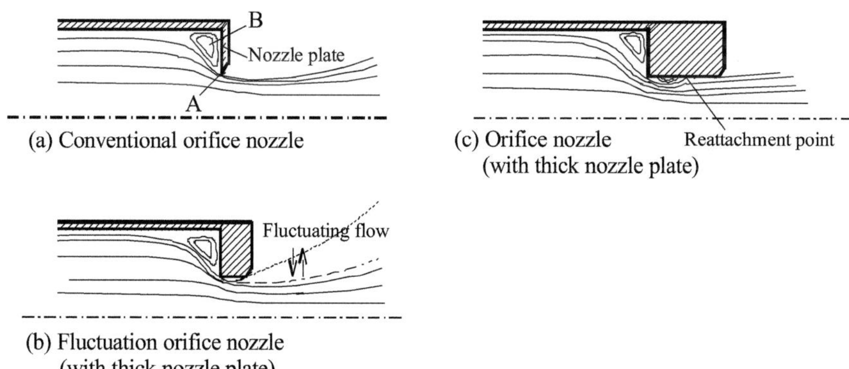

Fig. 64 Schematic flow pattern of orifice free jet at center cross section (only half region in radius direction, because of axisymmetric pattern)

Quinn (2006) pointed out that the effects of nozzle shape and outlet velocity distribution shape on flow development were not discussed. Furthermore, he experimentally investigated the development of jets from an orifice nozzle with a sharp outer edge and a contour nozzle with a gently reduced outlet in the field near the nozzle exit (initial region). The results showed that the jet from the orifice had greater mixing with the surrounding fluid than that from the contour nozzle. Lee and Lee (2000b) experimentally analyzed the effect of the nozzle shape of the orifice impinging jet on the heat transfer characteristics on the impingement plate. The results showed that in the case of an orifice nozzle with sharp corners, the heat transfer characteristics at the stagnation point are improved more than others due to the contraction effect. In addition, Shakouchi et al. (2008a, b) clarified the effects of the contraction area ratio of orifice nozzle on the flow and heat transfer characteristics of orifice free and impinging jets.

By the way, in order to increase the turbulence in the jet shear layer and increase the entrainment of surrounding fluid, that is, to improve the mixing and diffusion characteristics of the jet, the nozzle plate thickness of the orifice nozzle (hereinafter referred to as orifice plate thickness) may be considered to oscillate the flow as explained next, but there seems to be no such research. In other words, as shown in Fig. 64a, the flow in a normal orifice nozzle separates at the nozzle edge and causes contraction. However, when the nozzle plate thickness becomes considerably large (Fig. 64c), the flow separated at the nozzle edge forms an annular vortex region between the nozzle plate inner wall, reattaches to the nozzle plate inner wall, and flows out.

Here, we will clarify how the jet mixing and diffusion characteristics can be improved by the jet vibration phenomenon described above. In other words, the purpose is to control the flow characteristics of the orifice free jet by changing the orifice plate thickness, and the effect of the orifice plate thickness on the flow characteristics of the orifice free jet is visualized and observed, and the velocity distribution is measured. This will be experimentally clarified (Kito et al. 2008a, b).

2.7.1 Nozzle Configuration

Figure 65 shows the nozzle shape and coordinate system. The nozzle exit diameter, pipe diameter, pipe wall thickness, and straight pipe length are the same for all nozzles, $d_0 = 10.0$ mm, $d_i = 19.23$ mm, $t_p = 1.5$ mm, $L = 500$ mm ($L/d_0 = 50$). The contraction area ratio (nozzle exit cross-sectional area/original pipe cross-sectional area) of orifice nozzle was set to $CR = 0.27$ for a typical orifice nozzle. For comparison, a pipe nozzle with $d_o = 10.0$ mm, $t_p = 1.0$ mm, and $L = 500$ mm was also used. The coordinate system has the nozzle exit center as the origin O, the nozzle center downstream direction as the x coordinate, and the nozzle radial direction as the r coordinate.

Visualization and observation of the flow was performed as follows. In a sufficiently large water tank filled with water, the fluorescein sodium aqueous solution is jetted from the nozzle at a relatively low speed (Reynolds number $Re = u_m d_0/$

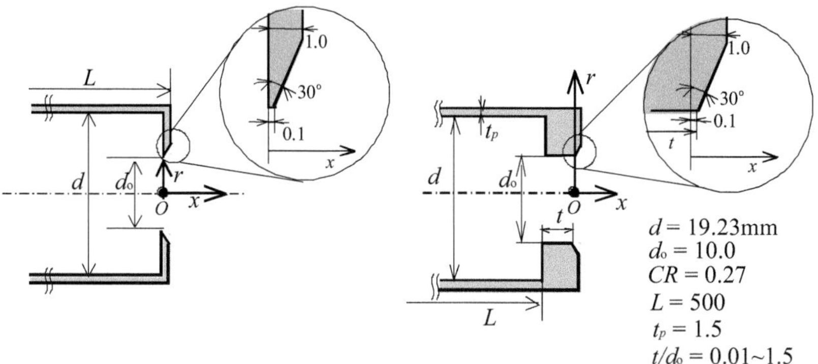

Fig. 65 Configuration of orifice nozzle ($CR = 0.27$)

$v = (1.0\text{~}3.0) \times 10^3$, u_m: mean velocity at the nozzle exit). A digital video camera was used to photograph and record the flow state at the center cross section of the jet, which was ejected and optically sectioned using a laser beam (Ar, 3W). Based on the results, we investigated the large-scale eddy structure and diffusion of the jet. The average and fluctuating velocity (turbulence strength) distribution of the jet was measured using a constant temperature hot wire anemometer/I type probe (hot wire: 5 μm diameter tungsten wire, prong spacing: 2.0 mm, sensor length: 0.8 mm). In addition, frequency analysis (FFT analysis) was performed on the centerline velocity near the nozzle exit to examine the fluctuation characteristics of the jet flow.

2.7.2 Visualized Flow Pattern

In order to visually examine the effect of the orifice plate thickness on the flow characteristics of the free jet, the orifice water jet was visualized and observed at $Re = (1.0\text{~}3.0) \times 10^3$.

Figure 66 shows a visualized image of the orifice water jet at $Re = 1.0 \times 10^3$. In a pipe jet (Fig. 66a), the velocity distribution at the nozzle exit has a thick boundary layer, the velocity gradient at the outer edge of the jet is not very large, and there is no turbulence, so the jet almost mixes with the surrounding fluid. It flows down in a belt shape. However, in the orifice jet, an axially symmetrical vortex ring is observed in all cases, and the location and frequency of its occurrence, as well as the state of diffusion, differ depending on the orifice plate thickness t. When t is small (Fig. 66b), the jet is strongly influenced by the contraction flow even after ejection, and a large-scale vortex ring is formed due to the large velocity gradient at the outer edge of the jet. The vortex ring develops downstream and then collapses, and the jet mixes and diffuses with the surrounding fluid. When t increases to $t/d_0 = 0.3$ (Fig. 66c), the flow state becomes as shown in Fig. 64b above, and radial diffusion is significantly

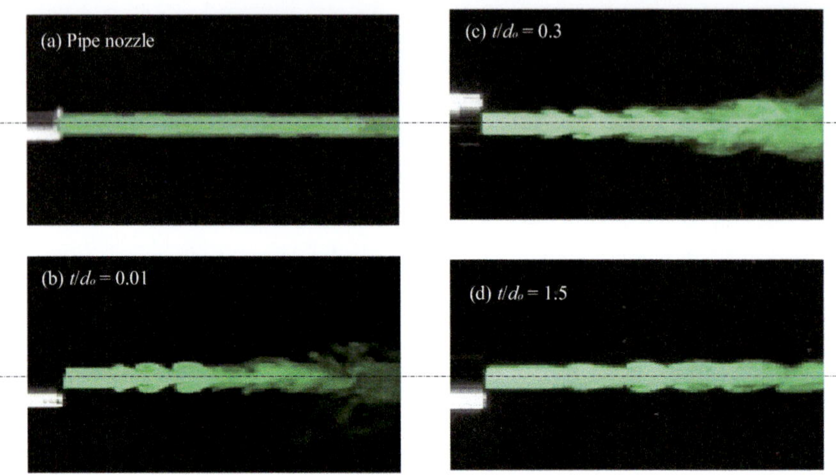

Fig. 66 Flow pattern of orifice free jet at $Re = 1.0 \times 10^3$ ($CR = 0.27$)

Fig. 67 Mean and fluctuating velocities at nozzle exit ($x/d_0 = 0.2$, $CR = 0.27$)

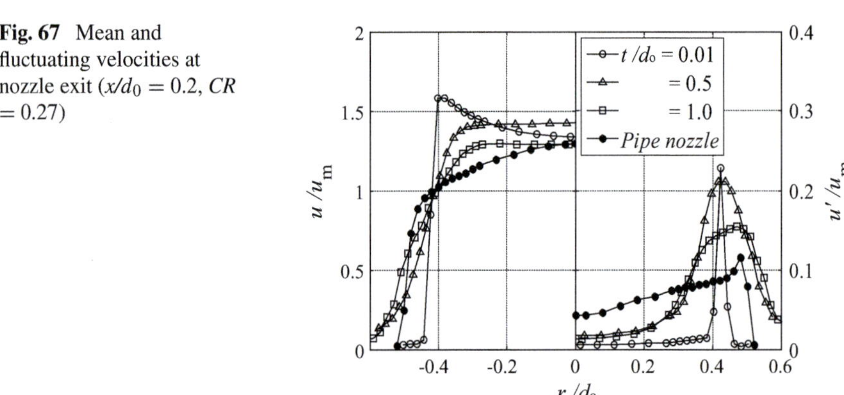

promoted. When t further increases to $t/d_0 = 1.5$ (Fig. 66d), the flow state becomes as shown in Fig. 64c, and the radial diffusion of the jet is not so large. It is thought that it approaches that of a pipe jet. In this way, the orifice plate thickness greatly influences the formation and development of a large-scale axisymmetric vortex ring.

2.7.3 Velocity Distributions at Nozzle Exit and on Jet Centerline

In the above, the effect of the orifice plate thickness on the flow characteristics of the orifice jet at low Reynolds numbers was investigated.

Here, we examine the effect of orifice plate thickness on the flow characteristics of the orifice air jet in the case of fully developed turbulence ($Re = 1.5 \times 10^4 =$ const.).

(a) **Velocity distribution at nozzle exit**

Figure 67 shows the orifice plate thickness $t = 0.1, 5.0$ and 10.0 mm, and the velocity distribution u/u_m (left half in the figure) and turbulence intensity distribution u'/u_m at the nozzle exit ($x/d_0 = 0.2$) of the pipe nozzle (right half in the figure) is shown. Since the distribution shape was axially symmetric, only half the area in the radial direction is shown. The pipe nozzle has a sufficiently long pipe length of $L/d_0 = 50$, so it exhibits a fully-developed velocity distribution shape. On the other hand, when $t/d_0 = 0.01$, the distribution shape has a maximum velocity at the outer edge of the jet, and the width of the jet is already quite small at the nozzle exit ($x/d_0 = 0.2$). Furthermore, when $t/d_0 = 0.1$, contraction was also observed and a similar distribution shape was obtained. As t increases, the velocity at the outer edge of the jet decreases and the jet width increases. At $t/d_0 = 1.0$, as mentioned above, the flow separated at the nozzle edge of the orifice reattaches to the inner wall while forming an annular vortex region between it and the inner wall of the nozzle plate, and then a boundary layer develops to some extent. In all cases, the turbulence intensity is maximum at the outer edge of the jet where the velocity gradient is greatest, and is minimum at the center of the jet. While the turbulence intensity at the outer edge of the jet at $t/d_0 = 0.01$ is caused by the velocity gradient, the turbulence intensity at $t/d_0 = 0.5$ is thought to be mainly due to vibrations occurring in the jet as described above. Note that the turbulence intensity near the center of the jet is greatest in the pipe jet and decreases as t decreases. Furthermore, the position where the turbulence intensity is maximum moves toward the outer edge of the jet as t increases.

(b) **Fluctuation properties at nozzle exit**

Figure 68 shows the turbulence intensity u'/u_m at the exit center of each nozzle ($x/d_0 = 0.2$, $r/d_0 = 0$). The turbulence intensity increases as t increases and reaches its maximum at $t/d_0 = 0.5$. The turbulent intensity at $t/d_0 = 0.5$ is approximately 2.3 times that at $t/d_0 = 0.01$. This is thought to occur because the flow separated from the nozzle edge forms a vortex region between it and the nozzle inner wall, repeating the situation in which it may or may not reattach to the nozzle exit end, as shown in Fig. 64b. In other words, this is thought to be due to the occurrence of oscillation phenomena. In the case of $Re = 1.0 \times 10^3$, this flow state occurred at $t/d_0 = 0.3$ (Fig. 66c), but at high Reynolds numbers, the annular vortex region formed between the separated flow and the inner wall of the nozzle plate increases, which is thought to occur at $t/d_0 = 0.5$. As t increases further, the orifice plate is thicker (the nozzle inner wall is longer), so the jet reattaches to the nozzle inner wall, and then the flow phenomenon approaches that of a pipe jet (the flow is not developed).

Figure 69 shows the dimensionless power spectrum Φ_u^* obtained by FFT analysis of velocity fluctuations at the center of the nozzle exit. f^* is the dimensionless frequency ($= f\, d_0/u_\mathrm{m}$). Here, the dimensionless power spectrum was obtained from the following equation, $\int \Phi_u^*(f*)df^* = 1$, according to Mi et al. (2001). Therefore, if the peak frequency Φ_u^* decreases, it means that the flow includes other frequency bands and the dominant frequency is no longer clear. When t is small ($t/d_0 \leq 0.1$), the center of the nozzle exit ($x/d_0 = 0.2$, $r/d_0 = 0$) contains various frequency components, and no clear dominant frequency can be seen. Mi et al. (2001) reported that

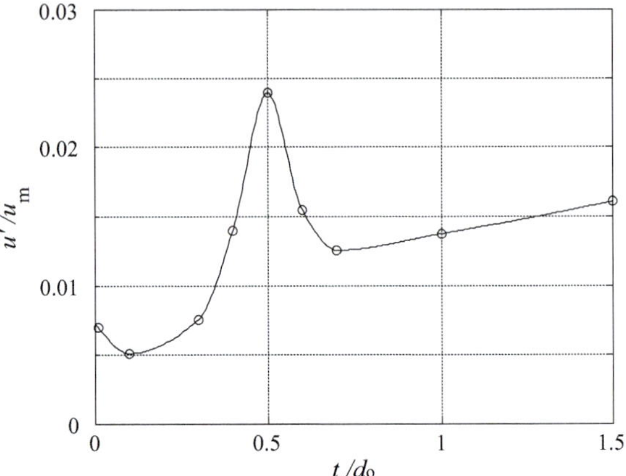

Fig. 68 Turbulent intensity on jet centerline at nozzle exit ($x/d_0 = 0.2$, $r/d_0 = 0$)

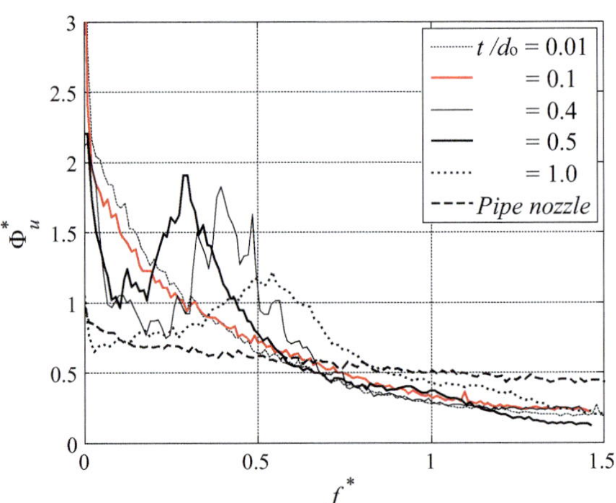

Fig. 69 Normalized power spectra at nozzle exit ($x/d_0 = 0.2$, $r/d_0 = 0$)

a normal orifice nozzle exhibits a dominant frequency $f^* = 0.7$ at $x/d_0 = 0$. In this experiment, when $t/d_0 \leq 0.1$, a similar dominant frequency $f^* = 0.7$ was observed after $x/d_0 \fallingdotseq 1.5$. Note that a clear dominant frequency appears when $t/d_0 \geq 0.4$, and when $t/d_0 \fallingdotseq 0.5$, Φ_u^* reaches its maximum at the dominant frequency $f^* \fallingdotseq 0$.

Figure 70 shows the relationship between the dimensionless frequency f^* and dimensionless power spectrum Φ_u^* of the velocity fluctuation at the center of the nozzle exit and the orifice plate thickness t/d_0. A dominant frequency appears as t/d_0 increases, and Φ_u^* reaches its maximum at $t/d_0 = 0.5$, clearly showing a dominant frequency. When $t/d_0 > 0.5$, Φ_u^* decreases, that is, various frequency components become visible. Also, the dominant frequency $f^* \fallingdotseq 0.3$ at $t/d_0 = 0.5$ is smaller than that for other jets.

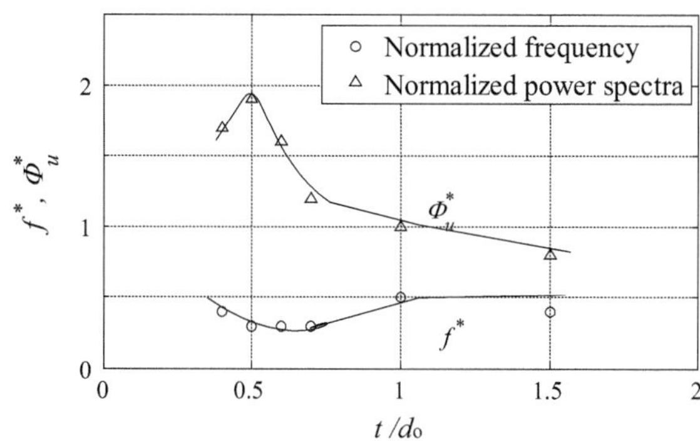

Fig. 70 Variation of f^* and Φ_u^* at nozzle exit ($x/d_0 = 0.2$, $r/d_0 = 0$)

(c) Flow characteristics on jet centerline

Figure 71a, b show the centerline flow velocity distribution u_c/u_m and turbulent intensity distribution u_c'/u_m, respectively. When $t/d_0 = 0.01$, as mentioned above, the velocity vector before the nozzle exit points toward the jet center, so the centerline velocity increases from the nozzle exit. After that, it passes through the potential core region, transition region, and reaches the development region. All distribution shapes in the developmental region are expressed as $u_c/u_m \propto (x/d_0)^{-a}$. Here, a is an experimental constant and $t/d_0 = 0.01, 0.5, 1.0$, and in the case of a pipe nozzle, a $\doteqdot$ $-1.0, -0.88, -0.92, -1.0$, respectively. Note that when t/do increases, no contraction occurs in the velocity distribution at the nozzle exit ($x/d_0 = 0.2$) (Fig. 67). Therefore, a potential core region is formed immediately after the nozzle exit, and it attenuates after passing through a transition region. Note that the length of the potential core region is $x/d_0 \doteqdot 6, 2.2$, and 3.5 when $t/d_0 = 0.01, 0.5$, and 1.0, respectively. Here, the length of the potential core region was set from the nozzle exit to the position where the centerline velocity was 95% of the nozzle exit velocity, according to Livingood et al. (1973). The potential core region length for $t/d_0 = 0.5$ is smaller than for other jets, and is about 0.4 times that for a normal orifice nozzle ($t/d_0 = 0.01$). In other words, it can be seen that the jets are rapidly mixed and diffused in the radial direction.

2.7.4 Velocity Distributions at Cross Section

(a) Velocity distribution

Figure 72a, b show the velocity distribution u/u_m and turbulence intensity distribution u'/u_m at each downstream cross section, respectively. When t/d_0 is small, contracted flow occurs, resulting in a large velocity gradient at the outer edge of the jet and a large shear force. As a result, the velocity at the center of the jet is large, and the entrainment of surrounding fluid becomes large. In addition, turbulence increases downstream due to the formation of strong shear layers. When $t/d_0 = 0.5$, the potential core region is shorter than that of other jets, and the jet center velocity decreases rapidly toward the downstream. In other words, when $x/d_0 \geq 3.0$, mixing and diffusion with the

Fig. 71 Jet centerline
velocities

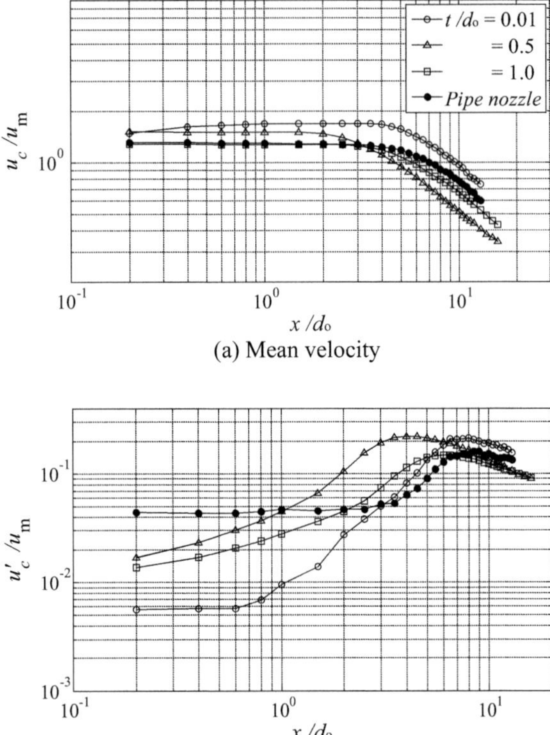

(a) Mean velocity

(b) Turbulenat intensity

surrounding fluid progresses, and the turbulence strength has a larger distribution
shape than others.

Figure 73 shows the velocity distribution in each cross section in dimensionless
form using the center velocity u_c and the half width $b_{1/2}$. The solid line in the figure
is the theoretical analysis result of Tollmien (1926) for an axisymmetric jet. When $t/d_0 = 0.01$, $r/d_0 > 5$, which almost agrees with Tollmien's analytical solution. When
$t/d_0 = 0.5$, $r/d_0 > 3.0$ almost matches Tollmien's analytical solution (Fig. 73b). Note
that when t/do is large ($t/d_0 \geq 1.0$), $r/d_0 > 5$, which almost agrees with Tollmien's
analytical solution.

(b) **Spreading of jet and volumetric entrainment flow rate**

Figure 74 shows the spread of the jet in terms of the half-width $b_{1/2}/d_0$ and the jet
width b/d_0. The jet width was determined by defining the radial position of 10% of
the maximum velocity in each cross section as the jet edge. When $t/d_0 = 0.5$, the
jet spreads greatly in the radial direction. Furthermore, as t/d_0 increases, both the jet
width and the half-width increase. he spread of the jet is large when the turbulence
intensity is $x/d_0 \geq 3.0$, and becomes largest when the peak frequency is small at t/d_0

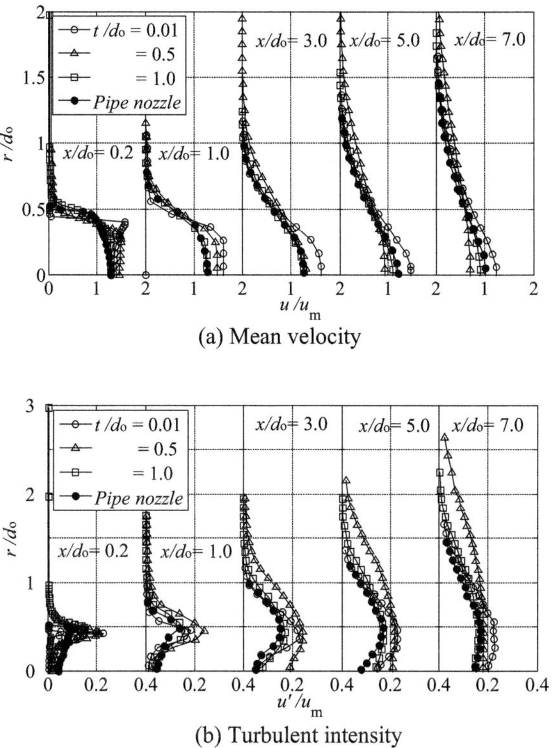

(a) Mean velocity

(b) Turbulent intensity

Fig. 72 Mean velocity and turbulent intensity at cross section

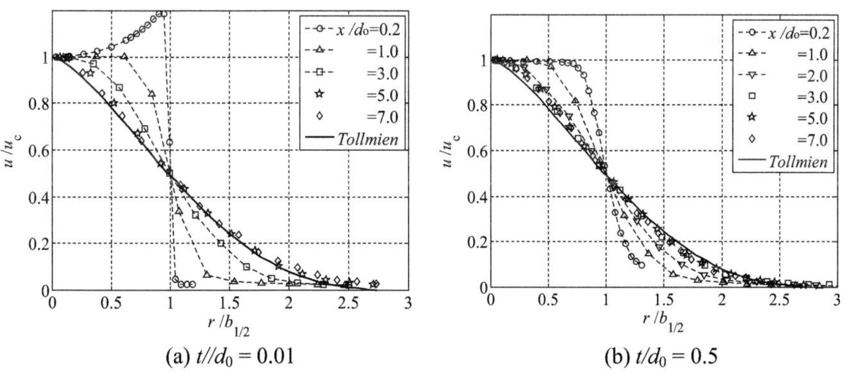

(a) $t/d_0 = 0.01$ (b) $t/d_0 = 0.5$

Fig. 73 Velocity profile

Fig. 74 Jet and half widths

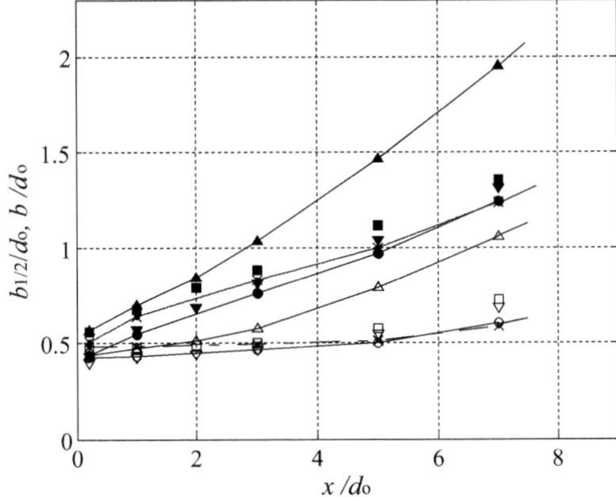

= 0.5. For example, the jet width at $x/d_0 = 7.0$ for $t/d_0 = 0.5$ is approximately 1.6 times that of the pipe jet and $t/d_0 = 0.01$.

Figure 75 shows the entrainment volumetric flow rate $(Q_x - Q_0)/Q_0$ determined from the velocity distribution. Q_x and Q_0 are the volumetric flow rates at arbitrary position x and nozzle outlet ($x/d_0 = 0.2$), respectively, and were calculated from the following equations.

$$Q = 2\pi \int_0^\delta urdr \tag{64}$$

Here, δ is the position in the radial direction that is 1% of the center flow velocity at the nozzle exit, as in Boguslawski et al. (1979). Figure 76 also shows the results of a pipe jet with a fully-developed turbulent velocity distribution in the pipe $[(Q_x - Q_0)/Q_0 = 0.183(x/d_0)]$ by Boguslawski et al. The pipe jet results of this experiment are in good

Fig. 75 Entrainment flow rate

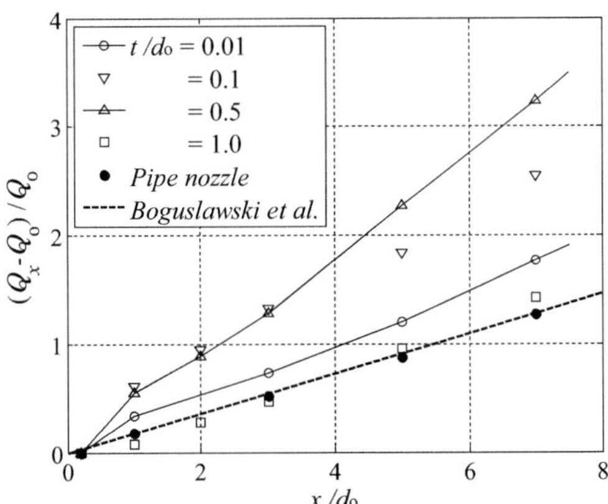

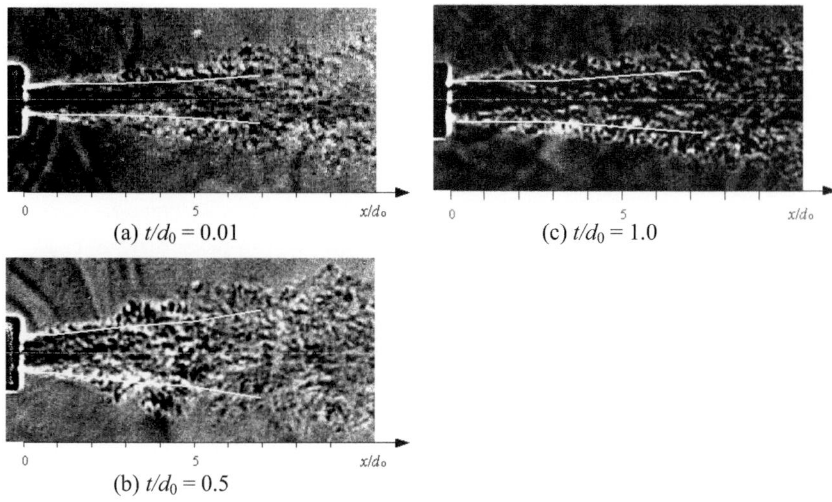

Fig. 76 Visualized flow pattern by Schlieren method ($CR = 0.27$, $Re = 1.5 \times 10^4$)

agreement with the results of Boguslawski et al. Also, when $t/d_0 = 1.0$, the results were almost the same as the pipe jet. The entrainment flow rate increases in the downstream direction as t increases, but the rate of increase in the entrainment flow rate also reaches its maximum at $t/d_0 = 0.5$, where the jet width is the largest. For example, the entrained flow rate at $x/d_0 = 7.0$ with $t/d_0 = 0.5$ is approximately 2.6 times and 1.8 times that of the pipe jet flow and $t/d_0 = 0.01$, respectively.

2.7.5 Diffusion of High Temperature Orifice Free Jet

Here, we use the Schlieren method to visualize and observe jets at high temperatures (60 °C) and high Reynolds numbers, and examine the mixing and diffusion characteristics due to jet fluctuation phenomena. Furthermore, we will examine the mixing and diffusion characteristics due to jet fluctuation phenomena, and examine the relationship between orifice plate thickness and fluctuation phenomena, and the effect of Reynolds number on mixing and diffusion characteristics, by measuring velocity distribution using a hot wire anemometer. We also clarify the relationship between the temperature distribution and velocity distribution of the jet.

(a) **Flow visualization by Schlieren method**

The nozzle Reynolds number is set to $Re = 1.5 \times 10^4 =$ const. using the kinematic viscosity coefficient of the high-temperature jet, and the mixing and diffusion characteristics due to the fluctuation phenomenon of the jet are clarified.

Figure 76 shows the Schlieren image. The white solid line in the figure indicates the position where $T - T_\infty = (1/2)\,(T_c - T_\infty)$, that is, the position of the half width R_{half} of the temperature distribution. When $t/d_0 = 0.5$, fluctuation phenomena

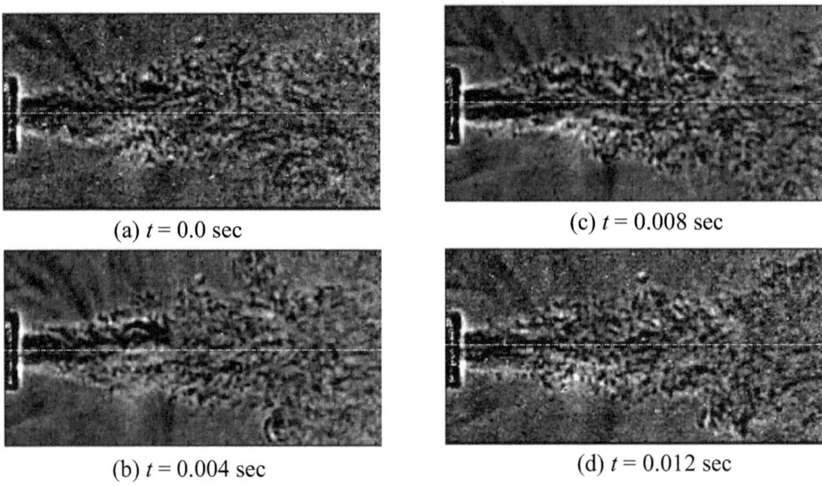

(a) $t = 0.0$ sec   (c) $t = 0.008$ sec

(b) $t = 0.004$ sec   (d) $t = 0.012$ sec

Fig. 77 Time series Schlieren images at 0.004 s images ($CR = 0.27$, $t/d_0 = 0.5$, $Re = 1.5 \times 10^4$)

that are not seen when $t/d_0 = 0.01$ and 1.0 occur, and the jet diffusion and mixing characteristics are significantly improved compared to other nozzles. For example, R_{half} at $x/d_0 = 3$ and 7 for $t/d_0 = 0.5$ is about 1.3 and 1.6 times, respectively, when $t/d_0 = 0.01$, and 1.4 and 1.5 times when $t/d_0 = 1.0$.

Figure 77 shows, as an example, a time-series Schlieren image at 0.004 s intervals when $t/d_0 = 0.5$. It can be seen that the jet flow fluctuates greatly. As mentioned above, this is thought to be caused by the fluctuation phenomenon caused by the flow separated from the nozzle edge reattaching to the nozzle outlet end or not. The fluctuation phenomena observed in the images were not periodic and sometimes behaved like a swirling jet. When the flow separated from the nozzle edge reattaches to the nozzle exit, the radial velocity increases, the radial diffusivity increases, and the turbulence within the jet increases significantly.

(b) **Mean and fluctuating velocity distributions on jet centerline**

Figures 78 and 79 show the mean velocity u_c/u_m and turbulence intensity distribution u'_c/u_m on the jet center line for nozzles with orifice plate thicknesses $t/d_0 = 0.01$ and 0.5. The mean velocity distribution becomes a stable turbulent state at high Reynolds numbers $Re = (0.70{\sim}2.5) \times 10^4$, and is almost the same in all cases. When $t/d_0 = 0.01$ (Fig. 78), contracted flow occurs at the nozzle exit, so the centerline velocity increases from the nozzle exit. After that, a potential core region where the velocity does not change is formed, and it decreases through a transition region. The centerline flow velocity distribution u_c/u_m is expressed as $u_c/u_m = (x/d_0)^a$ in all developed regions (a: experimental constant). When $Re = 3.0 \times 10^3$, the appearance of attenuation is different, but at high Reynolds number it is expressed as a $\doteqdot -1.0$. When $Re = (1.5{\sim}2.5) \times 10^4$, the turbulence intensity distribution is nearly identical and has one peak, but when $Re = (0{\sim}7.0) \times 10^3$, the distribution has two peaks. This distribution with two peaks is observed in the case of orifice jets with low Reynolds numbers, and is thought to be related to the generation of large-scale vortex structures. The turbulent intensity distribution (Figs. 78b and 79b) has two peaks in the case of low Reynolds number $Re = 3.0 \times 10^3$. The increase or decrease in turbulence is considered to be the effect of large-scale vortex structure generation. When $Re = (0.7\sim2.5) \times 10^4$, the experimental constant was a $\fallingdotseq$ −0.88. Furthermore, when t/d_0 increases further, the velocity distribution shape almost matches $Re = (0.3\sim2.5) \times 10^4$, and the experimental constant is a $\fallingdotseq$ −0.92. From this, the decrease of the temperature distribution and that of the velocity distribution shown in Sect. 2.7.4a show similar trends, the experimental constant a almost matches b, and when $t/d_0 = 0.5$, the attenuation on the center line is different from that of the velocity distribution. It is gentler than the jet flow. when $t/d_0 = 0.5$, the decrease of temperature and velocity on the jet centerline is gradual compared to other jets.

(c) **Potential core length**

As mentioned above, there is a potential core region in the jet region near the nozzle where the centerline velocity does not attenuate (in the case of an orifice jet, after the velocity increases as a result of contraction). There is a mixing region around it where the jet and surrounding fluid mix, and as the mixing region develops, the potential

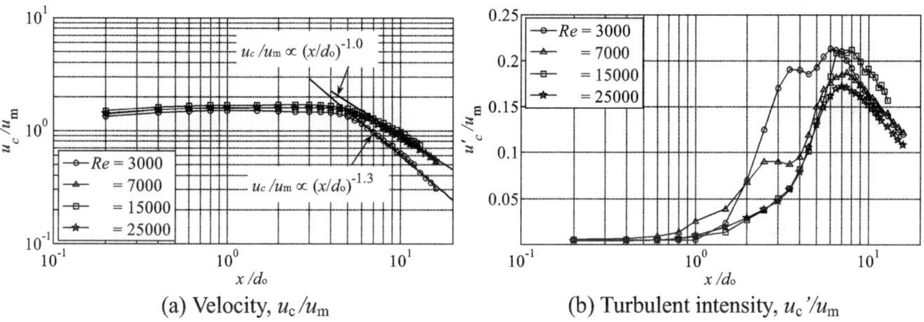

(a) Velocity, u_c/u_m (b) Turbulent intensity, u_c'/u_m

Fig. 78 Mean and fluctuating velocities on jet centerline ($t/d_0 = 0.01$)

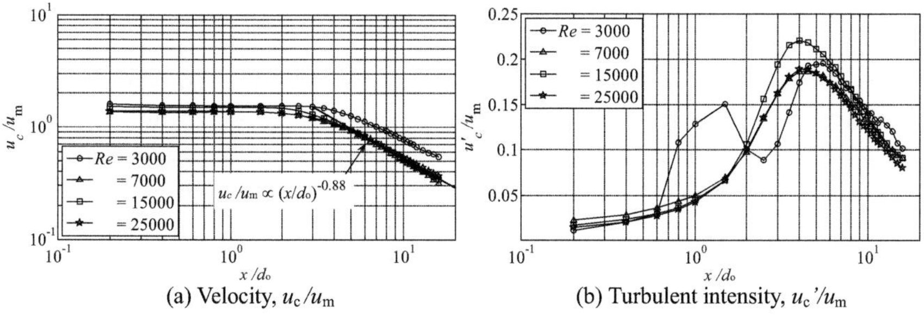

(a) Velocity, u_c/u_m (b) Turbulent intensity, u_c'/u_m

Fig. 79 Mean and fluctuating velocities on jet centerline ($t/d_0 = 0.5$)

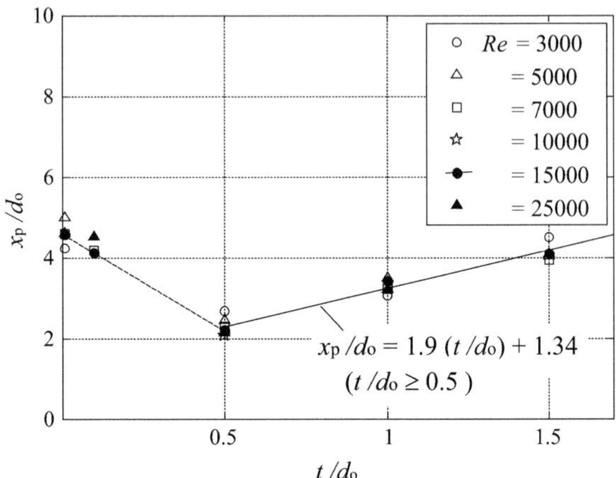

Fig. 80 Potential core length

core region decreases and eventually disappears. Therefore, in order to promote jet mixing and diffusion, it is desirable to eliminate the potential core region as quickly as possible. The potential core length x_p was set from the nozzle exit to the position where the centerline velocity attenuated to 95% of the nozzle exit velocity.

Figure 80 shows the influence of orifice plate thickness t/do on potential core length x_p at $Re = (0.30{\sim}2.5) \times 10^4$. x_p does not depend on the Reynolds number when $Re = (0.70{\sim}2.5) \times 10^4$, and becomes minimum when the orifice plate thickness $t/d_0 = 0.5$. The potential core length when $t/d_0 > 0.5$ and $Re > 0.70 \times 10^4$ is expressed by the following formula.

$$\frac{x_p}{d_0} = 1.9\left(\frac{t}{d_0}\right) + 1.34 \tag{65}$$

In addition, in the case of a pipe jet ($t/d_0 = \infty$), $x_p/d_0 \doteq 4.4$ was independent of the Reynolds number within the range of this experiment.

(d) Temperature distribution on jet centerline

Figure 81 shows the dimensionless temperature distribution $(T_c - T_\infty)/(T_0 - T_\infty)$ on the center line, taking into account room temperature fluctuations. When $t/d_0 = 0.5$, as is clear from the visualization image shown above, the jet has excellent radial diffusion. Therefore, for $t/d_0 = 0.5$, the temperature on the centerline starts to decay faster than for other jets. Note that the centerline temperature distributions for $t/d_0 = 0.01$ and 1.0 are almost the same. The centerline temperature of the pipe and orifice jet with $t/d_0 = 0.01$ and 1.0 is expressed by the following equation.

$$\frac{(T_c - T_\infty)}{(T_0 - T_\infty)} = \left(\frac{x}{d_0}\right)^b \tag{66}$$

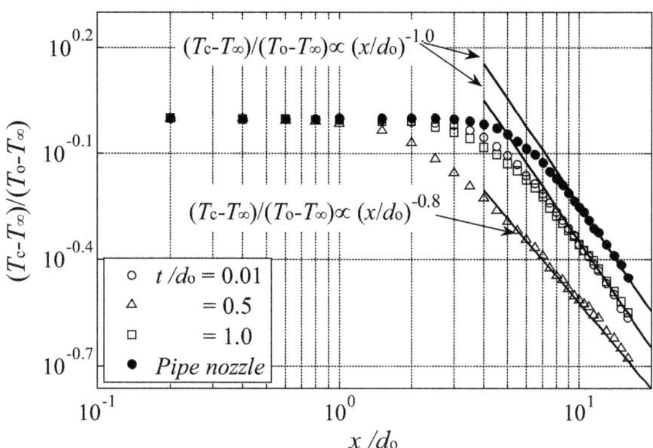

Fig. 81 Temperature distribution on jet centerline ($CR = 0.27$, $Re = 1.5 \times 10^4$)

Here, b is an experimental constant and $t/d_0 = 0.01$, 0.5, 1.0, and for pipe nozzles, $b = -1.0, -0.8, -1.0, -1.0$, respectively.

When $t/d_0 = 0.5$, the jet temperature on the center line decreases somewhat more slowly than other jets. Assuming that the potential core length of the temperature field is set to the position where the nozzle exit temperature is 95% as in the case of the velocity field, the potential core length is $x/d_0 = 4.59, 2.2, 41, 4.34$ for $t/d_0 = 0.01$, 0.5, 1.0, and for the pipe nozzle, respectively. From this, the potential core length of the temperature field is about 30% shorter than that of the velocity field, except for the pipe jet. This agrees with many past experimental results. Corrsin and Uberoi (1951) found that temperature fluctuations differ from velocity fluctuations in that they are distorted in a certain direction (skewness $\neq 0$), and associated this with the generation of vortices. The radial jet dispersion is caused by positive velocity fluctuations, which are thought to be induced by large-scale eddies.

On the other hand, temperature fluctuations are caused by the passage of large-scale eddies, which involve the low-temperature surrounding fluid, resulting in negative fluctuations. Therefore, temperature decay occurs faster than velocity decay, and the potential core length of the temperature field is shorter than that of the velocity field.

(e) **Temperature distribution at cross section**

Figure 82 shows a comparison of the distribution shapes of jet temperature and velocity u/u_{max} at each cross section. Since the distribution shape was axially symmetric, only half the area in the radial direction was shown. When $t/d_0 = 0.01$ (normal orifice jet), contracted flow occurs and the jet diameter becomes smaller than the nozzle exit diameter immediately after the nozzle exit, so the radial position where the jet temperature matches room temperature shifts to the jet center. When $t/d_0 = 0.5$, as seen in the central temperature distribution, when $x/d_0 > 1.0$, the jet spreads in the radial direction, but the distribution of temperature becomes larger than that of velocity. Furthermore, as t/d_0 increases, the spread of the temperature distribution becomes smaller than when $t/d_0 = 0.5$, similar to the velocity distribution.

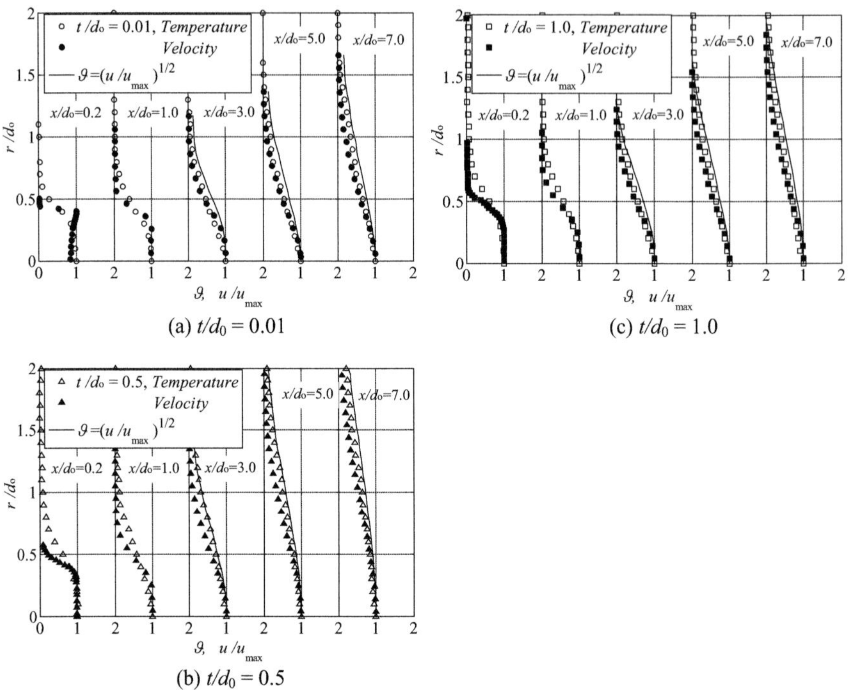

(a) $t/d_0 = 0.01$

(b) $t/d_0 = 0.5$

(c) $t/d_0 = 1.0$

Fig. 82 Temperature and velocity distributions

Now, assuming that the velocity and temperature at the center of the jet are approximately proportional to x^{-1}, the dimensionless temperature distribution is expressed by the following equation.

$$\vartheta = \frac{T - T_\infty}{T_c - T_\infty} = \left(\frac{u}{u_{max}}\right)^\delta \tag{67}$$

In Taylor's theoretical solution, $\delta = 1/2$, and the theoretical solution obtained based on the cross-sectional velocity distribution of each jet is shown in the solid line ($x/d_0 \geq 0$) in the figure. In either case, the experimental and theoretical results are almost in agreement. When $t/d_0 = 0.5$, the mixing region with the surrounding fluid develops quickly, so even when $x/d_0 = 0$, it is well represented by Taylor's theoretical solution.

(f) **Spreading of velocity and temperature distributions**

In Fig. 83, the radial position R_{half} where $\vartheta = 1/2$ is made dimensionless by the nozzle exit diameter d_0 and is shown as an open symbol. In addition, the half-width $b_{1/2}/d_0$ is also indicated with a black symbol. The diffusion of the temperature distribution shows the same tendency as the spread (half width) of the velocity distribution. R_{half} is not significantly affected by the orifice plate thickness near the nozzle exit ($x/$

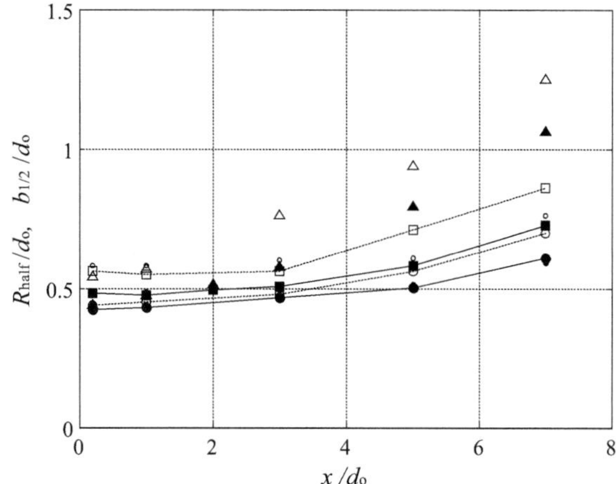

Fig. 83 Half widths of temperature and velocity distribution

$d_0 \leqq 1.0$). However, when $x/d_0 > 1.0$, the influence of the orifice plate thickness is seen, and when $t/d_0 = 0.5$, it spreads greatly in the radial direction. The jet spreads at a constant rate downstream. The gradient is about 0.08 for all other jets, but it is about 0.12 at $t/d_0 = 0.5$. Therefore, when $t/d_0 = 0.5$, the rate at which the jet spreads downstream is large.

In addition, the average value of the ratio of temperature to jet half-width $R_{\text{half}}/b_{1/2}$ is approximately 1.2 at $0 < x/d_0 \leqq 7.0$. This shows good agreement with the contour nozzle results of Drobiniak et al. (1998), and it can be seen that temperature diffusion has a slightly larger distribution shape than velocity diffusion.

When the orifice plate thickness increases, for example, at $Re = 1.5 \times 10^4$ and $t/d_0 = 0.5$, the flow separated from the nozzle edge forms a vortex region between the nozzle inner wall and the nozzle exit edge, and the jet oscillates. As a result, the turbulence intensity increases from the nozzle exit (about 2.3 times that of $t/d_0 = 0.01$), and the radial diffusion increases (for example, the jet width and entrained flow rate at $x/d_0 \doteqdot 7.0$ are about 1.6 and 2.6 times that of the pipe jet, respectively). In addition, the dominant frequency at $t/d_0 \doteqdot 0.5$ is smaller than that of the other jets, at about 0. The length of the potential core region at $t/d_0 = 0.5$ ($Re = 1.5 \times 10^4$) is shorter than that of the other jets, at $x/d_0 \doteqdot 2.2$, as a result of the above, and the attenuation of the maximum velocity is also quite large.

2.8 *Free Jer from Special Shaped Nozzle (Petal Shaped, Lobed, Nozzle)*

Here, the mixing and diffusion properties between the submerged free jet flow from the 6-petal shaped nozzle and the surroundings are examined experimentally by a flow visualization and mean and fluctuating velocity measurements. It is expected that these flow properties of the 6-petal shaped jet will be improved considerably because of the large wetting perimeter and generation of longitudinal large vortex structures (Shakouchi et al. 2014; Iriyama et al. 2012).

2.8.1 Petal Shaped Nozzle

Figure 84 shows the schematic diagram of the nozzle used of the 6-petal shaped double tube. It is consisted of the 6-petal shaped inner tube (hereafter, it is also called 6-p tube, or 6-p nozzle) and the outer tube. The inner tube is formed from 12.0 mm circular tube to the 6-p tube during 50 mm in length by a drawing process and then it becomes the 6-p tube of 200 mm in length. The outer tube is the straight tube of 250 mm in length with inner diameter of 20.1 mm, and it has an outer flow tube of diameter 10 mm set vertically on the outer tube. The inner and outer tubes are made by a pressure welding and their thickness is 0.8 mm.

Figure 85 shows the details of cross section of nozzle exit of the 6-petal shaped double tube. The exit cross section of the 6-p tube has the major axis, A-axis, of 18.5 mm in length and the minor axis, B-axis, of 7.0 mm in length, and the 6-petal shape with exchanging major and minor axes at every 30° around circumference. The area of the exit cross section of the 6-p nozzle is $A_i = 124$ mm^2, equivalent circle diameter is $d_{ei} \equiv d_e = 12.5$ mm. The area between the outer and inner tubes has 6 heteromorphic sectors, 6 outer nozzles, and total cross-sectional area is $A_0 = 108$ mm^2 and equivalent circle diameter is $d_{e0} = 11.7$ mm. The area ratio is $A_0/A_i = 0.871$. The jet flow from 6-p nozzle, 6-p jet, can be controlled by the outer jets from 6 outer nozzles, but in this study only the flow characteristics of the 6-p submerged free jet, 6-p jet, were examined. That is, no external jet was used. The results for the pipe nozzle and orifice nozzle are also shown for reference. The diameter and length of pipe nozzle are $d = d_0 = 10.0$ mm and $L/d = 50$, respectively. The coordinate system of which is the center of nozzle exit is the origin o, and in the downstream and radius directions are x and r axes, respectively, was used.

Fig. 84 Schematic diagram
of 6-petal shaped nozzle

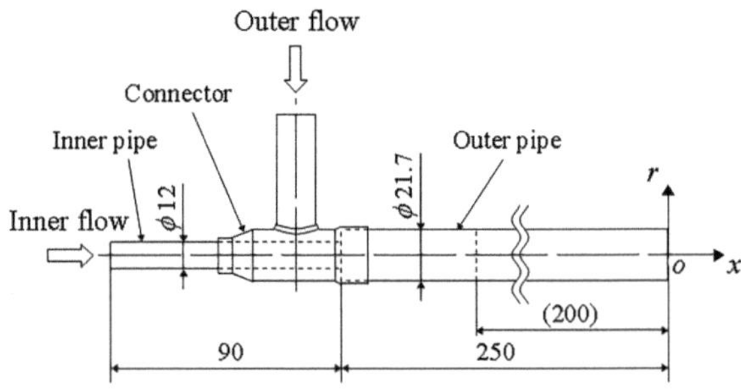

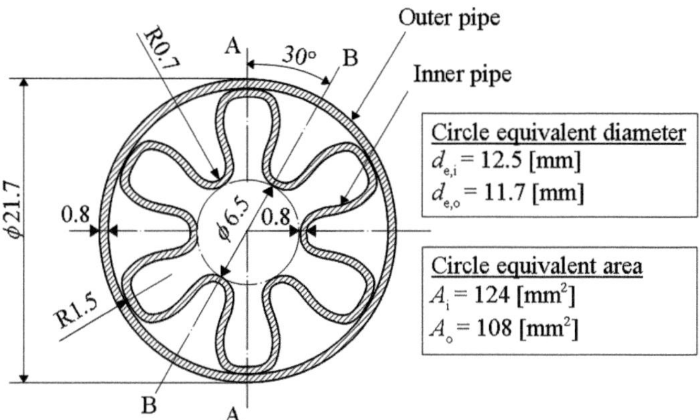

Fig. 85 Details of cross section of nozzle exit (6-petal shaped, lobed, nozzle)

2.8.2 Flow Pattern

The flow visualization and observation were carried out as follows. The nozzle was inserted vertically by the side wall under the water surface of the large test water tank which the water level is maintained a constant by the weir. The water solution of fluorescein sodium was issued from the nozzle into the still water and visualized by laser light (Spectra-Physics, Ar, 3W) sheet method. The visualized flow pattern was observed, and the brightness data was memorized and processed by a high-speed video camera (Photron, FASTCAM- 1024PCI, model 100 k), PIV (Photron, FACTCAM Viewer) and PC. The brightness distribution of the image will almost support the diffusive degree of the dye, water solution of fluorescein sodium. The flow visualization was carried out under the Reynolds number, $Re = u_m \, d \, /v = 3{,}000$ (u_m and v: mean exit velocity and kinematic viscosity, respectively).

Figure 86a shows the flow pattern at the center cross section of the pipe jet. First, there is a large non-disturbed area of the length of about $x/d_0 = 8$, and then the flow disturbance occurs from the edge area of the jet.

Figure 86b, c show the flow patterns at A- and B-sections of the 6-p jet. Since A-section is the major axis the jet width near the nozzle exit is large, and the disturbance occurs from about $x/d_e = 2.0$ near the jet edge and spreads to the downstream entraining the surrounding fluid. The jet width at A-and B-sections becomes almost same at about $x/d_e = 5.0$, and then it is thought that the cross-sectional shape of the jet edge approaches to a circle (see, Fig. 88). Compared with the pipe jet flow, the 6-p jet disturbs and spreads in the radius direction from much near the nozzle exit forming a large vortex structure, and then it is thought that 6-p jet has an excellent mixing and diffusion properties.

Figure 87 shows a three-dimensional, 3D, image of the 6-p jet. This image shows the outline of the jet, which is a stack of cross-sectional visualization images at $x/d_e = 0.4, 0.8, 1.2, 1.6, 2.4, 4.0$ ($x = 5, 10, 15, 20, 30, 50$ mm) stacked on the time axis. After $x/d_e \approx 1.2$, the jet begins to become turbulent and vibrates, and after about

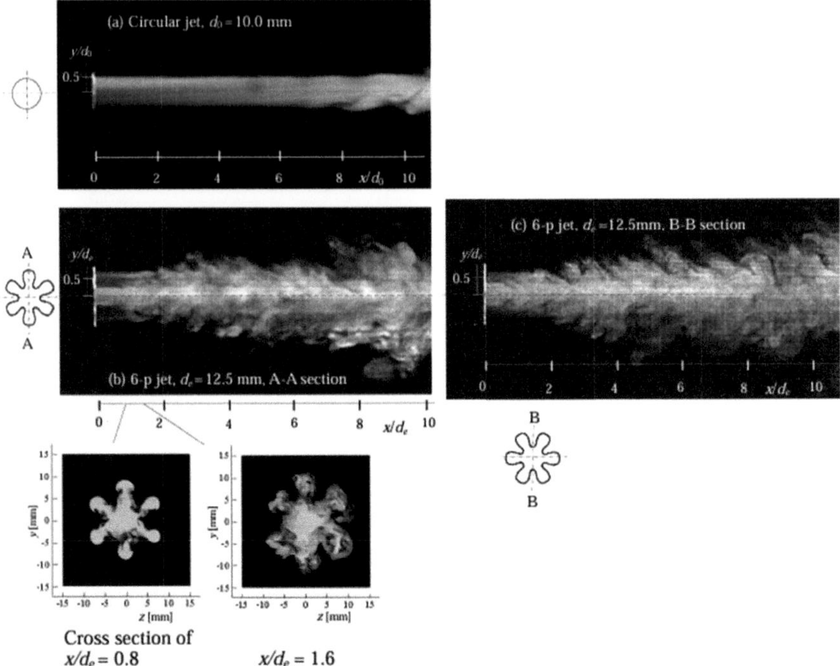

Fig. 86 Flow patterns of circular and 6-p submerged water jets, $Re = 3000$ ($d = d_0$ or d_e)

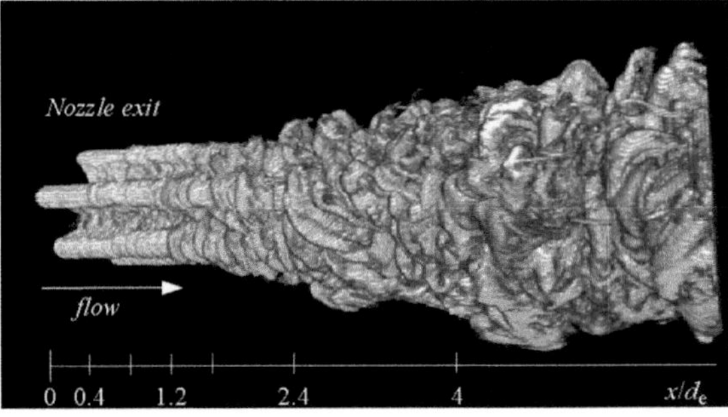

Fig. 87 Three-dimensional image, bird's eye view, of the jet was obtained by piling up the results ($Re = 3000$) (Shakouchi et al. 2014; Iriyama et al. 2012)

2.4, it spreads further in the radial direction. It can be seen that vortex masses of various sizes flow while intertwining in a complex manner, and the outline of the vortex structure on the outer surface of the jet can also be also seen.

Figure 88 shows the time mean brightness Br on the centerline of the visualized flow pattern of Fig. 88a, b. The brightness increases to the downstream from the nozzle exit and takes a maximum at $x/d_e \fallingdotseq 2.5$ and then decreases to the downstream following $Br \propto (x/d_e)^{-0.83}$ while fluctuating. It is thought that the mixing and diffusion characteristics can be estimated roughly by the brightness distribution.

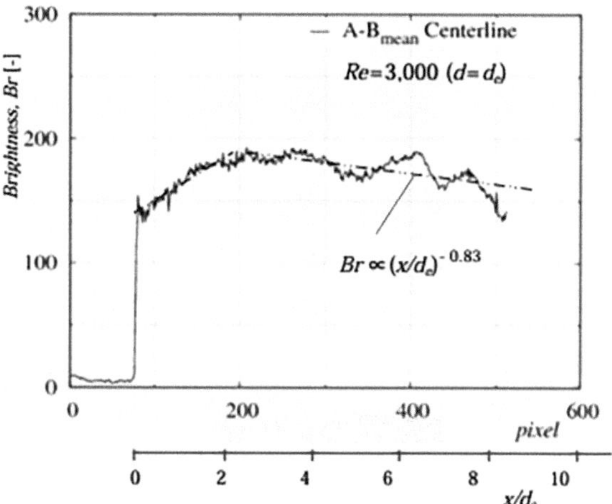

Fig. 88 Brightness Br [–] on the centerline of 6-p jet

Figure 89 shows the flow pattern and time mean brightness distribution at the cross section of $x = 10$, 15, 20, 30 mm ($x/d_e = 0.8$, 1.2, 1.6, 2.4). The mean brightness at A- and B-sections was obtained by averaging the data at three sections of the circumferential direction. Near the nozzle exit of $x = 10$ mm the flow pattern in the petal shaped area has a mushroom shape with twin counter rotating vortices by Prandtl's second kind of secondary flow. The outline of the flow is almost the shape of 6-p nozzle, and the flow oscillates irregularly and alternately in the circumference direction by a complex flow field with many large vortex structures including 6 twin longitudinal vortices (Figs. 89 and 90).

The profiles of the mean brightness agree mostly with the positions of the center and edge of the jet, and at A- section it takes a minimum value at around the root position of the petal of about $r/d_e = 0.4$. The jet spreads in the radius direction to the downstream, and the area of A- and B-sections mix each other (Fig. 89a–d). The two local maxima Br_{max} of the brightness decreases to the downstream following the mixing between the jet and the surroundings. During these mixing and spreading processes to the downstream, the jet oscillated irregularly and alternately in the circumference direction.

2.8.3 Vortex Structure

Figure 90 shows the sketch of the vortex structure. The mushroom shaped longitudinal twin counter rotating vortices called Prandtl's second kind of secondary flow are formed at the tip area of A-section and at the outer edge of B-section. As a result of the interaction of these vortices the jet flow oscillates irregularly and somewhat periodically in the circumference direction.

Figure 91 shows the details of the time change of the behavior of the vortices at $x/d_e = 0.8$ by the sketch of the flow. The mushroom shaped longitudinal twin vortices

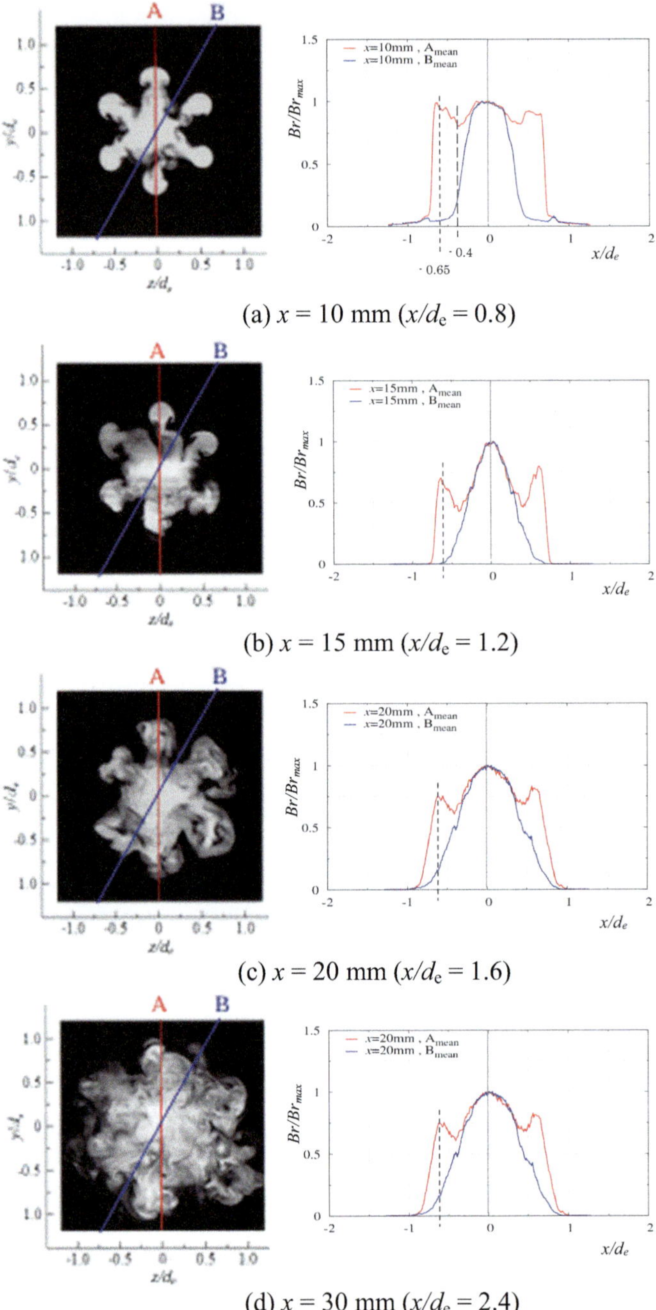

(a) $x = 10$ mm ($x/d_e = 0.8$)

(b) $x = 15$ mm ($x/d_e = 1.2$)

(c) $x = 20$ mm ($x/d_e = 1.6$)

(d) $x = 30$ mm ($x/d_e = 2.4$)

Fig. 89 Visualized flow pattern and mean brightness at the cross sections of 6-p jet ($Re = 3000$, $d = d_e$)

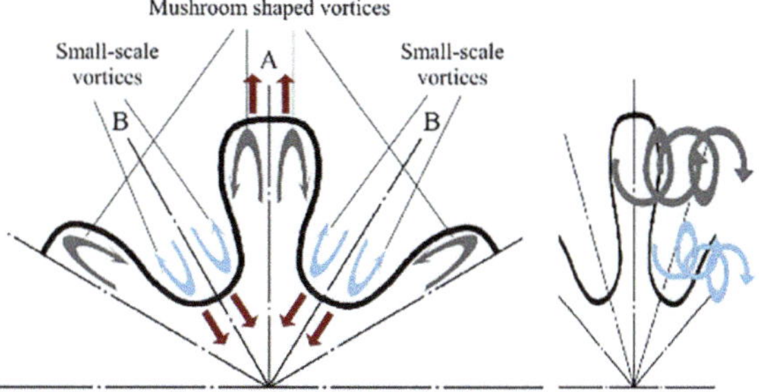

Fig. 90 Vortex formation

change their vorticities alternatively and then it oscillates irregularly and somewhat periodically in the circumference direction.

Figure 92a, b show the time change of the velocity at $x/d_e = 1.6$, $r/d_e = 0.64$ at B-section and FFT analysis, respectively. It looks like that the velocity oscillates

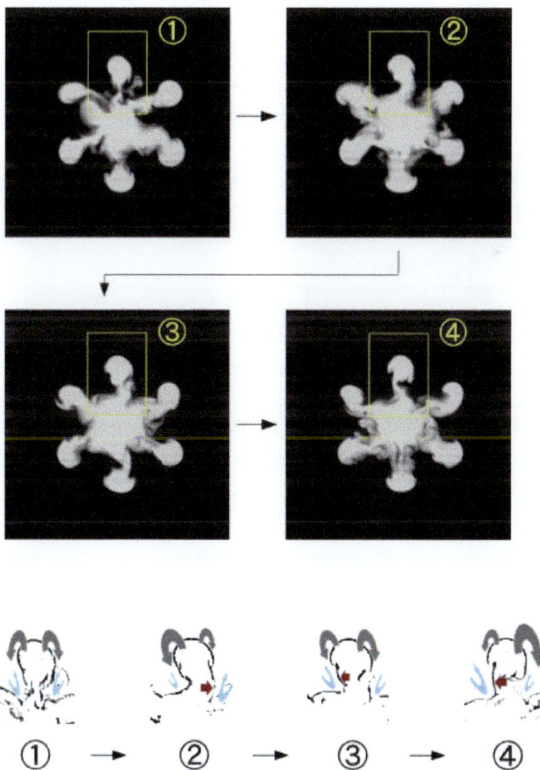

Fig. 91 Behavior of the vortices at the cross section of $x/d_e = 0.8$

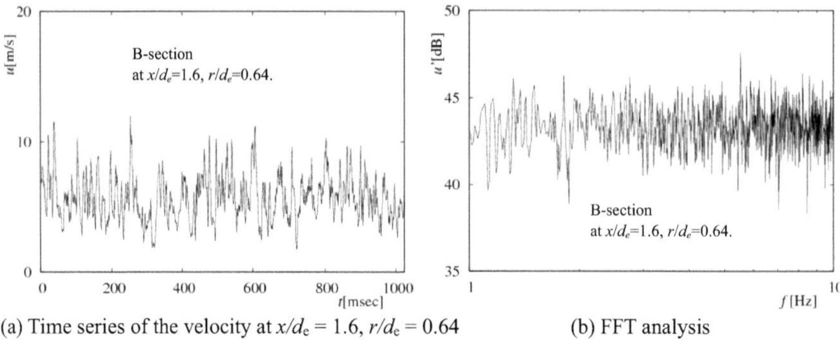

(a) Time series of the velocity at $x/d_e = 1.6$, $r/d_e = 0.64$ (b) FFT analysis

Fig. 92 Fluctuation of jet (velocity fluctuation)

somewhat periodically, and there are dominant frequencies at $f = 2, 5.5$ Hz, but it is not so clear.

2.8.4   Mean and Fluctuating Velocity Distributions on Jet Centerline

The mean and fluctuating velocities of air jet flow were measured by a hot wire anemometer (KANOMAX, IHW-100), single probe (wire: tungsten wire of 5 μm in diameter, prong distance: 2 mm, sensor distance: 1 mm) and PC. Here, the flow characteristics, flow pattern, mean and fluctuating velocity distributions of the 6-p jet and pipe jet were measured under the constant velocity of $u_m = 25$ m/s, and spread of jet and flow rate of the 6-p jet and pipe jet were measured under the constant operating power ($W = 1.72$ [W] = const.).

Figure 93a shows the mean jet centerline velocity u_c/u_m. The 6-p jet has a potential core region with a constant velocity, and then the velocity decreases to the downstream, $u_c/u_m \propto (x/d_e)^{-0.44}$. The centerline mean velocity of the 6-p jet decreases from an earlier position of about $x/d_e = 2.8$ than that of $x/d_e = 5.0$ for the pipe jet, because the mixing of the 6-p jet with the surrounding flow near the nozzle exit is larger than that of the pipe jet by a larger wetting perimeter. In the downstream the decreasing rate of u_c/u_m for the 6-p jet is smaller than that of the pipe jet and it decreases in proportion to $(x/d_e)^{-0.44}$.

Figure 93b shows the turbulent intensity profile u_c'/u_m. The 6-p jet has a larger turbulent intensity at the nozzle exit ($x/d_e = 0.2$) than that of the pipe jet. The turbulent intensity of the 6-p jet increases from about $x/d_e = 2.5$ and has the maximum value at the earlier position of $x/d_e = 4.5$ than that of about $x/d_e = 8.5$ for pipe jet. Then, it is thought that compared with the pipe jet, the 6-p jet has an excellent mixing and diffusion properties.

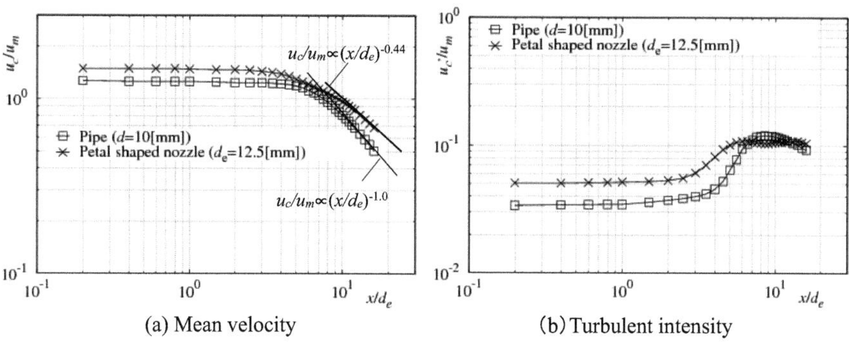

(a) Mean velocity   (b) Turbulent intensity

Fig. 93 Mean and fluctuating velocity distributions on jet centerline

2.8.5 Mean and Fluctuating Velocity Distributions at Cross Section

Figure 94a shows the velocity distribution u/u_m at the cross section. Since the profile
was axisymmetric only the results at the half region in the radius direction is shown.
The velocity distribution at A- and B-sections for the 6-p jet of $x = 2$ mm ($x/d_e =$
0.16) is entirely different depending on the nozzle geometry. At the petal area of
A-section the velocity is almost uniform but at the center area the velocity of A- and
B-sections is almost same and a maximum is on the center. The velocity at A-section
of $x = 10$ mm ($x/d_e = 0.8$) takes the local maximum at $r/d_e = 0.64$ and a maximum
at the center of $r/d_e = 0$. In comparison with A-section, the velocity distribution at
B-section takes the maximum on the center and changes smoothly in the profiles of
A- and B-sections are almost same.

Figure 94b shows the turbulent intensity u'/u_m at the cross section. The turbulent
intensity distributions at A- and B-sections of 6-p jet at $x = 2.0$ mm ($x/d_e = 0.16$)
are entirely different depending on the nozzle geometry. The turbulent intensity at
A-A section of $x = 10$ mm ($x/d_e = 0.8$) takes a local maximum at $r/d_e = 0.32$. In
comparison with A-section the turbulent intensity at B-section takes a local maximum
at $x = 10$ mm ($x/d_e = 0.8$, $r/d_e = 0.32$) and changes smoothly in the radius direction
mixing with the flow from the petal area. The jet spreads in the circumference and
radius directions to the downstream and at $x = 80$ mm ($x/d_e = 0.8$) the profiles at A-
and B-sections are almost same.

2.8.6 Spreading of Jet and Entrainment Flow Rate

Figure 95 shows the spread of jet, width b/d_e, of the 6-p jet. The experiments of the
spread and flow rate were carried out under the same operation power of $W = 1.72$
[W] $=$ const. ($u_m = 25$ [m/s]). Where, the jet edge was defined as the radial position
of $u/u_m = 0.1$. The jet width of the 6-p jet increases to the downstream after $x =$
20 mm. Near the nozzle exit until $x = 20$ mm the flow from A-section inflows to
B-section, and in the more downstream the spread of the 6-p jet is larger than that of

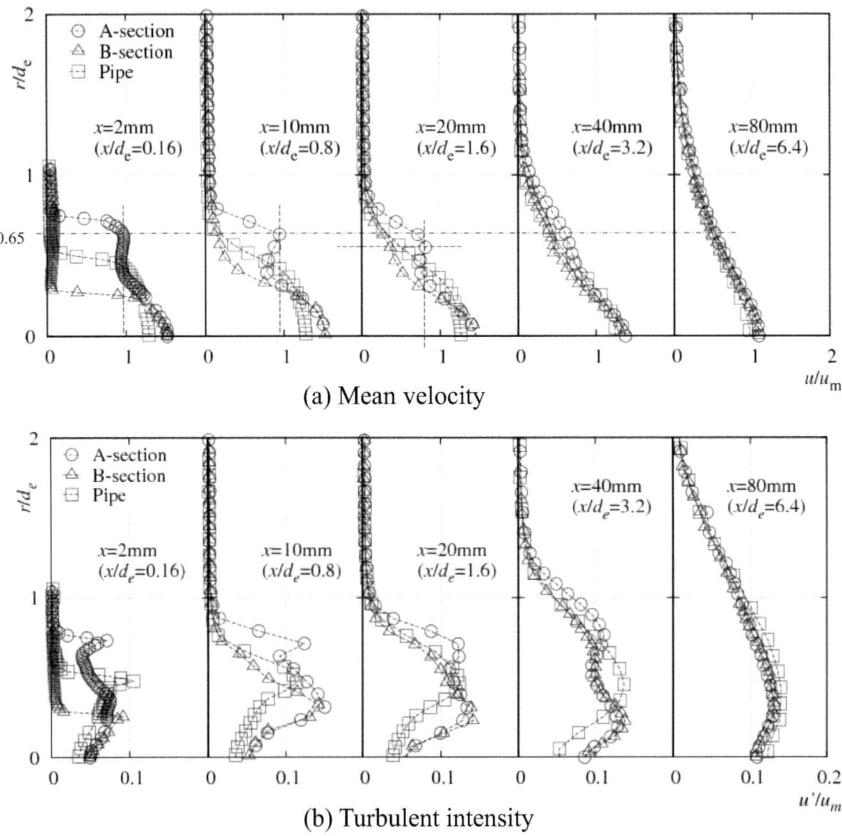

(a) Mean velocity

(b) Turbulent intensity

Fig. 94 Mean and fluctuating velocity distributions

the pipe jet. The mixing and diffusion of the 6-p jet with the surroundings are larger than that of the pipe jet. The switching phenomena which the major and minor axes of non-circular jet exchange depending on the deformation of the vortex ring could not be found in the range of $x/d_e > 8$.

Figure 96 shows the flow rate Q/Q_0, where, Q_0 is the flow rate at the nozzle exit. The flow rate of the 6-p jet was obtained approximately as the arithmetical mean of the flow rates calculated from the mean velocity profiles at A- and B-sections. The flow rate of the 6-p jet is larger than that of the pipe jet in the range of $x/d_e < 7$. This depends on the larger wetting perimeter of the 6-p nozzle than that of the pipe nozzle. It is known that the 6-p jet has a better mixing and diffusion characteristics than that of the pipe jet. For example, it is about 1.3 times of the pipe jet at $x/d_e = 1.6$.

Here, the flow characteristics, mixing and diffusion, of sub-merged free jet from 6 petal-shaped nozzle were examined experimentally. The 6-p jet has a good mixing and diffusion characteristics, and the flow rate of the 6-p jet is larger than that of the

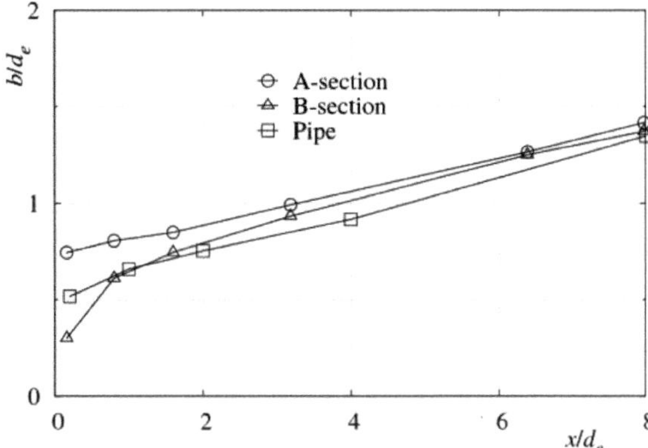

Fig. 95 Spread of jet, jet width b/d_e ($W = 1.72$ [W] = const., jet edge: y at $u/u_m = 0.1$)

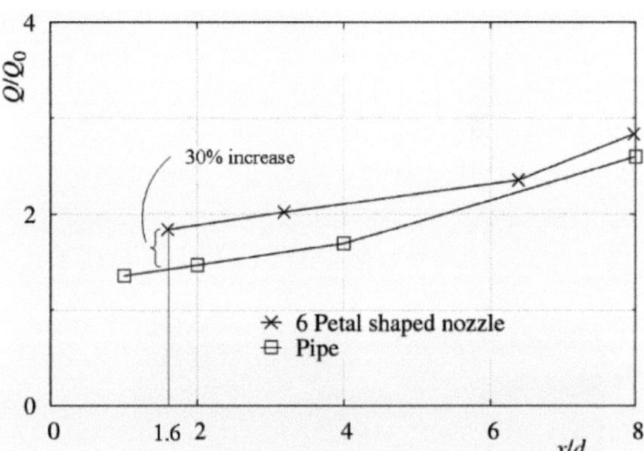

Fig. 96 Flow rate, $W = 1.72$ [W] = const. (jet edge: y at $u/u_m = 0.1$)

pipe jet in the range of $x/d_e < 7$ because of the larger wetting perimeter. For example, it is about 1.3 times of the pipe jet at $x/d_e = 1.6$ under the same operation power of $W = 1.72$ [W] ($u_m = 25$ [m/s]).

3 Impinging Jet Flows

Impinging jets have high heat and mass transfer properties near the stagnation point, so they are used in a wide range of industrial applications, such as cooling high-temperature objects, various heat sources, and electronic devices, drying paint films, and removing dirt and moisture from material surfaces. Therefore, clarifying the flow, heat transfer, and mass transfer characteristics of various impinging jets is not only interesting from an engineering standpoint, but also important for improving the performance of various related equipment.

Figure 97 shows an example of visualized submerged water jet issued from the orifice nozzle with contraction area ratio $CR = 0.11$ and impinged on the flat plate with the space of $H/d_0 = 5.0$ between nozzle exit and the plate. The nozzle diameter is $d_0 = 10.0$ mm and Reynold number is small and $Re = 1.0 \times 10^3$. The central cross-section (x-r plane) of an orifice submerged impinging water jet containing an aqueous sodium fluorescein solution was illuminated with a laser light sheet, and the visualized flow was observed and photographed. The result is shown in Fig. 97a and the sketch of (a). The submerged water jet of $Re = 1.0 \times 10^3$ from the orifice nozzle passes through the laminar shear layer, and a vortex ring is generated due to the instability of the flow. As it goes downstream, the vortex ring develops, rolls up, and reaches the impingement plate. The vortex ring generated when $x/d > 2.5$ impinges with the flat plate and flows down left and right along the wall surface. Figure 97b shows a visualization of the flow seen from the back side of a transparent impingement plate. On the flat plate, a group of vortex rings can be seen, following the spacing of the vortex rings created in the jet before the impingement.

There have been many studies on the flow and heat transfer characteristics of two-dimensional and three-dimensional circular impinging jets. Some attempts to improve the flow and heat transfer characteristics of impinging jets include the following. In other words, there are methods for manipulating the large-scale vortex

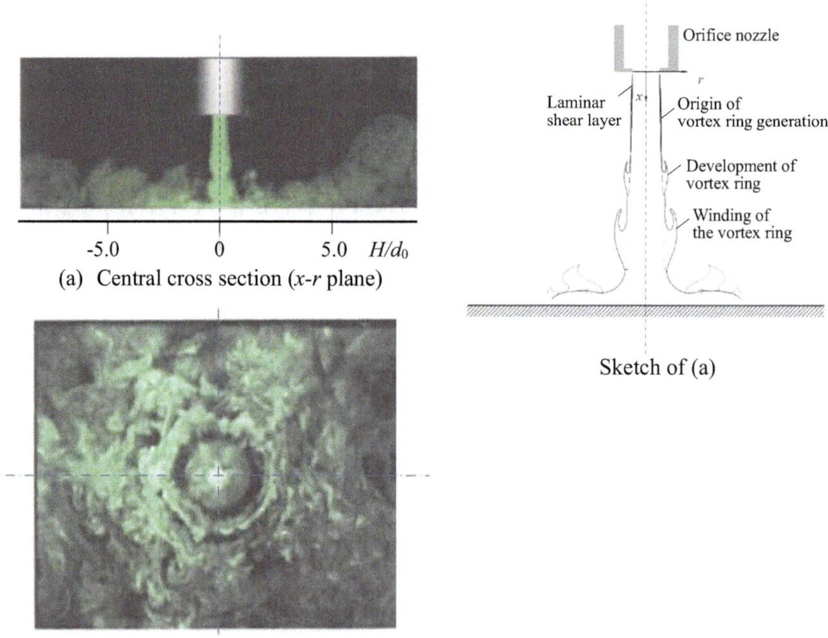

(a) Central cross section (x-r plane)

Sketch of (a)

(b) On the plate (View from the back of a transparent impingement plate)

Fig. 97 Visualized flow pattern of impinging submerged water jet from orifice nozzle (Contraction area ratio, $CR = 0.11$, $d_0 = 10.0$ mm, $Re = 1.0 \times 10^3$)

structure that occurs in the turbulent shear layer of the jet by using tabs, ribs, or non-circular nozzles, and for exciting the fluctuating velocity of the flow using sound waves, micro-actuators, plasma actuators, etc. Vertical impinging jets have also been the subject of much research, as they are used for surface cleaning, roughening, and jet cutting.

In this section, we will mainly discuss the flow and heat transfer characteristics of vertical impinging jets and their improvement and enhancement when jets are ejected from two-dimensional and three-dimensional circular nozzles into the same type of stationary fluid.

3.1 Flow Model of Impinging Jet

Figure 98 shows a flow model of a jet from a two-dimensional nozzle with a nozzle width of b_0 and a three-dimensional circular nozzle with a nozzle diameter of d_0 impinging perpendicularly on a flat plate. When the distance H between the nozzle exit and the flat plate is larger than a certain degree, the jet issued from the nozzle can be divided into three main regions: the free jet region, which is not affected by the impingement flat plate; the impinging jet region or stagnation region; and the wall jet region, where the jet flows along the flat plate after impinging the flat plate.

In the Region A or initial region in Fig. 98 the jet issued from the nozzle has a potential core region in which the velocity remains constant and equal to the nozzle exit velocity. In the region B or developed region, the jet is characterized by a decay of the jet centerline velocity and a spreading of the jet in the r direction by entraining

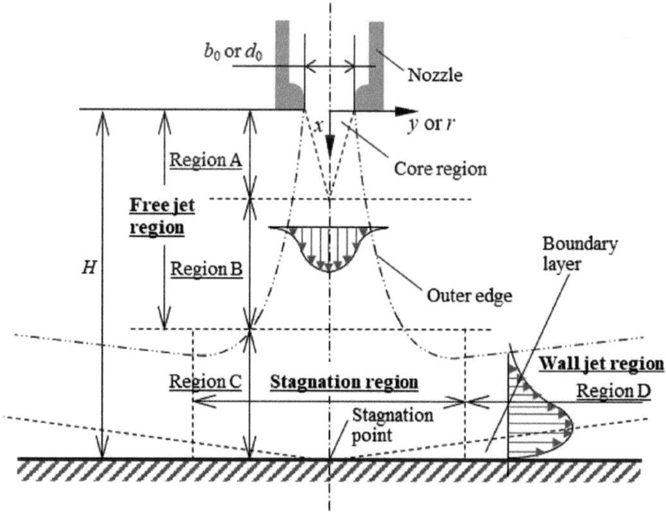

Fig. 98 Flow model of two- or three-dimensional impinging jet

the surrounding fluid. In the Region C or stagnation region, the jet is deflected from the x axis in the r direction. The Region D is known as the wall jet region, where the boundary layer on the flat wall is developed along the wall surface.

3.2 Two-Dimensional Impinging Jet

3.2.1 Flow Characteristics

Figure 99 summarizes the maximum flow velocity u_m, half-width $y_{1/2}$, turbulence intensity $\sqrt{\overline{u'^2}}$, etc. in the free jet region of a two-dimensional impinging jet (Wooly and Karamcheti 1973). The core region exists up to about $6b_0$, and the centerline maximum flow velocity u_m and half-width $y_{1/2}$ are expressed as a function of H by the following equation. Here, um0 is the maximum velocity at the nozzle exit.

$$\frac{u_\mathrm{m}}{u_\mathrm{m0}} = 2.45\left(\frac{H}{b_0}\right)^{-1/2} \tag{68}$$

$$y_{1/2} = 0.114H \tag{69}$$

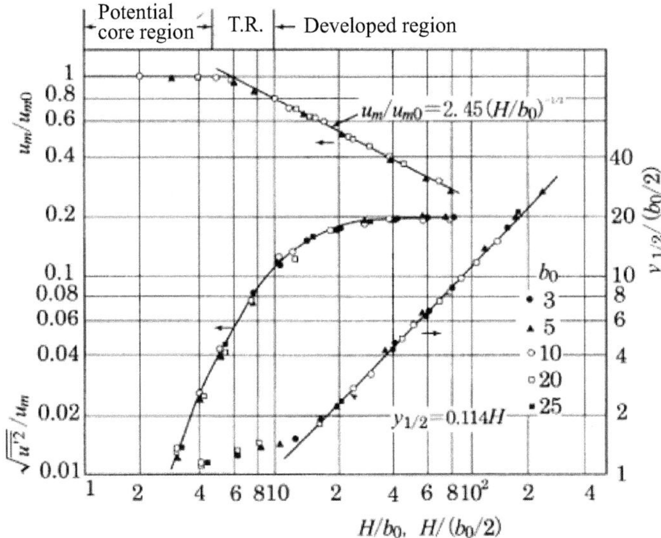

Fig. 99 Flow characteristics of two-dimensional impinging jet (by Wooly and Karamcheti 1973)

3.2.2 Heat Transfer

(a) Stagnation point

According to Gardon and Akfirat (1966), the heat transfer coefficient (Nusselt number, Nu_b) of two-dimensional impinging jet at the stagnation point of the impingement plate is:

$$Nu_0 = 0.535 Pr^{0.4} Re_b^{0.5} \quad H/b_0 \leq 4 \tag{70}$$

$$Nu_b = 1.42 Pr^{0.43} Re_b^{0.58} (H/b_0)^{-0.62} \quad H/b_0 \geq 8 \tag{71}$$

where, $10^4 < Re_b < 2 \times 10^5$,

$Nu_0 = hb_0/k$ (h : heat transfer coefficient, k : thermal conductivity)
   Pr : Prandtle number, $Re_b = u_0 b_0/v$
   Also, the maximum heat transfer coefficient is obtained at $H/b_0 \fallingdotseq 9$. This is because at $H/b_0 \fallingdotseq 9A$, the impinging velocity (maximum flow velocity) is not significantly attenuated and turbulence intensity increases. In general, the heat transfer coefficient of an impingement plate depends on the impinging velocity, the fluctuation velocity caused by large-scale eddies, the fluctuation period, the turbulence intensity, etc.

(b) Wall jet region

For the local Nusselt number in the wall jet region, the following relationship has been determined from experimental results of mass transfer in air flow (based on the analogy between mass transfer and heat transfer, the law of similarity).

$$\frac{Nu_x}{Nu_0} = 0.049 Re_b^{0.22} \left(\frac{x}{H}\right)^{-0.37} \tag{72}$$

where, $2 \times 10^4 < Re_b < 2 \times 10^5$

3.3 Three-Dimensional Circular Impinging Jet

3.3.1 Flow Characteristics

Figure 98 also shows a flow model of an impinging jet from a three-dimensional circular nozzle with nozzle diameter d_0. The free jet region, stagnation region (impinging jet region), and wall jet region are generally similar to those of a two-dimensional impinging jet.

(a)   **Jet centerline velocity and pressure distributions on impingement plate**

Figure 100 shows the centerline velocity u_m/u_{m0} of the jet until it impinges the flat plate (Tani and Komatsu 1964). The jet centerline velocity is not affected by the flat plate up to the vicinity of the flat plate and follows that of a free jet, after which it rapidly decreases and impinges on the flat plate. The free jet region of the impinging jet $\bar{x}$ can be expressed experimentally as follows.

$$\frac{\bar{x}}{d_0} \fallingdotseq 0.95\frac{h}{d_0} - 3.0 \quad \left(3 \leqq \frac{h}{d_0} \leqq 30\right) \tag{73}$$

where, $\bar{x} = h - z$

Figure 101 shows the pressure distribution with pressure coefficient on the impingement plate. The pressure has the maximum at the stagnation point, and it decreases in the radius direction and the Increase of H/d_0.

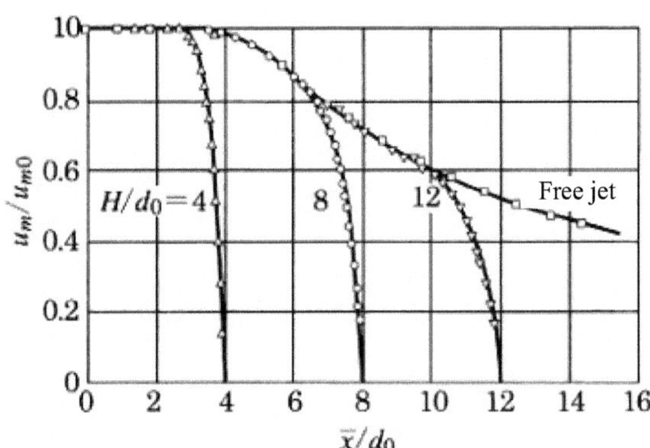

Fig. 100   Jet centerline velocity ($\bar{x} = H - z$)

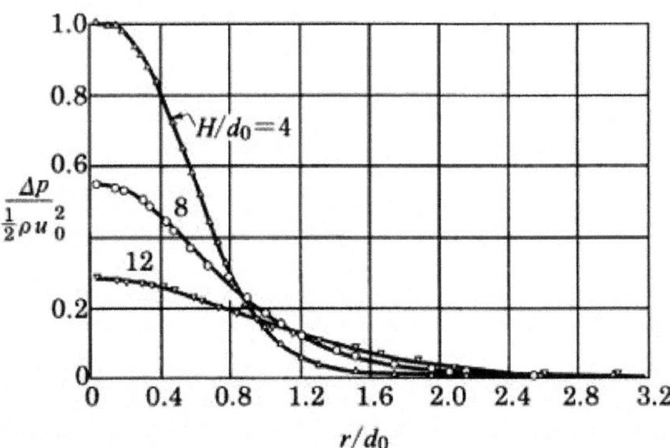

Fig. 101   Pressure distribution on flat plate

(b) **Numerical analysis**

In order to numerically analyze the flow and heat transfer characteristics of the above circular impinging jet with $H/d_0 = 2.0$, we used a standard k-ε model and solved the Navier equation and continuity equation using the finite difference method.

Figure 102 shows the results of numerical analyses, and since all results were axisymmetric and only half region in the r direction. Velocity vector, contour maps of pressure, contour maps of velocity u and v, turbulence kinetic energy k and eddy viscosity are shown in the left and right halves of Fig. 102a–c, respectively. The pressure, $p/(\rho u_0{}^2/2)$. Appeared in the right half of Fig. 102a has a maximum at the stagnation point. The velocity u/u_{me} and v/u_{me} appeared in the right and left halves, respectively, of Fig. 102b. Where, u_{me} is mean velocity at the nozzle exit. The jet centerline velocity decreases rapidly as the plate approaches.

(c) **Velocity distribution in radial wall jet region**

The velocity distribution u/u_{m} and maximum velocity u_{m} of a radial wall jet is given by the following equations.

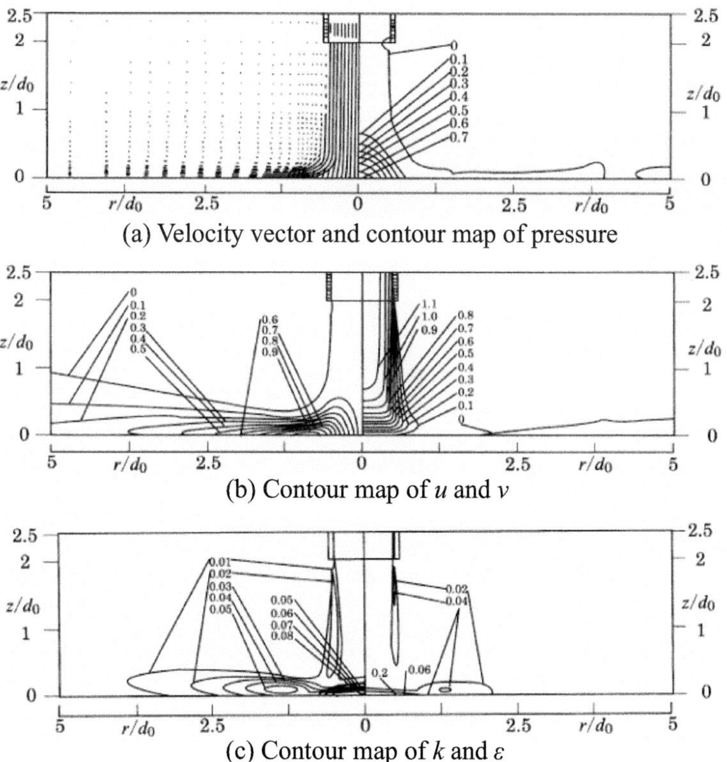

Fig. 102 Flow characteristics of impinging jet ($d_0 = 10.0$ mm, $Re = 2.67 \times 10^4$)

$$\frac{u}{u_{\mathrm{m}}} = \exp\left[-0.693\left(\frac{y-\delta}{y_{1/2}}\right)^2\right] \tag{74}$$

$$u_{\mathrm{m}} = 3.5\left[\frac{(Hr_0)^{1/2}}{r}\right]u_0 \tag{75}$$

However, here, we will use the profile method to find the velocity distribution. Now, assume that the velocity distribution of radial wall jet is similar everywhere and is given by the following equation.

$$\frac{u}{u_{\mathrm{m}}} = f\left[\left(\frac{z}{\delta}\right)^n\right] \tag{76}$$

Moreover, the boundary conditions are as follows.

$$U = 0 \text{ at } z = 0,\ u/u_{\mathrm{m}} = 0.1 \text{ and } \partial u/\partial z = 0 \text{ at } z = \delta \tag{77}$$

Considering these boundary conditions and expressing Eq. (76) as the simplest polynomial equation, we get

$$\frac{u}{u_{\mathrm{m}}} = C_1\left(\frac{z}{\delta}\right)^n + C_2\left(\frac{z}{\delta}\right)^{n+1} + C_3\left(\frac{z}{\delta}\right)^{n+2} + C_4 \tag{78}$$

where, $C_1 \sim C_4$: experimental constants.

From Eqs. (77) and (78) we get

$$\frac{u}{u_{\mathrm{m}}} = \left(\frac{z}{\delta}\right)^n\left\{C_3\left[1-\left(\frac{z}{\delta}\right)^2\right] + 0.1n\left[1-\left(\frac{z}{\delta}\right)\right] + 0.1\right\} \tag{79}$$

Now, if we put $z_{\mathrm{m}} = \delta/k$ (k: const, $k > 1$), and $u/u_{\mathrm{m}} = 0.1$ and $\partial u/\partial z = 0$ at $z = z_{\mathrm{m}}$,

$$\frac{u}{u_{\mathrm{m}}} = \left(\frac{z}{\delta}\right)^n\left[\frac{k^n - 0.1n(1-1/k) - 0.1}{1-1/k}\left(1-\frac{z}{\delta}\right)^2 + 0.1n\left(1-\frac{z}{\delta}\right) + 0.1\right] \tag{80}$$

From a comparison with the experimental results, in the boundary layer region near the wall, the experimental constants are given as $n = 1/4$, $k = 8.48$, and $C_3 = 2.063$, and as a result, the velocity distribution is given by the following equation.

$$\frac{u}{u_{\mathrm{m}}} = \left(\frac{z}{\delta}\right)^{1/4}\left[2.063\left(1-\frac{z}{\delta}\right)^2 + 0.025\left(1-\frac{z}{\delta}\right) + 0.1\right] \tag{81}$$

In addition, in the mixing region, the experimental constants are given as $n = 1/5$, $k = 10.317$, and $C_3 = 1.811$.

$$\frac{u}{u_m} = \left(\frac{z}{\delta}\right)^{1/5}\left[1.811\left(1 - \frac{z}{\delta}\right)^2 + 0.02\left(1 - \frac{z}{\delta}\right) + 0.1\right] \tag{82}$$

Now, if we give the half width of the jet (the larger value of z that makes $u/u_m = 0.5$) by $z_{1/2} = \delta/k'$ (k': const., $k' > 1$) and rewrite Eq. (80), in the boundary layer region the constants are given as $n = 1.4$, $k' = 2.0$ and $C_3 = 4.307$, and the following equation is written as obtain.

$$\frac{u}{u_m} = \left(\frac{z}{z_{1/2}}\right)^{1/4}\left[4.307\left(1 - \frac{z}{z_{1/2}}\right)^2 + 0.025\left(1 - \frac{z}{z_{1/2}}\right) + 0.1\right] \tag{83}$$

Also, in the mixed region, the constants are given as $n = 1/5$. $K' = 2.0$ and $C_3 = 4.307$, and the following equation is obtained.

$$\frac{u}{u_m} = \left(\frac{z}{z_{1/2}}\right)^{1/5}\left[4.155\left(1 - \frac{z}{z_{1/2}}\right)^2 + 0.02\left(1 - \frac{z}{z_{1/2}}\right) + 0.1\right] \tag{84}$$

These results will be discussed later in comparison with experimental results.

Figure 103a, b respectively show the velocity distribution u/u_m and turbulence intensity distribution u'/u_m at each radial position of an air jet (radial wall jet) ejected perpendicularly to the heating plate from a pipe nozzle with an inner diameter of $d_0 = 10.0$ mm, an outer diameter of 12.0 mm and a length of $L_n/d_0 = 50$. The mean velocity at the nozzle exit is $u_0 = 40$ m/s ($Re = 2.67 \times 10^4$) and the distance between the nozzle exit and plate is $H/d = 0.1\sim2.0$.

As H/d_0 becomes smaller, both the velocity and turbulence intensity increase rapidly, but the difference becomes much smaller at far distances of r/d_0.

Figure 104a, b respectively show dimensionless velocity distribution in the case of $H/d_0 = 0.2$ and 2.0 as examples. In either case, the velocity profile downstream at $r/d_0 = 4.0$ and 8.0 are similar, and are well expressed by Eqs. (83) and (84) shown by solid lines in the figure.

3.3.2 Heat Transfer of Impinging Jet

There are many experimental equations for the heat transfer coefficient (Nusselt number) at the stagnation point of a circular impinging jet, but for example, it is expressed by the following equation.

$$Nu_0 = 0.94Pr^{0.4}Re_d^{0.5} \tag{85}$$

$$Nu_0 = 11.6Pr^{0.5}Re_d^{0.5}\left(\frac{H}{d_0}\right)^{-1} \tag{86}$$

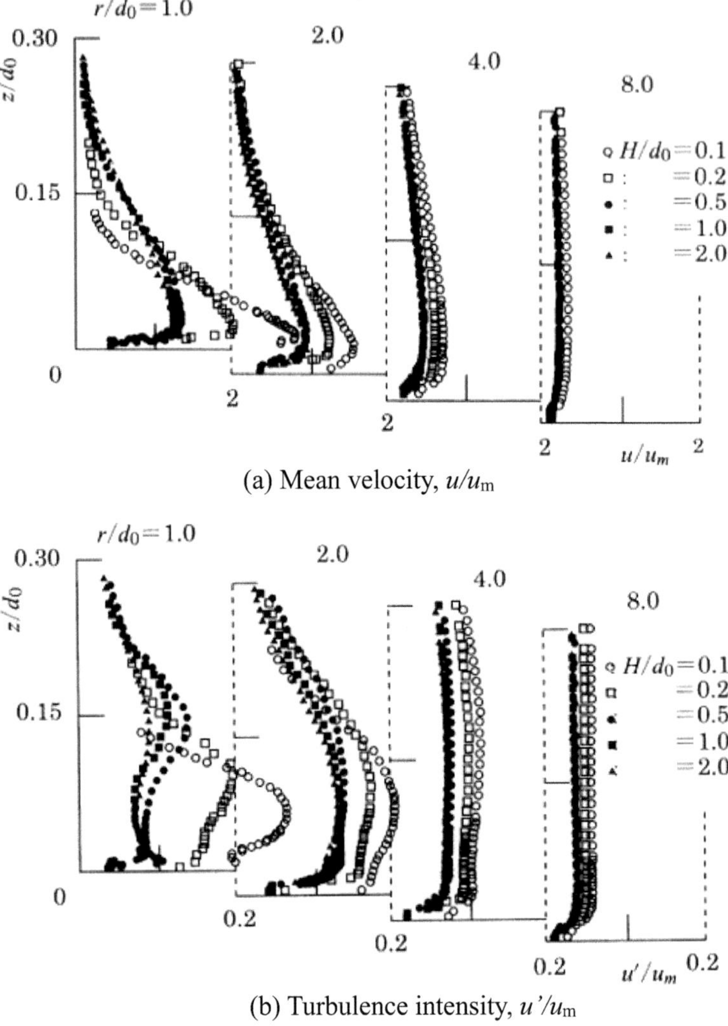

(a) Mean velocity, u/u_m

(b) Turbulence intensity, u'/u_m

Fig. 103   Mean velocity and turbulence intensity distribution

where, Pr: Prandtl number

According to Martin (1977), the mean Nusselt number $\overline{Nu}$ is expressed by the following equation.

$$\overline{Nu} = G\left(\frac{d_0}{r}, \frac{H}{d_0}\right) F(Re_d) Pr^{0.42} \tag{87}$$

where,

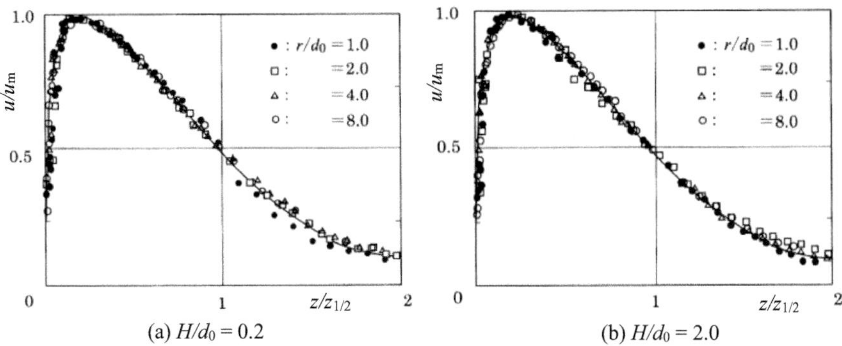

(a) $H/d_0 = 0.2$ (b) $H/d_0 = 2.0$

Fig. 104 Nondimensional velocity distributions of radial wall jet

$$G = \left(\frac{d_0}{r}\right)\frac{1 - 1.1d_0/r}{1 + 0.1(H/d_0 - 6)d_0/r}$$

$$F = 2Re_d^{0.5}\left(1 + Re_d^{0.55}/200\right)^{0.5}$$

for $2 \times 10^3 \leqq Re_d \leqq 4 \times 10^5$, $2.5 \leqq r/d_0 \leqq 7.5$, $2 < H/d_0 < 12$

Figure 105 shows the heat transfer characteristics of impinging jet. The curved lines in the figure show the numerical analyses. The numerical analyses were carried out using standard a standard k-ε model and solved the governing equations using the finite difference method. Since the profile was axisymmetric only half region in the r direction is shown. The heat transfer coefficient, Nusselt number, increases rapidly as decreasing H/d_0 and it takes the maximum value at $r/d_0 = 0.5$ when $H/d_0 < 0.2$. The numerical results represent the experimental results fairly well.

Table 5 summarizes the heat transfer coefficient and Nusselt number Nu_0 at the stagnation point of the three-dimensional circular impingement jet.

Fig. 105 Heat transfer characteristics of impinging jet ($d_0 = 10.0$ mm, $Re_d = 2.67 \times 10^4$)

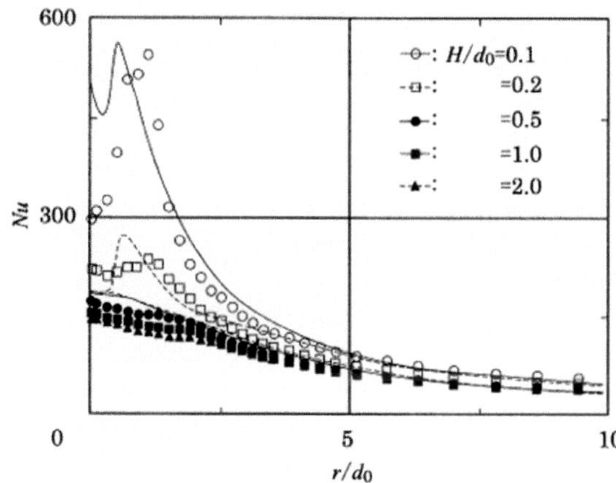

Table 5 Correlation on Nusselt number, Nu_0, at the stagnation point (Three-dimensional circular impinging jet)

Authors	Re	d_0 [mm]	H/d_0	Correlation
Liu et al. (1991)	$(2.0 - 4.0) \times 10^3$	3.8, 9.5		$Nu_0 = 0.715 Re^{0.5} Pr^{1/3}$
Liu and Sullivan (1996)	$(1.2 - 1.51) \times 10^4$	12.7	$1 - 2$	$Nu_0 = 0.585 Re^{0.5} Pr^{0.4}$
Lytle and Webb (1991)	$(0.36 - 2.76) \times 10^4$	$7.8 - 10.9$	$0.1 - 6$	$Nu_0 = 0.726 Re^{0.53} (H/d_0)^{-0.191}$
Mohantry and Tawfek (1993)	$(0.48 - 1.53) \times 10^4$	5, 7	$9 - 39.6$	$Nu_0 = 0.388 Re^{0.696} (H/d_0)^{-0.345}$
Ozman and Byadar (2008)	$(3,0 - 7.0) \times 10^4$	12.6	$1 - 10$	$Nu_0 = 0.092 Re^{0.69} (H/d_0)^{0.019}$
Katti and Prabhu (2008)	$(1.2 - 2.8) \times 10^4$	7.35	$0.5 - 8$	$Nu_0 = a Re^{0..5} Pr^{1/3} (H/d_0)^{-0.11}$ a: depends on H/d_0

3.3.3   Mechanism of Heat Transfer Enhancement

When a jet is ejected from a nozzle into a stationary fluid of the same type, the jet entrains the surrounding fluid through shear mixing and spreads. At this time, a large-scale vortex structure exists in the flow near the nozzle exit, resulting in a complex flow state. Kataoka et al. (1987) explained in detail what happened during this time based on flow visualization and observation results, and showed the relationship between the large-scale vortex structure and the heat transfer characteristics of the impinging jet. He also showed the relationship between the heat transfer coefficient around the stagnation point and the distance H/d_0 between the nozzle exit and the impingement plate (Fig. 106), and showed that the heat transfer coefficient at the stagnation point is maximum at $H/d_0 = 6{\sim}8$. This position corresponds to the position where the centerline velocity of the impinging jet is less reduced than that at the nozzle exit and, moreover, the turbulence is maximally developed.

It is thought that the first factor that increases the heat transfer of impinging jet is the nozzle Reynolds number and the impinging velocity to the heat transfer surface, and the second factor is the interface renewable frequency and the turbulence intensity of the turbulent flow.

Figure 107 shows the heat transfer enhancement model. In this model, the heat transfer enhancement rate due to turbulence is

$$\varepsilon_0 = Nu_0/(Nu_0)_f \tag{88}$$

where, $(Nu_0)_f$: Nu_0 in undisturbed flow condition.

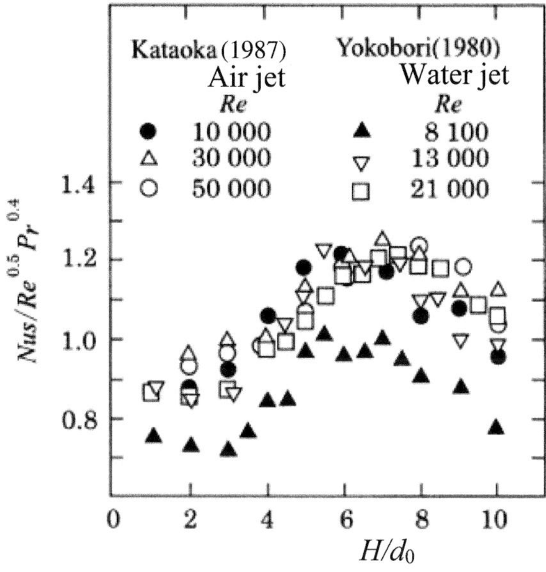

Fig. 106 Heat transfer coefficient at stagnation point (Kataoka et al. 1987)

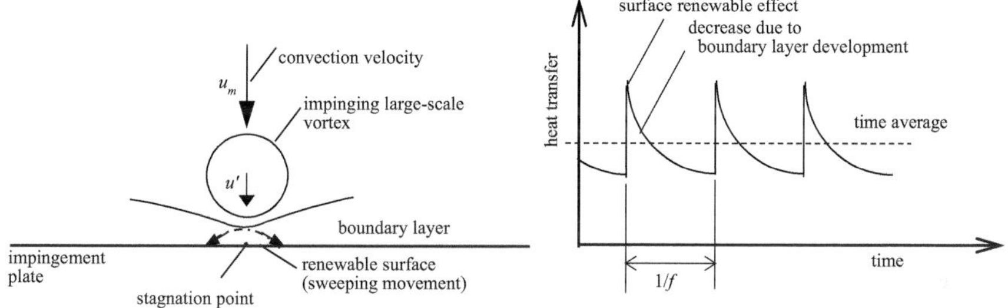

Fig. 107 Heat transfer enhancement model

The interface renewable intensity is the turbulence intensity at the center line of the free jet at the location of the heat transfer surface, and its frequency is evaluated by conditional statistical analysis of the temporal variation of the center velocity (Fig. 108). Here, the interface renewable parameter is defined as follows.

$$SR = \frac{\left(\overline{u'^2}\right)^{1/2}}{\overline{u_0}} \cdot \frac{fd_0^2}{\nu} = Re_t St \tag{89}$$

This can also be thought of as a product of the Reynolds number of the turbulence and the Strouhal number. As shown in Fig. 109, the heat transfer enhancement rate ε_{h0} at the stagnation point of various impinging jets is related to the interface renewable parameter SR.

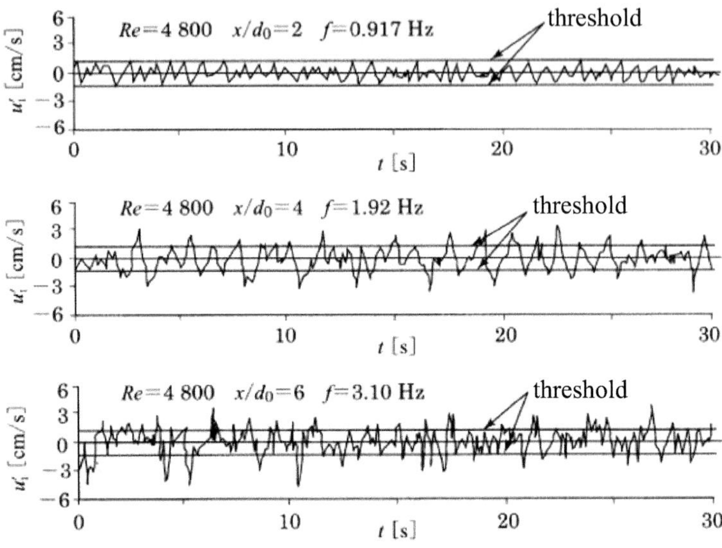

Fig. 108 Analysis of fluctuating velocity (Kataoka et al. 1987; Kataoka 1990)

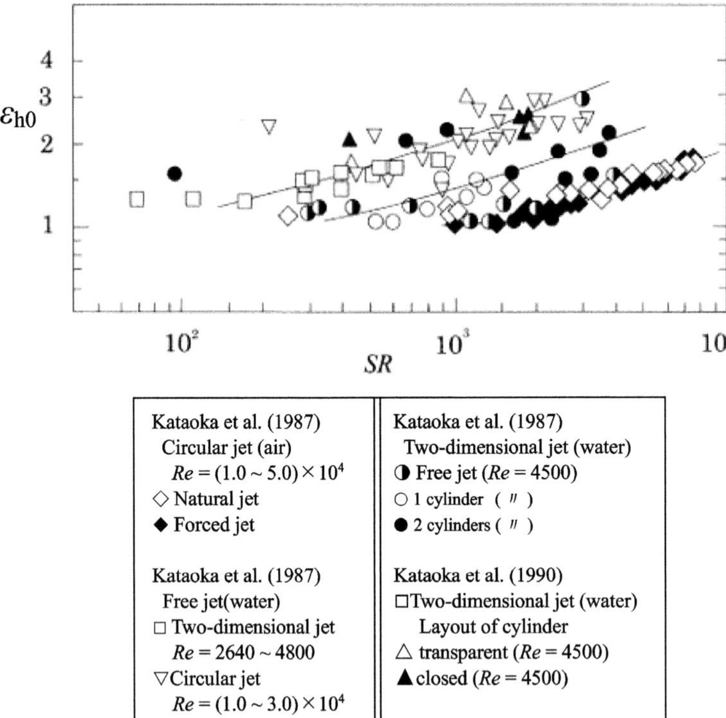

Fig. 109 Heat transfer enhancement rate and interface renewable parameter, ε_{h0} vs. SR (Kataoka et al. 1987; Kataoka 1990)

By the way, parameters that are considered to affect the heat transfer of the impinging jet include the impinging velocity, the fluctuation velocity and period caused by large-scale eddies, and the turbulence intensity, etc. However, according to this theory by Kataoka et al. (1987), Kataoka (1990) it is unclear how much influence each of them has.

3.4 Impinging Jet with Small Nozzle-Plate Spacing

Figure 110 shows the flow model of impinging jet. Since the flow is axisymmetric, only half the area was shown in the radial direction. In the impinging jet, when the distance, spacing, H between the nozzle and the flat plate become smaller, it is generally called a radial wall jet in $H/d_0 < 1$.

$H/d_0 > 1$: Impinging wall jet.

$H/d_0 < 1$: Radial wall jet.

Here, we will call it "Impinging wall jet".

*** Heat transfer and pressure drop of radial wall jet with small spacing**

It is known that in a circular impinging jet, the heat transfer coefficient of the impinging plate increases rapidly when the nozzle-plate distance becomes considerably small.

The relationship between the Nusselt number distribution Nu on the impingement plate and nozzle-plate spacing H/d_0 for a circular jet impinging perpendicularly to a heated plate with a constant heat flux has already been shown in Fig. 105. When H/d_0 becomes smaller, for example, the maximum value of Nusselt number at $H/d_0 = 0.1$ becomes more than twice that at $H/d_0 = 0.2$. This is because when h/d_0 becomes considerably small, the flow passing between the nozzle exit end and the flat plate is rapidly accelerated, and as a result, the turbulence component of the flow also increases. Furthermore, when h/d_0 becomes smaller, the maximum value of Nusselt number appears in addition to the stagnation point, and its position r/d_0 is expressed by the following equation.

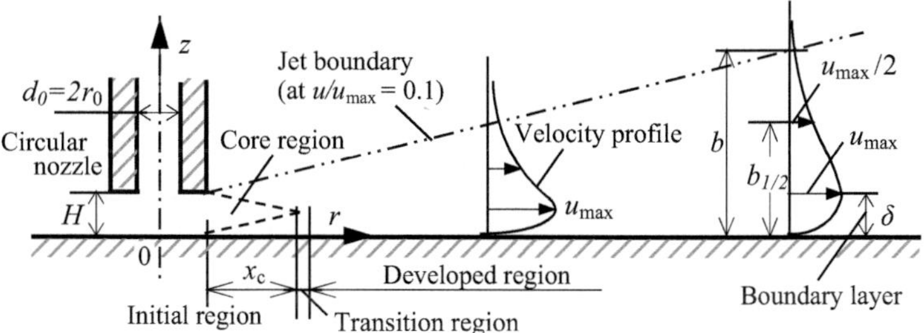

Fig. 110 Flow model of radial wall jet ($H/d_0 < 1$)

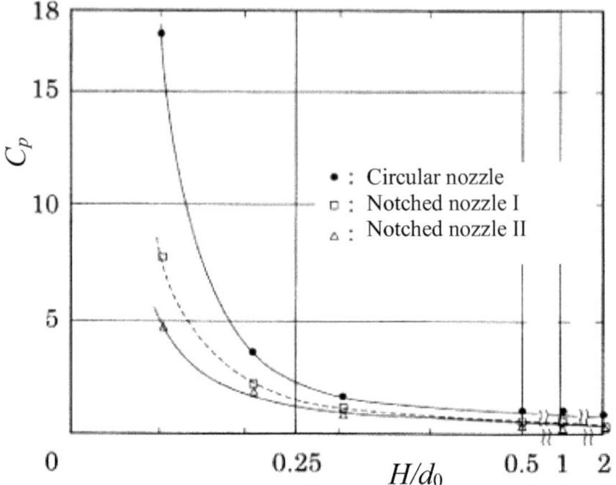

Fig. 111 Flow resistance of nozzle-impingement plate system

$$\frac{r}{d_0} = 0.188 Re^{0.241} \left(\frac{H}{d_0}\right)^{0.224} \tag{90}$$

$$\text{for} \quad 1.1 \times 10^4 < Re < 2.76 \times 10^4, \quad 0.1 < H/d_0 < 1.0$$

Figure 111 shows the pressure drop Δp of the circular nozzle-impingement plate system using the pressure coefficient C_p. In either case, the pressure drop increases rapidly as H/d_o decreases.

When the nozzle-plate distance is quite small, it is important to utilize its excellent heat transfer properties without significantly increasing flow losses.

3.5 Improvement of Heat Transfer of Impinging Wall Jet

3.5.1 Using of Notched Pipe Nozzle

When H/d_0 is quite small, one possible way to reduce the flow resistance and increase the turbulence of the flow is to provide a notch at the nozzle exit end (Shakouchi et al. 2011).

Figure 112 shows a special nozzle with notches installed at four circumferential locations at the nozzle exit end. The nozzle inner diameter and wall thickness are $d_0 = 10.0$ mm and $t = 1.0$ mm, respectively, and the depth of the notch is $N_n = 1.0$, 2.0 mm and the width is $N_w = 2.0$ mm. Hereafter, the notched nozzle with $N_n = 1.0$ mm and 2.0 mm will be referred as notched nozzle I and II, respectively. The length of a circular nozzle is $L_n/d_0 = 50$ and then the flow at the nozzle exit has a fully developed velocity profile. In case of a notched nozzle, since the area between the nozzle exit and the plate increases the flow resistance Δp decreases considerably.

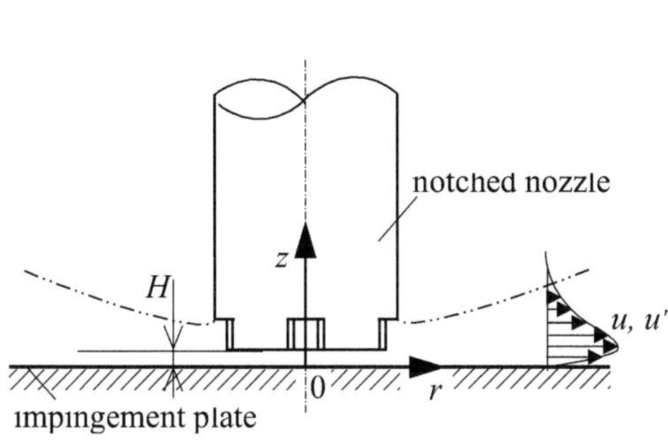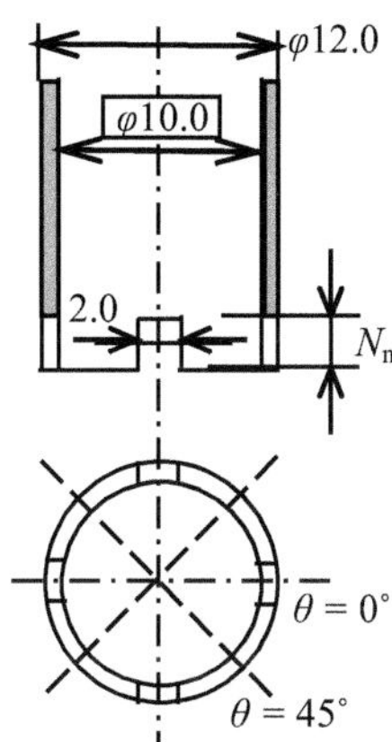

$* N_n = 1.0$ mm : Notched nozzle I
$\quad = 2.0 \quad$: Notched nozzle II

Fig. 112 Notched pipe nozzle

(a) Flow resistance of notched pipe nozzle-plate system

In Fig. 111, the flow resistance of the notched nozzle-plate system is also shown. In case of notched nozzle I, the flow resistance for $h/d_0 = 0.1$ and 0.2 decrease about 0.62 and 0.6 times that of the circular nozzle, respectively. Also, in the case of notched nozzle II, they decrease about 0.38 and 0.58 times that of the circular nozzle, respectively.

(b) Heat transfer

Figure 113 shows Nusselt number Nu distribution for a circular nozzle, notched nozzle I and II when the mean velocity at the nozzle exit is $u_0 = 40$ m/s ($Re = 2.67 \times 10^4$) and the nozzle-plate spacing $H/d_0 = 0.2$. Sinde the profile was axisymmetric, only the half region in the radius direction r is shown. In the figure, $\theta = 0$ and $45°$ show respectively the positions on the r-axis at $z = 0$ and on the center between r and y axes. In case of notched nozzle I and when $\theta = 0°$, Nu in the region of $r/d_0 \leqq 2.5$ is larger than that for the circular nozzle, and in the outer region, it is somewhat larger in $\theta = 0°$ of notched nozzles I and II, but the distribution shape is almost the same in all cases. In addition, the maximum value $Nu_m = 267$ occurs at $r/d_0 = 1.1$ near where the outer edge of the jet impinges on the flat plate, and is approximately 1.14 times that for a circular nozzle, and at the stagnation point, Nusselt number $Nu = 259$, it is approximately 1.12 times that for a circular nozzle. In case of notched nozzle I and when $\theta = 45°$, Nu distribution is larger in the range of $r/d_0 \leqq 1.9$ than in the case of a circular nozzle, but smaller than in the case of $\theta = 0°$ in almost the entire range of r/d_0.

In this way, when the notched nozzle I is used, the flow resistance in the nozzle-flat plate system can be reduced and the heat transfer characteristics of the impinging jet can be improved compared to when a circular nozzle is used. In case of the notched

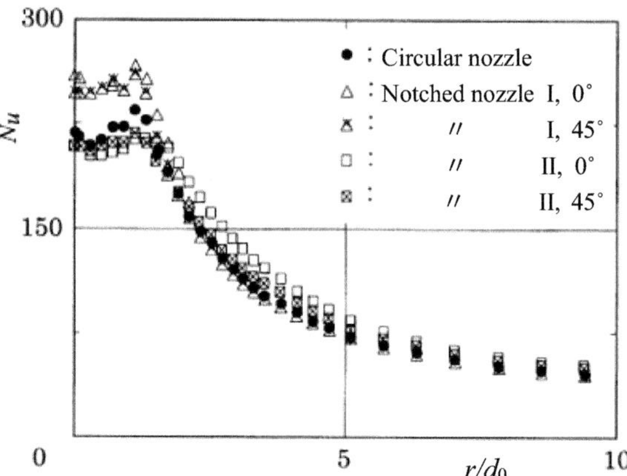

Fig. 113 Local Nusselt number ($H/d_0 = 0.2$)

nozzle II when $\theta = 0$ and 45°, Nu distribution are smaller than that of the circular nozzle in the region of $r/d_0 \leq 1.4$ and 1.7, respectively, but it is larger than that of the circular nozzle in the outer region. Next, we will consider the amount of heat transferred per unit operating power, that is, the heat transfer efficiency.

Figure 114 shows Nu of notched nozzle I. The nozzle-flat plate distance was $H/d_0 = 0.2$, and the operating power of the nozzle-flat plate system was made the same as that of the circular nozzle. For $\theta = 0°$, Nu is larger than for the circular nozzle over the entire range of r/d_0. The maximum Nusselt number $Nu_\mathrm{m} = 296$ is about 1.26 times that of the circular nozzle and the Nu at the stagnation point is about 1.32 times that of circular nozzle. In case $\theta = 45°$, the same thing as above can be said, although Nu at the stagnation point is somewhat smaller than in case $\theta = 0°$.

Figure 115 shows the mean Nusselt number $\overline{Nu}$ distribution for Fig. 114, calculated using the following equation.

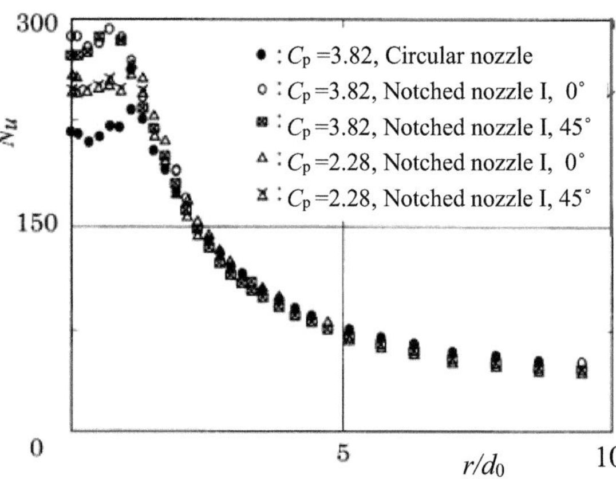

Fig. 114 Local Nusselt number for same operation power ($H/d_0 = 0.2$)

$$\overline{Nu} = \frac{1}{\pi r^2} \int Nu(2\pi r)dr \tag{91}$$

$\overline{Nu}$ of notched nozzle I is larger than that of a circular nozzle in the entire range of r/d_0, for example, the $\overline{Nu}$ for $\theta = 0$, 45° and $r/d_0 = 0\sim1$ is about 1.31 times that of the circular nozzle. The notched nozzle I can be improved the heat transfer, $\overline{Nu}$, about 31% than that of the circular nozzle when $H/d_0 = 0.2$, $\theta = 0°$, 45° and $r/d_0 = 0\sim1$ (Fig. 115).

Figure 116 shows the results for $H/d_0 = 0.1$. The $\overline{Nu}$ for notched nozzle I and II is smaller than that of the circular nozzle in the entire range of r/d_0. This is because when $H/d_0 = 0.1$, the distance between the nozzle exit and the impingement flat plate becomes extremely small, so the flow passing between the nozzle exit end and the flat plate is rapidly accelerated, creating a strong shear flow, which increases flow resistance, but it is thought that the heat transfer performance will be greatly improved. Thus, when $H/d_0 = 0.1$, notched nozzle I can reduce the flow resistance of the nozzle-flat plate system by about 32%, but cannot improve the heat transfer

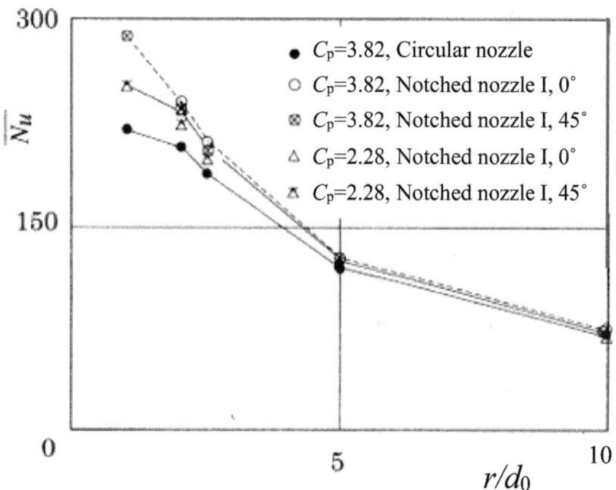

Fig. 115 Mean Nusselt number for same operation power ($H/d_0 = 0.2$)

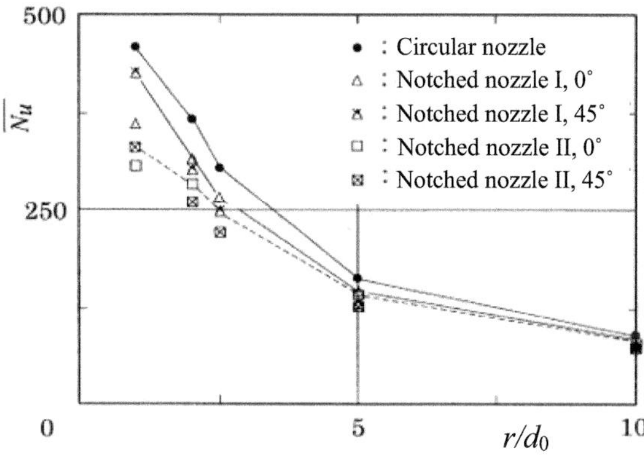

Fig. 116 Mean Nusselt number for same operation power ($H/d_0 = 0.1$)

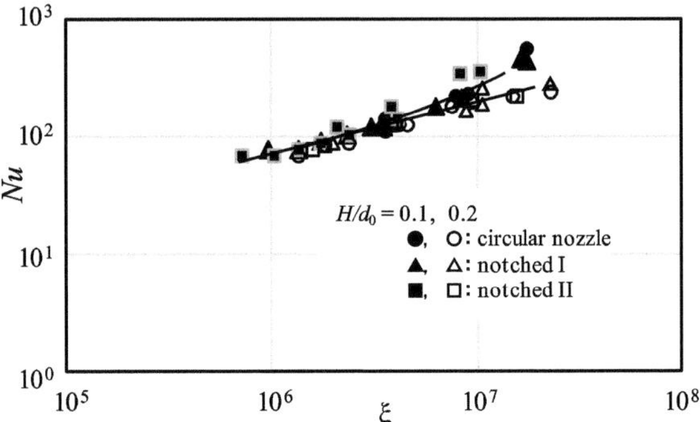

Fig. 117 Heat transfer enhancement parameter, ξ

performance. For example, $\overline{Nu}$ in the range of $r/d_0 = 0\text{~}1$ for $\theta = 0$ and $45°$ is about 0.86 and 0.93 times that of the circular nozzle, respectively.

Furthermore, in the case of $H/d_0 = 0.3$, where the nozzle-plate distance is larger than in the above case, when the driving power is the same as that of the circular nozzle, $\overline{Nu}$ was almost the same as that of the circular nozzle over the entire range of r/d_0 for $\theta = 0$ and $45°$.

In this way, the heat transfer performance of the impinging jet when using a notched nozzle is improved within a limited range of H/d_0.

(c) **Mechanism of heat transfer enhancement**

Kataoka et al. (1987), Kataoka (1990) describes the heat transfer characteristics of an impinging jet using the following parameters:

$$\xi_k = \left(u'/u_m\right)\left(f_e d^2/v\right) = Re'St \tag{92}$$

However, when the nozzle-plate spacing is extremely small, the frequency f_e of generation of large-scale vortices in the shear layer at the outer edge of the jet is thought to be of little relevance.

$$Nu \propto F\left[(ud/v)\cdot(v/a)\cdot\left(u'h/v\right)\right] \equiv F\left(Re\cdot Pr\cdot Re'\right) \equiv F(\xi) \tag{93}$$

where, a: thermal conductivity, $Re' = u'H/v$

From this, we predict that Nu is a function of ξ. ξ is heat transfer enhancement parameter.

Figure 117 shows the relation between Nu and ξ. When calculating ξ, the values measured at $z/d_0 = 0.005$, the closest neighbor to the plate surface, were used for u and u'. Nu increases with increasing ξ and the proposed heat transfer enhancement parameter by Kataoka et al. (1987), Kataoka (1990) well represents the heat transfer characteristics of the impinging jet.

3.5.2 Using of Thick Pipe Nozzle Plate

In a circular impinging jet that impinges perpendicularly on a flat plate, when the nozzle-plate spacing H/d_0 becomes considerably small, its heat transfer characteristics are greatly improved. For example, the maximum Nusselt number when $H/d_0 = 0.1$ is about 0 and 2.2 times larger than when $H/d_0 = 6.0$ and 0.2, respectively. However, at that time, the flow loss (operation power) of the nozzle-plate system also increases rapidly. Furthermore, especially in close-range impingements, the shape of the nozzle exit is considered to have a large effect on the jet flow and heat transfer characteristics. Here, when the nozzle-plate spacing H/d_0 is quite small, we will change the wall thickness and sheath shape of the nozzle exit and examine the relationship between the flow, heat transfer characteristics, and operation power (Shakouchi et al. 2003a, b).

(a) **Nozzle-plate system in close range impingement**

Figure 118a shows the nozzle-plate system in close-range impingement and the nozzle-plate spacing H can be changed. The details of nozzle exit are shown in Fig. 118b, and since the nozzle shape was axisymmetric and only the half region in the y direction is shown. In Fig. 118b, (1) shows a standard nozzle with nozzle wall thickness $t = 1.0$ mm (hereinafter referred to as st-nozzle), and (2) shows a nozzle with t varied from 2.0 to 30.0 mm (tw-nozzle). Furthermore, as a simple method for suppressing flow separation at the corner of the inner wall of the nozzle exit, a nozzle (hereinafter referred to as step-nozzle) provided with a step of 1 mm in height and 1 mm in width is shown in (3), and (4) shows a nozzle with a radius of 1 mm (r-nozzle).

(b) **Velocity distribution**

To measure the r-direction velocity u and turbulence u' (rms value) on the impingent plate, a single probe inserted from the back side of the impingement plate as shown in Fig. 119, and hot wire anemometry with the single probe of 5 μm diameter was used. The distance from the plate to the hot wire is variable in the range of 0 to 2 mm. In addition, the entire plate was moved in the r direction, and the velocity u and turbulence u' at arbitrary positions in the r direction were measured.

Figure 120a shows the velocity distribution u/u_{me} at $z/d_0 = 0.05$ for $H/d_0 = 0.1$ and Re $= 2.37 \times 10^4$. It seems that the velocity u/u_m directly below the nozzle exit at $r/d_0 = 0 \sim 0.5$ does not strictly indicate a velocity component in the r direction. The distribution shape for st-nozzle and tw-nozzle with $t/d_0 = 0.5$ reaches a maximum at $r/d_0 \fallingdotseq 0.4$, then rapidly decreases to a second maximum value, and after that, it gradually increases, reaches a local maximum value, and then gradually decreases. Since the st- or tw-nozzle has a right-angle corner inside of the nozzle exit, then the flow separates there. As shown in Fig. 118b, (1) and (2), the flow path between the separation stream line and the impingement plate narrows, and the velocity increases to the downstream.

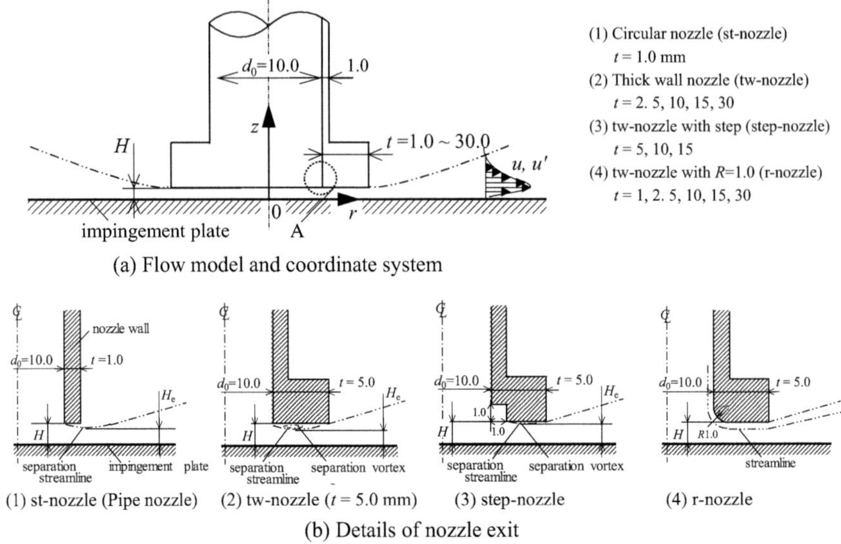

Fig. 118 Nozzle-plate system in close-range impingement

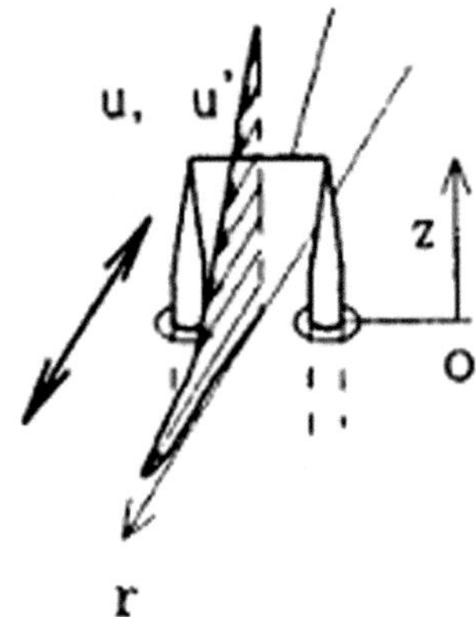

Fig. 119 Single probe, hot wire anemometry

Figure 120b shows the jet turbulence intensity u'/u_{me} at $z/d_0 = 0.05$. u'/u_{me} of the st-nozzle increases from $r/d_0 = 0.5$ when the flow is accelerated and the velocity gradient is large, and reaches its maximum around $r/d_0 = 0.8 \sim 1.2$ after the jet is issued into free space. Furthermore, the position where this maximum value occurs moves toward the center axis of the nozzle as t/d_0 increases, and its value decreases approximately 0.68 times in the tw-nozzle compared to the st-nozzle. This is because in the tw-nozzle, the separation region on the lower surface of the nozzle wall is suppressed in the z direction as t/d_0 increases.

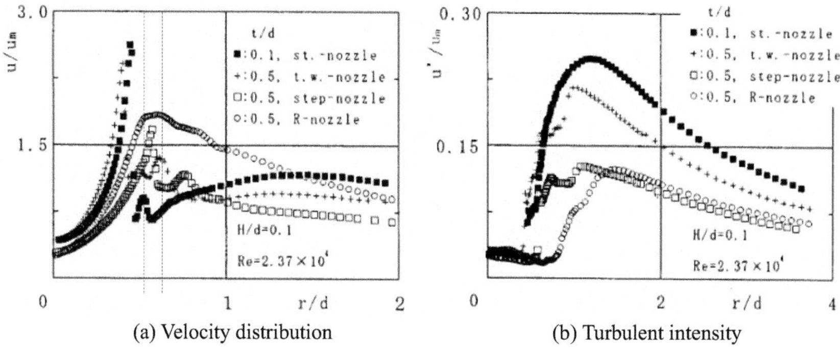

(a) Velocity distribution (b) Turbulent intensity

Fig. 120 Velocity distributions, $z/d\_0 = 0.05$

(c) **Heat transfer**

The heat transfer characteristics of this impinging jet were investigated as follows. First, a 50 μm-thick stainless steel foil attached to a Bakelite impingement surface was directly heated by electricity to create a heat transfer surface with constant heat flux. Next, the temperature distribution on the surface of the impingement plate of the impinging jet was measured using a number of Cu–Co thermocouples installed on the back side of the stainless-steel foil, and the Nusselt number was calculated. The mean Nusselt number was calculated by the following equation.

$$\overline{Nu} = \frac{1}{\pi r^2} \int_0^r Nu(2\pi r)dr \tag{94}$$

The operation power $W[\mathrm{W}]$ of the nozzle-impingement plate system was calculated by the following equation.

$$W = \Delta P \frac{\pi d^2}{4} u_{\mathrm{m}} \tag{95}$$

where, ΔP: differential pressure between the static pressure hole provided at $15d_0$ upstream of the nozzle exit and atmospheric pressure

Figure 121a shows the Nusselt number distribution for $H/d_0 = 0.1$, $Re = 2.37 \times 10^4$, and st-nozzle and tw-nozzle with $t/d_0 = 0.5, 0$. Since the profile was axisymmetric, only the half region in r-direction was shown. Nusselt number increases from the stagnation point, reaches a maximum at $r/d_0 \fallingdotseq 0.7$, and then decreases gradually. Nusselt number also decreases as t/d_0 increases, and the maximum values of tw-nozzle with $t/d_0 = 0.5$ and tw-nozzle with $t/d_0 = 0$ decrease to approximately 0.9 and 0.45 times that of st-nozzle, respectively. Figure 121b shows the results for st-nozzle, and tw-, step-, and r-nozzle with $t/d_0 = 0.5$. Nusselt number also takes a maximum at $r/d_0 \fallingdotseq 0.7$, and then decreases gradually. In step- and r-nozzles, Nu decreases significantly at the stagnation point compared to st-nozzle, with the respective maximum

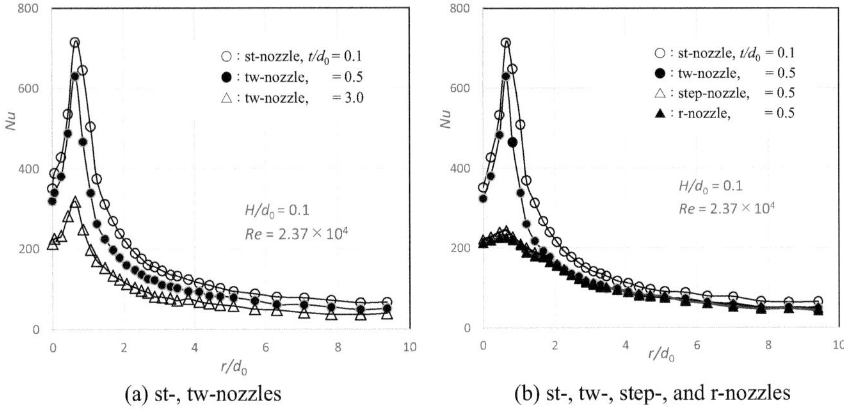

(a) st-, tw-nozzles   (b) st-, tw-, step-, and r-nozzles

Fig. 121 Nusselt number distribution

values decreasing to approximately 0.35 and 0.32 times st-nozzle. This is thought to be because in step- and r-nozzles, flow separation from the inner corner of the nozzle exit is suppressed (the separated vortex area is small), the effective passage area does not decrease much, and the velocity and turbulence intensity decreases, resulting in a decrease in heat transfer characteristics.

(d)   **Operation power and heat transfer**

Figure 122a shows the relationship between the mean Nusselt number $\overline{Nu}$ of the tw-nozzle with $t/d_0 = 0.2$ to 0 up to $r/d_0 = 10$ and the operation power W of the nozzle-plate system. $\overline{Nu}$ and W decrease with increasing t/d_0. This is thought to be because as t/d_0 increases, flow separation on the underside of the nozzle wall is suppressed, the effective passage area increases, flow resistance decreases, and the operating power of the nozzle-plate system decreases. Furthermore, the velocity and turbulence intensity also decrease, so the heat transfer characteristics are thought to decrease.

Figure 122b, c show the results for step- and r-nozzles. In these cases, $\overline{Nu}$ and W also decrease with increasing t/d_0.

(e)   **Heat transfer efficiency**

Figure 123 shows the heat transfer efficiency $\eta = (\overline{Nu}/W)/(\overline{Nu}/w)_{\text{st}}$ of the tw-, step-, and r-nozzles, that is, the ratio of the amount of heat transferred per unit operation power to the st-nozzle. η of tw-nozzle is approximately unity in the range of $t/d_0 = 0.2 \sim 0$. That is, it is the same as that of st-nozzle. η of the step-nozzle is about 6 at $t/d_0 = 0$, and η of the r-nozzle almost increases with increasing t/d_0 and it is about 2.3 at $t/d_0 = 0$.

Providing a step or radius on the corners of the inner wall at the nozzle exit suppresses flow separation, increases the effective passage area under the nozzle, and reduces flow resistance and operating power. In addition, the velocity and turbulence

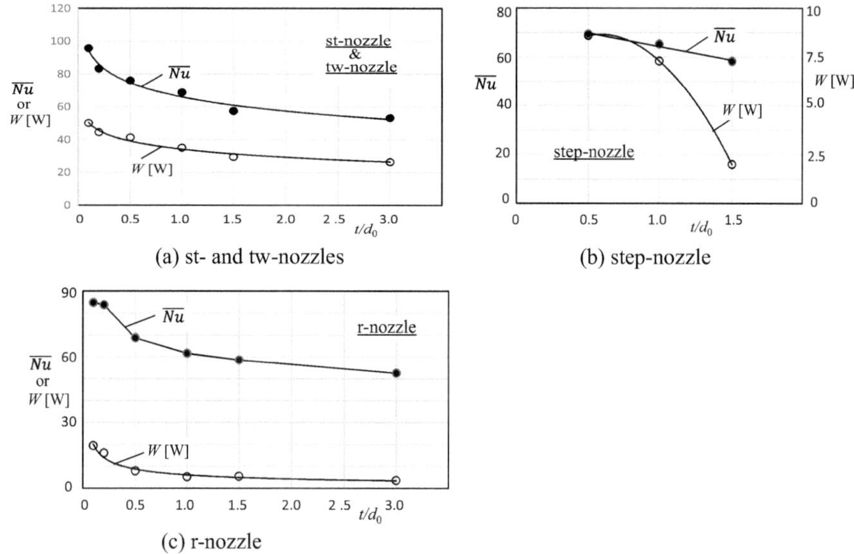

(a) st- and tw-nozzles   (b) step-nozzle

(c) r-nozzle

Fig. 122 Mean Nusselt number $\overline{Nu}$ and operation power W

Fig. 123 Heat transfer efficiency

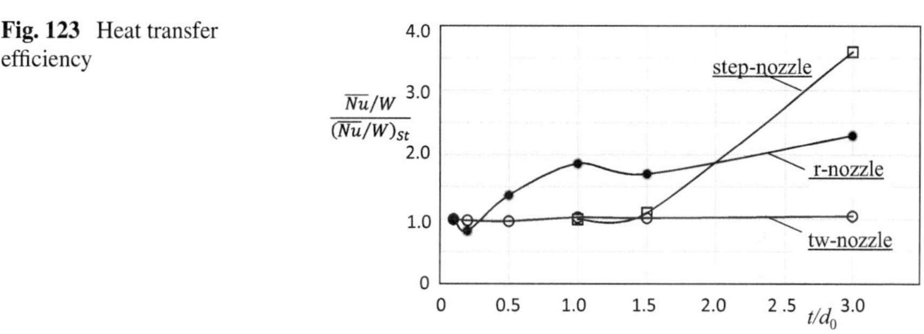

components between the underside of the nozzle and the impingement plate are reduced, so heat transfer characteristics deteriorate. However, the operating power is reduced more than this, so heat transfer characteristics taking operating power into account improve.

3.5.3 Using of Notched Orifice Nozzle

In this section, we consider the heat transfer characteristics of the impinging jet from the notched orifice nozzle shown in Fig. 54 in this chapter, Sect. 2.6.1.

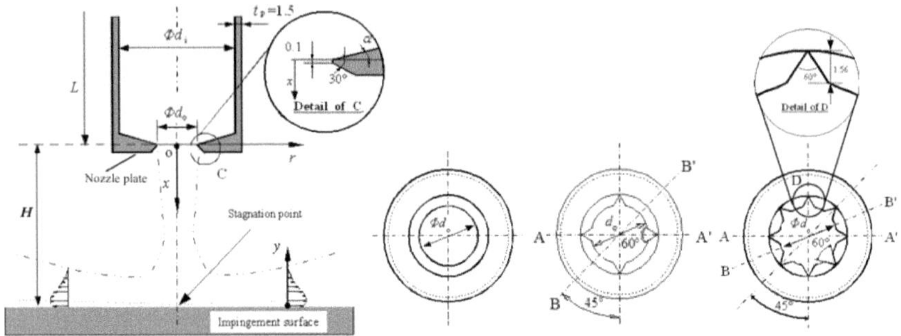

(a) Orifice nozzle – impingement plate   (b) Orifice nozzle and with 4- and 8-notches

Fig. 124   Schematic diagram of experimental set-up and nozzle geometry

(a) **Nozzle geometry and procedure**

A schematic of the nozzle geometry is shown in Fig. 124. The impingement plate is made of Bakelite with a thickness of 20 mm. A thin 30 μm stainless steel foil applied to this plate was directly heated by a uniform heat flux (2.5 kW/m^2) regulated by a slide trance. A total of 25 thermocouples (Cu–Co) were attached to the back of the foil and the temperature distribution was measured simultaneously. The local Nusselt number [$Nu = hd_0/\lambda a = q(T - T_a)d_0/\lambda_a$, where h: heat transfer coefficient, λ_a: thermal conductivity of air, q: heat flux, T: temperature, T_a: ambient temperature] was calculated from the temperature distribution and the average Nu was calculated from the following formula.

$$\overline{Nu} = \frac{1}{\pi r^2} \int_0^r Nu(2\pi r)dr \tag{96}$$

The operating power W was calculated by

$$W = Q\left(\Delta P + \rho u_{pm}^2/2\right) \tag{97}$$

where, Q: flow rate, ΔP: difference in pressure between system and atmosphere.

Static pressure was measured with a manometer at $20d_0$ upstream from the nozzle exit. Reynolds number was $Re = 1.5 \times 10^4$ unless otherwise stated. An orifice nozzle with an exit diameter of 10.0 mm nozzle and a pipe inner diameter of 29.75 mm [contraction area ratio is $CR = (d_0/d_i)^2 = 0.27$, see Fig. 124b] was used unless otherwise stated. Figure 124b shows an orifice nozzle with 8 notches having 60° openings and 1.56 mm depths. The inside of the 8 notched orifice nozzle is conically tapered at the angle $\alpha = 0, 5, 10, 15°$. Pipe nozzle with an inner diameter of 10 mm was also used for comparison. All nozzles were 500 mm long and 1.5 mm thick. The origin O of the coordinate system was placed at the center of the jet at the narrowest point of the nozzle opening, and the r and x axes were oriented radially and axially. The y-axis had its origin at the impingement plate surface and was oriented in the $-x$ direction.

(b) **Effects of notch on the flow characteristics**

Flows issuing from the orifice and notched orifice nozzle were visualized to demonstrate the effects of the notches on the flow characteristics. Flow visualization was performed in a sufficiently large water tank. Water jets seeded homogeneously with uranine (fluorescein) issued from the nozzle into the water tank. The vortex structures on the transparent impingement plate were visualized using a laser light sheet (Ar-laser, 3W) and xenon lamp and were recorded from the backside with a digital video camera.

Figure 125a–d show the flow patterns of the impingement jets from pipe-, orifice-, 4-notched orifice-, and 8-noched orifice-nozzles, respectively. The nozzle-plate distance is $H/d_0 = 2.0$ and the Reynolds number is $Re = 1000$. The temperature distribution of the impingement plate measured by thermography is shown on both sides. In the case of orifice-impinging jet, the flow fluctuates at the nozzle exit edge and spreads radially to produce a ring-shaped flow pattern that intermittently impinges on the plate, indicating a transition from laminar to turbulent flow. The spacing of the ring-shaped pattern corresponds to the frequency of the vortex emission. It is noteworthy that a longitudinal eddy structure was observed around the ring-shaped flow.

Figure 125c, d show the flow patterns of the 4- and 8-notched orifice jet, respectively. It is clearly observed that the presence of notches strongly affects the flow pattern. Fluctuations at the nozzle edge generate large vortex structures downstream that intermittently impinge on the plate, as observed in orifice impinging jets. However, the large vortex structures in notched orifice jets are distorted by the small notches. The transition from laminar to turbulent flow occurs slowly, and because the impingement area in the A-A' direction is larger than that in the B-B'

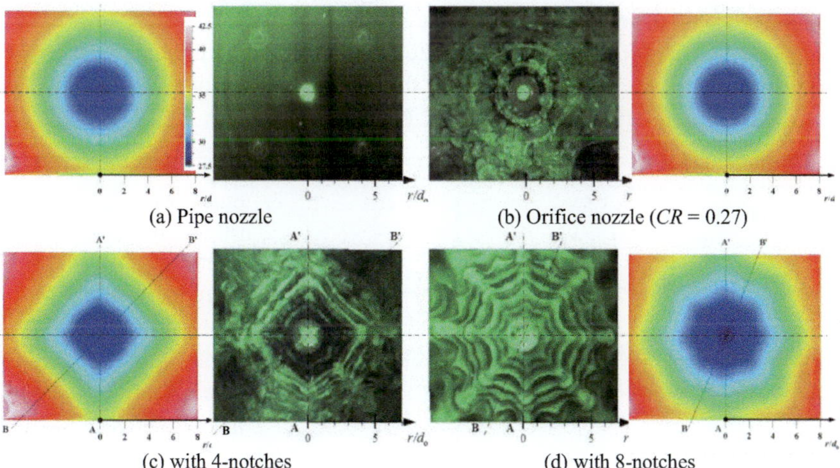

(a) Pipe nozzle   (b) Orifice nozzle ($CR = 0.27$)

(c) with 4-notches   (d) with 8-notches

Fig. 125 Visualized flow pattern and temperature distribution (by thermography) of impinging jet ($H/d_0 = 2, Re = 1000$)

direction at the eight notches, vortex up winding occurs downstream. This results in an octagonal star pattern corresponding to the position of the small notch at the nozzle outlet.

(c)   Reference length

The nozzle exit diameter is generally selected as a reference length to represent the dynamical characteristics of jet flows. However, it is difficult to represent the universal dynamical characteristics for notched orifice nozzles because there is little similarity in flow patterns and characteristics even though an equivalent diameter based on the wet length and cross-sectional area is used. In this paper, the operating power was calculated based on the flow resistance of each orifice jet under the same operating power ($W = 1.41[W] =$ const.) as that of the orifice jet operated at $Re = 1.5 \times 10^4$.

Figure 126 shows the flow rate Q_0 at the nozzle exit under the same operating power. The flow rate Q_0 of the notched orifice jet increases while the flow resistance decreases compared to that of the orifice jet without notches because of the increase in the cross-sectional area at the nozzle exit. The flow rate Q_0 increases linearly as the taper angle α increases. The Q_0 of 8-notched orifice jets with $\alpha = 0, 5, 10, 15°$ increased by 10.9, 11.3, 11, 15%, respectively, compared with that of the orifice jet.

(d)   Flow characteristics

The centerline velocity u_c under the same operating power is shown in Fig. 127. For orifice nozzles, the centerline velocity increases from the nozzle exit due to the vena contracta effect, becoming nearly constant in a region called the potential core, and then reaching the fully developed region. A similar increase in centerline velocity was observed for the notched orifice jet, but the effect of vena contracta weakened as the nozzle exit size became larger. The centerline velocity of the 8-notched orifice jet increased with increasing conical angle α.

Figure 128a shows the mean and fluctuating velocity distributions in the A-A' direction at the nozzle exit $x/d_0 = 0.2$ for the orifice, notched orifice, and pipe jets.

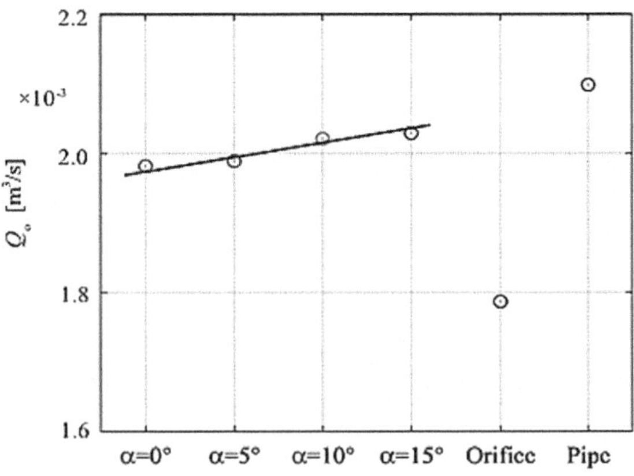

Fig. 126 Flow rate at the nozzle exit under the same operation power

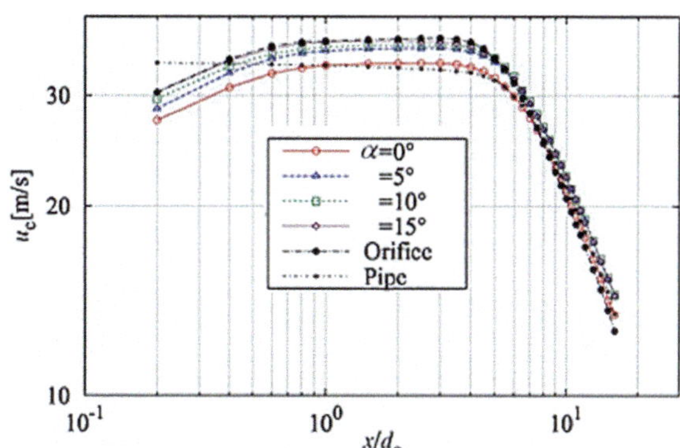

Fig. 127 Mean velocity along jet centerline

The mean exit velocity u/u_m is plotted on the left-hand side of Fig. 128 and the fluctuating velocity u'/u_m is on the right-hand side because they were axisymmetric. The results in the B-B' direction at the nozzle exit is plotted in Fig. 128 in the same manner. The saddle-backed profile observed for the orifice jet is caused by the vena contracta effect. The maximum fluctuating velocity of the orifice jet occurring at the jet edge is attributed to the exit velocity vector inside the nozzle toward the center because of the orifice nozzle contraction. A similar profile can be seen for the notched orifice jets, especially in Fig. 128b. The jet widths of the notched orifice jets increase because of the notches in the A-A' plane, while the vena contracta effects appear in the B-B' plane. The maximum velocity occurs at $r/d_0 = 0.45$, which is smaller than the nozzle edge due to the vortex generated by the notches. The fluctuating velocity of the notched-orifice jets increases in the A-A' plane. In the case of the pipe jet, the mean velocity profile of the pipe jet matched a fully developed turbulent pipe flow because the 500 mm long pipe is long enough for the boundary layer to grow completely. The fluctuating velocity reflecting the steep velocity gradient near the center of the jet is higher than that of the notched orifice jets that are discharged toward the center of the jet and accelerated. The turbulence intensity near the nozzle exit is larger than that of the other jets because the pipe jet has a fully developed turbulent velocity profile; however, the growth of the turbulence intensity is gradual and the turbulence intensity decreases at $x/d_0 \geq$

(e) **Mean and fluctuating velocity distributions on the impingement plate**

Figure 129a, b plot the mean wall-jet velocity, u, and fluctuating velocity u' normalized by the mean velocity, um, on the impingement plate for $H/d_0 = 2.0$ at $r/d_0 = 2.0$. The presence of the notches increases the velocity in the A-A' direction in relation to that of the orifice jet. On the other hand, the velocity in the B-B' direction is smaller than those of the other two jets. The velocity difference between in the A-A' and in the B-B' directions could be attributed to the turbulence formation. A maximum fluctuating velocity occurs at $y/d_0 = 0.07\text{~}0.1$, where there is a large velocity gradient.

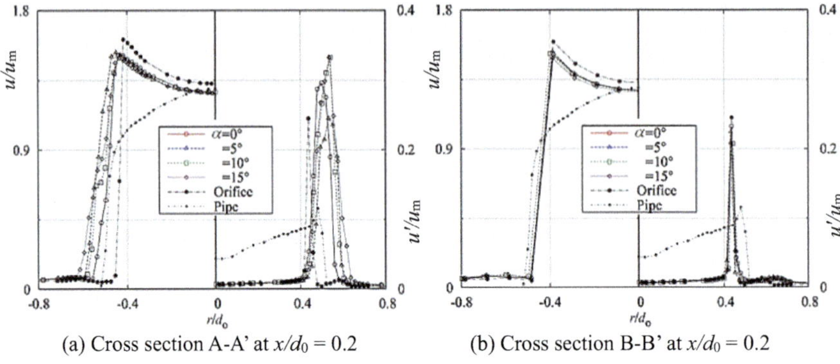

(a) Cross section A-A' at $x/d_0 = 0.2$ (b) Cross section B-B' at $x/d_0 = 0.2$

Fig. 128 Mean and fluctuating velocity distributions at nozzle exit ($x/d_0 = 0.2$)

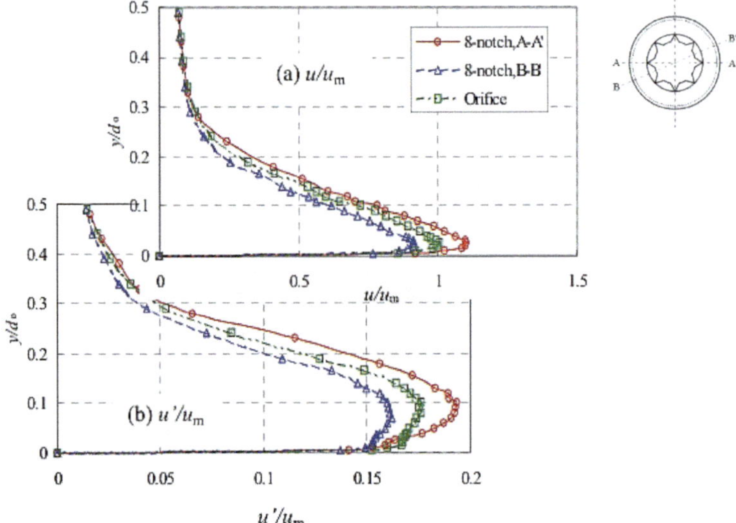

Fig. 129 Velocity distribution in the y-r plane at $r/d_0 = 2.0$

Figure 130 plots the velocity distribution in Fig. 130a normalized by the maximum velocity, u_{max}, and the half jet width, $y_{1/2}$ (the larger value). The solid line in Fig. 131 is a value calculated using the profile method in the following equation. Every jet is well matched with the equation.

$$\frac{u}{u_{max}} = \left(\frac{y}{\delta}\right)^{1/5}\left[1.811\left(1 - \frac{y}{\delta}\right)^2 + 0.02\left(1 - \frac{y}{\delta}\right) + 0.1\right] \tag{98}$$

Figure 131a, b present the results at $r/d_0 = 5.0$ in the same manner as Fig. 129. The maximum velocities are seen at about $y/d_0 = 0.1$ Considering that the maximum

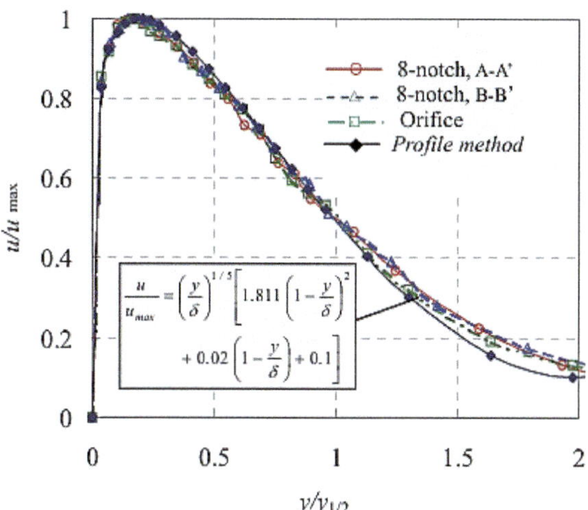

Fig. 130 Velocity distribution, $u/u_{\max} - y/y_{1/2}$

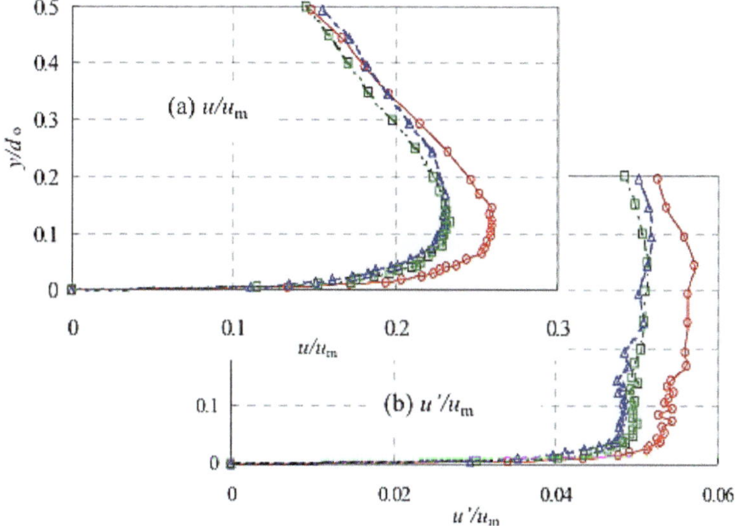

Fig. 131 Velocity distribution in the y–r plane at $r/d_0 = 5.0$

velocities occur at about $y/d_0 = 0.025$ in Fig. 131a, an increase of the boundary layer is observed from the location where the maximum velocity occurs. Note that the normalized velocity distributions ($u/u_{\max} - y/y_{1/2}$) for the velocity profiles at $r/d_0 = 5.0$ were also well fitted with Eq. (98).

(f) Heat transfer

The heat-transfer characteristics of the orifice, the notched orifice, and the pipe-impinging jets under the same operating power are shown in this section.

Figure 132 presents the local Nusselt number, Nu, distribution on the impingement plate at $H/d_0 = 2.0$. Due to the axisymmetry of the profile, only half of the radial distribution from the stagnation point is plotted. The Nu reaches a maximum near the stagnation point and decreases outward. The Nu distribution of the notched orifice jet is larger than that of the other jets, while that of the pipe jet is smaller, corresponding to the larger centerline velocity (impinging velocity) and the larger turbulent intensity due to the notch.

Figure 133a, b show the mean Nusselt number, $\overline{Nu}$, increasing ratio IR in relation to that of the pipe jet for the orifice and the notched orifice jets with taper angle α in the A-A' and the B-B' plane, respectively.

$$IR = \frac{\left(\overline{Nu} - \overline{Nu}_{\mathrm{pipe}}\right)}{\overline{Nu}_{\mathrm{pipe}}} \tag{99}$$

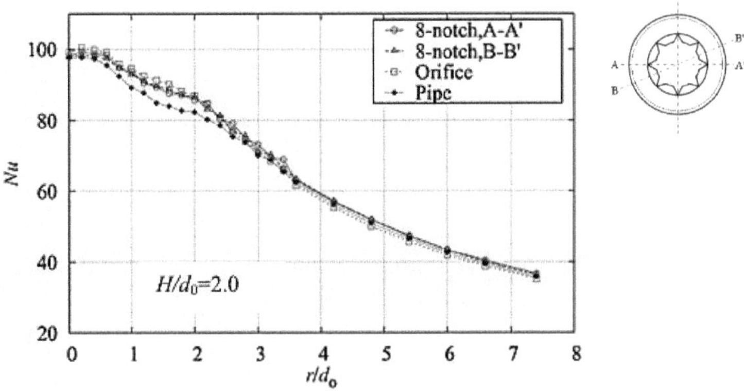

Fig. 132 Nu distribution of orifice nozzle

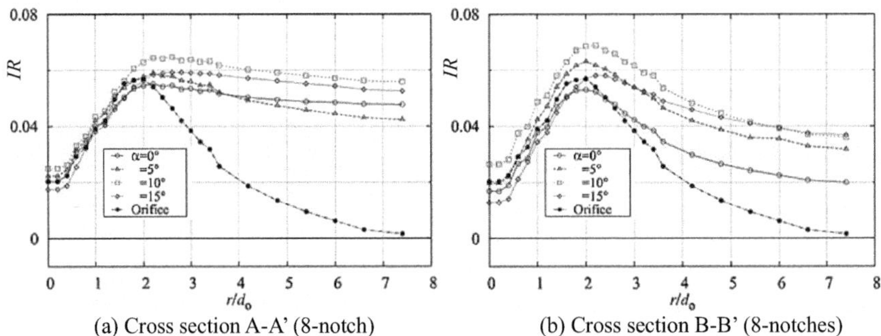

(a) Cross section A-A' (8-notch)   (b) Cross section B-B' (8-notches)

Fig. 133 Increasing ratio of mean Nu to the pipe jet

The mean Nusselt number, $\overline{Nu}_m$, was obtained approximately by summing and averaging the local Nusselt number calculated from the temperature distribution in the A-A' and the B-B' directions considering the measured temperature as the reference temperature in the range of $\theta = 2\pi/16$ [rad.]. The mean Nusselt number for the notched orifice jets with tapering significantly increase at $r/d_0 = 2\sim4$, especially in the A-A' direction because the notched orifice jets have a larger jet width, improving their mixing performance with ambient fluid, and increasing the flow rate and turbulent intensity.

Figure 134 shows the increasing ratio IR_m [IR when $\overline{Nu}$ is replaced by $\overline{Nu}_m$ in Eq. (99)] of the mean Nusselt number $\overline{Nu}_m$ to that of the pipe jet under the same operating power. The mean Nusselt number $\overline{Nu}_m$ was obtained by averaging the mean Nusselt number in the A-A' and the B-B' directions in Fig. 124. The enhanced Nusselt number is observed in the stagnation region due to the increase of jet center line velocity by vena contracta effect, but because of the decrease of jet width the heat transfer characteristics at the downstream becomes bad. However, in the case of the orifice jet, the heat transfer coefficients decrease more the others in the downstream region because of the size of the nozzle exit. The $\overline{Nu}$ for the notched orifice jet increases at $r/d_0 \fallingdotseq 2.2$ compared to that of the pipe jet. The ratio of $\overline{Nu}_m$ to that of the pipe jet was attained at $\alpha = 10°$ and reached 6.7% at maximum. In addition, the heat transfer characteristics for $\alpha = 0, 5, 10, 15°$ were enhanced by about 4, 7, 4.6, 4.5%, respectively, in the downstream at $r/d_0 = 7.4$.

For reference, in Fig. 135 the increasing ratio of the mean Nusselt number $\overline{Nu}$ to that of the orifice jet of $CR = 0.11$, $\alpha = 15°$, $H/d_0 = 2.0$ under the same operating power is shown. The result of a cone orifice jet having $\alpha = 15°$ is also plotted in Fig. 135 for comparison with the notched orifice jets of $CR = 0.11$. The heat transfer characteristics are enhanced in the impingement region by a maximum of 7.8% for the cone orifice jet and then decrease monotonically in the downstream. The ratios of the $\overline{Nu}$ of the notched orifice jets with tapering in the downstream $r/d_0 > 3$ were

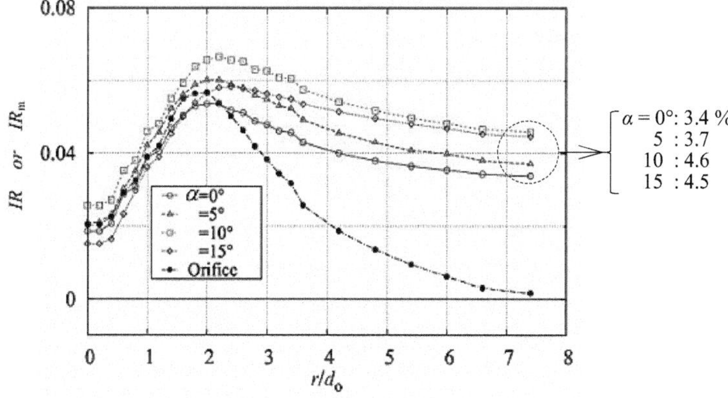

Fig. 134 Increasing ratio of mean Nu_m to the pipe jet

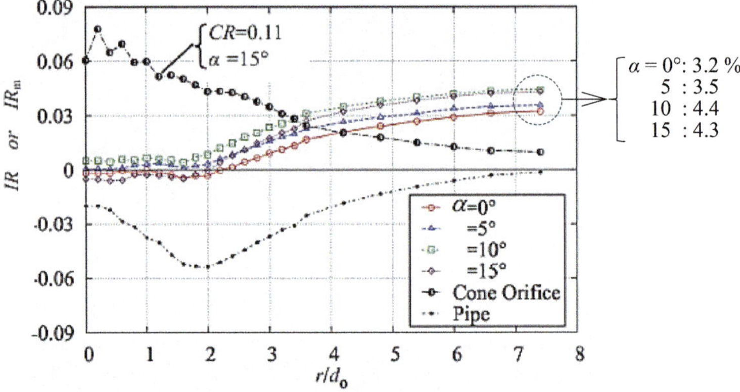

Fig. 135 Increasing ratio of mean Nu_m to the orifice jet

larger than that of the orifice jet. In the case of $CR = 0.27$ almost all same results were obtained.

Experimental measurements were made to obtain mean flow velocity, fluctuating flow velocity, and temperature distribution on an impingement plate to investigate flow and heat transfer characteristics of notched orifice jets with different taper angles.

(1) The notched orifice jet reduced the nozzle resistance and operating power as well as increased turbulence at the nozzle exit compared to the existing orifice jet.
(2) The heat transfer characteristics of the tapered notched orifice jet increased significantly in the A-A' direction at $H/d_0 = 2.0$ due to the larger turbulent intensity generated by the notch.
(3) Under the same operating power, the heat transfer characteristics of an 8-notch, $\alpha = 10°$ conical notched orifice jet increase by up to about 6.7% at $r/d_0 \fallingdotseq 2.2$ and about 4.6% at $r/d_0 = 7.4$ compared to a pipe jet.

The proposed tapered notched-orifice nozzle would be of great value in improving heat transfer characteristics in impinging jet systems.

4   Wall Jet Flows

Jets ejected along solid boundaries (walls), so-called wall jets, are used for boundary layer control, film cooling of high-temperature walls, heat insulation, etc., and it is important to understand their flow characteristics.

Here, we will discuss the flow characteristics of a wall jet flowing along a flat plate.

4.1 Two-Dimensional Wall Jet

Figure 136 shows a flow model of a two-dimensional turbulent wall jet. In any case, a wall jet flows with a boundary layer (wall shear layer) on the flat wall and a free boundary (free shear layer) outside the jet. When both shear layers intersect, the potential core disappears, and the flow becomes a fully developed flow after this cross section.

4.1.1 Dimensional Analysis

Now, assuming that the momentum M_0 of the jet is approximately conserved and that the effects of viscosity are ignored, the maximum jet velocity $u_{\max}$, the jet width b, and the wall shear stress τ_0 are functions of M_0, ρ, and x, respectively.

$$U_{\max} = F_1(M_0, \rho, x), \ b = F_2(M_0, \rho, x), \ \tau_0 = F_3(M_0, \rho, x) \tag{100}$$

Using the π-theorem,

$$\frac{u_{\max}}{U_0} = C_1 / \sqrt{\frac{x}{b_0}} \tag{101}$$

$$b = C_2 \cdot x \tag{102}$$

$$c_f = \tau_0 / \frac{\rho U_0^2}{2} = C_3 / \frac{x}{b_0} \tag{103}$$

where, c_f is skin friction coefficient.

From Eq. (102), we have

$$c_f = \tau_0 / \frac{\rho u_{\max}^2}{2} = \frac{C_3}{C_1} \tag{104}$$

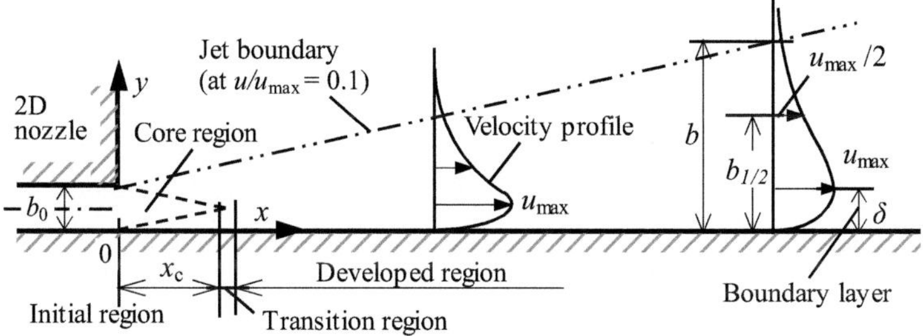

Fig. 136 Two-dimensional turbulent wall jet

4.1.2 Velocity Distribution for Turbulent Two-Dimensional Wall Jet

Foerthmann (1934) measured first the velocity distribution of the two-dimensional wall jet. Figures 137 and 138 show it with dimensions and non-dimensionalized with maximum velocity u_m and half-width $b_{1/2}$. The velocity distribution in the development region is similar, and when non-dimensionalized it is represented by a single curve.

Figure 139 show the experimental results of Verhoff (1963) and the following empirical formula are shown.

$$\frac{u}{u_{\max}} = f(\eta) = 1.479\eta^{1/7}[1 - \mathrm{erf}(0.67758\eta)] \tag{105}$$

where, $\eta = y/b_{1/2}$.

Figure 140 shows the measurements of velocity distribution of u by Yamada et al. (1980) and compared with the Wilson and Goldstein (1976) and Schwarz and Cosart (1961). The height (aspect ratio) of the nozzle and the parallel part are $b = 10.0$ mm

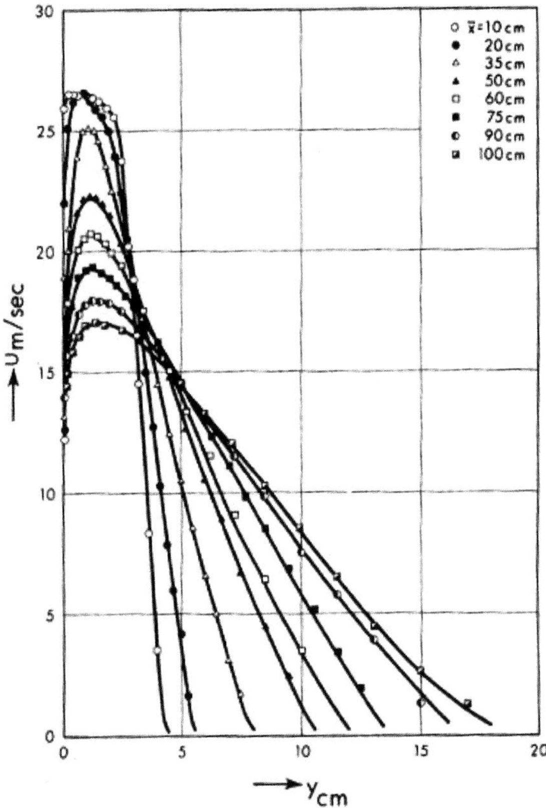

Fig. 137 Velocity profiles, plane wall jet (Foerthmann 1934)

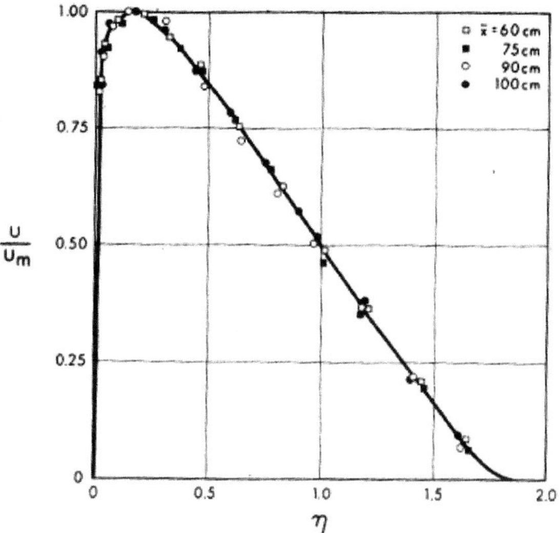

Fig. 138 Nondimensional velocity profiles, plane wall jet (Foerthmann 1934)

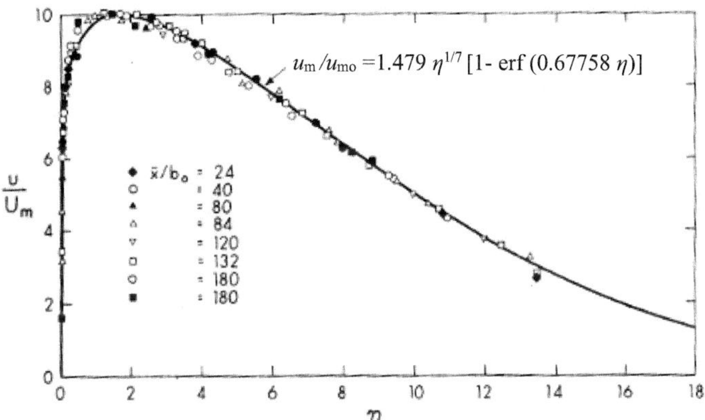

Fig. 139 Verhoff's Eq. (6) and experiments (Verhoff 1963)

$(AR = 30)$ and 50.0 mm, respectively. The velocity distribution was measured at $Re = u_o\, b/v = 2.0 \times 10^4$ using a constant temperature hot wire anemometer and an x-type probe. The measurements are good agreement with those of Wilson and Goldstein and Schwarz and Cosart. The mean velocity v in the y-direction was calculated from the continuity equation using the measured value u.

$$v = \int_0^y (\partial u/\partial x)\,dy \tag{106}$$

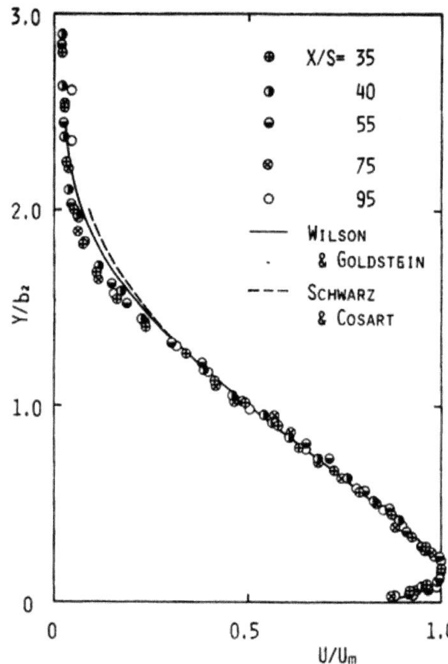

Fig. 140 $u/u_\mathrm{m}-y/b_{1/2}$
(Yamada et al. 1980)

Figure 141 shows the velocity distribution v. Wilson and Goldstein (1976) were also obtained v with the same method and reported that the distribution shape of v is similar in the case of a plane wall jet. The measurements are a little different with the Wilson and Goldstein.

4.1.3   Maximum Velocity

Figure 142 shows the decay of the maximum velocity u_m in the x direction. $u_\mathrm{m}/u_\mathrm{mo}$ decreases in proportion to $x^{-1/2}$ and it can be expressed by the following equation in Sect. 3.3.1c.

$$\frac{u_\mathrm{m}}{u_\mathrm{m0}} = 3.5\left(\frac{b_0}{x}\right)^{1/2} \; for\left(\frac{x}{b_0} < 100\right) \tag{107}$$

4.1.4   Spreading of Jet and Half Width

Figure 143 shows the results of several experiments expressing the spread of the two-dimensional wall jet in terms of its half-width (the larger value), $b_{1/2}$. The half-width increases linearly downstream but the results of Sigalla (1958) are somewhat

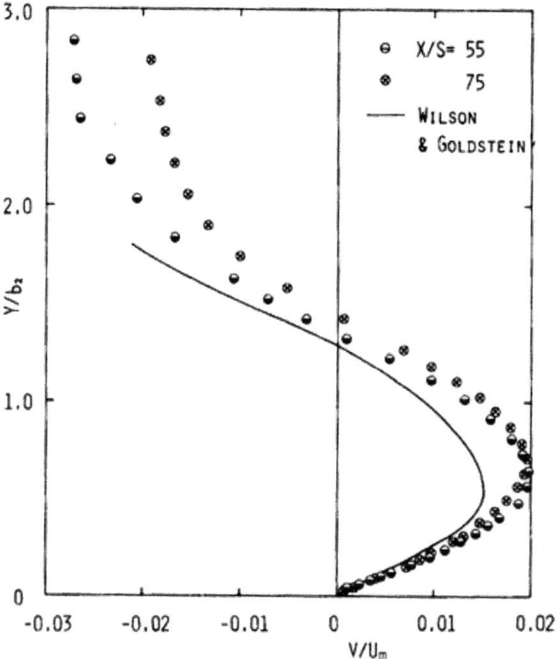

Fig. 141 $v/u_\mathrm{m}-y/b_{1/2}$ (Yamada et al. 1980)

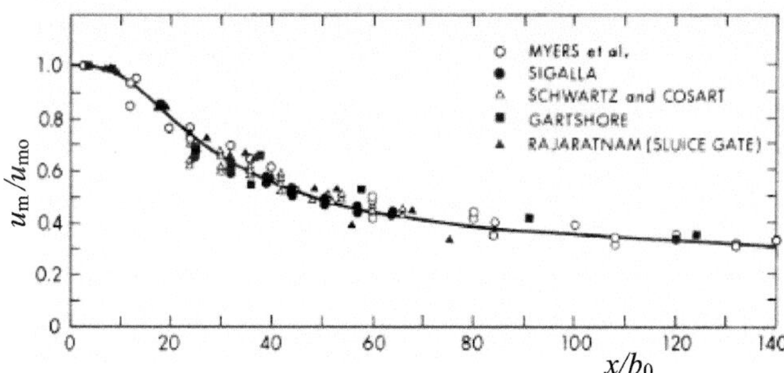

Fig. 142 Jet centerline velocity, $u_\mathrm{m}/u_\mathrm{mo}$ (Rajaratnam and Subramanya 1967)

larger than others. If the virtual origin is located at $x/b_0 = -10$, the half-width can be approximately expressed by the following equation.

$$b_{1/2} = 0.068\,x \tag{108}$$

The spread of the two-dimensional wall jet, expressed in terms of half-width, is about 0.7 times that of the two-dimensional free jet described above.

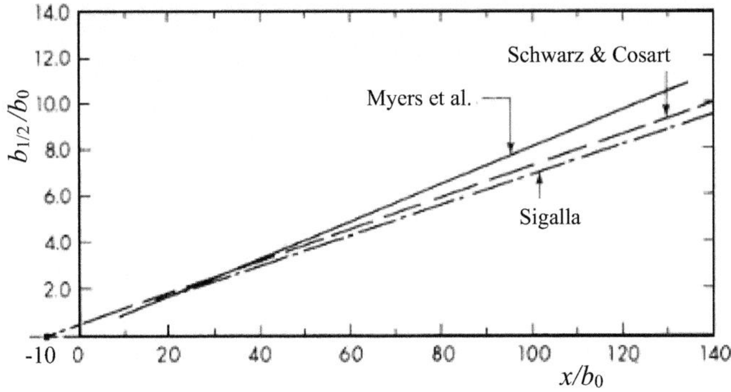

Fig. 143 Spread of two-dimensional wall jet, half width (Rajaratnam 1976)

4.1.5   Entrainment Flow Rate

The entrainment flow rate per unit length in the wall jet is

$$Q = \int\limits_{0}^{\infty} u\,dy = u_{max}b_{1/2} \int\limits_{0}^{\infty} f\,d\eta \tag{109}$$

Using velocity distribution $f(\eta)$, Q is calculated as follows.

$$\frac{Q}{Q_0} = 0.248\left(\frac{x_c}{b_0}\right) \tag{110}$$

where, Q_0 is flow rate per unit length at the nozzle exit.

The entrainment velocity v_e can be expressed by the following equation using Eq. (107)

$$v_e = dQ/dx = 0.035 u_{max} \tag{111}$$

4.1.6   Wall Shear Stress

Myers et al. (1963) measured the velocity distribution near the wall using the hot film method and applied it to the Clauser's chart to calculate the wall shear stress τ_0. T_0 decays in proportion to x^{-1} ($\tau_0 \propto x^{-1}$), and the wall friction stress coefficient is expressed by the following equation.

$$C_f = \frac{\tau_0}{\rho u_0^2/2} = \frac{0.1976}{(x/b_0)(u_0 b_0/v)^{1/2}} \tag{112}$$

Sigalla (1958) obtained τ_0 using Preston tube and showed the following experimental formula.

$$C_\mathrm{f} = \frac{\tau_0}{\rho u_m^2/2} = \frac{0.0565}{(u_m\delta/v)^{1/4}} \tag{113}$$

4.1.7 Turbulent Properties

Figure 144 shows the measurements of the turbulent intensity, $\sqrt{\overline{u'^2}}/u_\mathrm{m}$, in the x-direction by Yamada et al. (1976). The solid lines in the figure are the results of Wilson and Goldstein. The measurements are good agreement of the Wilson and Goldstein. The profile.

If the $\sqrt{\overline{u'^2}}/u_\mathrm{m}$ of each cross section is made dimensionless by its maximum value $\sqrt{\overline{u'^2}}_\mathrm{max}$ it will be similar and be expressed by almost a single curve. Also, the distribution profile is almost the same as the Wilson and Goldstein.

Figure 145 shows the nondimensional Reynold stress, $\overline{u'v'}/\overline{u'v'}_\mathrm{max}$. In this case, the profile is also similar shape.

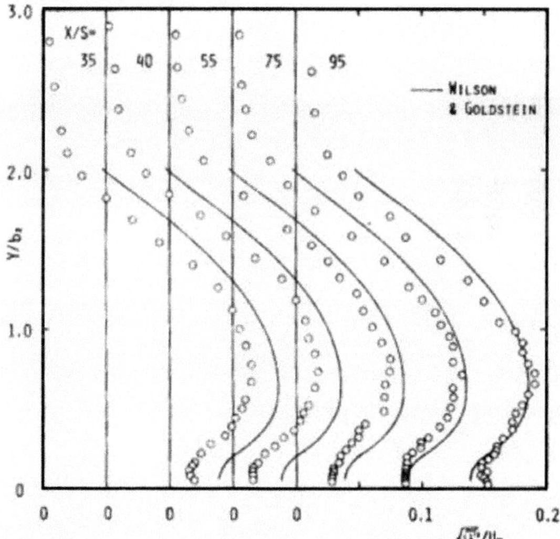

Fig. 144 Turbulent intensity, $\sqrt{\overline{u'^2}}/u_\mathrm{m}$ (Yamada et al. 1976)

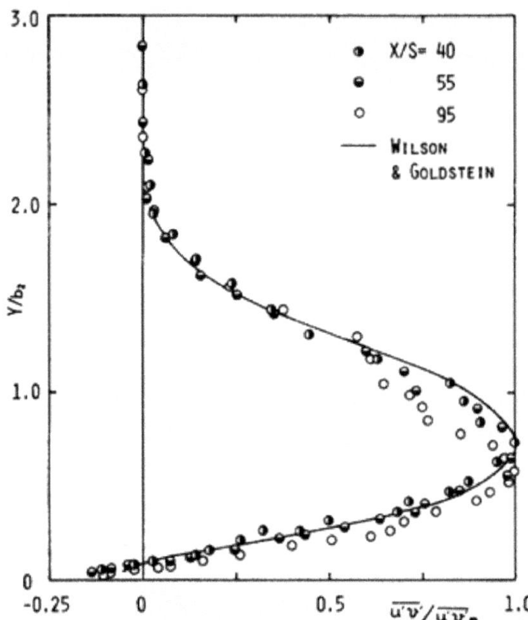

Fig. 145 Reynolds stress, $\overline{u'v'}$ (Yamada et al. 1976)

4.2   *Impinging Wall Jet and Radial Wall Jet Flows*

A part of the flow characteristics of impinging wall jet (radial wall jet) has been examined in Sect. 3 in Chap. 3. Here we examine other aspect. The flow model of imping wall jet ($H/d_0 > 1$) and radial wall jet ($H/d_0 < 1$) is shown in Figs. 98 and 110.

4.2.1   Pressure Distribution on Impingement Plate

Baltaos and Rajaratnam (1974) showed the pressure distribution p_w on the impinging plate can be expressed by

$$\frac{p_w}{p_s} = \exp\left[-114\left(\frac{r}{H}\right)^2\right] \quad for\ (H/d_0 > 8) \tag{114}$$

where, p_s: stagnation pressure, p_s and p_w: gauge pressure

Furthermore, it was shown that the wall shear stress τ_0 is also expressed by the following equation.

$$\frac{\tau_0}{\tau_{\mathrm{m}}} = 0.18\left\{\frac{1 - \exp\left[-114\left(\frac{r}{H}\right)^2\right]}{r/H}\right\} - 9.43\left(\frac{r}{H}\right)\exp\left[-114\left(\frac{r}{H}\right)^2\right] \quad for\ (H/d_0 > 8)$$

$$(115)$$

where, τ_{m} is maximum wall shear stress and can be expressed by

$$\tau_{\mathrm{m}} = 0.16\rho U_0^2\left(\frac{H}{d_0}\right)^{-2} \quad \text{at}\ (r/H) \doteq 0.14 \quad \text{for}\ (H/d_0 > 8) \tag{116}$$

4.2.2 Velocity Distribution

Figure 146 shows the flow model and notations of impinging wall jet.

Figure 147a shows the theoretical analyses of velocity distribution by Glauert (1956), Abramovich (1963) and Goertler (1967). In order to obtain a solution, Glauert was considered by dividing the inner layer and the outer layer on the wall side with the position of the maximum flow velocity as the boundary. At that time, the velocity distribution of the inner layer was assumed to have a 1/7 power law, and the eddy viscosity coefficient was assumed to be constant for that of the outer layer.

Poreh et al. (1967) and Abramovich (1963) introduced the reference boundary velocity u_{rb} where $u_{\mathrm{rb}} = \text{Const.}/r$ to determine the velocity distribution of the radial wall jet. It is obtained by extrapolating the velocity profile of the outer region of the jet to the wall (Fig. 146). The velocity profile by Goertler (1942) for the outer region is

$$\frac{u}{u_{\mathrm{rb}}} = 1 - \tanh^2 \mathrm{C}\frac{z}{r} \quad \text{where, C : integral const.} \tag{117}$$

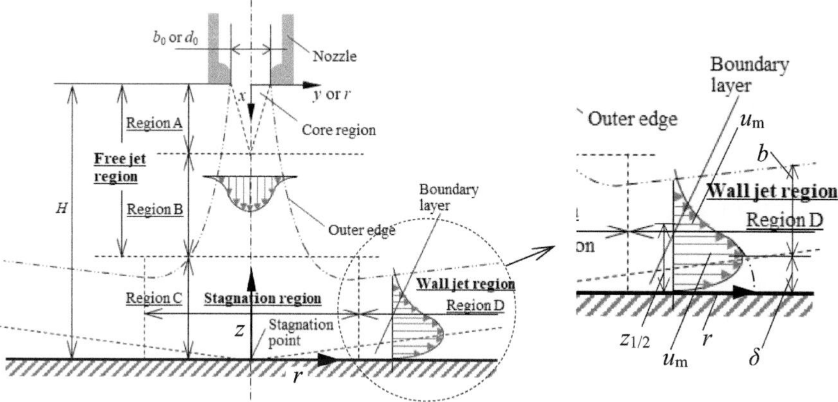

Fig. 146 Flow model of radial wall jet and notations

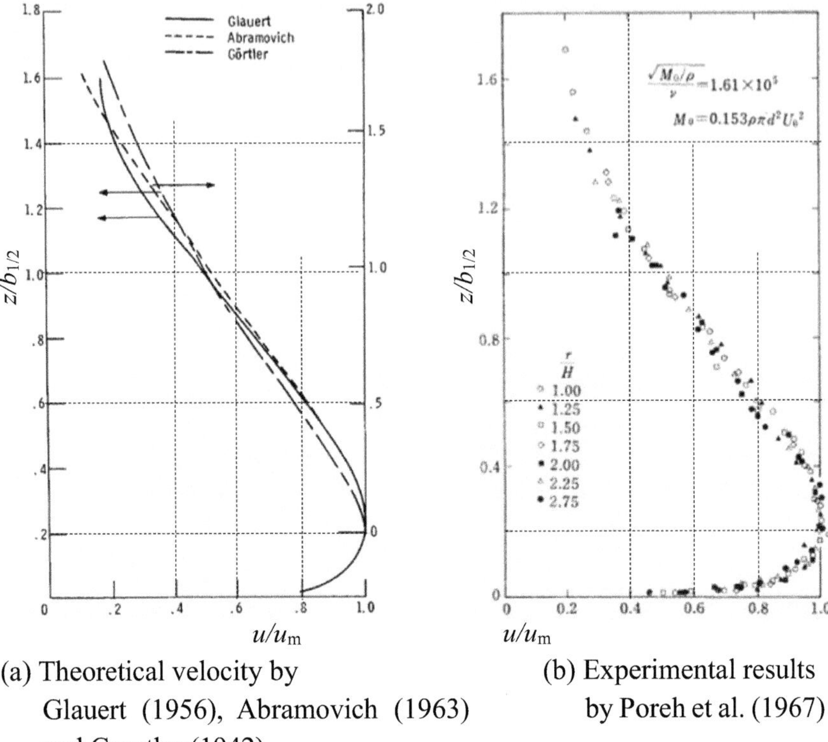

(a) Theoretical velocity by
Glauert (1956), Abramovich (1963)
and Goertler (1942)

(b) Experimental results
by Poreh et al. (1967)

Fig. 147   Velocity distribution of wall jet

Abramovich used the following relation

$$\frac{u}{u_\mathrm{m}} = \left(1 - \varsigma^{2/3}\right)^2 \quad \text{where,} \ \varsigma = \frac{z - \delta}{b'} \tag{118}$$

The three theoretical results in Fig. 147a and experimental results in Fig. 147b show fairly good agreement.

4.2.3   Wall Shear Stress

In general, the following relationships hold for radial wall jet:

$\tau_0 \propto 1/r^2$, i.e., $u^* \propto 1/r$, where, u^* is friction velocity $(=\sqrt{\tau_0/\rho})$.

Poreh et al. (1967) showed the relation in Fig. 148 between $1/u^*$ and r and they are in a linear relationship.

Now, we have that the wall shear stress τ_0 is the function of jet momentum M_0, ρ, H and r, and then

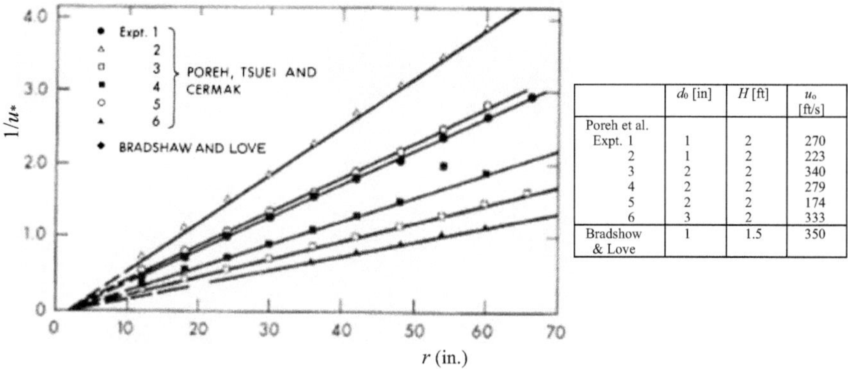

	d_0 [in]	H [ft]	u_0 [ft/s]
Poreh et al. Expt. 1	1	2	270
2	1	2	223
3	2	2	340
4	2	2	279
5	2	2	174
6	3	2	333
Bradshow & Love	1	1.5	350

Fig. 148 Wall shear stress in radial wall jets (u^*: friction velocity $= \sqrt{\tau_0/\rho}$)

$$\frac{\tau_0}{M_0/H^2} = F\left(\frac{r}{H}\right) \propto \frac{1}{(r/H)^2} \tag{119}$$

Using the relation in Fig. 148, we get

$$c_f = \frac{\tau_0}{\rho u_0^2/2} = \frac{c_f\prime}{(r/d_0)^2} \tag{120}$$

$c_f\prime$ is a function of Reynolds number and is represented by the following equation.

$$c_f\prime = \frac{0.238}{Re^{0.54}} \tag{121}$$

In Table 6 the flow characteristics such as velocity distribution, jet width, shear stress on surface etc. of two- and radial wall jet flows and impinging jet flow are summarized.

Table 6 Submerged Turbulent wall jet flows and impinging jet (Blevins 1984)

Jet flow characteristics ($x > x_c$)	Two-dimensional, plane, wall jet	Radial wall jet flow ($H/d_0 < 1$)
a. Length of core region, x_c	$x_c = 3.2b_0$	–
b. Velocity profile, u/u_m	$u/u_m = \exp\{-0.693[(y-\delta)/b_{1/2}]^2\}$ for $y \geqq \delta$ or $u/u_m = 1.479(y/b_{1/2})^{1/7}[1-(0.6776y/b_{1/2})]$	$u/u_m \doteq \exp[-0.693[(y-\delta)/b_{1/2}]^2]$
c. Maximum velocity, u_{max}	$u_{max} = 3.5(b_0/x)^{1/2}u_0$	$u_{max} = 3.5[(Hr_0)^{1/2}/r]u_0$
d. Shear stress on surface, τ	$\tau = 0.2(b_0/x)[v/(u_0b_0)]^{1/2}(\rho u_0^2/2)$	$\tau \doteq [(Hr_0)^{1/2}/r](\rho u_0^2/2)$
e. Boundary layer thickness, δ	$\delta = 0.014x$	$\delta = 0.016r$
f. Half width, $b_{1/2}$	$b_{1/2} = 0.068x$	$b_{1/2} = 0.078r$
g. Entrainment velocity, v_e	$v_e = 0.035u_{max}$	$v_e = 0.081u_{max}$
h. Volumetric flow rate, Q	$Q = 0.248xQ_0/b_0$	$Q = 0.201rQ_0/(Hr_0)^{1/2}$

Jet flow characteristics ($x > x_c$)	Impinging jet flow ($H/d_0 > 1$, $r/H > 0.22$)
a. Length of core region, x_c	–
b. Velocity profile, u/u_m	$u/u_m \doteq \exp\{-0.693[(y-\delta)/b_{1/2}]^2\}$
c. Maximum velocity, u_m	$u_m = 2.1(r_0/r)u_0$
d. Shear stress on surface, τ	$\tau = 0.655[v/(u_0r_0)]^{0.54}(r_0/r)^2(\rho u_0^2/2)$
e. Boundary layer thickness, δ	$\delta = 0.02r$
f. Half width, $b_{1/2}$	$b_{1/2} = 0.087r$
g. Entrainment velocity, v_e	–
h. Volumetric flow rate, Q	–

5 Reattached Jet Flows (to Flat and Curved Wall Surfaces)

Here, we will consider the so-called reattached jet, a phenomenon in which a jet flows while reattaching to a flat surface and a curved wall surface installed in a still space near the nozzle exit.

5.1 Two-Dimensional Reattached Jet to Adjacent Flat Plate by Coanda Effect

If a flat wall surface or sidewall exists near a two-dimensional, 2D, incompressible jet, the fluid between the outer edge of the jet on the sidewall side is caught up and carried away downstream as a result of the velocity gradient and fluid viscosity, resulting in a decrease in pressure and the jet flowing toward the sidewall. It flows while reattaching to the side wall while curving. This phenomenon is called the Coanda effect after discoverer Henri Coanda. At this time, the center line of the reattached jet is approximated by a circular arc, and the jet flows steadily in a state where the force due to the pressure difference between the inside and outside of the jet and the centrifugal force are balanced (Fig. 149). Borque and Newman used momentum theory to find the attachment distance, x_R, and the pressure within the bubble region, P_b, etc. under some assumptions.

Their assumptions are (1) the jet is 2D and incompressible flow, (2) the centerline of reattached jet can be expressed the arc of radius R, (3) the pressure within bubble region is uniform, P_b, and (4) the entrainment flow rate and the velocity distribution are similar to those of a two-dimensional free jet.

Now, the velocity distribution of a 2D free jet is

$$u = \left[\frac{3K\sigma}{4(x + x_0)} \right]^{1/2} \left[1 - \tanh^2 \left(\frac{\sigma y}{x + x_0} \right) \right] \tag{122}$$

where, x: coordinate along jet centerline, distance along jet axis from the nozzle exit,
x_0 : distance from nozzle exit to hypothetical origin, σ : spreading parameter ($=$ 7.67 for free jet by Reichardt), $K := J/\rho$, J : jet momentum per unit depth ($= \rho b_0 u_0^2$)

The volumetric flow rate Q between the jet centerline and reattaching streamline ($y = y'$) is constant and can be expressed by the following equation.

$$Q = \int_0^{y'} u\, dy \tag{123}$$

Substituting the equation of velocity distribution and integrating, we get

$$u_0 b_0 = \left[\frac{3J(x + x_0)}{4\rho\sigma} \right]^{1/2} \tanh \left[\frac{\sigma y'}{x + x_0} \right] \tag{124}$$

Since $J = \rho b_0 u_0{}^2$, so we get

$$\left[\frac{3(x + x_0)}{4b_0\sigma}\right]^{1/2} \tanh\left[\frac{\sigma y'}{x + x_0}\right] = 1 \tag{125}$$

Now, if we take $t = \tanh\left[\sigma y/(x + x_0)\right]$, the equation that gives the reattachment streamline is as follows.

$$\frac{3x}{\sigma b_0} = \frac{1}{t^2} - 1 \tag{126}$$

* First theory (Impingement point flow model).

Next, considering the momentum balance at the reattachment point (impingement point flow model), for the side wall of $\alpha = 0°$ (parallel plate),

$$J\cos\theta = J_1 - J_2 = \int\limits_{-\infty}^{y'} \rho u^2 \mathrm{d}y - \int\limits_{y'}^{\infty} \rho u^2 \mathrm{d}y \tag{127}$$

where, J_1: Momentum of the flow flowing out along the flat plate after the reattachment point,

J_2: momentum of the refluxing flow in the vortex region along the flat plate from the reattachment point (Fig. 149).

From this, the following equation is obtained.

$$\cos\theta = \frac{3t}{2} - \frac{t^2}{2} \tag{128}$$

From the geometrical relationship, we obtain

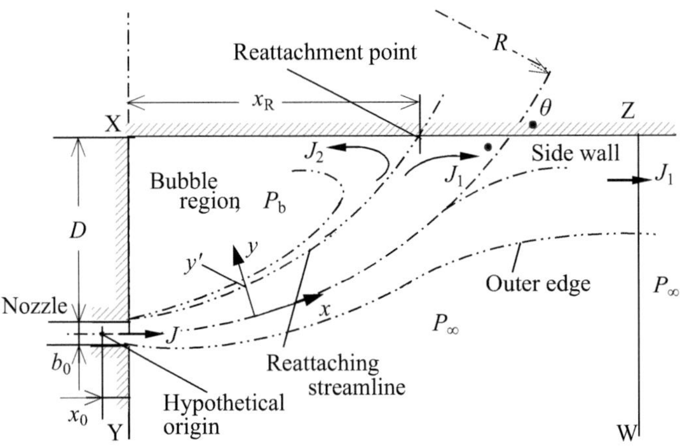

Fig. 149 Two-dimensional, incompressible reattached jet

$$\frac{D}{b} = \sigma\left(\frac{1}{t^2} - 1\right)\left(\frac{1 - \cos\theta}{3\theta}\right) - \frac{1}{2} \tag{129}$$

where, D: offset distance (Fig. 149).

The pressure difference C_{ps} on both sides of the jet is expressed by the following equation.

$$C_{pc} = \frac{P_\infty - P_b}{P_0 - P_\infty} = -\frac{2b}{r} \tag{130}$$

where, P_0: supplied pressure, P_b: bubble pressure, P_∞: ambient pressure

Therefore, the reattachment distance x_R is

$$\frac{x_R}{b_0} = \sigma\left(\frac{1}{t^2} - 1\right)\frac{\sin\theta}{3\theta} - \frac{\tanh^{-1}t}{3t^2\sin\theta} \tag{131}$$

where, t and θ are expressed as function of D/b_0 from Eqs. (128) and (129) or Eqs. (133) and (129). Equation (128) is the result regarding the impingement flow model and Eq. (133) is the result regarding the control volume flow model. Therefore, reattachment distance x_R and bubble pressure C_{pb} can be found as a function of D/b_0.

* Second theory (Control volume flow model)

In addition, from the balance of forces applied to the control volume ABCD shown in Fig. 149 the following equation is obtained.

$$J - J_1 = (P_\infty - P_b)\left(D + \frac{b}{2}\right) \tag{132}$$

Now, it's $P_\infty - P_b = J/R$, so

$$\cos\theta = \frac{1}{2} + \frac{3t}{4} - \frac{t^3}{4} \tag{133}$$

Equation (133) replaces Eq. (128) in the first theory and the rest equation are used without change.

Figure 150 shows the experimental and theoretical calculation results for the reattachment distance x_R obtained by Borque and Newman (1960). The reattachment distance x_R/b_0 increases almost linearly with increasing offset distance D/b_0. The theoretical calculation results for x_R better represent the experimental results when using impingement flow model (First theory) with $\sigma = 12$.

Figure 151 shows the contour map of pressure distribution, and there is a pressure distribution in the bubble region, and there is a pressure increase where the jet impinges on the side wall, and the minimum pressure position on the side wall corresponds to the minimum pressure position within the bubble region.

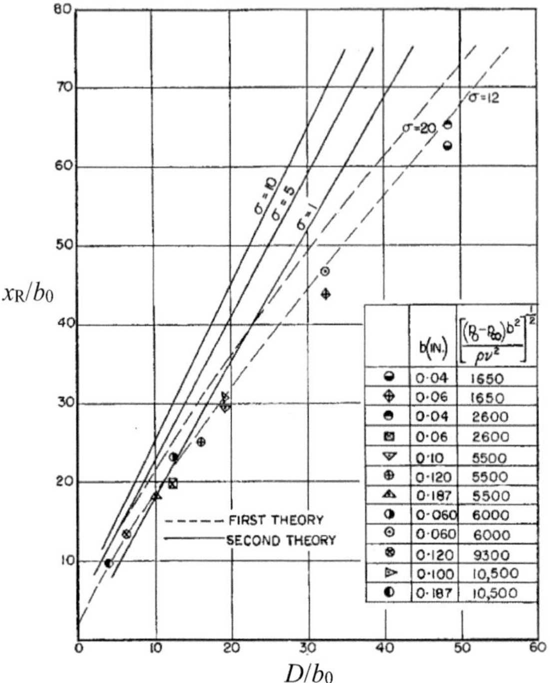

	b(IN.)	$\left[\dfrac{(p_0-p_\infty)b^2}{\rho v^2}\right]^{\frac{1}{2}}$
◓	0·04	1650
⊕	0·06	1650
◑	0·04	2600
⊠	0·06	2600
▽	0·10	5500
⊕	0·120	5500
▲	0·187	5500
◐	0·060	6000
⊙	0·060	6000
⊗	0·120	9300
▷	0·100	10,500
◑	0·187	10,500

Fig. 150   Reattachment distance, x_R

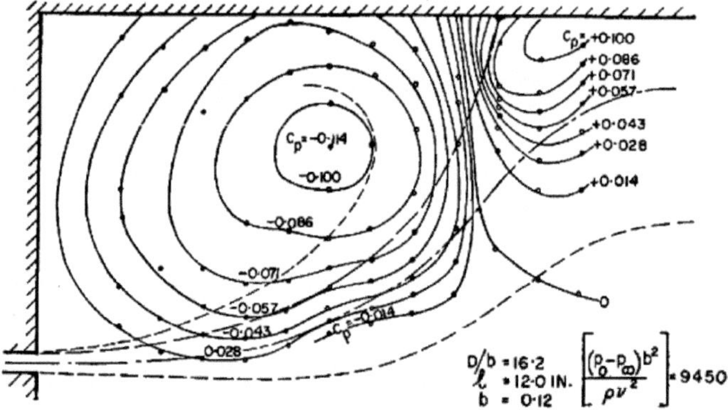

Fig. 151   Pressure distribution of reattached jet

Based on comparison with experimental results, the impingement flow model is considered better in case $D/b_0 > 3$. In addition, Borque and Newman (1960) shows a similar analysis for an inclined sidewall with an offset distance of $D = 0$, and in the case of $\alpha > 25°$, the control volume flow model is considered to be better.

5.2 Reattachment of Two-Dimensional Jet to Curved Wall Surface

The two-dimensional, 2D, and three-dimensional, 3D, jets flowing along the curved wall surface due to the Coanda effect are important to clarify its flow characteristics in relation to boundary layer control on the curved wall surface, that is, increasing the lift of the wing (Takahashi 1986; Maita 1986) and the film cooling of the turbine blade, cyclone separators, centrifugal concentration of uranium, and classification of fine solid particles etc.

Here, we will particularly examine the flow characteristics of 2D and 3D circular jets flowing along the wall surface of a cylinder, which are basic shapes.

5.2.1 Two-Dimensional Reattached Jet to Cylindrical Wall (Approximate Theory)

When a two-dimensional jet is issued along a cylindrical wall, the jet attaches to the wall due to the Coanda effect, flows under the effect of centrifugal force, and separates from the wall downstream (Fig. 152). In the following, we will discuss the approximate theory for determining the velocity distribution, pressure distribution, etc. of a cylindrical wall jet (Shakouchi et al. 1988, 1992a, 1990; Shakouchi and Onohara 1989).

(a) Momentum integral equation

Considering a two-dimensional, incompressible, turbulent jet flow on a cylindrical wall, taking the test volume as ABCD in Fig. 153 and deriving the momentum integral equation, we get the following equation.

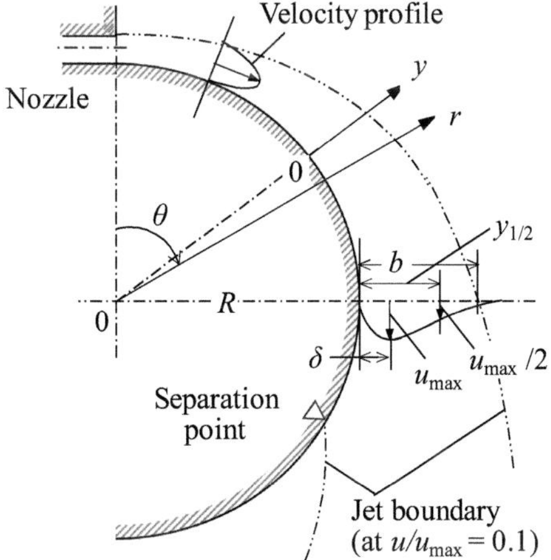

Fig. 152 Two-dimensional cylindrical wall reattached jet

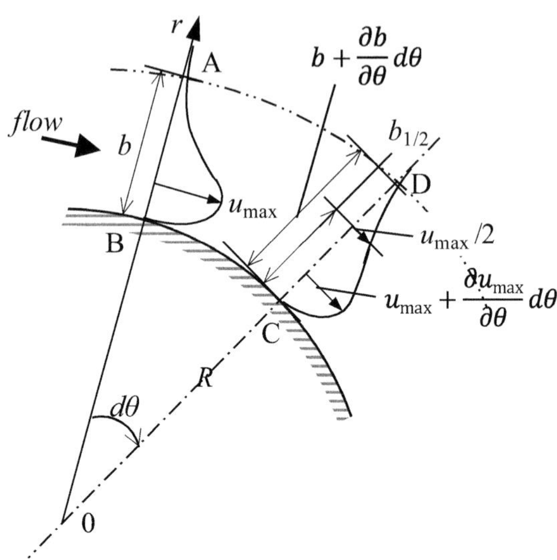

Fig. 153 Control volume, ABCD

$$\frac{\partial}{\partial \theta} \int_0^b \rho u^2 \mathrm{d}y - v_e \frac{\partial}{\partial \theta} \int_0^b \rho u \mathrm{d}y = -\frac{\partial}{\partial \theta} \int_0^b p \mathrm{d}y + p_\infty \frac{\partial b}{\partial \theta} - \tau_w R \qquad (134)$$

where, b: jet width, p: pressure, p_∞: ambient pressure, (r, θ): polar coordinate, u: circumferential velocity, y : coordinates in the r direction with the origin at the cylinder wall, v_e: velocity at which the surrounding fluid is entrained into the jet from surface BC.

(b) Velocity distribution

To solve Eq. (134), it is necessary to give the velocity distribution of the jet in the circumferential direction. Here, we will calculate u using the profile method.

First, it is assumed that the velocity distribution shape is similar at any position x (distance along the cylinder wall from the nozzle exit) and is expressed by the following equation.

$$\frac{u}{u_{\max}} = F\left[\left(\frac{y}{b}\right)^n\right] \qquad (135)$$

where, F: function symbol, n: constant

Next, the simplest functional form that satisfies the following four boundary conditions, $u = 0$ at $y = 0$, $u = U$ and $\partial u/\partial y = 0$ at $y = \delta$, and $\partial^2 u/\partial^2 y = 0$ at $y = 0$ when $\partial p/\partial y = 0$, is a polynomial in y that includes four arbitrary constants $C_1 \sim C_4$.

$$\frac{u}{u_{\max}} = C_1\left(\frac{y}{b}\right)^n + C_2\left(\frac{y}{b}\right)^{n+1} + C_3\left(\frac{y}{b}\right)^{n+2} + C_4 \qquad (136)$$

Therefore, if C is constant,

$$\frac{u}{u_{\max}} = C\left(\frac{y}{b}\right)^n\left[1 - \left(\frac{y}{b}\right)\right]^2 \tag{137}$$

Next, if $u = u_{\max}$ and $\partial u/\partial y = 0$ at $y_{\mathrm{m}} = b/k$ (k: const.),

$$\frac{u}{u_{\max}} = \left[\frac{(n+2)^{n+2}}{4n^2}\right]\left(\frac{y}{b}\right)^n\left[1 - \left(\frac{y}{b}\right)\right]^2 \tag{138}$$

$n = 1/6$ was obtained from the velocity distribution measurements. From this, the values of $u/u_{\max}$ and k are given by the following equation.

$$\frac{u}{u_{\max}} = 1.8\left(\frac{y}{b}\right)^{1/6}\left[1 - \left(\frac{y}{b}\right)\right]^2 \tag{139}$$

$$k = 13 \tag{140}$$

Also, if we give $b_{1/2} = b/k'$ (k': const., $k' > 1$) the half-width of the jet and rewrite Eq. (139), we get the following equation. The half width is the distance from $y = 0$ to the larger value of y at $u/u_{\max} = 0.5$.

$$\frac{u}{u_{\max}} = 0.3\left(\frac{y}{b_{1/2}}\right)^{1/6}\left[2.3 - \left(\frac{y}{b_{1/2}}\right)\right]^2 \tag{141}$$

$$k' = 2.3 \tag{142}$$

The following equation is obtained from the above velocity distribution Eqs. (141) and (142).

$$y_{\mathrm{m}} = \frac{b}{13}, \quad b_{1/2} = \frac{b}{2.3} \tag{143}$$

(c) Maximum velocity

Here, the maximum flow velocity $u_{\max}$ is calculated using the velocity distribution Eq. (141) and momentum integral Eq. (134) that were obtained previously. Now, the jet width b is given by the following equation from the velocity distribution measurements.

$$\frac{b}{b_0} = A_1\exp(A_2 x) \tag{144}$$

where, $A_1 = 4.42/b_0 = $ const., $A_2 = 0.0465/b_0 = $ const.

In addition, the entrainment velocity v_{e} is calculated as follows, where Q is the volumetric flow rate of the jet flow.

$$v_e = \frac{\partial Q}{R \partial \theta} = \frac{\partial}{R \partial \theta} \int_0^b u \, dy \tag{145}$$

The radial pressure gradient is

$$\frac{\partial p}{\partial r} = \frac{\rho u^2}{r} \quad \text{or} \quad \frac{\partial p}{\partial y} = \frac{\rho u^2}{R + y} \tag{146}$$

From this, the attraction p at any position is

$$p = p_\infty + \int_0^y \frac{\rho u^2}{R + y} dy - \int_0^b \frac{\rho u^2}{R + y} dy \tag{147}$$

In addition, the wall friction coefficient C_f is calculated from the velocity distribution measurement results.

$$\frac{C_f}{2} = \frac{\tau_w}{\rho u_{max}^2} = 0.0225 \left(\frac{u_{max} y_m}{\nu} \right)^{-0.223} \tag{148}$$

the above results in Eq. (134), we get the following.

$$\begin{aligned}
0.203 b^2 &\left(\frac{d u_{max}}{d\theta} \right)^2 - 2 u_{max} b \left[0.324 - 0.068 \left(\frac{b}{R} \right) + 0.0217 \left(\frac{b}{R} \right)^2 - 0.0087 \left(\frac{b}{R} \right)^3 \right. \\
&\left. - 0.203 \left(\frac{db}{R d\theta} \right) \right] \left(\frac{d u_{max}}{d\theta} \right) \\
&- u_{max}^2 \left[0.324 - 0.137 \left(\frac{b}{R} \right) + 0.0651 \left(\frac{b}{R} \right)^2 - 0.0348 \left(\frac{b}{R} \right)^3 \right. \\
&\left. - 0.203 \left(\frac{db}{R d\theta} \right) \right] \left(\frac{db}{d\theta} \right) - 0.0399 (u_{max})^{1.78} b^{-0.223} \nu^{0.223} R = 0
\end{aligned} \tag{149}$$

By solving the above equation using the Runge–Kutta method, the relationship between u_{max} and the distance x along the cylinder wall from the nozzle exit can be found.

(d) **Pressure distribution**

Based on the above results, the pressure distribution of the flow field is determined as follows. Now, assuming that the velocity distribution is expressed by the following equation,

$$\frac{u}{u_{max}} = F(\eta) = 1.8 \eta^{1/6} (1 - \eta)^2 \tag{150}$$

where, $\eta = y/b$.

Equation (146) can be expressed by

$$p = p_\infty + \int_0^\eta \frac{\rho u_{\max}^2}{R + b\eta} f^2(\eta) b d\eta - \int_0^1 \frac{\rho u_{\max}^2}{R + d\eta} f^2(\eta) b d\eta \tag{151}$$

By substituting Eqs. (150) into (151) and using the jet width b and the maximum flow velocity $u_{\max}$ determined earlier, the pressure distribution around the cylinder can be determined.

5.2.2 Two-Dimensional Reattached Jet to Cylindrical Wall (Experimental Results)

An air jet with a mean velocity of $u_{m0} = 40$ m/s at the nozzle exit is ejected from a two-dimensional, 2D, nozzle with a nozzle width $b_0 = 2.0$ mm and an aspect ratio $AR = 20$ in the tangential direction of a cylindrical wall surface with a diameter $D = 80.0$ mm (Fig. 152). The shape of the nozzle is a contour nozzle with a short parallel section (Fig. 154), and the velocity distribution u in the circumferential direction at the nozzle exit $[x = 0\ (\theta = 0°),\ z = 0]$ has a relatively thin boundary layer thickness and small turbulence intensity ($\sqrt{\overline{u'^2}}/u_{mo} = 0.0015$ at $y/b_0 = 0.5$) as shown in Fig. 155. u and $\sqrt{\overline{u'^2}}$ were measured using a hot wire anemometry and a single probe, and the separation pint was measured using boundary layer pitot tube with tube thickness of 0.36 mm.

(a) Velocity distributions

Figure 156a shows the velocity distribution u in the circumferential direction at the center height of the channel ($z = 0$, Fig. 152). The average flow velocity at the nozzle exit is $u_0 = 40$ m/s, and the separation point position of the reattached flow is $x/b_0 = 41.9\ (\theta = 120°)$. In addition, Fig. 156b shows the above results rendered dimensionless using maximum flow velocity $u_{\max}$ and half width $b_{1/2}$. The velocity distribution shape is similar in the developed region ($9.2 \leq x/b_0 \leq 41.9$) from the core region ($x/b_0 \leq 9.2$) up to the separation point, and matches well with Eq. (141)

Fig. 154 Nozzle configuration

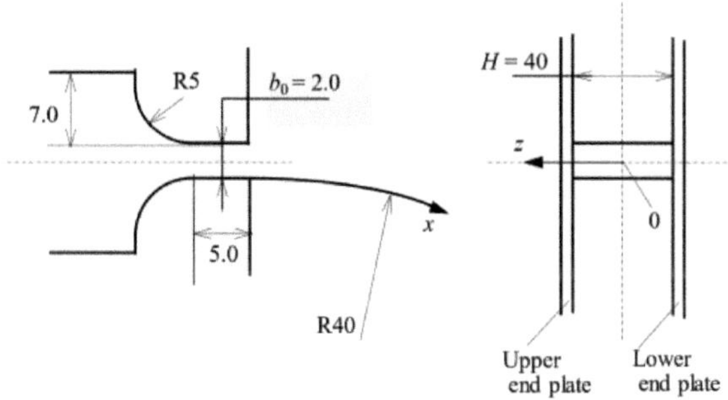

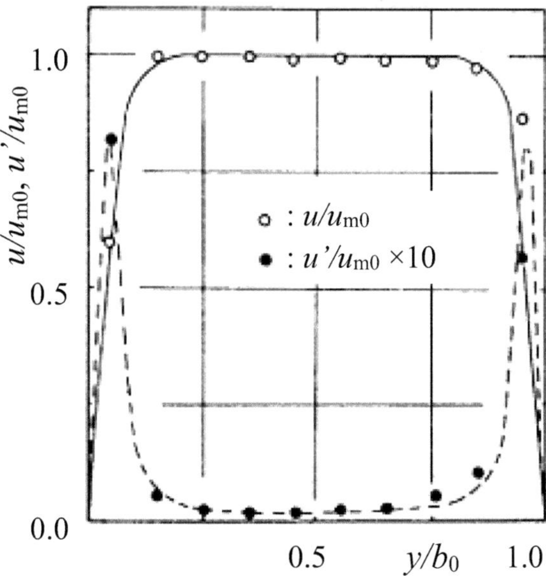

Fig. 155 Velocity distribution at nozzle exit

shown by the solid line in Fig. 156b. For reference, the figure also shows Verhoff's empirical equation (Verhoff 1963) for a 2D wall jet flowing along a flat plate.

$$\frac{u}{u_{\mathrm{m}}} = 1.479\left(\frac{y}{y_{1/2}}\right)^{1/7}\left\{1 - \mathrm{erf}\left[0.6776\left(\frac{y}{y_{1/2}}\right)\right]\right\} \tag{152}$$

The velocity distribution of the cylindrical wall jet is relatively close to that of the two-dimensional, 2D, wall jet.

Figure 157 shows the turbulent intensity $\sqrt{\overline{u'^2}}/u_{\mathrm{max}}$ at $z = 0$. The turbulence intensity, which was maximum at $y/y_{1/2} \fallingdotseq 1$ near the nozzle exit, shifts to a smaller position $y/y_{1/2} < 1$ as it goes downstream.

Figure 158 shows the maximum velocity $u_{\mathrm{max}}/u_{\mathrm{m0}}$ (u_{m0}: maximum velocity at the nozzle exit). The broken line in the figure shows the theoretical results mentioned above, and for reference, the results of 2D-wall and free jets are also shown by a dash-dot line and a two-dot chain line, respectively. The length of core region is $x/b_0 \fallingdotseq 9$ and it is larger than that of the free jet $x/b_0 \fallingdotseq 5.3$ and smaller than that of the wall jet $x/b_0 \fallingdotseq 10$. In the developed region of $9.2 \leq x/d_0 \leq 41.9$, the maximum velocity decreases in proportion to $(x/b_0)^{-1/2}$, as in the case of the 2D free and wall jets. The calculated results represent the experimental results well.

Figure 159 shows the wall friction coefficient C_f calculated using the measurements of velocity distribution and Clause's chart. Re' is the Reynolds number expressed by $Re' = u_{\mathrm{max}}\,\delta/\nu$. The results can be expressed by

$$C_f = 0.045 Re'^{\,-0.223} \tag{153}$$

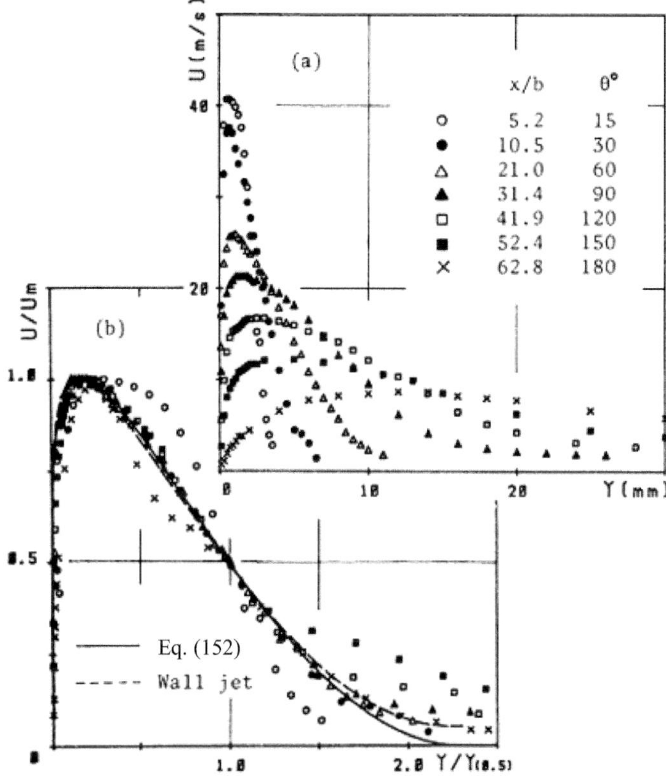

Fig. 156 Velocity distribution ($u_0 = 40$ m/s)

Fig. 157 Turbulent intensity

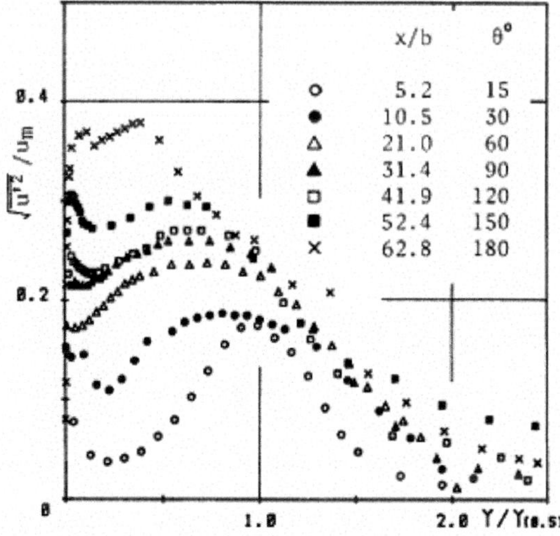

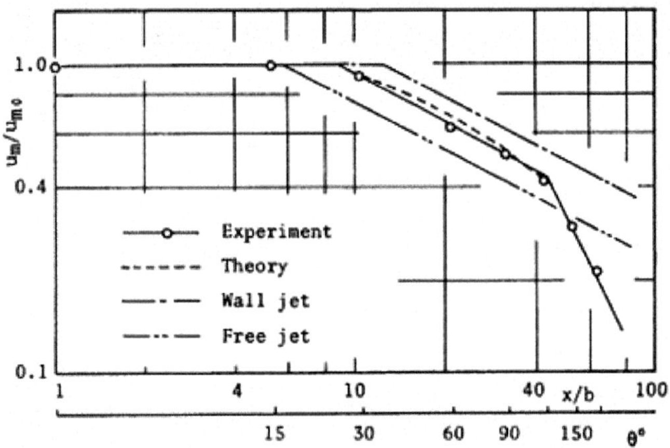

Fig. 158 Maximum velocity

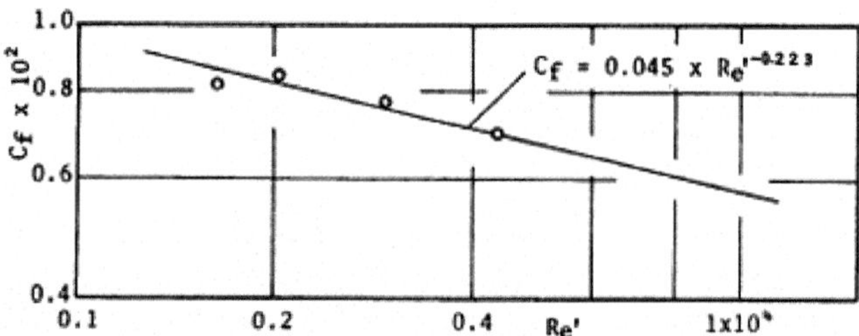

Fig. 159 Wall friction coefficient

From this, the wall shear stress is calculated as follows.

$$\tau_{\mathrm{w}} = 0.0225 \rho u_{max}^2 \left(\frac{u_{max}\delta}{\nu} \right)^{-0.223} \tag{154}$$

This value is close to the wall shear stress on a smooth flat plate as shown in the following equation.

$$\tau_{\mathrm{w}} = 0.0225 \rho u_{\infty}^2 \left(\frac{u_{\infty}\delta}{\nu} \right)^{-0.25} \tag{155}$$

where, u_{∞}: mainstream velocity, δ: boundary layer thickness

(b) **Flow rate, momentum and kinetic energy**

The volumetric flow rate Q, momentum M, and kinetic energy E of the cylindrical wall jet can be obtained using the following equations.

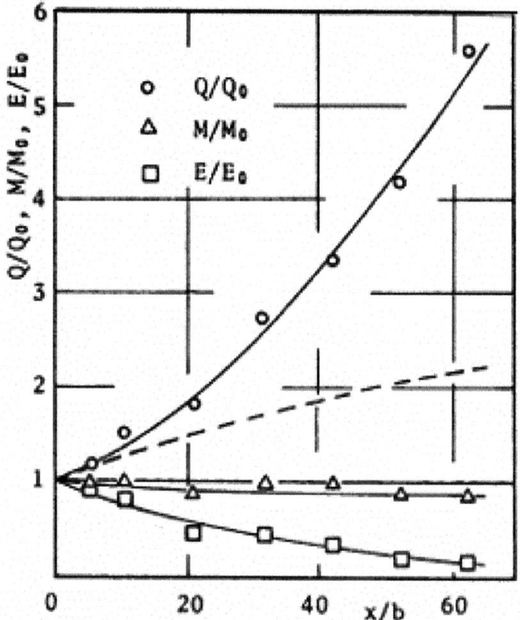

Fig. 160 Flow rate, momentum and kinetic energy

$$Q = \int_{0}^{b} u\,dy, \quad M = \rho \int_{0}^{b} u^2\,dy, \quad E = \frac{1}{2}\rho \int_{0}^{b} u^3\,dy \tag{156}$$

The results are shown in Fig. 160, and the broken line is the result for the 2D wall jet by Rajaratnam (1976). The flow rate Q increases rapidly to the downstream and much larger than that of the wall jet. In the case of a free jet, the momentum M is constant in the downstream direction, but it decreases slightly. These quantities are thought to be influenced by R/b_0 and u_0, but unless R/b_0 is particularly small, the qualitative trends are thought to be the same as the above results.

(c) Half width $b_{1/2}$, jet width b and position of maximum velocity δ

Figure 161 shows changes in the jet half-width $b_{1/2}$, jet width b, and maximum flow velocity position δ in the downstream direction calculated using the velocity distribution measurement results described above. The half width decreases rapidly and can be expressed by the following experimental equation.

$$\frac{b_{1/2}}{b_0} = 0.96\exp\frac{0.046x}{b_0} \tag{157}$$

The broken line in the figure shows the two-dimensional wall jet by Rajaratnam (1976). In the case of 2D wall jet, $b_{1/2}$ increases slowly with increasing x but in the case of a cylindrical wall jet in a centrifugal force field, the increase is even more rapid.

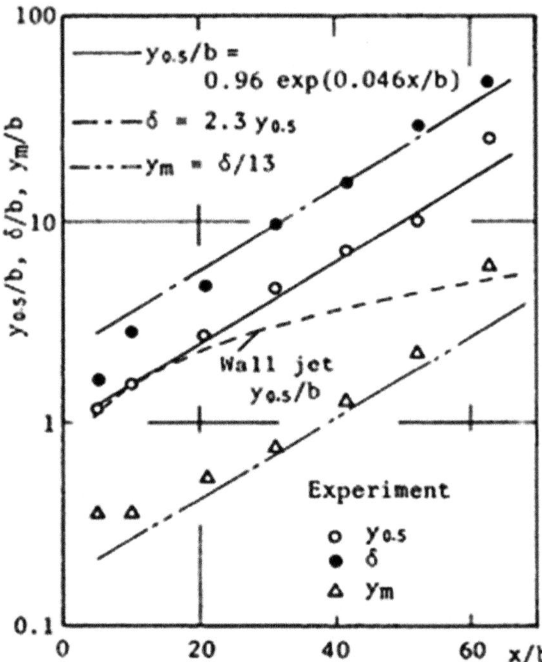

Fig. 161 Half width, jet width and position of maximum velocity

The half width and jet width also show similar changes, which are well expressed by Eq. (143) obtained by the approximate theoretical calculation described above. The dash-dot line and two-dot chain line in the figure are the results by Eq. (143)

(d) Reattachment distance

Figure 162 shows the relation between the reattachment distance x_R, distance along the cylinder wall from nozzle exit to separation point, and the nozzle exit mean velocity u_0. X_R increases rapidly with increasing u_0 and approaches to a constant $x_R/b_0 \fallingdotseq 52$ at $u_0 \fallingdotseq 80$ m/s ($Re \fallingdotseq 1.1 \times 10^4$). The separation point at this time is $\theta = 150°$.

(e) Pressure distribution

The pressure distribution along the cylinder surface is shown in Fig. 163. Since the flow from the nozzle flows along the cylindrical wall under the influence of centrifugal force, the pressure on the wall becomes a negative value. Furthermore, at the downstream, the pressure recovers and approaches the surrounding pressure (atmospheric pressure, $C_p = 0$). In cases $u_0 = 20$ m/s and $u_0 = 40$ m/s, there is a region where the value of C_p is quite different, but in $u_0 \geqq 40$ m/s it seems that the flow state is almost similar. The solid line in the figure shows the results by the approximate theory mentioned above at $u_0 = 40$ m/s can express well the experimental results in the area up to the separation point ($x/b_0 = 41.9$, $\theta = 120°$).

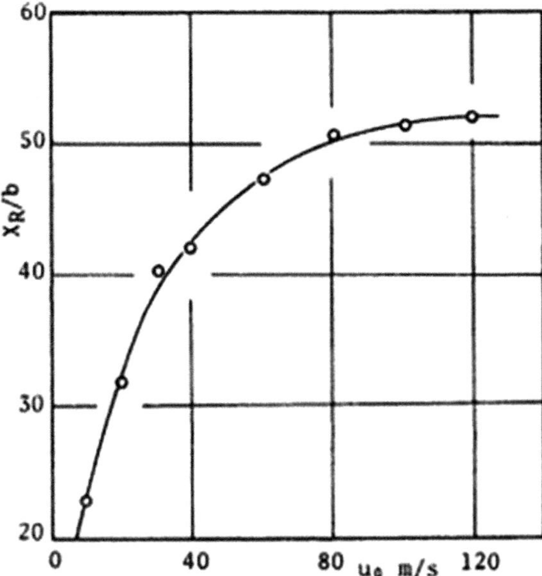

Fig. 162 Reattachment distance, x_R

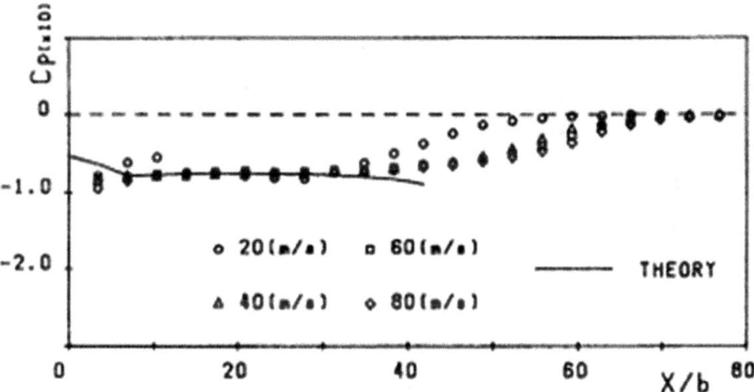

Fig. 163 Pressure distribution on the cylinder wall

The pressure distribution in the flow field is in Fig. 164. The solid line and broken line show the results by experiments and approximate theory mentioned above, respectively. The theoretical results represent the experimental results well, except downstream after $\theta \fallingdotseq 100°$.

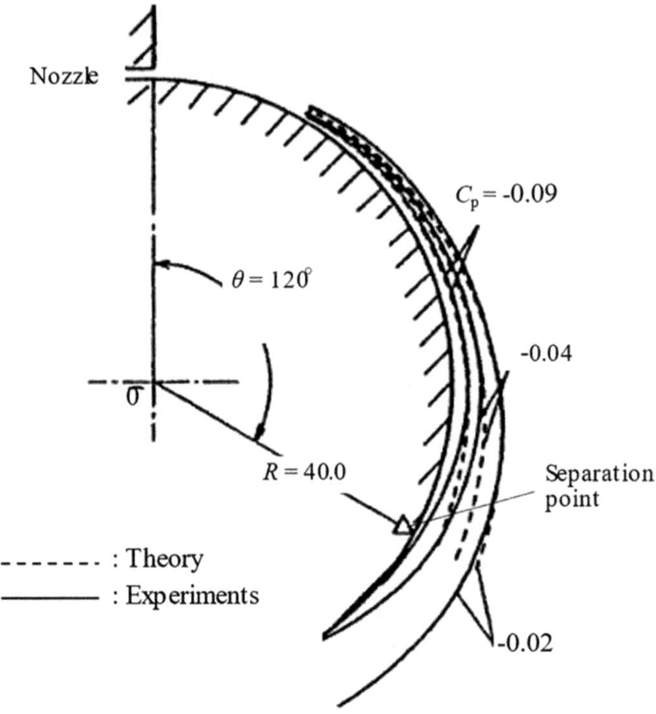

Fig. 164 Pressure distribution in flow field

5.3   Reattachment of Three-Dimensional Circular Jet to Cylinder Surface

Here, we will examine how a circular jet flows along a cylindrical wall. When a circular jet is ejected in the tangential direction of the cylindrical wall, the contact area between the jet and the cylindrical wall near the nozzle exit is small, so the jet hardly attaches to the cylindrical wall and flows without being deflected much. On the other hand, as shown in Fig. 165, if the ejection angle α of the jet is changed so that it impinges on the cylinder, the contact area between the jet and the wall increases, and the jet attaches to the cylinder due to the Coanda effect. In the following, flow characteristics such as velocity distribution, pressure distribution, jet spreading, attachment distance, etc. of a three-dimensional circular jet flowing along a cylindrical wall surface are shown, and examine the effect of the jet angle α on the flow characteristics (Shakouchi et al. 1992b, 1993).

Figure 165 shows the circular nozzle—cylinder system. A jet flow is issued from a contour nozzle with a diameter of 10.0 mm onto a cylindrical wall surface with a diameter $D = 215$ mm at an ejection angle α as shown in Fig. 165. For example, an ejection angle of $\alpha = 0°$ indicates that the jet is ejected in the tangential direction of

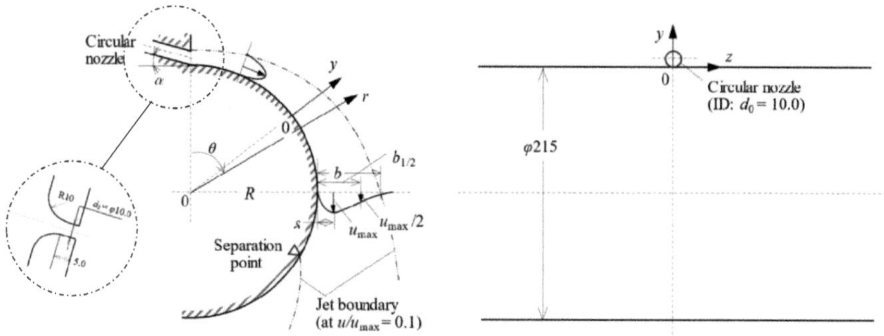

Fig. 165 Circular nozzle—Cylinder system

the cylinder (Fig. 165). The x axis and angle θ were taken from the nozzle exit along the cylinder wall surface, and the y and z axes were taken with the cylinder surface as the origin.

5.3.1 Mean and Fluctuating Velocities

(a) Mean velocity

Figure 166a, b show the velocity distributions at x–y plane ($z = 0$) of ejection angle of $\alpha = 0$ and $20°$, respectively, and Fig. 167 shows the position of the maximum velocity. The velocity distribution shape at the nozzle exit is a rectangular shape with a thin boundary layer thickness. In case $\alpha = 0°$, the contact area between the jet and the cylinder wall near the nozzle exit is small, so the jet flows without much attachment to the cylinder wall and with almost no deflection. On the other hand, in case B, the jet collides with the cylindrical wall and the contact area increases, so the jet adheres to the wall due to the Coanda effect. The attachment distances in each case are $x_R/d_0 = 11.5$ ($\theta = 61.3°$) and $x_R/d_0 = 22.0$ ($\theta = 117°$), respectively.

Figure 168a, b respectively show the results of Fig. 166 in dimensionless form using the jet flow's half-width $b/_{1/2}$ in the y direction and the maximum flow velocity. In both cases $\alpha = 0°$ and $\alpha = 20°$, the velocity distribution shapes on the x–y plane at $z = 0$ are not similar. The solid line and broken line in Fig. 168b are the theoretical curve by Glauert (1956) and the experimental result by Bakke (1957) for a two-dimensional wall jet, respectively. In addition, the dash-dotted line is the result of the following equation (Shakouchi et al. 1988) for a two-dimensional wall jet, and is shown for reference.

$$\frac{u}{u_{\max}} = 0.3\left(\frac{y}{b_{1/2}}\right)^{1/6}\left[2.3 - \left(\frac{y}{b_{1/2}}\right)\right]^2 \tag{158}$$

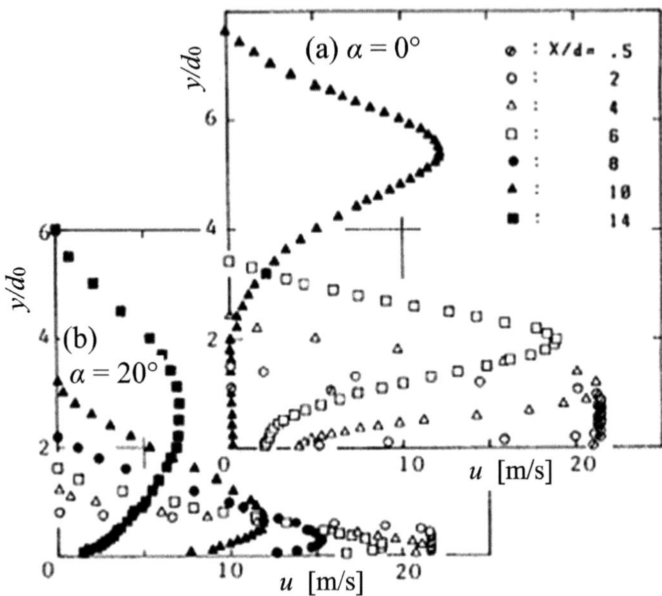

Fig. 166 Velocity distribution, $u-y/d_0$

Fig. 167 Position of
maximum velocity

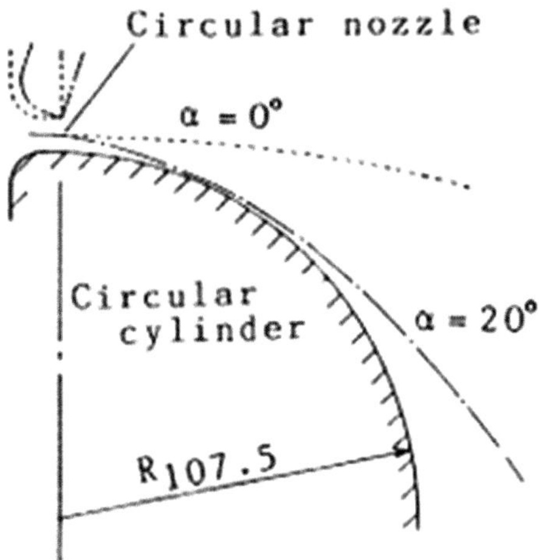

Figure 169a, b show the velocity distributions in the x–z plane ($y = \delta$) of $\alpha =$ $0°$ and $\alpha = 20°$, respectively, rendered dimensionless by the half-width $z_{1/2}$ in the z direction and the maximum flow velocity. The solid line in the figure is the theoretical calculation result for a circular free jet by Goertler (Schlichiting 1979). In both cases, the distribution was symmetrical with respect to $z = 0$, so only the results for the region of $z \geq 0$ are shown. In the case of $\alpha = 0°$, the velocity distribution shapes are

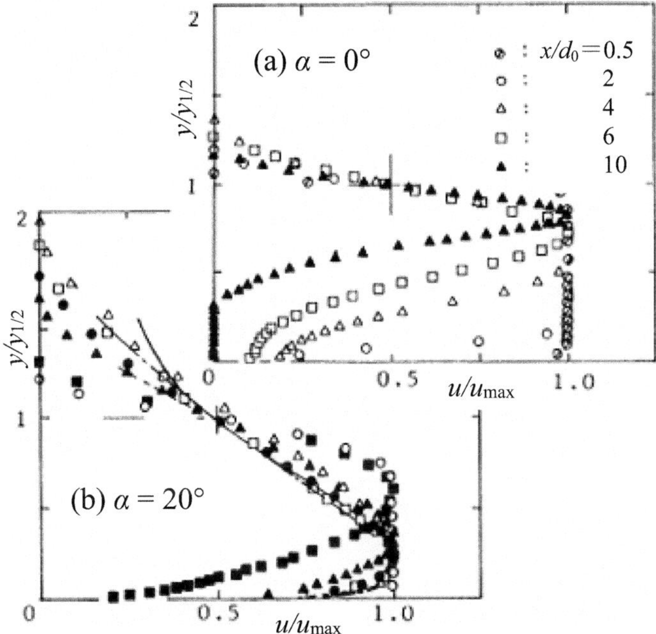

Fig. 168 Nondimensional velocity distribution ($u/u_{max} - y/b_{1/2}$)

almost similar when $x/d_0 < 6$, but these are not well represented by the theoretical calculation results of Goertler. In the case of $\alpha = 20°$, the velocity distribution shape is not similar in the outside $z/z_{1/2} \geq 1$ of the jet. This is because the jet impinges on the cylinder wall and is spread in the z direction.

(b) Turbulent intensity

Figure 170a, b show the turbulent intensity u'/u_{max} (u': RMS) in the mainstream direction in the x–y plane ($z = 0$) at $\alpha = 0°$ and $\alpha = 20°$, respectively. The broken line in the figure is the result of a two-dimensional wall jet at $x/d_0 = 6$, and is shown for reference. The turbulent intensity u'/u_{max} corresponds to the velocity distribution and increases where the velocity gradient is large, and its maximum value increases with the ejection angle α downstream. Moreover, in either case, the distribution shapes are not similar.

Figure 171a, b show the u'/u_{max} (u': RMS) in the x–z plane (at $y = \delta$) at $\alpha = 0°$ and $\alpha = 20°$, respectively. Since the distribution shape was axially symmetrical with respect to the z-axis, only the region where $z \geq 0$ was shown. Although it generally increases toward the downstream, it does not have a similar distribution shape.

(c) Maximum velocity

Figure 172 shows the maximum velocity u_{max}. The broken line in the figure is the result of a two-dimensional free jet and is shown for reference. The core region is

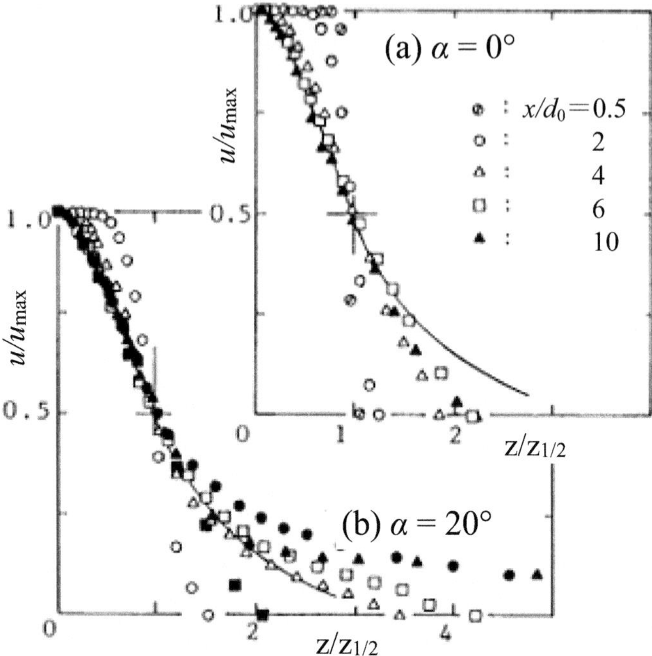

Fig. 169 Nondimensional velocity distribution ($u/u_{\max} - z/z_{1/2}$)

in the range of $0 \leqq x/d_0 < 3.5$ in each case, the transition region is in the range of $3.5 < x/d_0 < 6$ for the flat wall jet and the $\alpha = 20°$, and in the range of $3.5 < x/d_0 < 7$ for the $\alpha = 0°$. In each case, the maximum velocity $u_{\max}$ in the developed region decreases with distance x/d_0 in the form of a power function:

$$\frac{u_{\max}}{u_{m0}} \propto \left(\frac{x}{d_0}\right)^n \tag{159}$$

where, n is in the following Table 7

In addition, in the case of a cylindrical wall jet, since the jet tends to spread in the y direction, the decay of the maximum flow velocity is considerably larger than in the case of a flat plate wall jet.

5.3.2   Reattachment Distance

Figure 173 shows the relationship between the reattachment distance x_R and the ejection angle α. x_R increases almost linearly with increasing α. As α becomes larger, the jet is strongly pressed against the cylinder immediately after the nozzle exit, and is expanded in the z direction to become a planar attached jet. That is, as a result of

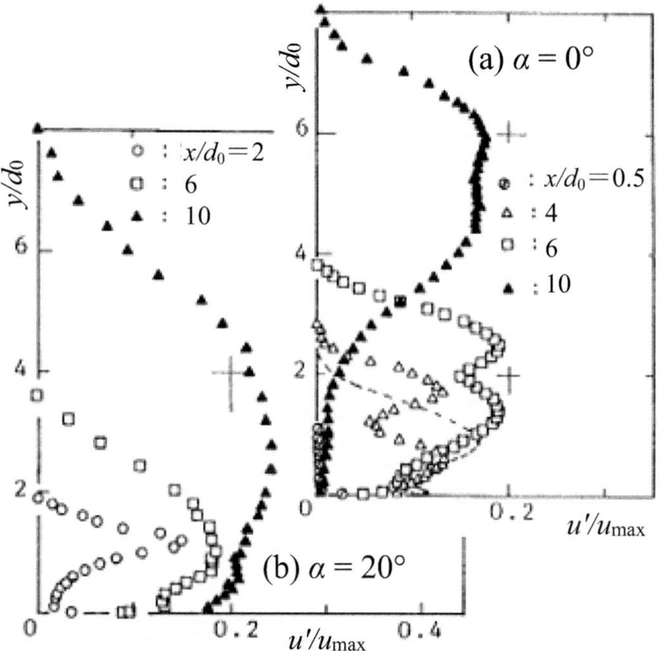

Fig. 170 Turbulent intensity (x–y plane) ($u'/u_{max}-y/d_0$)

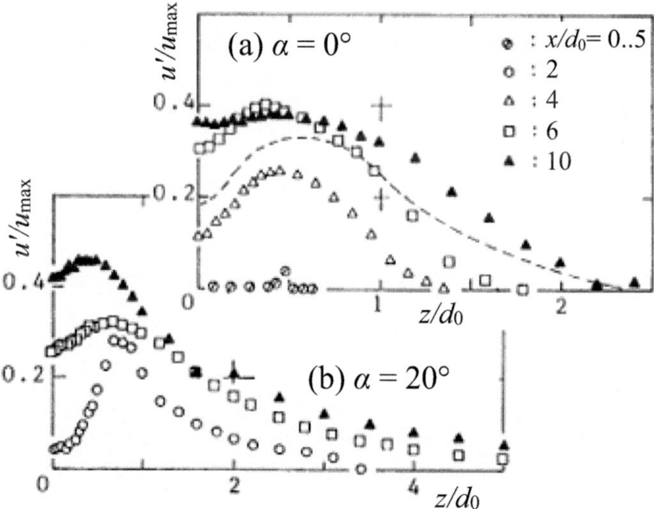

Fig. 171 Turbulent intensity (x–z plane) ($u'/u_{max}-z/d_0$)

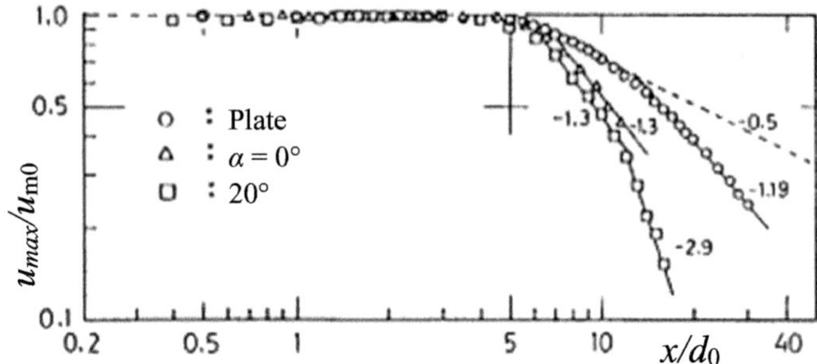

Fig. 172   Maximum velocity

Table 7   Constant n in Eq. (38)

	$x/d_0 < 12$	$12 < x/d_0$
Plate wall jet	$n = -0.5$	$n = -1.19$
$\alpha = 0°$	-1.3	
$= 20°$	-1.3	-2.9

the increase in the contact area between the jet flow and the cylindrical wall near the nozzle exit, the attachment phenomenon due to the Coanda effect becomes noticeable and the attachment distance x_R increases. Their relationship is given by the following equation.

$$\frac{x_R}{d_o} = 0.492\,\alpha + 12 \tag{160}$$

Fig. 173   Reattachment distance, x_R

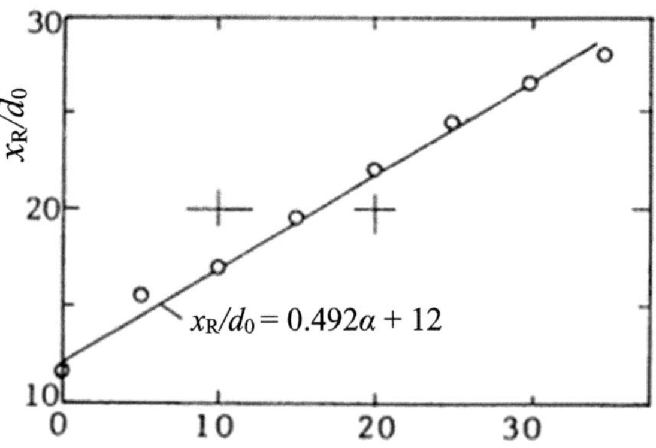

5.3.3 Spreading of Jet

Figure 174 shows the distance δ from the wall to the position of maximum flow velocity and the half-width $b_{1/2}$ of the cylindrical wall jet and flat plate wall jet in cases $\alpha = 0$ and $20°$. In the case of a plate wall jet, δ and $b_{1/2}$ are $\delta/d_0 \fallingdotseq 0.25$ and $b_{1/2}/d_0 \fallingdotseq 1.21$, respectively, and are approximately constant, unlike in the case of a plate wall jet, except in the core region where the maximum flow velocity does not change. This is because the jet also spreads in the z direction. On the other hand, in the case of a cylinder wall jet, both δ/d_0 and $b_{1/2}/d_0$ increase in the form of an exponential function shown by the following equation as the distance x/d_0 increases, and are much larger than in the case of $\alpha = 0°$, which hardly attaches to the cylinder. However, in the case of $\alpha = 20°$, the jet impinges on the cylinder wall near the nozzle exit, so these values are small in the range of $0 \leqq x/d_0 < 4$.

$$\frac{\delta}{d_0} \text{ and } \frac{b_{1/2}}{d_0} = C_1\exp\left(\frac{C_2 x}{d_0}\right) \tag{161}$$

where, C_1 and C_2 are as shown in the following Table 8

Figure 175 shows the half width $z_{1/2}$ in the z direction. In the case of a wall jet and a cylindrical wall jet of $\alpha = 0°$, $z_{1/2}$ increases linearly in the range of $0 \leqq x/d_0 < 6$. In case $\alpha = 20°$, the jet impinges on the cylinder near the nozzle exit, so it takes a maximum value at $x/d_0 \fallingdotseq 4$. Moreover, when $x/d_0 > 7$, both increase almost linearly. These relationships are given by the following equation.

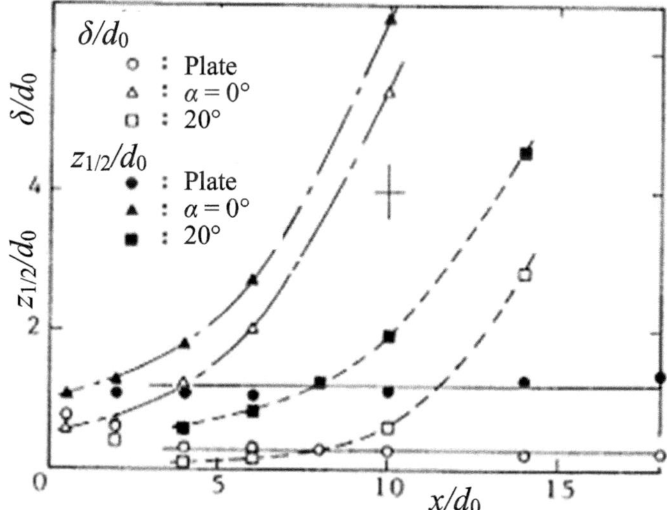

Fig. 174 Spreading of jet (δ/d_0, $z_{1/2}/d_0 - x/d_0$)

Table 8 Constants C_1 and C_2 in Eq. (40)

$\alpha = 0°\ (x/d_0 > 2)$	C_1	C_2
δ/d_0	0.42	0.26
$b_{1/2}/d_0$	0.72	0.22
$\alpha = 20°\ (x/d_0 > 6)$	C_1	C_2
δ/d_0	0.013	0.39
$b_{1/2}/d_0$	0.22	0.21

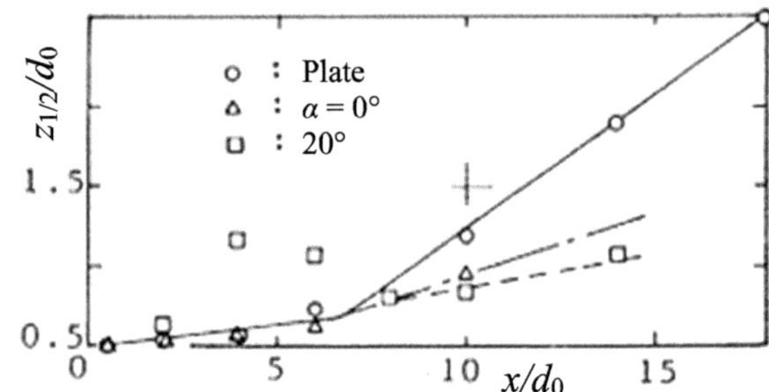

Fig. 175 Spreading of jet $(z_{1/2}/d_0 - x/d_0)$

Table 9 Constants C_1' and C_2' in Eq. (40)

	Plate wall jet	$\alpha = 0°$	$\alpha = 20°$
C_1'	0.16	0.075	0.048
C_2'	−0.35	0.19	0.39

$$\frac{z_{1/2}}{d_0} = C_1'\left(\frac{x}{d_0}\right) + C_2' \tag{162}$$

where, C_1' and C_2' are in the following Table 9.

In addition, in the case of a plate wall jet, the jet tends to spread in the y direction, so $z_{1/2}$ is considerably larger than in the case of a plate wall jet.

5.3.4   Pressure Distribution

Figure 176 shows the pressure distribution on the cylinder wall surface in terms of pressure coefficient $C_P\ [=\ (p-p_\infty)/(\rho u_0^2/2)]$ when a circular jet is ejected onto the cylinder at an angle $\alpha = 0\text{–}35°$. In any case other than $\alpha = 0°$, the jet impinges won the cylinder wall near the nozzle exit, so C_p reaches its maximum value at that position. Furthermore, since the jet reattaching to the cylinder is under the action of centrifugal force, the pressure on the cylinder wall becomes a negative value, and becomes equal to the surrounding pressure (atmospheric pressure) downstream.

Figure 176b shows the relationship between the maximum pressure $C_{p\text{max}}$ and minimum pressure $C_{p\text{min}}$ found based on the results in Fig. 176a, and their positions

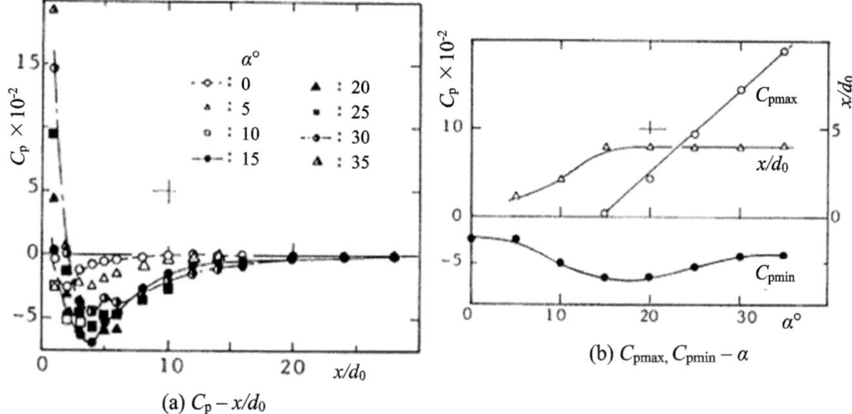

Fig. 176 Pressure distribution on cylinder wall

x/d_0 and α. C_{pmax} increases with increasing α. On the other hand, C_{pmin} occurs in the range of $\alpha = 15{\sim}20°$. Furthermore, the position where C_{pmin} occurs increases with α up to the range of $\alpha = 15{\sim}20°$, but remains at a nearly constant value ($x/d_0 \fallingdotseq 4$) within the range of $\alpha > 20°$.

6 Some Practical Applications

Hear, some practical applications of jet flow are shown.

6.1 Globular Forming of Fine Solid Particle Using High Temperature Air Jet Flow

High temperature air jets are used widely in the various fields of industry, such as heating, drying, powder treatment and others, but the heat transfer characteristics from high temperature air jet to fine particles have not been sufficiently understood yet. In general, in order to produce powder/particles of a few micro-meter the jet grinding method (Chemical Eng. Handbook 1988) is used. Particle laden jet is made impinged on target plate, and then the particle is pulverized. The shape of grinded fine particle is not spherical nor uniform. In order to improve the efficiency of the powder, surface fusing or globular forming of the toner satisfying these requirements includes a chemical reaction, polymerization method. However, the problems for this method are to use organic solvent and high production cost. For a simple, desiccative and continuous process, the following method is considered to solve such problems. Injecting the fine particles of low melting temperature into a high temperature air jet flow, the particles will be melted instantaneously and become globular by their own surface tension. However, in this process for a high temperature air jet flow, large amount of heat rate is needed, and the loading ratio of fine particle is not possible to make so large. In addition, it is desired to improve the energy efficiency.

Here, one of the purposes is to improve the performance of a surface fusing, globular forming system for thermoplastic fine particles.

First, the flow and thermal characteristics of normal and high temperature air jet flows are examined experimentally. Next, the behavior of particle of a few micronmeter is examined by a numerical analysis and flow visualization, and the relation between the heat transfer rate from a high temperature jet flow to fine particle and the globular forming characteristic are clarified experimentally. Using these results, in order to improve the performance of globular forming system (Shakouchi et al. 2003; Nakamura et al. 2003; Nakamura and Shakouchi 2004) for thermoplastic fine particles, a new globular forming system is proposed.

In the former globular forming system, fine particles were introduced into a high temperature air jet flow from around it (Fig. 177a). On the other hand, the new system (Fig. 177b) has the three-concentric-nozzle, and the particles are introduced into the center of the high temperature air jet flow. The effect of nozzle shape on the performance of globular forming is also examined.

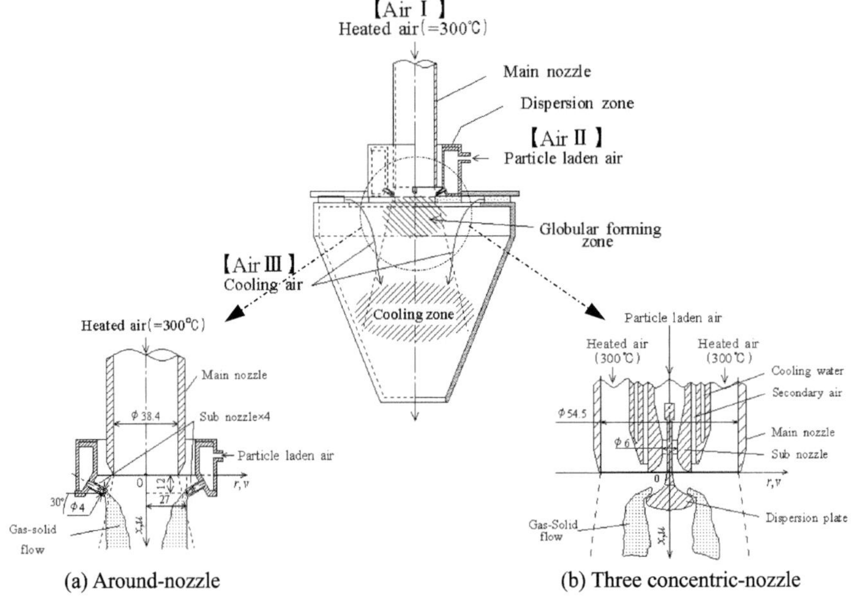

Fig. 177 Globular forming systems, (**a**), (**b**): nozzle part

6.1.1 Method of Globular Forming

Figure 177 shows the former and new globular forming systems, called the Around-nozzle (a) and the Three-concentric-nozzle (b), respectively. These systems have no moving parts, and they have three regions for dispersing fine particles for forming the particles globular by heat where the particles are introduced into a high temperature air jet flow, and for cooling the particles after globular forming.

Figure 178 shows the schematic diagram of experimental set up. After passing through the globular forming system and the cyclone ⑫ the high temperature air jet flow from the main nozzle (Air I: $T = 300$ °C), the particle laden jet flow (Air II: $T = 20$ °C) and the cooling air (Air III: $T = 20$ °C) are exhausted to outside by a blower after passing through a dust collector.

For the around-nozzle, the raw material, powder, fed as constant amount by the feeder ⑩ enters into the ejector nozzle ⑪, dispersion zone, and then into the globular forming zone from four sub nozzles which are set on the circumference of high temperature air jet flow.

Fine particles in well dispersed state accelerated quickly in the sub nozzles are introduced into the high temperature air jet flow in the globular forming zone.

For the three-concentric-nozzle, in order to create the flow and temperature fields in which all particles pass through the high temperature region, the raw material is injected by the sub nozzle which is placed at the center of the main nozzle. In order to have the best dispersion state of fine particle right before injecting into the high temperature region, the sub nozzle is narrowed near the exit. The contraction will give a large shearing force to the fine particles. A dispersion plate is set to the downstream of nozzle exit in order to attain better dispersion state of fine particle and better mixing of high temperature main jet flow and normal temperature sub jet flows. The fine particles are accelerated quickly in the sub nozzle and collided with the dispersion plate.

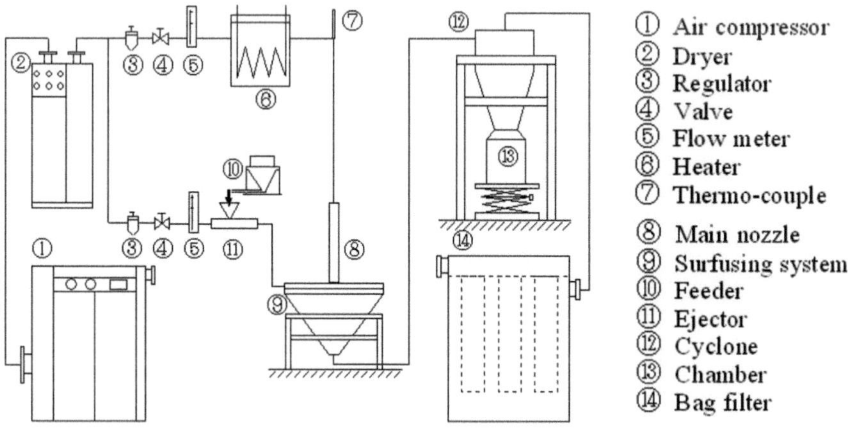

Fig. 178 Schematic diagram of set up for globular forming

In the three-concentric-nozzle, the raw material, powder, fed with a constant rate by the feeder ⑩ passes through the ejector nozzle ⑪, and then is accelerated quickly in the sub nozzle, is made collide with the dispersion plate and results in well dispersion.

After that fine particles change the trajectory in the r direction, and it is dispersed more by the large shearing stress due to the secondary air from surroundings. In a best dispersion state, they are introduced to the globular forming zone of high temperature air jet flow.

The temperature of fine particles is beyond the melting point (over 120 °C), so they get melted. The melted particles are fused by their own surface tension, and the particles are cooled and solidified quickly by the cooling air from surroundings.

The particles are exhausted continuously and collected by the cyclone.

6.1.2   Numerical Analysis

Numerical analysis for jet flow characteristics from the main nozzle and flying path of fine particle was carried out. The turbulent flow transport equations of equation of continuity, Navier–Stokes' equation and $k - \varepsilon$ turbulent model were solved by the finite difference method. The equation of motion of particle was also solved by Runge–Kutta method. The turbulent components were estimated by random simulation. The influence of particle on air flow is negligible because the loading ratio, mass flow rate of particle to air flow, is very small, in this study.

Figure 179 shows the flow model, coordinate system and some of the symbols used. Only the positive half region in the radius direction r is shown, because the flow pattern is axisymmetric. The mean velocity of air jet flow from the main nozzle of diameter $d = 38.4$ mm is $u_m = 33$ m/s. The sub nozzle of diameter $d_s = 4.0$ mm was set at four places on the circumference. The position was $(x, r) = (12, 27$ mm) at the angle of 30°. Gas–solid two-phase flow was introduced at $u_s = 33$ m/s in the main jet flow. The particle's diameters are $d_p = 2, 5$ and 10 μm, and the density $\Box\rho_p$ $= 600$ kg/m^3. The flying path was estimated by solving the following equation, the equation of motion of the particle.

$$m_p \frac{du_p}{dt} = C_D \pi d_p^2 \rho_a (u_a - u_p) |u_p - u_a| + m_p g \left(1 - \frac{\rho_a}{\rho_p}\right) \tag{163}$$

The particle Reynolds number Re_p is in the Stokes region, and then the drag coefficient of the particle is

$$C_D = \frac{24}{Re_p} \tag{164}$$

The initial velocity of particle at the sub nozzle exit is assumed to be the same as the velocity of air jet. The Runge–Kutta method was used to solve the equation of motion. The calculation area of $(x \times r) = (576 \times 269)$ mm was divided to the unequal intervals of 155×269 lattice.

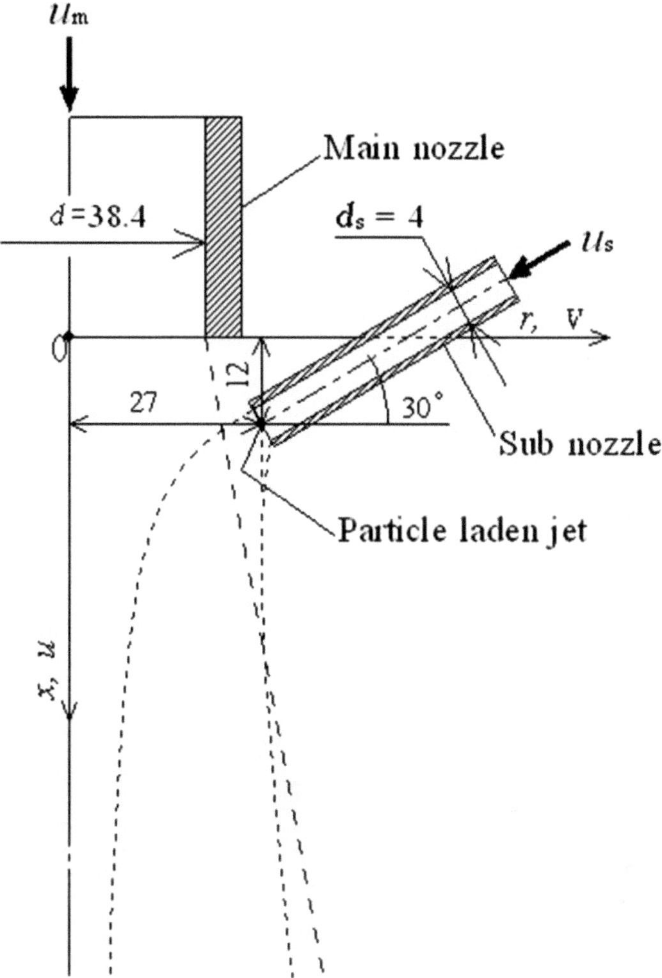

Fig. 179 Flow model and coordinate system

6.1.3 Set up and Procedure

(a) **Relationship between heat transfer to fine solid particle and globular forming characteristics**

In order to obtain the heat transfer rate to fine particle which was required to form the particle globular, an annular nozzle was used as shown in Fig. 180. The raw material used here is polyester toner I. The mean roundness, 50% diameter, density and melting point of the toner, are $R_p = 0.919$, $d_{p50} = 8.8\ \mu m$, $\rho_p = 600\ kg/m^3$, $T_{mel} = 120\ °C$, respectively.

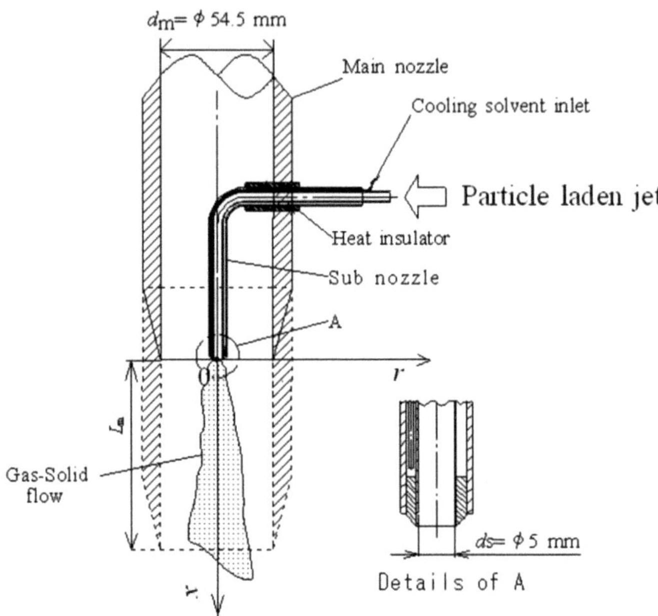

Fig. 180 Annular nozzle

Here, the main nozzle has diameter $d = 54.5$ mm. The mean velocity of high temperature air jet flow at the nozzle exit was kept constant at $u_m = 21$ m/s, while three different temperatures are used for the high temperature air jet flow; $T = 150$, 250, 300 °C, which are measured by thermo-couple as shown in Fig. 178. Fine particles were introduced into the center region of high temperature air jet flow by dispersion sub nozzles of diameter $d_s = 5.0$ mm. The mean nozzle-exit velocity at the sub nozzle was $u_s = 21$ m/s. In order to control the residence time of fine particles in the high temperature region, the length of nozzle edge L_m was changed from 0 to 400 mm. The Globular forming performance of fine particle was estimated by the mean roundness. The mean roundness was measured by the particle analyzer "FPIA-2100" (Sysmex Co., Ltd). The mean roundness R_p was defined by the ratio of the perimeter of a non-uniform particle of projection image to the perimeter of a spherical particle with the same area of a non-uniform particle. SEM shows only two-dimensional data, but since more than ten thousands of particles were managed for one picture of SEM as the result three-dimensional roundness can be estimated.

For calculating the heat transfer rate to the particle of $d_p = 8.8$ μm, the following equation can be used. It was assumed that the particle velocity is the same as the gas velocity.

$$Q_1 = \pi k d_p \left(T - T_p\right)\left(2 + 0.6 Re_p^{0.5} Pr^{0.33}\right) \tag{165}$$

where, Re_p is the particle Reynolds number, which is given by

$$Re_p = \frac{(u_m - u_p)d_p}{\nu} \tag{166}$$

Q_2 is the heat transfer rate for the residence time t, which is given by

$$Q_2 = Q_1 t \tag{167}$$

(b) Temperature distribution

The temperature of air jet flow cannot be visualized directly by a thermograph since the infrared ray is not radiated from the air. A thin rubber plate with thickness of 1 mm, width of 100 mm and length of 400 mm was set on the center-plane of main nozzle. The infrared rays radiated from the rubber plate was received by the thermograph after the temperature profile of the heated rubber plate became steady. The temperature profile measured by the thermograph was calibrated by comparing with the temperature which was obtained from the thermo-couple at a horizontal section of nozzle exit.

(c) Globular forming of fine solid particle

The globular forming performances of the Around-nozzle (Fig. 177a) and the Three-concentric-nozzle (Fig. 177b) were examined experimentally. Raw materials used here are polyester toner I and II. The mean roundness, 50% diameter, density and melting point of the polyester toner II, are $R_p = 0.925$, $d_{p50} = 7.4\ \mu m$, $\rho_p = 600\ kg/m^3$, $T_{mel} = 120\ °C$, respectively. In the case using the polyester toner I, the flow rate of high temperature air jet flow was $F_m = 1.0\ Nm^3/min$. The maximum temperature at the nozzle exit for the Around-nozzle was $T_{max} = 300\ °C$. On the other hand, the maximum temperatures for the Three-concentric-nozzle were $T_{max} = 250, 300\ °C$. In the case using the polyester toner II, the flow rate of high temperature air jet flow was $F_m = 1.0\ Nm^3/min$. The maximum temperature at the nozzle exit for the Around-nozzle was $T_{max} = 390\ °C$, while, the maximum temperatures for the Three-concentric-nozzle were $T_{max} = 250, 300\ °C$. The mean velocity at the sub nozzle exit for the Around-nozzle and the Three-concentric-nozzle were $u_s = 33\ m/s$. The flow rates of secondary air for the Three-concentric-nozzle were $F_{se} = 0.16, 0.24\ Nm^3/min$. For the Around-nozzle, the feed rate of polyester toner I was changed; $G = 0.3, 1.3, 2.7\ kg/hr$ by the feeder. For the Three-concentric nozzle, the feed rate was $G = 6.0\ kg/hr$. For the Around-nozzle, the feed rate of polyester toner II was changed; $G = 1.5, 0\ kg/hr$. For the Three-concentric-nozzle, the feed rate was $G = 0\ kg/hr$.

Globular forming of fine particle was evaluated by the degree of mean roundness and coupling Δd_p, the ratio of the mean diameter of globular-formed particle to the mean diameter of raw material.

6.1.4   Relationship Between Heat Transfer Rate to Fine Solid Particle and Globular Forming Characteristics

Figure 181 shows the relation between the mean roundness of fine particles and the residence time in high temperature region. The residence time t is given by

$$t = \Delta t_1 + \Delta t_2 \tag{168}$$

where, Δt_1 is the passing time in the main nozzle, which is given by

$$\Delta t_1 = L_\mathrm{m}/u_\mathrm{s} \tag{169}$$

where, Δt_2 is the time effecting on globular forming from the main nozzle exit, and it is given by the mean of roundness $R_\mathrm{p} = 0.919$ of raw material at the crossing point of the curved line $T = 150$ °C. This gives $\Delta t_2 \cdot = .1.4 \times 10^{-2}$ s. It is assumed that the time Δt_2 is the same for the cases of $T = 250, 300$ °C.

For $T = 150$ °C, the mean roundness increases a little with increasing the residence time t; however, for $T = 250, 300$ °C it increases rapidly up to $R_\mathrm{p} \fallingdotseq 0.97$.

Figure 182 shows the relation between the mean roundness and the heat transfer rate to a fine particle of $d_\mathrm{p} = 8.8$ μm. Globular forming will be influenced by not only heat transfer rate but also surface tension. In this study, for the simplicity an approximate estimation about the relation between roundness and heat transfer under a constant surface tension will be shown. The pictures taken by SEM are also shown in Fig. 182. The mean roundness is improved by increasing the heat transfer rate to fine particle. These results are clear in the pictures by SEM. The relation between the mean roundness and the heat transfer rate to a fine particle is given by

$$R_\mathrm{p} = \left(-4.84 \times 10^9\right)Q_2^2 + 36000Q_2 + 0.91 \tag{170}$$

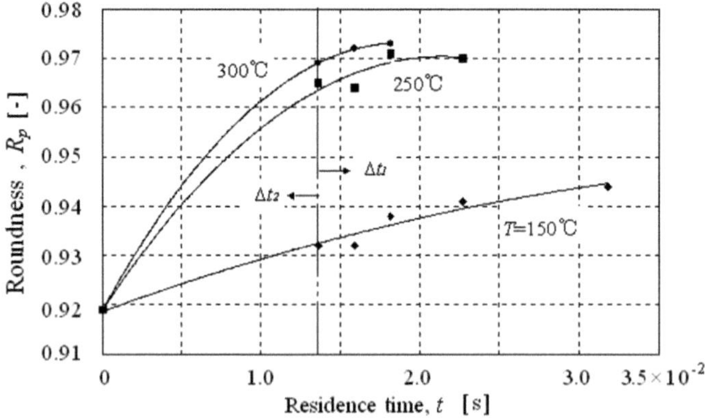

Fig. 181 Roundness and residence time

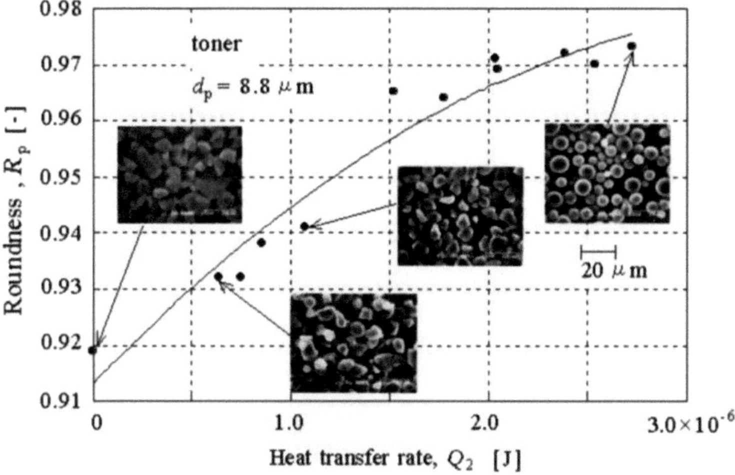

Fig. 182 Roundness and heat transfer to a fine particle

where, R_p is the mean roundness.

6.1.5 Temperature Profile

Figures 183, 184 and 185 show the temperature profiles of high temperature air jet flow for the Around-nozzle and the Three-concentric- nozzle, respectively. The flow rate of a high temperature air jet flow from the main nozzle is $F_m = 1.0\ \text{Nm}^3/$ min. The maximum temperature at the main nozzle exit is $T_{max} = 300\ °C$. The mean velocity at the sub nozzle exit is $u_{sv} = 33$ m/s. The flow rates of secondary air for the Three-concentric-nozzle are $F_{se} = 0.16$ and $0.24\ \text{Nm}^3$/min in Figs. 184 and 185, respectively. For the Around-nozzle, the jet flows from the main and sub nozzles collide with and interact with each other, and the main jet flow contracts somewhat in the centre direction. The centreline temperature T_c decreases rapidly to the downstream because the high temperature air jet flow is mixed with the jet flow from the sub nozzle. In this study, the high temperature region is set to $T_c >$ 120 °C because the melting point of toner used here is $T_{mel} = 120\ °C$. Then the high temperature region becomes $0 < x/d < 3$. For the Three-concentric-nozzle shown in Figs. 184 and 185, the main jet flow flows downward annularly due to the sub jet flow and the dispersion plate placed at the center of the main nozzle. The sub and secondary jet flows spread most at $x/d = 0.5$ and 0.6, respectively. The maximum centreline temperatures for the flows are $T_c = 180$ and $150\ °C$ at x/d \fallingdotseq 3.0, ˙ = . 0, respectively. The high temperature region becomes $0 < x/d < 4$. In Figs. 183 and 184, the sketches of fine particle's flying trajectories in the high temperature field which are mentioned above are also shown. For the Around-nozzle, the trajectories of some particles reach to the center at $x/d = 1.4$, and some of them flow the outside

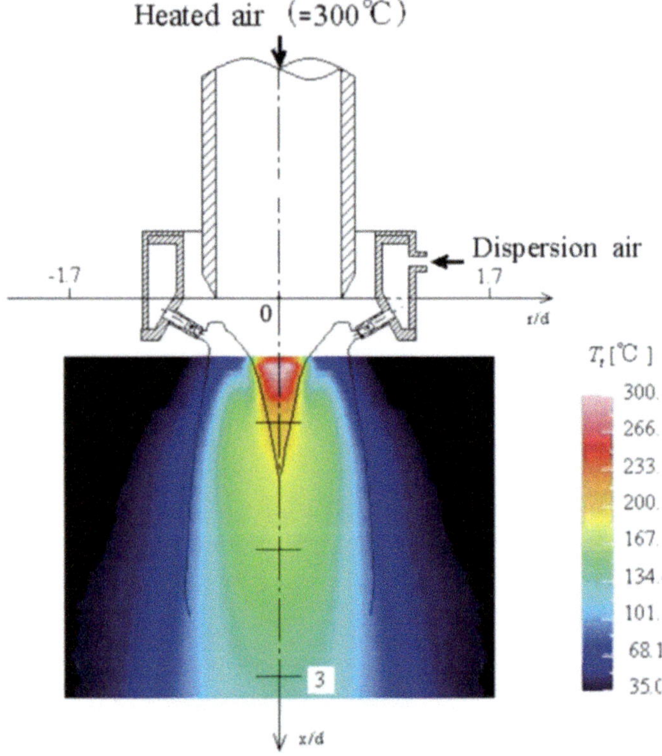

Fig. 183 Temperature profile and particle trajectories of the Around-nozzle

of main jet flow. For the Three-concentric-nozzle type, the width of trajectory of fine particle spread most in the radius direction at $x/dy \fallingdotseq 0.3$, then the trajectory follows the high temperature air jet flow.

6.1.6 Numerical Analysis

Figure 186 shows the numerical results for the flow characteristics, the streamline and contour of pressure for the main and sub nozzle's exit velocity of $u_m = u_s = 33$ m/s. The numbers in Fig. 186 show the stream function ψ and pressure coefficient C_p. As mentioned before, the jet flows from main and sub nozzles collide and interact with each other, and the main jet contracts some to the center. The pressures in the collision and interact areas of both jet flows increase.

Figure 187 shows the flying trajectories of fine particles of $d_p = 2, 5, 10$ μm. The trajectory is deflected to the downstream by the main jet flow, and then follows the flow. The trajectory of the small particle of diameter $d_p = 2$ μm is considerably deflected by the main jet flow and does not reach the center because of the particle's

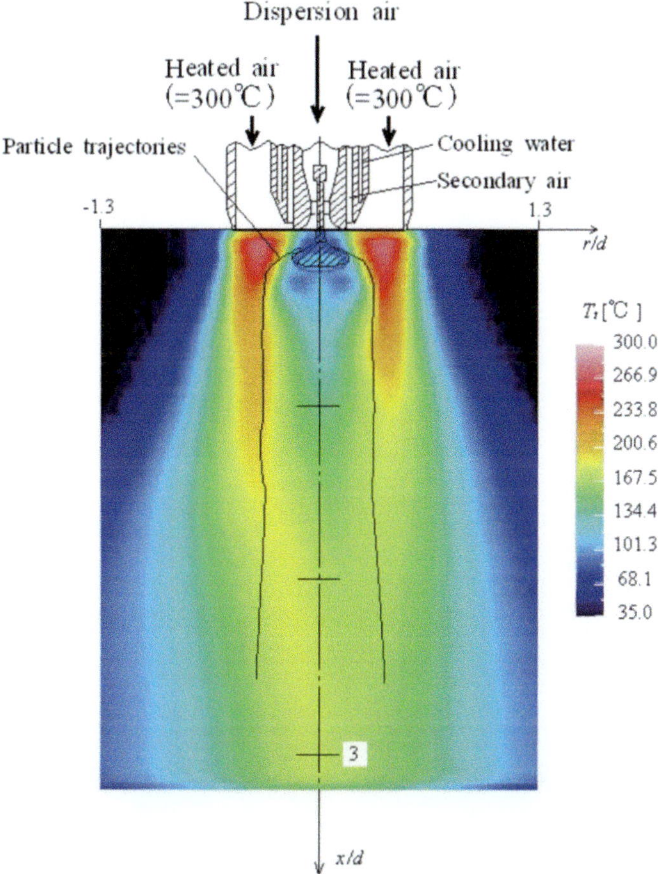

Fig. 184 Temperature profile and particle trajectories of the Three-concentric-nozzle

small inertia. The trajectory for large particle of diameter $d_p = 10$ μm follows the flow, and it does not reach to the center, either. That is, the trajectories of particles up to $d_p = 10$ μm do not reach the center of the main jet flow. Here, the mean diameter of raw material is $d_{p50} = 8.8$ μm, and the proportion of the particles which are larger than $d_p = 10$ μm is approximately 10%. Most of the particles fly outside the high temperature main jet flow, namely in the low temperature region. For the Three-concentric-nozzle type, the width of trajectory of fine particle spread most in the radius direction at $x/d = 0.3$, then it follows the high temperature air jet flow. Numerical analysis and visualization for the trajectory of fine particle, in the around- nozzle, most of the particles fly outside ($T < 120$ °C) of main jet flow, on the other hand, in the three-concentric-nozzle, most of the particles fly in the region ($T > 200$ °C). As a result, it is confirmed that the new globular forming system can introduce most of the particles to the high temperature region of $T > 200$ °C.

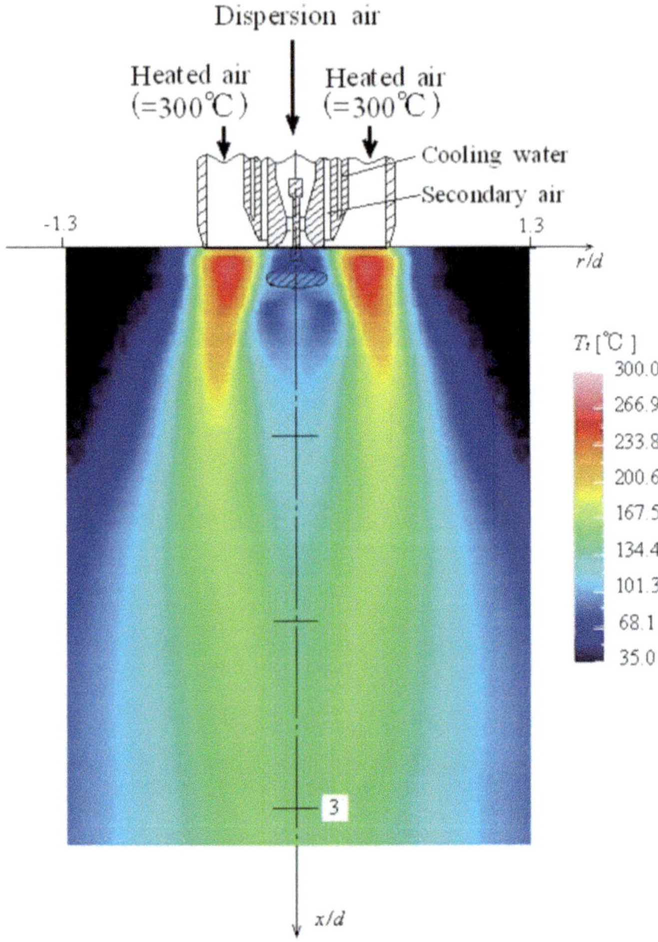

Fig. 185 Temperature profile of the Three-concentric-nozzle

6.1.7   Surface Fusing of Fine Solid Particle

Figure 188 shows the globular forming performance of polyester toner I in terms of the mean roundness and coupling. For the Around-nozzle, the mean roundness decreases with increasing the feeding rate, while the coupling increases. It suggests that this decrease in roundness is caused by the decrease in the heat transfer rate to a particle because the number density of particle increases in the high temperature region. About the coupling, it suggests that the collision of each particle will be increased because the number density of particle increases in the high temperature region. For the Three-concentric-nozzle, with increasing the temperature of high temperature air jet flow, both the mean roundness and the coupling increase. It suggests that the increase in roundness depends on the increase in the heat transfer

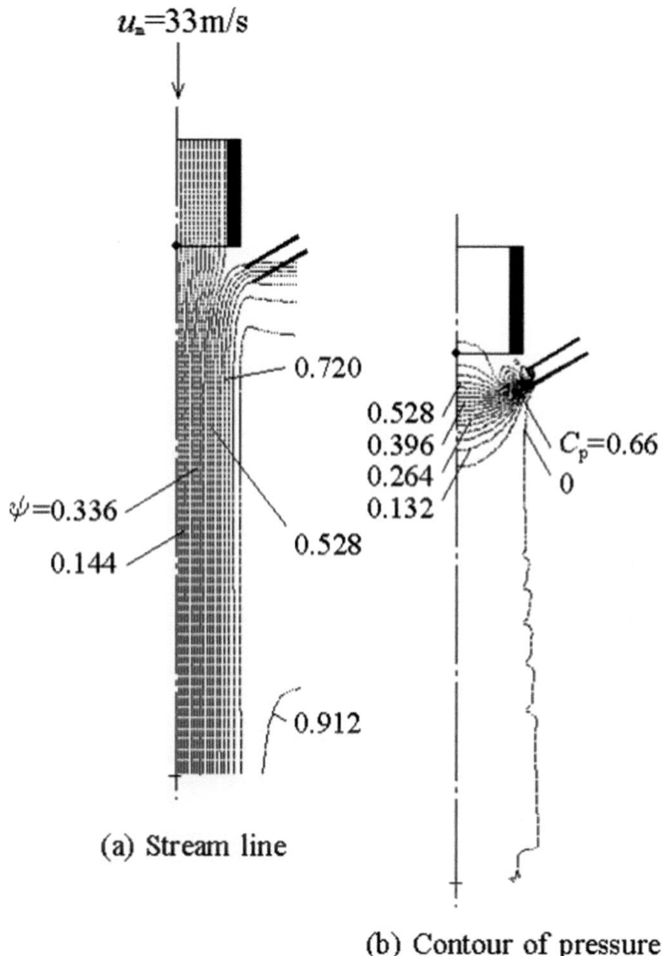

Fig. 186 Flow characteristics

rate to a particle. It also suggests the increase in coupling depends on the increase in collision of particle because the long high temperature region is wider to the downstream. The globular forming performances of both point of the solid and broken lines in Fig. 188. That is, in order to get the performance of roundness $R_p = 0.953$ and the coupling $\Delta d_p = 0.27$ μm at the crossing point for the Three-concentric-nozzle, the temperature of high temperature air jet flow and the feed rate should be $T^{\cdot} = .\,260\,°C$ and $G = 6.0$ kg/hr, respectively. The energy efficiency in this case for both globular forming systems, η_r and η_{th} [kg/kJ] can be defined by

$$\eta = G/mC_p t \tag{171}$$

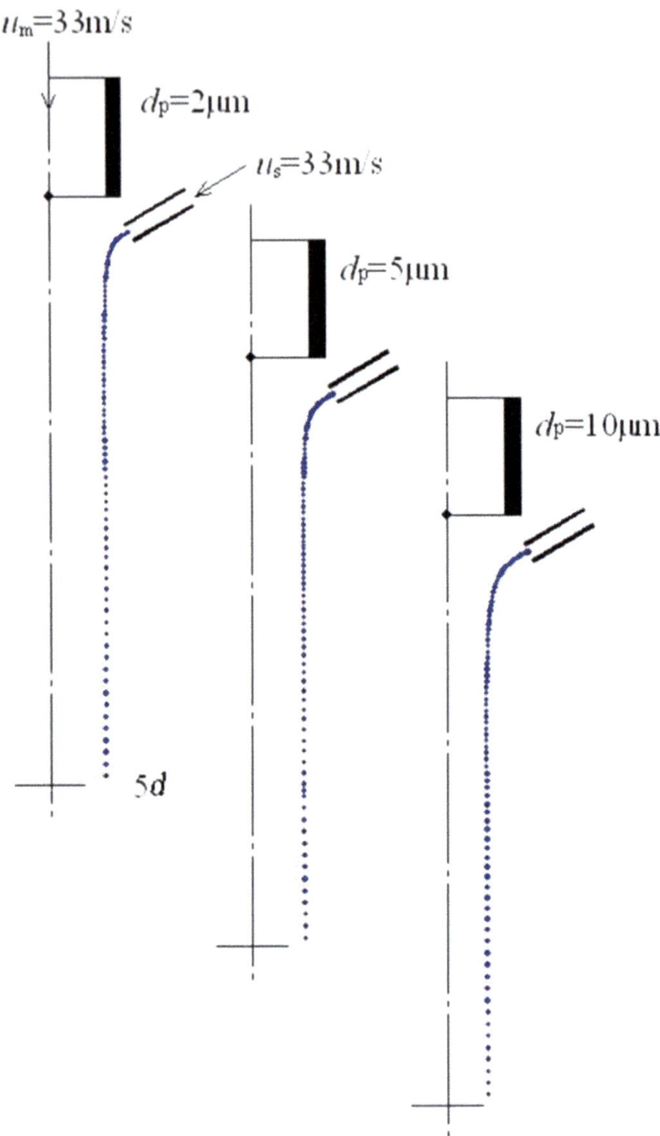

Fig. 187 Particle trajectory (calculation)

where, m: the flow rate of high temperature air jet flow, C_p: the specific heat, subscript $_r$ and $_{th}$ mean the around-nozzle and the three-concentric-nozzle, respectively.

The energy efficiency ratio σ [–] is defined by

$$\sigma = \eta_{th}/\eta_r \tag{172}$$

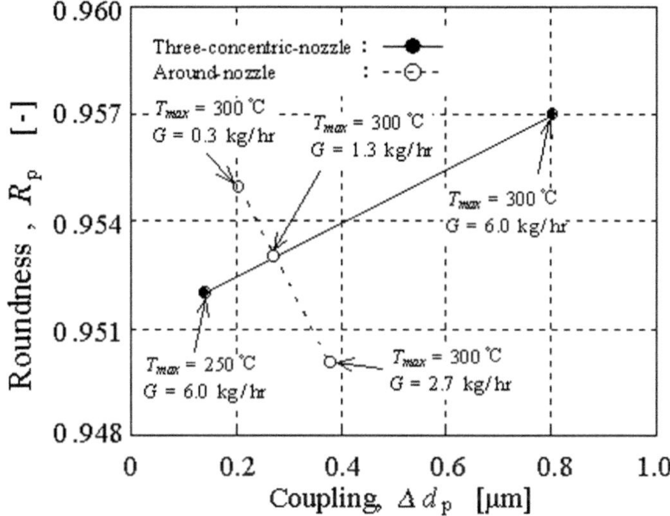

Fig. 188 Roundness and coupling (Toner I)

Using these equations, the energy efficiency ratio σ is approximately 5.

Table 10 and Fig. 189 show the globular forming performance of polyester toner II and the SEM photos by electron microscope. The SEM photographs confirm that the fine particles are made globular in both nozzles. The Three-concentric-nozzle has a very high performance on globular forming of fine particles. Comparing with the Around-nozzle, the mean roundness R_p and coupling Δd_p for the concentric nozzle type are improved; $R_p = 0.972$ and $\Delta d_p = 0.50\,\mu m$, $R_p = 0.962$ and $\Delta d_p = 0.50\,\mu m$, respectively. Moreover, the number density of the particle whose roundness <0.9 for the Three-concentric-nozzle is improved, comparing with the around-nozzle, from 3.93% to 7.33%.

The globular forming performances of both systems can be obtained by the crossing point of the solid and broken lines in Fig. 188. In order to get the performance of roundness $R_p = 0.966$ and the coupling $\Delta d_p = 0.43\,\mu m$ at the crossing point by the Around-nozzle and the Three-concentric-nozzle, the temperature of the high temperature air jet flow and the feed rate should be $T^{\cdot} = .\ 390\,°C$ and $T^{\cdot} = .$

Table 10 Globular forming performance

Type	Feed rate	Temperature	Roundness		Coupling
			Mean	<0.9	
	[kg/hr]	[°C]	[−]	[%]	[μm]
Around-nozzle	1.5	390	0.966	7.39	0.5
Three-concentric-nozzle	3.0	300	0.972	3.98	0.5

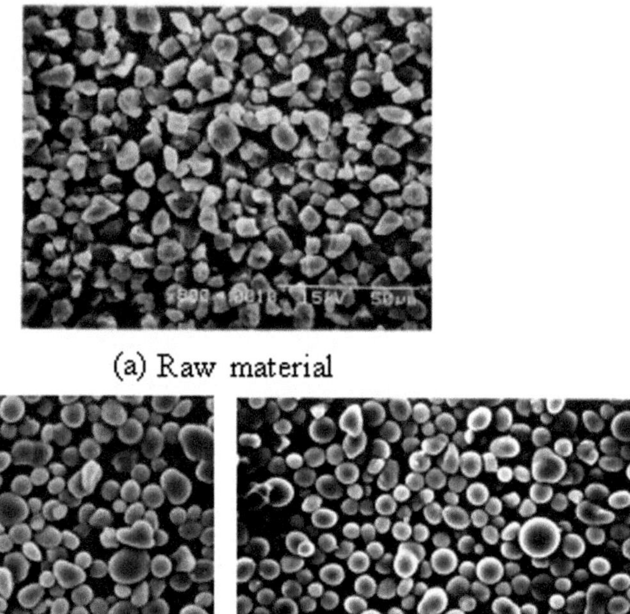

(a) Raw material

(b) Around-nozzle   (c) Three-concentric-nozzle

Fig. 189 Globular forming products

300 °C, $G = 1.35$ kg/hr and $G = 0$ kg/hr, respectively. Then, using Eqs. (171) and (172), the energy efficiency ratio turns to be approximately 2.9.

Figure 190 shows the globular forming characteristics of polyester toner II in terms of the mean roundness and the coupling, when the flow rate of secondary air for the Three-concentric-nozzle is changed to $F_{se} = 0.16, 0.24$ Nm^3/min at $T_{max} = 250$ °C. As increasing the flow rate of secondary air the mean roundness and coupling decrease. It suggests that the decrease in both roundness and coupling is caused by the decrease in the heat transfer rate to a particle, because the high temperature region decreases as shown in Fig. 191.

The new three-concentric-nozzle globular forming device proposed here allows fine particles to flow into a higher temperature region than the around-nozzle type. The energy efficiency is greatly improved, approximately 5.3 times that of the around-nozzle type.

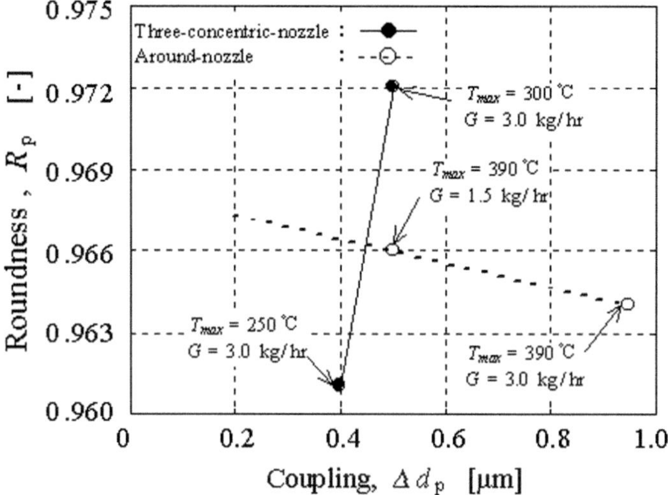

Fig. 190 Roundness and coupling (Toner II)

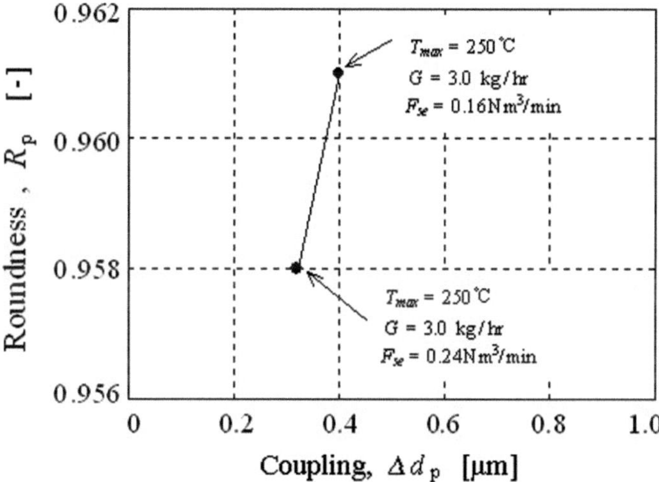

Fig. 191 Effect on secondary air (Toner II)

6.2 *Jet Pump*

A jet pump is a pump in which a driving fluid is ejected from a nozzle at high speed, entraining the surrounding driven fluid and discharging the mixed fluid. Its structure is generally simple, with no moving parts. Due to this characteristic, jet pumps have been used in boiling water reactors (BWRs) to obtain the recirculation flow of coolant required to cool the reactor core under the high temperatures and pressures

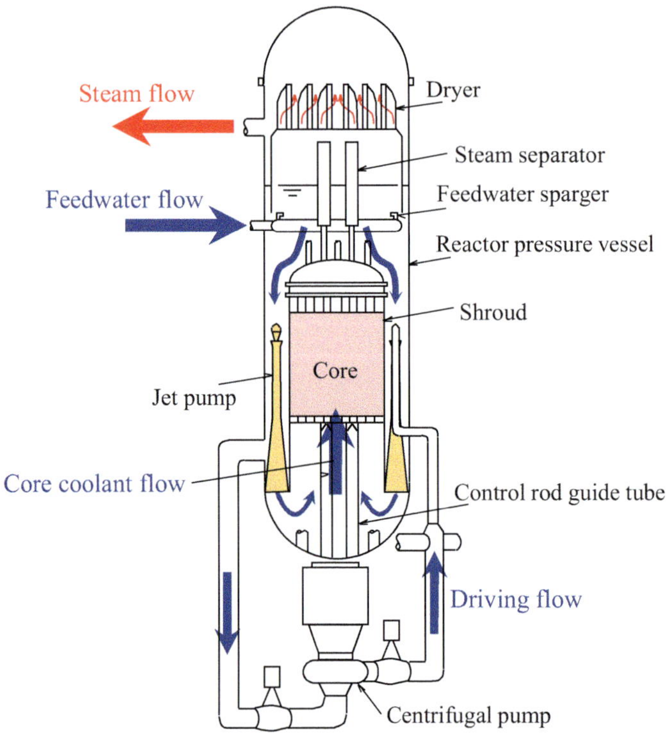

Fig. 192 Schematic of Boiling Water Reactor

(approximately 280 °C and 7 MPa) inside the reactor pressure vessel. Figure 192 shows a schematic cross-section of a BWR. In a BWR, part of the coolant inside the reactor pressure vessel is directed through a recirculation loop consisting of large-diameter pipes and gate valves with a nominal diameter of about 600–700A installed outside the reactor pressure vessel, pressurized by a centrifugal pump, and supplied as the driving fluid for the jet pump (Figs. 192 and 193).

Jet pumps are generally known to be less efficient than centrifugal pumps, and BWRs use jet pumps with an efficiency of approximately 35–40%.

6.2.1   Effect of Surface Roughness on Jet Pump Performance

One main reason why jet pump efficiency is low occur in the 1arge friction loss in the throat, which needs enough length to mix the two fluids. And the optimum throat length exits for each jet pump specification to make its efficiency improve. Therefore, increasing the surface roughness of the throat, which causes friction loss, affects jet pump performance whose throat length was optimized.

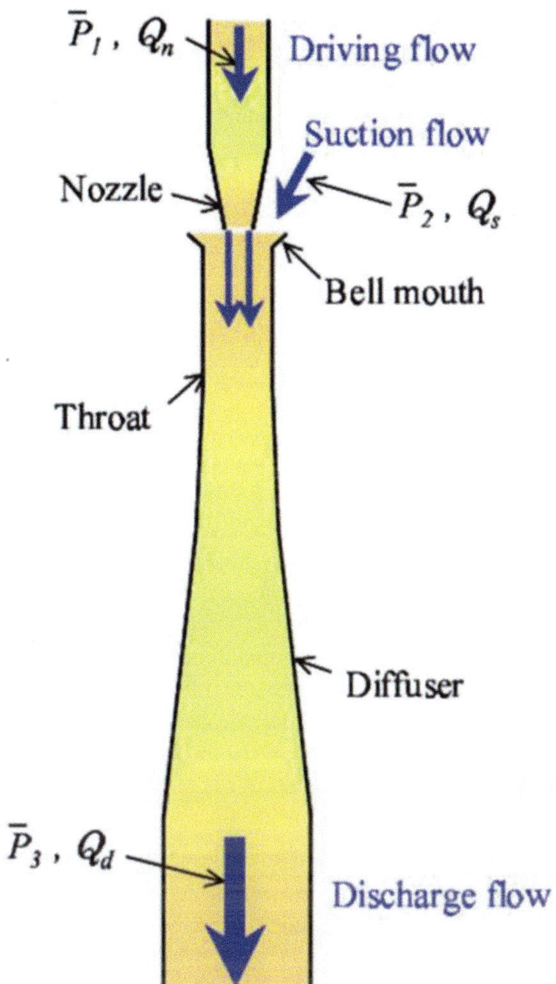

Fig. 193 Schematic of jet pump

It is known that in BWRs, metal ions of Zn injected for the purpose of reducing radiation exposure, and Cr and Fe from the piping materials, form cladding due to static electricity and adhere to the inner surface of the jet pump. This increases surface roughness and increases frictional resistance to the flow.

To obtain fundamental knowledge of the effect of surface roughness on jet pump performance, experimental studies were performed for several surface roughness of a typical single nozzle jet pump using water at room temperature. Experimental values of efficiency for the flow rate ratio were compared to a one-dimensional theoretical prediction equation, and the effect of roughness on the frictional resistance coefficient in the throat for each roughness was investigated.

In a typical single-nozzle jet pump for a BWR, it is necessary to identify an effective cleaning location for removing crud that has adhered to the inner throat surface and restoring its performance. That is, the effect of the axial position of the rough surface caused by crud adhesion on the inner throat surface and its surface roughness (roughness) on the performance of the jet pump is experimentally clarified. Furthermore, in order to evaluate the degree of crud adhesion from the amount of efficiency reduction, the amount of efficiency reduction due to changes in the roughness of the inner throat surface is experimentally shown (Yamazaki et al. 2005, 2006; Narabayashi et al. 2006).

(a)   **Set-up and procedure**

The experimental set-up of' the 1/5 scale jet pump performance test loop for BWR plant is shown in Fig. 194, The working fluid was water at room temperature, High pressure water pumped by a centrifugal pump flows into the nozzle through the magnetic flow meter from the chamber and is discharged to the throat inlet in the shape of bell-mouth, as the driving jet flow. The driving jet and induced flow are mixed in the throat and flow out after recovering pressure in the diffuser, they return to the chamber through another magnetic flow meter in the discharge line. In this study, driving flow rate Q_n was fixed at 6.5×10^{-3} m^3/s (nozzle Reynolds number $Re_n = V_n \times D_n/v = 4.8 \times 10^5$), and flow rate ratio M $(= Q_s/Q_n)$ was changed from 0.5 to 1.8 (Q_s is the induced flow rate). The chamber was pressurized at about 0.35 MPa to prevent the occurrence of cavitation in the jet pump threat.

Figure 195a, b show the configuration of jet pump nozzle and throat, respectively. The cross section of a jet nozzle is circular, and its exit diameter is $D_n = 17$ mm, The throat is a $L_t = 440$ mm straight pipe whose diameter is $D_t = 35$ min. The distance of the nozzle and the inlet of the throat are $s = 0$ mm. The rough surface inside the throat is shaded area shown in Fig. 195. Coated abrasives affixed inside the throat were substituted for the inside surface roughness of the jet pump. Here, the length

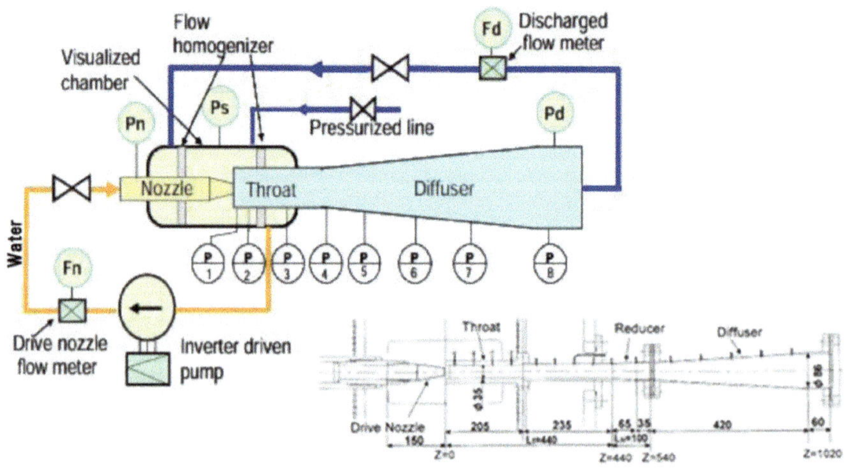

Fig. 194 Schematic diagram of the 1/5 scale jet pump test loop for BWR plant (Narabayashi et al. 2006)

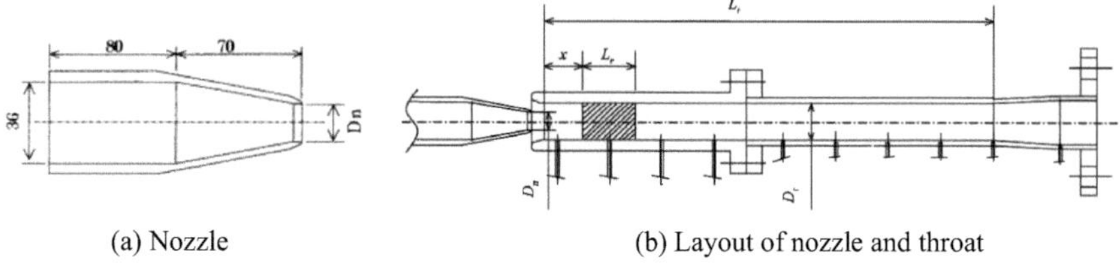

(a) Nozzle   (b) Layout of nozzle and throat

Fig. 195 Configuration of jet pump, nozzle and throat

of the coated abrasives was $L_r = 50$ mm, and the location was varied from $x = 12$ mm (at the throat inlet except bell-mouth) to $x = 155$ mm in the throat by affixing new coated abrasives after peeling off the previous ones, where x is the streamwise position from the throat inlet.

(b) Jet pump parameters

Jet pump performance is generally considered to be a function of the parameters defined as follows. Flow rate ratio M is the ratio of induced flow to the driving flow defined by

$$M = \frac{Q_s}{Q_n} \tag{173}$$

Pressure ratio N is the ratio of the specific energy increase of the induced flow energy to the specific energy decrease of the driving flow defined by

$$N = \frac{\overline{P}_d - \overline{P}_s}{\overline{P}_n - \overline{P}_d} \tag{174}$$

where, Q is the flow rate, $\overline{P}$ is the total pressure, and subscript n, s, and d refer to the driving flow, induced flow, and discharged flow, respectively.

Jet pump efficiency η is defined by

$$\eta = M \times N \tag{175}$$

(c) Surface roughness parameters

The surface roughness inside the throat is changed by affixing commercial coated abrasives. Here, three sizes of coated abrasive grains were used: P150, P240, and P1200 (based on ISO International Standards). The grain size of the 50% frequency distribution is standardized to be 98 ± 8 μm for P150, 58.5 ± 2.0 μm for P240, and 15.3 ± 1.0 μm for P1200. Here, equivalent roughness k_e is defined as grain size in the 50% frequency distribution.

On the other hand, surface roughness k_e is of the acrylic throat affixing no coated abrasives is defined as hydraulically smooth. Its equivalent roughness is presumed to be $k_e = 1.5$ μm, the same as the drawn pipes.

Figure 196 shows visual roughness was about the 3-D surface profile of the optical inspection using the same lens of magnification ratio 450 for all coated abrasives. The surface roughness (peak to peak) measured by the roughness gauge is about 92 μm P-P for P150, 61 μm P-P for P240, and 10 μm P-P for P1200. These results were almost the same as the standardized grain sizes of the 50% frequency distribution. Therefore, the three kinds of relative roughness are defined as $k_e/D_t = 0.0028 \pm 0.0002, 0.0017 \pm 0.0001$, and 0.0004 ± 0.00003, respectively. The relative roughness of the hydraulically smooth is $k_e/D_t = 0.00004$. Using the $L_r = 50$ mm long coated abrasives $(0.12\ L_r)$ rough surface are located on the entire inside race from $x/L_t = 0$ (at the throat inlet except for bell-mouth) to $x/L_t = 0.12$.

To investigate the effect of rough location on jet pump performance, rough location was changed from $x/L_r = 0$, $x/L_t = 0.20$, and $x/L_t = 0.32$ in the throat on which the induced flow was greatly accelerated, using the relative roughness $k_e/D_r = 0.0028$. In addition to that, the roughness at the inlet of the diffuser was also investigated for reference.

(d)   Effects of rough location on jet pump performance

Figure 197 shows the effect of rough location in the throat using coated abrasives P150 (relative roughness $k_e/D_t = 0.0028$) on jet pump performance.

Figure 198 shows flow rate ratio Mb at the best efficiency point and best efficiency ratio $\eta_b/\eta_{b,hs}$, where $\eta_{b,hs}$ is best efficiency of the hydraulically smooth jet pump. From these figures, flow rate ratio M_b at the best efficiency point, and best efficiency η_b were decreased more than the hydraulically smooth $(k_e/D_t = 0.00004)$ results as rough location x/L_t was closer to the throat inlet. Surface roughness great [y affected jet pump efficiency η in the higher flow rate ratio M region. In cases of rough location in the diffuser, the effect of roughness was smaller than in the throat,

From the velocity profile measurement results in case of hydraulically smooth throat, local skin friction coefficient C_f tended to decrease toward the throat exit and then increase after the transition to the turbulent boundary layer. Therefore, apparently roughness in the throat inlet greatly affected jet pump efficiency because C_f nearer the throat inlet was larger. This study revealed that the surface roughness located nearer the throat inlet had a greater effect on jet pump efficiency.

(e)   Effects of relative roughness

Flow rate ratio M_b at the best efficiency point, and best efficiency η_b are decreased more than the hydraulically smooth $(k_e/D_t = 0.00004)$ results as relative roughness k_e/D_t increases. Identical to effects of rough location on jet pump performance in Fig. 195, relative roughness k_e/D_t greatly affects jet pump efficiency η in the higher flow rate ratio M region.

Figure 199 shows the effect of relative roughness k_e/D_t located from $x/L_t = 0$ to 0.12 in the throat, where jet pump performance was extremely affected, on jet pump

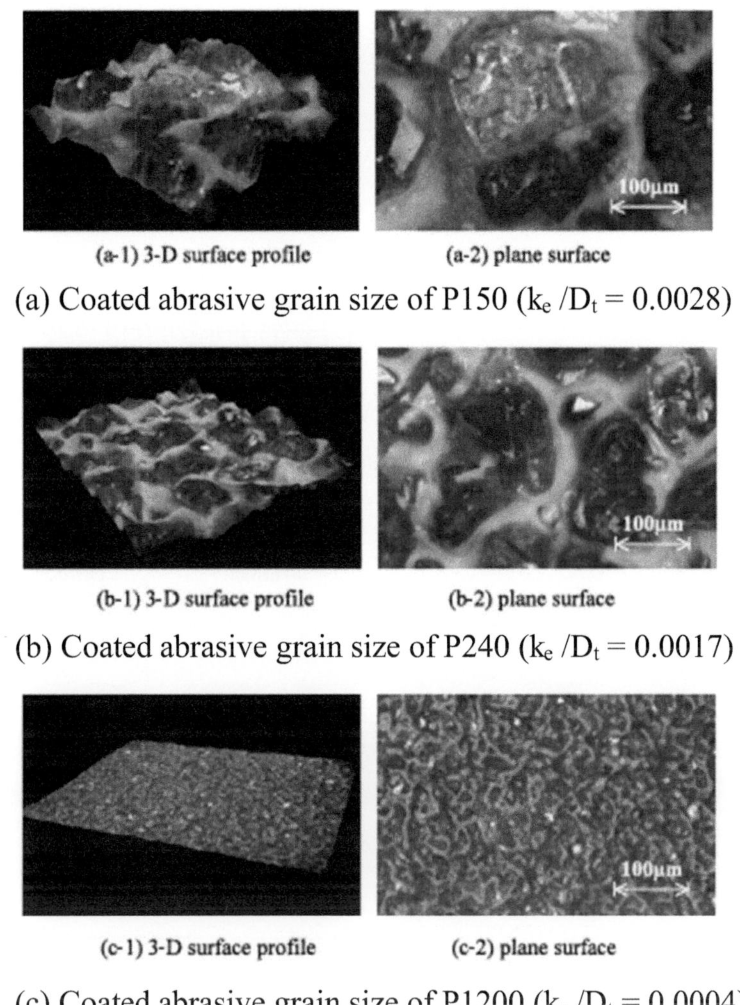

(a) Coated abrasive grain size of P150 (k_e /D_t = 0.0028)

(b) Coated abrasive grain size of P240 (k_e /D_t = 0.0017)

(c) Coated abrasive grain size of P1200 (k_e /D_t = 0.0004)

Fig. 196 Three-dimensional surface profile and plane surface of coated abrasives

efficiency. Flow rate ratio M_b, at the best efficiency point and $\eta_b/\eta_{b,hs}$ is linearly decreased as relative roughness k_e /D_t increases.

(f) **Theoretical prediction**

Regarding the effect of relative roughness k_e/D_t located near the throat inlet on jet pump performance, the effect on frictional resistance coefficient λ in the throat was investigated for each roughness by a non-dimensional theoretical equation which was consisting of an application of energy, momentum, and continuity equations across the jet pump. The energy equations at the driving nozzle, suction inlet, and diffuser are

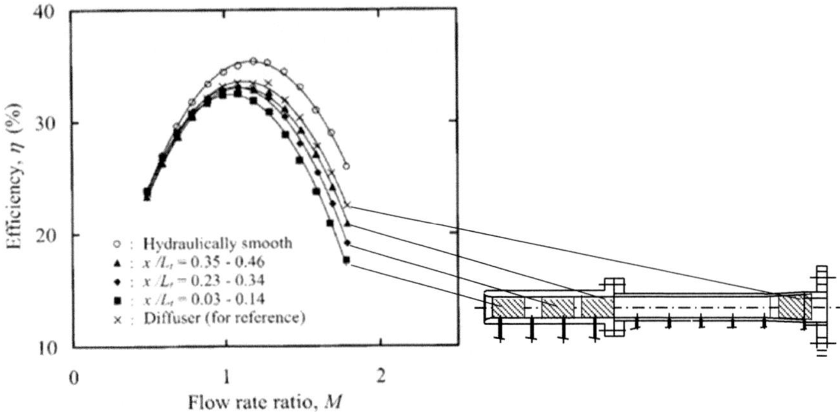

Fig. 197   Effect of rough location on jet pump performance ($k_e/D_t = 0.0028$)

Fig. 198   Effect of rough
location on best efficiency
($k_e/D_t = 0.0028$)

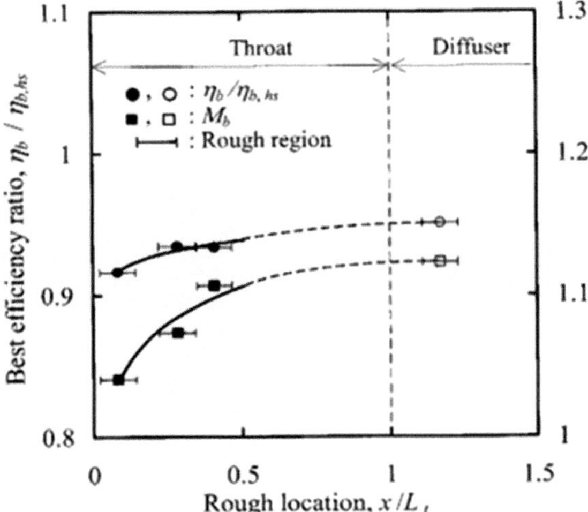

$$\overline{P}_n - \overline{P}_{ti} = k_n \frac{\rho v_n^2}{2} \tag{176}$$

$$\overline{P}_s - \overline{P}_{ti} = k_s \frac{\rho v_s^2}{2} \tag{177}$$

$$\overline{P}_{t0} - \overline{P}_d = k_d \frac{\rho v_i^2}{2} \tag{178}$$

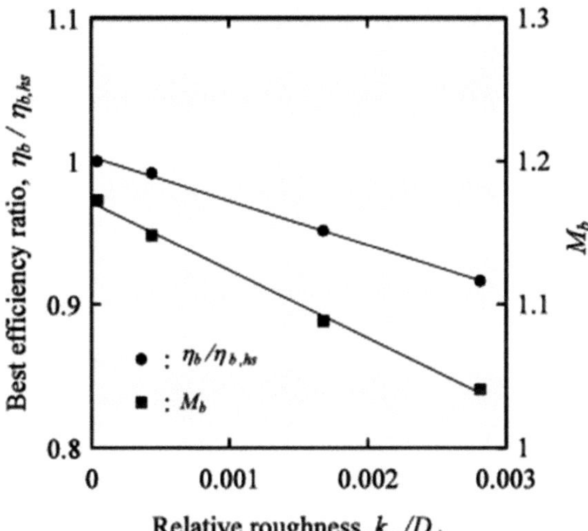

Fig. 199 Effect of relative roughness on best efficiency (rough location, $x/L_t = 0 \sim 0.12$)

The momentum equation which is applied for the throat is

$$P_{t0} - p_{ti} = \rho\left[RV_n^2 + (1-R)V_s^2 - V_t^2\right] - k_t\frac{\rho v_t^2}{2} \tag{179}$$

Equation (178) assumes that momentum is fully transferred in the throat. Mean velocity V at suction inlet and throat are described as follows using continuity equations.

$$V_s = \left(\frac{R}{1-R}\right)M \times V_n \tag{180}$$

$$V_t = R(1+M) \times V_n \tag{181}$$

Therefore, a one-dimensional theoretical equation of jet pump efficiency η becomes

$$\eta = M \times \frac{\overline{P}_d - \overline{P}_s}{\overline{P}_n - \overline{P}_d} =$$
$$- M \times \frac{(1+k_s)\left(\frac{RM}{1-R}\right)^2 - 2R - \frac{2R^2M^2}{1-R} + R^2(1+M)^2(1+k_l+k_d)}{(1+k_n) - 2R - \frac{2R^2M^2}{1-R} + R^2(1+M)^2(1+k_l+k_d)} \tag{182}$$

Although friction loss coefficient k is generally defined by $k = \lambda \times (L/D)$ for straight pipe flow, here friction loss coefficient k_t in the throat, where flow is not fully developed, is defined by

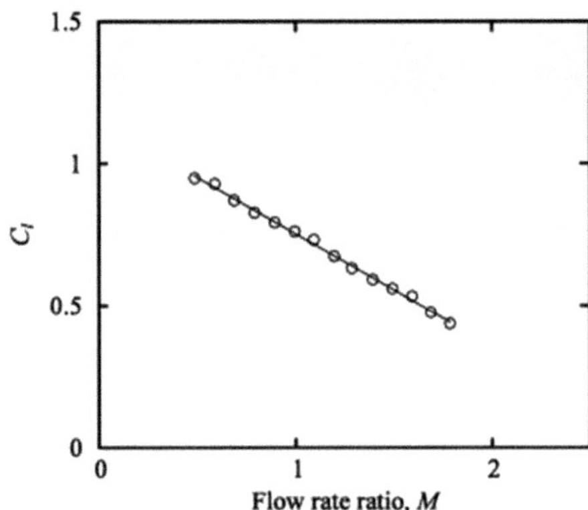

Fig. 200 C_1 distribution for hydraulically smooth jet pump

$$k_t = \lambda \times \frac{C_1 L_t}{D_t} \tag{183}$$

where, C_1 is the equivalent frictional length coefficient for a hydraulically smooth jet pump, λ is the frictional resistance coefficient obtained from Colebrook's equation of the Moody diagram using k_e/D_t, and throat Reynolds number Re_t $(= V_t \times D_t/\nu$, V_t is the mean velocity of mixed flow rate in the throat). The values of k_n, k_s and k_d in Eq. (182) are 0.01, 0.02 and 0.16, respectively, which are obtained by the experimental result and general literature.

Figure 200 shows C_1 distribution which are calculated by Eqs. (182) and (183) and the experimental result. C_1 increases linearly as M increases. This means that as increasing M the derivation from the friction resistance coefficient λ for fully developed pipe flow increases. To investigate how the frictional resistance coefficient λ is affected by relative roughness, λ for three relative roughness are evaluated by Eqs. (182) and (183) and C_1 in Fig. 200.

In Fig. 201, λ for three relative roughness are drawn on Moody Diagram. Figure 201 also has λ which are obtained by Colebrook's equation in case of the fully developed pipe flow for $k_e/D_t = 0.0028, 0.0017, 0.0004$ and 0.00004.

From λ distribution in Fig. 201, λ is not affected very much by relative roughness in lower Re_t. However, λ increased as Re_t increased and becomes almost the same as the dashed line, which is Colebrook's equation calculated from Re_t, and k_e/D_t for fully developed pipe flew. Although the rough surface is a part of the throat, the roughness located nearer the throat inlet affected as the same increases of λ as the entire arca roughness of the throat in larger Re_t.

This revealed that the frictional resistance coefficient in the throat for each roughness became obvious by fitting a one-dimensional theoretical prediction equation to the experimental results.

To obtain fundamental knowledge of the effect of surface roughness in the throat on jet pump performance, experimental studies were performed for a typical single nozzle jet pump by affixing commercial coated abrasives onto the throat's inner surface.

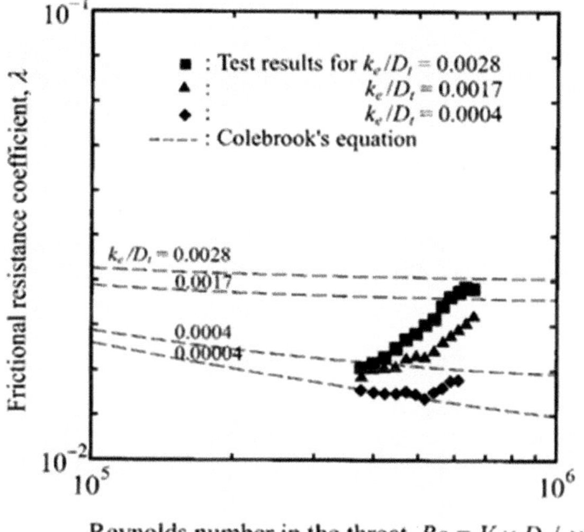

Fig. 201 Effect of relative roughness on jet pump throat (rough location, $x/L_t = 0.03 \sim 0.14$ in the throat)

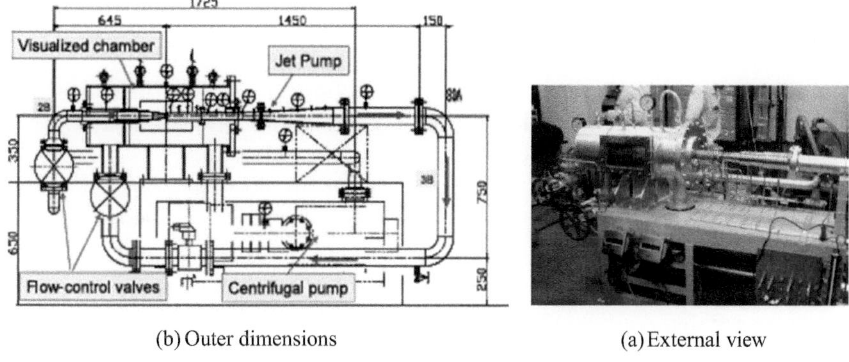

(b) Outer dimensions   (a) External view

Fig. 202 Experimental loop of jet pump

(1) Surface roughness located nearer the throat inlet has a greater effect on the jet pump efficiency because the local skin friction coefficient nearest the throat inlet is the larger.

(2) The flow rate ratio at the best efficiency and best efficiency decrease linearly as surface roughness increases, especially surface roughness greatly affects jet pump efficiency in the higher flow rate ratio region.

(3) The frictional resistance coefficient in the throat each roughness becomes obvious by fitting one-dimensional theoretical prediction equation to experimental results.

6.2.2   Effects of Throat Shape on Jet Pump Performance

The throat of a jet pump needs to have an appropriate length to mix the driving fluid and the driven fluid and transfer the momentum of the driving fluid to the driven fluid. However, due to its length, the fluid inside the throat is subject to wall friction, resulting in energy loss, which is known to have a large impact on the efficiency of the jet pump.

Here, we consider how to improve and enhance the efficiency of the jet pump by making the throat a diverging pipe with a small diverging angle, slowing the average velocity and reducing friction losses.

(a)   Normal throat and modified diverging throat

The experimental analysis was carried out using the 1/5 scale model mentioned before (Figs. 194 and 202).

Figure 203a, b show the normal throat and modified diverging throat, respectively. The normal throat consists of the length $L_t = 440$ mm straight part and a diffuser part ($2\theta_u = 6.0$ deg, $L_u = 100$ mm). The diverging throat consists of $L_t = 130$ mm straight part and $2\theta_u = 1.0$ deg, $L_u = 410$ mm diverging part. The diameter of each straight part is $D_t = 35.0$ mm.

(b)   Effects of throat shape

Figure 204 shows the effect of throat shape on the relationship between efficiency η and flow rate ratio M. The efficiency η of diverging throat is significantly improved compared to the normal throat, especially as the flow rate ratio M increases. The maximum efficiencies of normal and diverging throats are $\eta = 34.0\%$ at $M = 1.1$ and $\eta = 35.4\%$ at $M = 1.3$, respectively.

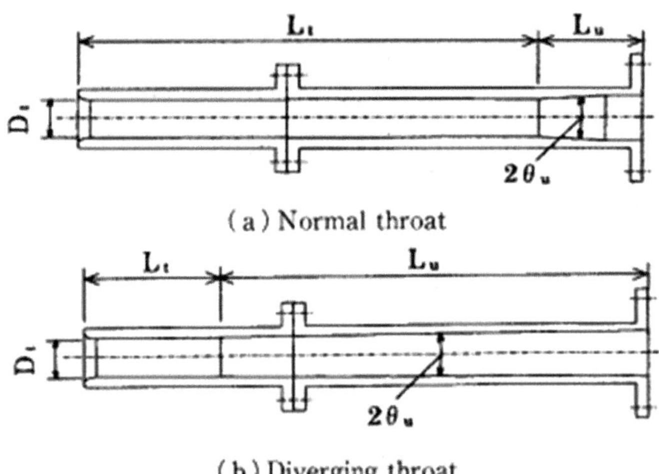

Fig. 203   Normal throat and modified diverging throat ($2\theta = 1.0°$)

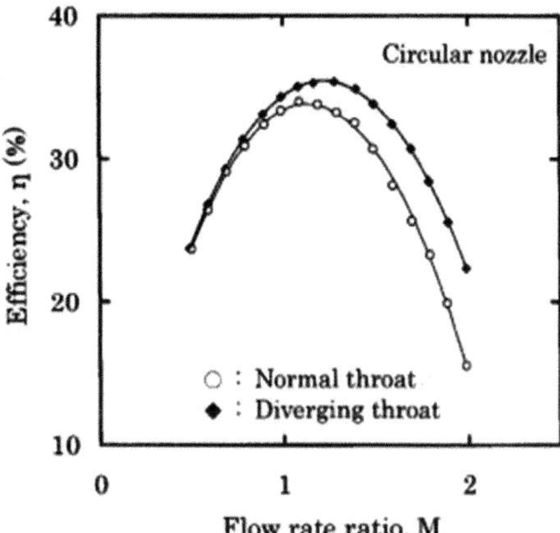

Fig. 204 Effect of throat shape on jet pump efficiency

Below, the reason for the improvement in efficiency by the diverging throat is considered from the measurement results of the pressure distribution and velocity distribution in the throat.

Figure 205 shows the effect of throat shape on local pressure ratio N_L. The figure shows the results for $M = 1.1$ and 1.3, which are the highest efficiencies for the normal throat and the diverging throat. In the case of the diverging throat, static pressure recovery is slower than in the normal throat, and the peak moves downstream. The static pressure at the throat exit is greater in the diverging throat.

Next, the velocity distribution in the throat was measured using air as the working fluid for $Q_n = 9.2 \times 10^{-3}$ m^3/s ($Re_n = 4.5 \times 10^4$), $M = 1.3$. A cylindrical single-hole Pitot tube with a diameter of 1.0 mm and a hole diameter of 0.4 mm was used to measure the velocity distribution. The deviation S_v between the result and the 1/7 power law was calculated and is shown in Fig. 206. In the diverging throat, the asymptotic trend to the 1/7 power law is slower than in the standard throat, and the

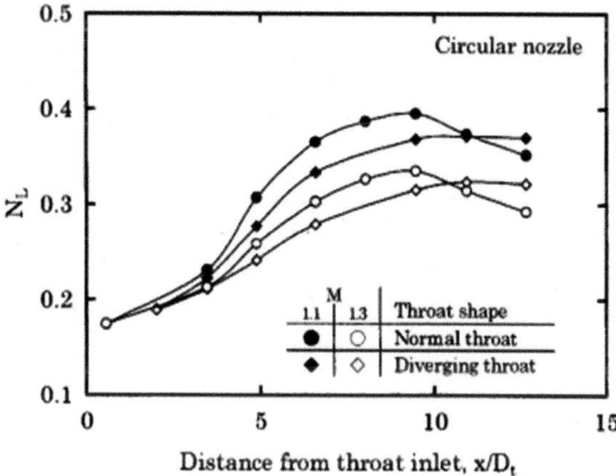

Fig. 205 Effect of throat shape on pressure distribution

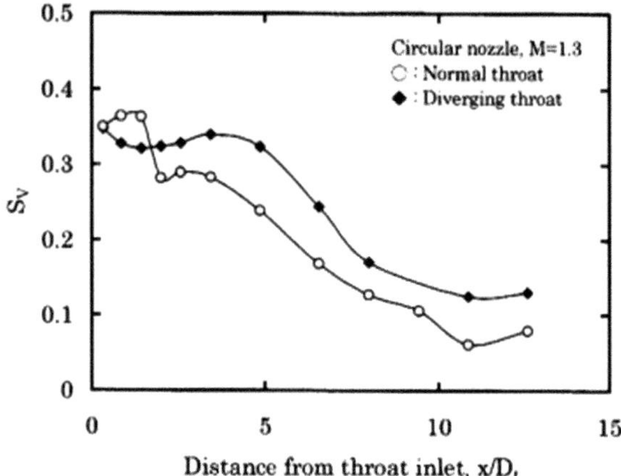

Fig. 206 Effect of throat shape on Deviation of velocity

mixing of the driving fluid and the driven fluid is shifted downstream. This is thought to be the reason why the static pressure recovery peak shifts downstream in Fig. 205.

Here, the 1/7 power law distribution at each cross-section x/D_t of the expansion throat is expressed by the following equation using the radius of the expansion part $r_u(x)$ and the average velocity $V_u(x)$, where the maximum flow velocity of the 1/7 power law distribution is $1.22V_u(x)$.

$$v_{1/7}(r) = 1.22V_u(x) \times \left[\frac{r}{r_u(x)} \right]^{1/7} \tag{184}$$

Figure 207 shows the local friction coefficient C_f calculated from the velocity distribution near the wall at $M = 1$. With a standard throat, the boundary layer on the wall transitions to turbulence when $x/D_t \fallingdotseq 3$, and C_f increases rapidly, but with an diverging throat, no such increase in C_f is observed.

The effect of throat shape on the performance of jet pump is summarized as follows.

In the normal throat, mixing of the driving fluid and the driven fluid is promoted approximately in proportion to the distance from the throat inlet over a distance of approximately $11D_t$, and the static pressure is restored. From the intermediate stage where mixing of the two fluids is at $x/d_t \fallingdotseq 3$, the boundary layer transitions to turbulence downstream, and C_f increases. Therefore, the amount of recovery of static pressure decreases downstream from $x/D_t \fallingdotseq 3$, and the static pressure decreases due to friction resistance downstream from $x/D_t \fallingdotseq 9$, where the mixing of the driving fluid and the driven fluid is almost completed. In the case of a diverging throat, compared to a normal or straight throat of the same length, the promotion of mixing between the driving fluid and the driven fluid is inferior, and as a result, the recovery of static pressure is insufficient and the peak of static pressure recovery is lower. On the other hand, the boundary layer does not transition to turbulence, and friction resistance is smaller than that of a normal throat of the same length. Therefore, even though the

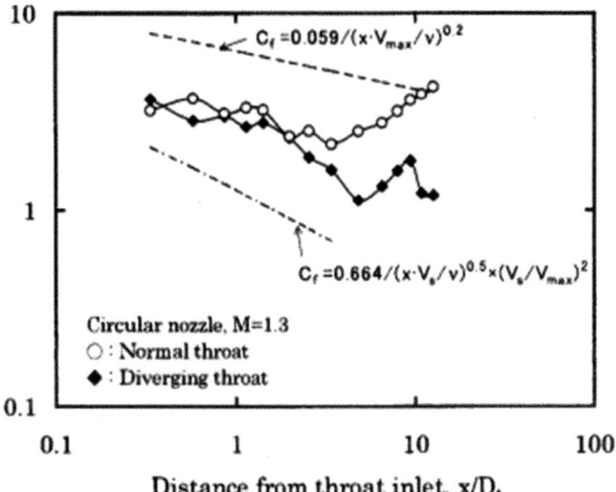

Fig. 207 Effect of throat shape on C_f distribution ($M = 1,3$)

Distance from throat inlet, x/D_t

recovery of static pressure is insufficient in the diverging throat, the amount of static pressure drop due to friction resistance is small. As a result, it is thought that the efficiency of the jet pump will be higher.

The maximum efficiencies of the jet pump with normal and diverging throats were $\eta = 34.0\%$ at $M = 1.1$ and $\eta = 35.4\%$ at $M = 1.3$, respectively. The expansion angle of the expansion section is extremely gentle at $2\theta = 1°$, but this shape has a large effect on the mixing process of the two fluids.

6.3 Air Curtain Jet Flow

Open refrigerated showcases without glass doors are widely used in supermarkets to display fresh foods, refrigerated foods, and frozen foods. In order to maintain a low temperature inside the case, air curtain jets are used to block the outside air, but the insulation performance is not sufficient.

Here, we examine the behavior of the air curtain jet from the results of numerical analysis, and from flow visualization and observation using the dye tracer method. Furthermore, we consider the cooling performance of an open show-case, and prediction of temperature distribution from dye diffusion (concentration distribution).

6.3.1 Test Open Refrigerated Showcase and Numerical Model

Figure 208a, b show the numerical analysis model of the open show case and the 1/5 scale experimental model of the actual machine, respectively.

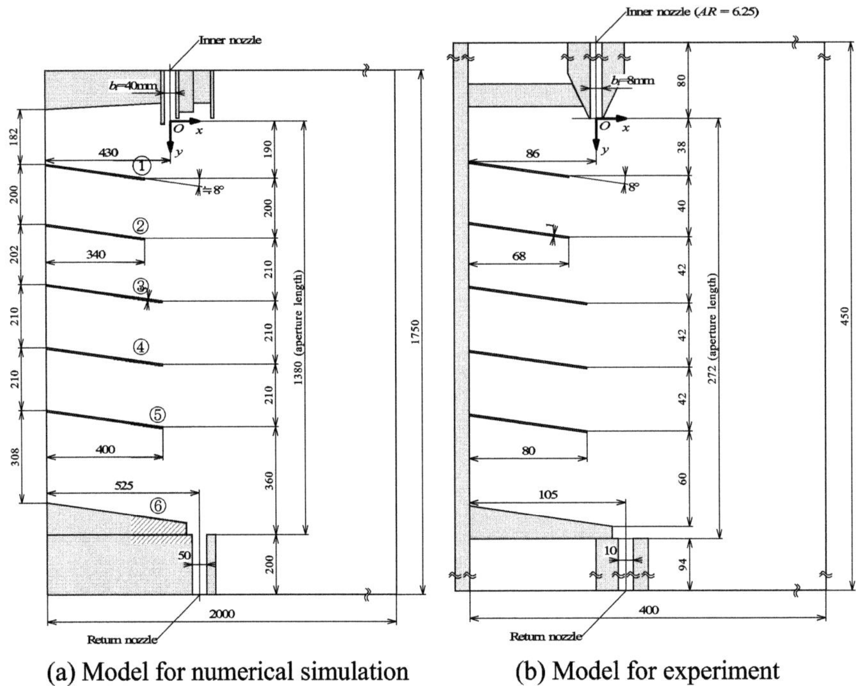

(a) Model for numerical simulation   (b) Model for experiment

Fig. 208   Test models of open showcase for numerical simulation and experiment

The width of the inner nozzle (hereafter referred to as In-Noz) of the experimental model is $b_i = 8.0$ mm, the aspect is $AR = 6.25$, and it has end plates on the front and back. In addition, there are a slit and an outer nozzle with widths of 1.6 and 8.0 mm on the inside and outside of the In-Noz., respectively, which can be used to control the jet from the In-Noz.

6.3.2   Velocity and Temperature Field (Numerical Analysis Results)

Figure 209 shows the velocity vector in the y-direction obtained from the numerical analysis (two-dimensional calculation) for the flow path in Fig. 208a. The calculation was performed using CFD2000 software, which uses a standard k-ε turbulence model. An air jet with a temperature of 3 °C and a uniform velocity of $V_i = 1.5$ m/s (nozzle Reynolds number, $Re = V_i b_i / v = 4400$) is ejected from the In-Noz ($b_i = 40.0$ mm) into a field with an ambient temperature of 27 °C, and ambient air is sucked in at uniform velocity $V_r = 1.5$ m/s from the lower return nozzle with a width of 50.0 mm. The flow passes through a flow channel with a heat exchanger installed on the backside of the showcase to the upper nozzle, where it is ejected again as a jet at 3 °C. The jet flows down along the edge of each shelf, entraining the surrounding fluid; as it does so, part of the jet flows into the shelf from the edge and flows along

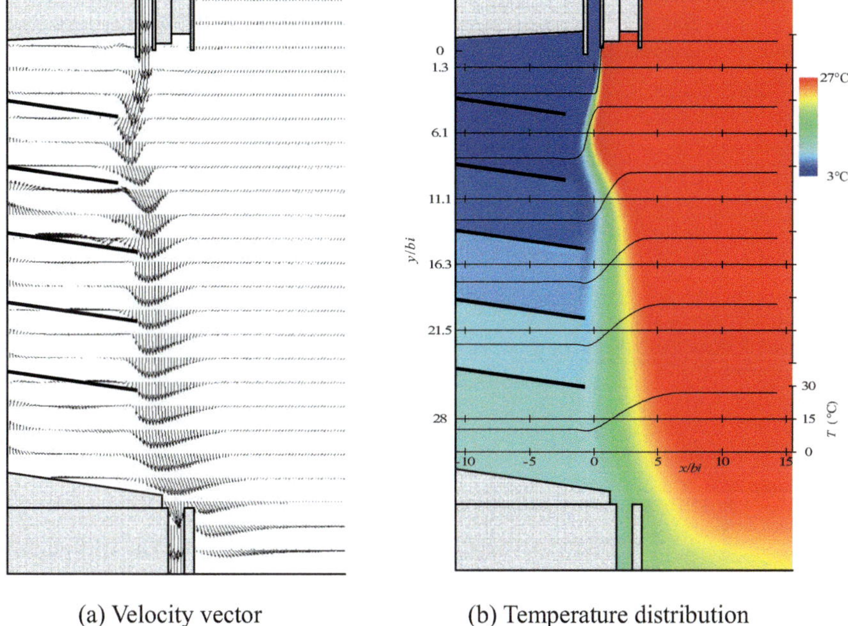

(a) Velocity vector   (b) Temperature distribution

Fig. 209 Flow and heat characteristics

the wall, creating a clockwise swirling flow that is entrained in the jet at the top. Figure 209b shows the contour map of temperature distribution. The temperature distribution at each cross section in the y-direction is also shown in the figure. The temperature in the x-direction of the jet at each cross-section changes from low on the shelf side to high around the surroundings. The temperature of the jet decreases to the downstream y-direction, and the temperature of the shelves also drops from 3 °C at the top shelf to 10 °C at the bottom shelf as the jet flows, entraining the hot surrounding fluid.

6.3.3 Performance of Open Refrigerated Showcase

The nozzle of the actual machine is equipped with a main nozzle (inner-nozzle), and slit- and outer-nozzles on both sides to control the mixing characteristics of the jet from the inner-nozzle with the surrounding fluid (Fig. 210).

Figure 211 shows the temperature distribution T °C (experimental values are the results of the actual machine) at the center of the back of each shelf. The speed and temperature of the inner-, slit-, and outer-nozzles are (1.5 m/s, 3 °C), (2.0, 3), and (1.0, 10), respectively. The temperature of each shelf decreases downstream, and the experimental results qualitatively match the numerical analysis results, but there are quantitative differences. The experimental and numerical analysis results are expressed by the following equation, and the two match via a correction coefficient.

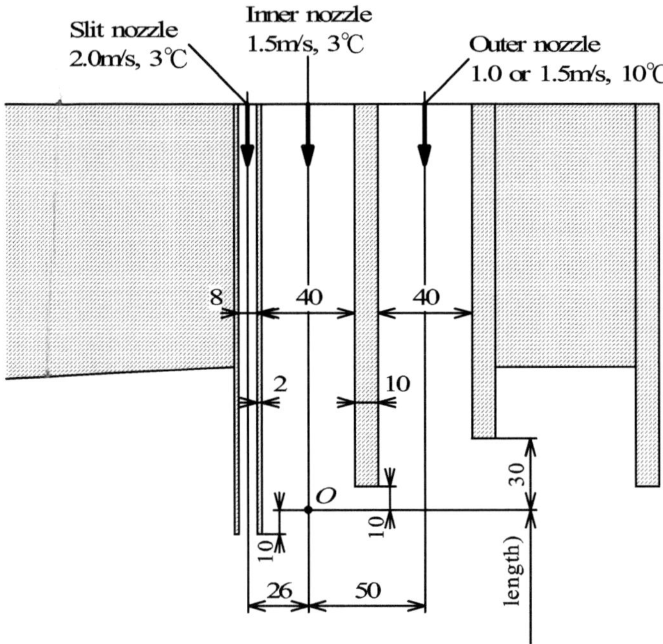

Fig. 210 Details of configuration of nozzle part (scale shown is 1/5 of actual machine)

$$T = 0.0065\left(\frac{y}{b_i}\right)^2 + 0.021\left(\frac{y}{b_i}\right) + 4.58 \tag{185}$$

$$T = 0.0038\left(\frac{y}{b_i}\right)^2 - 0.024\left(\frac{y}{b_i}\right) + 3.03 \tag{186}$$

Figure 211 also shows the calculation results for $V_i = 1.5$ m/s. Compared to the previous results with the slit- and outer-nozzle, the cooling performance inside the case is significantly reduced.

6.3.4 Dye Diffusion and Temperature Distribution

Figure 212 shows a conceptual diagram of predicting temperature distribution from the diffusion (concentration distribution) of dye (water blue). First, for the two-dimensional free water jet visualized with dye shown in Fig. 212a [the device was a two-dimensional free jet realized by removing unnecessary parts in Fig. 208b], the relationship between the concentration distribution (transmitted light intensity) C on the center line and the numerical analysis result of the temperature distribution T shown in Fig. 212b (calibration curve) is obtained. Next, the concentration distribution of a refrigeration model installed under the water surface of a large aquarium

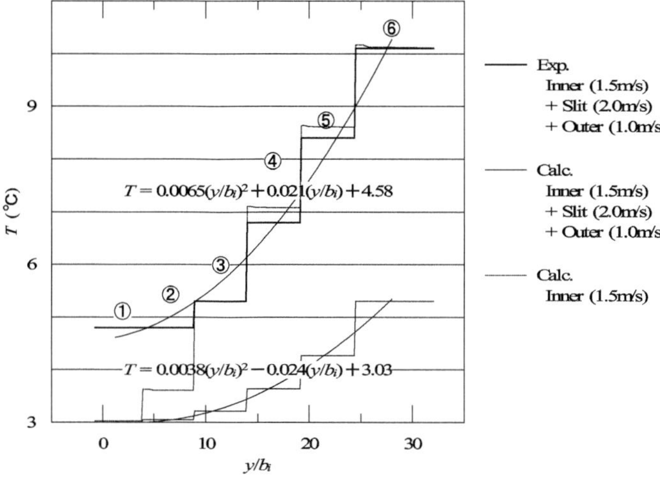

Fig. 211 Temperature distribution at the center of the back of each shelf

is measured (Fig. 212c), and the temperature distribution is obtained using the calibration curve (Fig. 212d). In actuality, a low-temperature air jet of 3 °C is sprayed into a space of 27 °C, so in this experiment, a saltwater water jet corresponding to the density difference between them was used.

Figure 213 shows the relationship (calibration curve) between temperature $T^*[= (T - T_\infty) / (T_0 - T_\infty)]$ and concentration $C^*[= (C - C_\infty) / (C_0 - C_\infty)]$. The inflow velocity of the water jet was set to Reynolds number $Re = 4{,}400$, which is the same as that of the actual machine (air, $V_i = 1.5$ m/s, 3 °C). The temperature and concentration distributions match well when $C^* > 0.785$. However, when $C^* < 0.785$, the difference with T^* increases as C^* decreases, but the relationship between them is well expressed by the following equation.

$$T^* = 1.04(C^*)^{2.03} \tag{187}$$

6.3.5 Prediction and Actual Temperature Distribution

Figure 214 shows the temperature distribution (thin line) calculated using the calibration curve Eq. (187) from the dye concentration distribution inside the case at $V_i = 1.5$ m/s (dashed line in the figure). The thin line roughly represents the experimental temperature distribution results (thick solid line).

Here, the flow and heat transfer characteristics of an open refrigerated showcase were examined using numerical analysis and experimental results, and the results showed that: *

<u>Free jet</u>

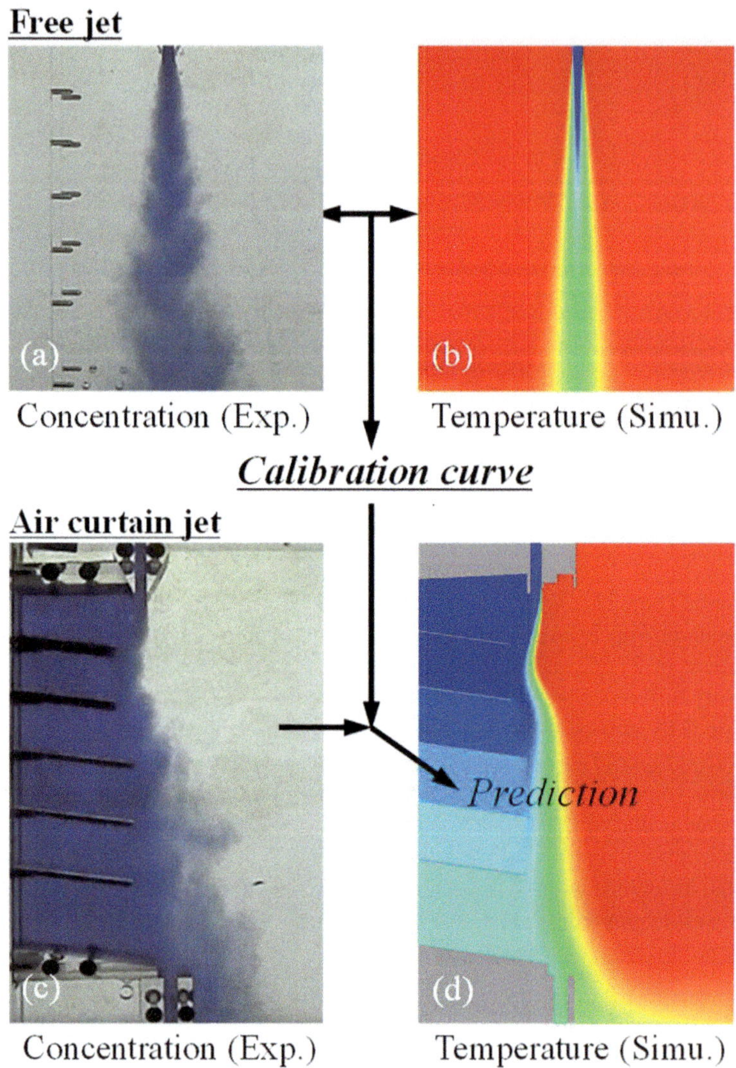

Fig. 212 Dye concentration and temperature distribution

(1) Cooling performance is greatly improved when a slit- or outer-nozzle is provided.
(2) Temperature distribution can be predicted from the dye diffusion and concentration distribution.

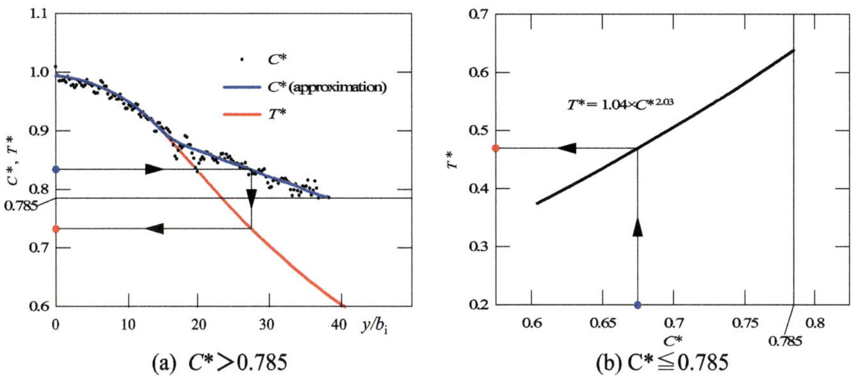

Fig. 213   Calibration curve of C* and T*

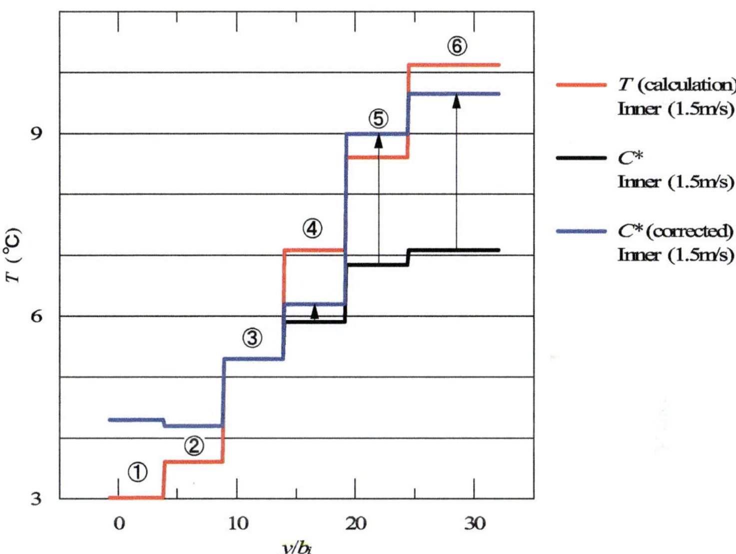

Fig. 214   Temperature distribution

7   Jet Stability and Oscillation Phenomena

The jet issued from the nozzle has flow fluctuations of various wave numbers and
disturbance components. If a disturbance component of a certain wave number is
amplified temporally or spatially in some way, the jet fluctuates greatly and oscillates
regularly (self-excited oscillation).

In this chapter, regarding the oscillation phenomena that occur in jet flows, we
will mainly focus on the edge tone oscillation phenomenon, and discuss the oscillation mechanism based on stability theory and the mechanism for determining the

oscillation frequency. Fluidic oscillation phenomena will also be discussed as other examples of jet oscillation (Shakouchi et al. 1985, 1986).

7.1 Edge Tone Oscillation Phenomenon

7.1.1 Mechanism of Oscillation

When a two-dimensional jet impinges on a wedge-shaped object (hereinafter referred to as an edge) placed at a certain distance from the nozzle outlet on its center line, the phenomenon that regular vibrations begin in the direction directly below the center transition is, which is accompanied by the generation of sound when the working fluid is gas, has long been known as an edge tone oscillation phenomenon. Incidentally, the sound production of various types of flutes is based on the edge tone oscillation phenomenon.

Figure 215 shows the schematic of edge tone oscillation. The advancing velocity u_a of the meandering motion of the jet is approximately 0.4 times the jet exit velocity, $u_a \doteqdot 0.4u_0$, and the pressure increase generated when the jet impinges on the edge is fed back to the root of the jet and changes the direction of the jet there is deflected in the y direction as shown in Fig. 215a. A large vortex forms at the root of the jet and moves in the x direction with velocity $u_v \doteqdot 0.4u_0$. Figure 215b shows the deflected jet with opposite phase.

Conventionally, the following are typical theories regarding the oscillation mechanism of the edge tone oscillation phenomenon.

(1) **Vortex theory by Curle** (1953)

The cause of the oscillation was thought to be the vortex arrays formed at the edges of the jet synchronously on both sides. That is, due to the presence of vortex rows on both sides of the edge, the ejected vortex formed by the jet that crosses the tip

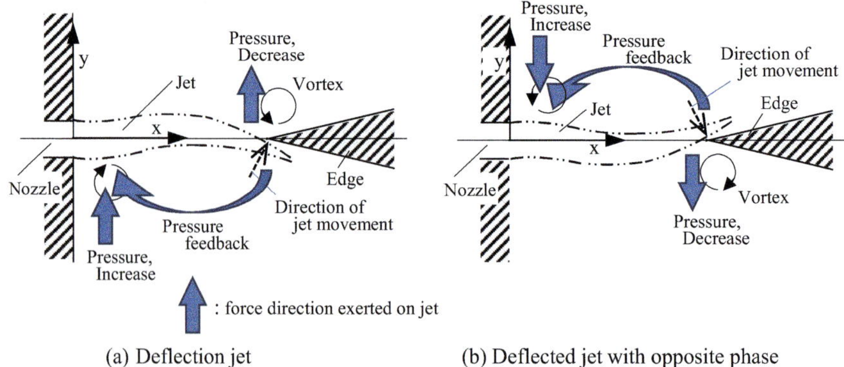

(a) Deflection jet   (b) Deflected jet with opposite phase

Fig. 215 Schematics of edge tone oscillation

of the edge while swinging conversely bends the direction of the jet at the upstream end of the jet (nozzle outlet). He also proposed a type of self-excitation mechanism (feedback mechanism) in which the newly created small vortices are amplified as they reach the edge position due to the inherent instability of the jet.

(2) **Jet instability theory by Powell** (1961)

First, he inferred that oscillations occur at frequencies close to sinusoidal disturbances, which is the frequency at which jets are most likely to become unstable. He also considered that the jet oscillation caused by the disturbance creates pressure fluctuations with opposite phases on both sides of the edge, and that the jet is bent at the nozzle exit due to this pressure fluctuation, resulting in amplified pressure fluctuations at the edge position. This is an oscillation mechanism based on feedback.

(3) **Pressure feedback theory by Nyborg** (1954)

He proposed that the behavior of a jet consists of independent (non-continuous) fluid particles passing moment by moment through the center of the nozzle in a pressure field based on alternating pressure sources formed at the edges by the oscillating jet. On the assumption that particles can be described as motion, he proposed a differential equation that includes the feedback effect of the pressure source on particle motion.

(4) **Theory based on flow stability theory by Wooly and Karamcheti** (1973)

This model can be considered to be based on the most reliable hydrodynamic foundation in the sense that Orr-Sommerfeld equation, which describes the stability of the flow, is used as the starting point for discussion. However, this paper only offers speculations about the mechanism that determines the oscillation frequency that is essential to phenomena.

7.1.2 Experimental Verification of Mechanism of Oscillation

We will attempt to experimentally verify the mechanism of the edge-tone oscillation phenomenon below.

(a) Experimental apparatus and test section

Figure 216 shows the experimental apparatus, jet–edge system (model). The model consists of a nozzle ① with a long parallel section, nozzle wall ② perpendicular to ①, an edge ③ with a wedge shaped tip angle 20° which is set on the centerline of the nozzle, an upper plate ④ and lower plate ⑤. Each section except for the brazen edge is made of a transparent acrylic plate. The model is set under water in a large water table with a constant water head, and water from a head tank or pump is conducted to the nozzle. The nozzle width and aspect ratio are $b_0 = 10.0$ mm and $AR = 6$, respectively. The edge distance h can be varied within 300 mm by rack and pinion ⑥. The injection pipes ⑦, ⑦′ and ⑦″ of dye for flow visualization are set on the centerline of the nozzle ($z = 0$), and at $y = \pm 5$ mm ($z = 0$) of nozzle exit, respectively. The inner and outer diameters of injection pipe are 0.6 and 1.0 mm,

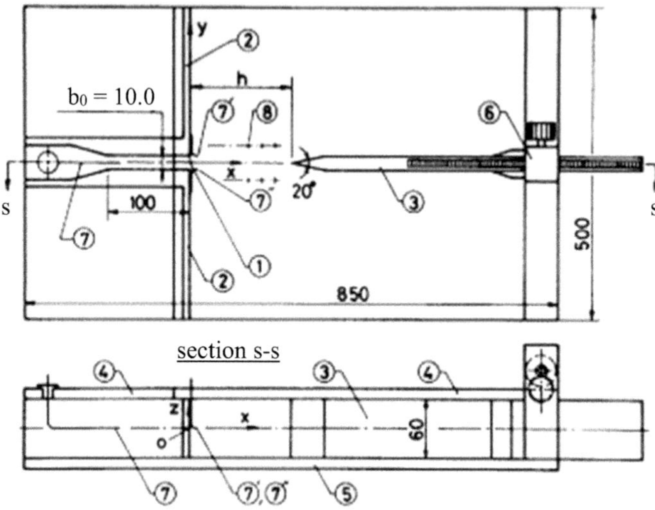

Fig. 216 Experimental apparatus, jet–edge system (model)

respectively. Static pressure taps of 1.0 mm in diameter are set at every 10 mm along
the lines of $y = \pm 15$ mm parallel to the x-axis.

(b) Effects of pressure feedback

The following experiment was conducted to confirm the effect of pressure feedback
near the nozzle outlet. As shown in Fig. 217, using an upper plate with holes of
10 mm in diameter at four locations A, B, A', and B', for the following three cases,
the pressure fluctuations in the flow field were measured at 4 points (only two points
① and ② are shown) every 30 mm on the σ axis (origin 0″), which vertically bisects
the straight line connecting A and B. Three cases are (1) Connect A and B' and A'
and B with vinyl pipes (inner diameter 8 mm, length 140 mm) filled with water,
keeping the holes watertight, (2) A and B For the three cases of connecting A' and
B' (parallel connection), and (3) normal state where each hole is plugged,
  In the Fig. 217, P_a is the pressure amplitude at each measurement point from ①
to ④, and P_{an} is the pressure amplitude at point ① in the case of (iii). In the case
of cross-connection (i) in which anti-phase feedback is added to normal feedback,
pressure fluctuations in the flow field are almost eliminated. Furthermore, in the
case of parallel connection (ii), an increase in pressure fluctuation is observed (the
frequency remains almost unchanged). The results of this experiment suggest that
pressure feedback near the jet exit is the mechanism by which edge tone oscillation
occurs. In addition, this experiment was conducted by changing the channel height
to $H/b_0 = 1$ in order to make it easier to see the effect of pressure feedback via the
vinyl pipe.

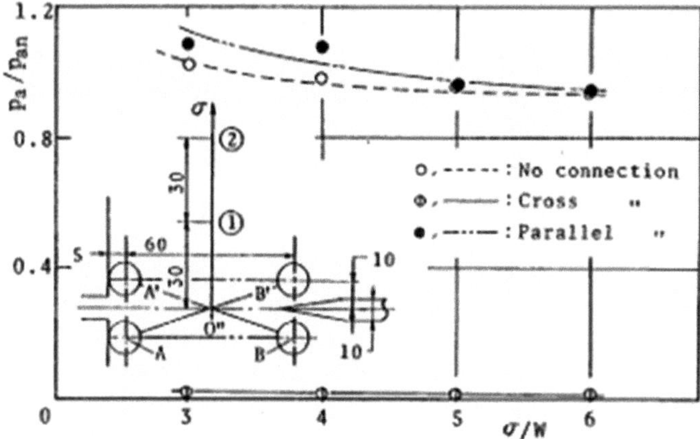

Fig. 217 Effect of pressure feedback ($h/b_0 = 6$, $Re = b_0 u_m/v = 710$, $f = 38$ Hz, mode I)

7.1.3 Oscillatory Frequency of Edge-Tone

Here, we will discuss the mechanism for determining the oscillatory frequency.

(a) Experimental observation

(a.1) Visualized flow pattern of free jet

An element having upper and lower end plates with a nozzle width $b_0 = 10.0$ mm and an aspect ratio $AR = 6$ of the parallel part of the nozzle is installed below the water surface in a large constant head water tank, and a submerged water jet is issued from the nozzle into stationary water.

Figure 218 shows the behavior of the jet when the average flow velocity u_0 (Reynolds number, $Re = u_0 b_0/v$) at the nozzle exit is changed by visualizing streak lines starting from the center of the nozzle exit and both ends of the nozzle exit at $z = 0$. The jet flow was visualized using an aqueous solution of sodium fluorescein as a tracer.

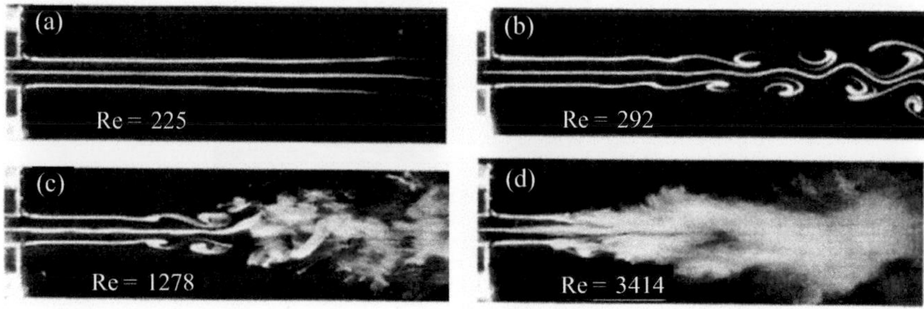

Fig. 218 Visualized flow pattern of two-dimensional submerged free jet (Effects of Re number)

At $Re = 225$, the jet flows down with almost no turbulence, but when the Reynolds number increases and $Re = 292$, vortex arrays arranged in a staggered manner occur on both sides of the jet. When the Reynolds number further increases and $Re = 1278$, vortex arrays are generated that are symmetrically arranged on both sides of the jet. At $Re = 3414$, the jet becomes highly turbulent when $x/b_0 > 2.5$.

(a.2) **Visualized flow pattern of edge-tone oscillation**

Figure 219 shows the behavior of the jet flow when an edge is installed on the nozzle axis with a distance $h/b_0 = 6$ from the nozzle exit to the edge in a flow of $Re = 292$ (Fig. 218b). Figure 219a–e show streak lines at 1 s intervals, with state (a) as the origin of time, $t = 0$. The behavior of the jet is completely different from that without an edge; the streak lines oscillate with a certain width at the edge tip position ($x = h$), and a noticeable large vortex region occurs at the outer edge of the jet. The oscillatory period is T $= 4.5$ s. When the edge distance h is small, the jet flow is divided into two parts symmetrically with respect to the nozzle axis. In Fig. 219f, only the streak lines emanating from the center of the nozzle are visualized, and the flow pattern is recorded every 0.4 s using a strobe device, and the jet moves regularly and symmetrically with respect to its center line.

Figure 220 shows an example of the imaging results of streak lines when $h/d = 9$ for the same Reynolds number. In this case, at $T = 2$ s, the stage mode of oscillation

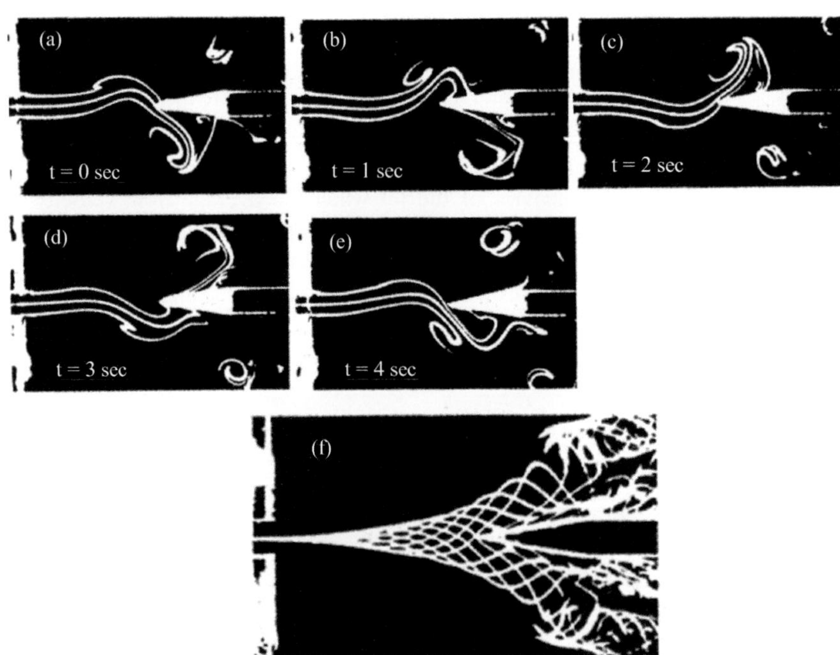

Fig. 219 Visualized flow pattern of streak lines of edge-tone oscillation ($b_0 = 10.0$ mm, $h/b_0 = 6$, $Re = 292$, $T = 4.5$ s, mode I)

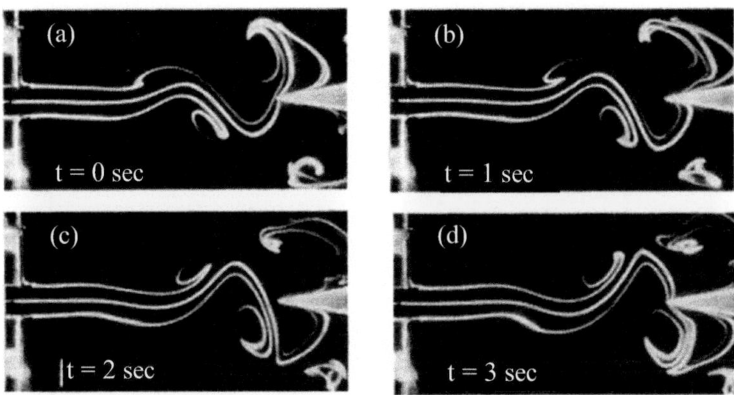

Fig. 220 Visualized flow pattern of streak lines of edge-tone oscillation ($b_0 = 10.0$ mm, $h/b_0 = 9$, $Re = 292$, $T = 3.1$ s, mode II)

is significantly different from that in Fig. 219, and two nodes are seen in the flow direction. The oscillation modes shown in Figs. 219 and 220 are called modes I and II, respectively.

(a.3) Oscillatory frequency

Figure 221 shows an example of measuring the frequency f when the nozzle-edge distance h is changed quasi-statically by rack and pinion mechanism at a relatively small Reynolds number ($Re = 250 = $ const.). The white circles and black circles in the figure are the measured values when the value of h was gradually increased and decreased, respectively.

The followings were found.

(1) The jet oscillation phenomenon occurs at edge distances h greater than a certain value.
(2) A jet flow has multiple modes of vibration (stage mode, mode I, mode II etc.), and a jump in frequency with a hysteresis appears between adjacent modes of vibration.

Figure 222 shows the measured value of f (mode I results only) for a relatively large Reynolds number. f is proportion to u_0. In addition, when u_0 was gradually increased while keeping h constant, a jump from mode I to mode II was observed for all h shown in Fig. 222 in the range of $Re \lesssim 1.0 \times 10^3$.

(b) Theoretical model of jet oscillation

As mentioned above, Woolly and Karamcheti inferred that the oscillatory frequency f was determined in term of the fluctuation with the wave number having a spatial growth in which the amplitude was maximum at an edge. If the theoretical model was made of a parallel jet without spreading under this inference, f was decided independently of edge position. However, this fact is in conflict with the experimental

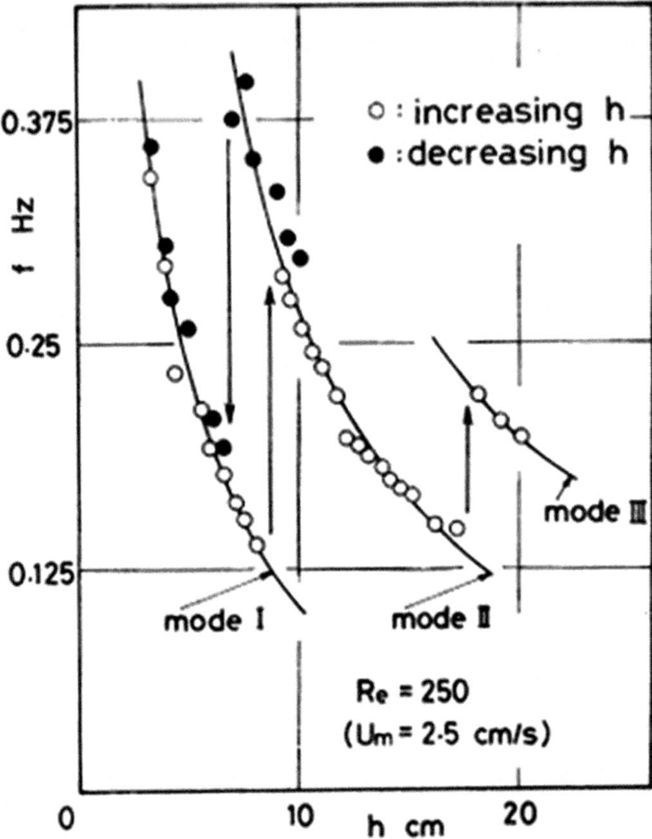

Fig. 221   Oscillation frequency ($Re = 250$)

results (Fig. 221). They proposed a theoretical model considering a spreading of the jet by the reasoning above mentioned. But they did not verify the theoretical results for f experimentally. Now, when we make a theoretical model for the oscillation of a jet, they also summarize that the oscillation of a jet can be explained by the occurrence of fluctuation. But the present investigations entirely disagree with the opinion of Woolley and Karamcheti about the mechanism to decide the oscillation frequency which is fundamental fact on this phenomenon. The explanations for the qualitative characteristic of oscillation and the mechanism to decide the oscillatory frequency can be well obtained from the investigation about the fluctuation under the parallel jet without a spreading.

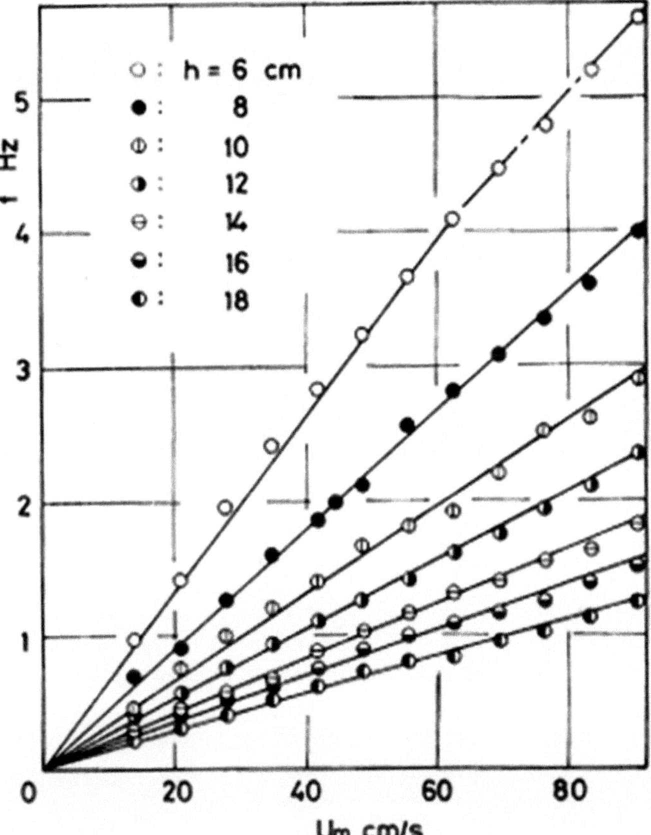

Fig. 222 Oscillation frequency ($Re = 1.4 \times 10^2$–9×10^3)

(b.1) **Preliminary considerations for model configuration**

When the oscillation of the jet in the nozzle edge system is examined by dimensional analysis on the assumption that the nozzle and edge shapes are geometrically similar, the following equation is obtained.

$$St = F(Re, h/b_0) \tag{188}$$

where, St: Strouhal number ($= fb_0/u_\mathrm{m}$)

From the results of Eq. (188) and Fig. 221, that is, the proportional relationship between f and u_m, it can be seen that St is hardly affected by Re in a relatively large

Reynolds number range. Therefore, the effect of working fluid viscosity on f appears to be secondary. Naturally, consideration of viscosity is necessary when considering critical conditions for oscillation occurrence. Therefore, the following theoretical model is constructed on the assumption that the working fluid is inviscid.

(b.2)   Consideration of parallel jet

(b.2.1)   Waveform disturbance in flow field

Under the assumption that the working fluid is inviscid and incompressible, the flow field of a two-dimensional parallel jet is expressed as follows, using Euler's equation of motion and the equation of continuity as the fundamental equation group.

$$u = \overline{u}(y) + \tilde{u}(x, y, t) \tag{189}$$

$$v = \tilde{v}(x, y, t) \tag{190}$$

$$p = \tilde{p}(x, y, t) \tag{191}$$

where, u, v: velocity components in the x and y directions, respectively, $\overline{u}(y)$: velocity component of the basic flow in the x direction, p: kinematic pressure ($=$ pressure/density), $\sim$: disturbance.

Equations (189)–(191) are used in the above group of basic equations, and assuming that the magnitude of the disturbance is a much small amount relative to that of the main flow, a set of linear differential equations for the disturbance components such as $\tilde{u}$, $\tilde{v}$, and $\tilde{p}$ can be obtained. The above equation group is converted to non-dimensional form using the following dimensionless quantities:

$$\overline{u}_n = \frac{\overline{u}}{U_0}, \hat{u}_n = \frac{\hat{u}}{U_0}, \hat{p}_n = \frac{p}{U_0^2}, \alpha_n = \alpha\delta(\alpha : \text{wave number}),$$

$$c_n = \frac{c}{u_0}(c : \text{phase velocity}), x_n = \frac{x}{\delta}, y_n = \frac{y}{\delta}, t_n = \frac{tU_0}{\delta}$$

where, U_0: maximum velocity of jet, δ: characteristic width of jet, (for example, momentum thickness)

Furthermore, assuming the following complex solution for the disturbance,

$$\hat{u}_n = U_n(y_n) \cdot e^{i\alpha_n(x_n - c_n t_n)} \tag{192}$$

$$\hat{v}_n = V_n(y_n) \cdot e^{i\alpha_n(x_n - c_n t_n)} \tag{193}$$

$$\hat{p}_n = P_n(y_n) \cdot e^{i\alpha_n(x_n - c_n t_n)} \tag{194}$$

If these are used for the non-dimensional linear equations mentioned above, the following equations are obtained.

$$i\alpha_n(\overline{u}_n - c_n)U_n + (d\overline{u}_n/dy_n)V_n + i\alpha_n P_n = 0 \tag{195}$$

$$i\alpha_n(\overline{u}_n - c_n)V_n + (dP_n/dy_n) = 0 \tag{196}$$

$$i\alpha_n U_n + (dV_n/dy_n) = 0 \tag{197}$$

where, $^\wedge$ means a complex solution, and the subscript n represents a dimensionless quantity. Also, U_n, V_n are abbreviations for $U_n(y_n)$, $V_n(y_n)$, respectively. Furthermore, α_n, c_n, and i are dimensionless wave number of disturbance, dimensionless phase velocity of disturbance, and imaginary unit, respectively.

Next, from Eqs. (195)–(197), by eliminating U_n and P_n, we obtain the following equation.

$$\frac{d^2 V_n}{dy_n^2} = \left[\left(\frac{1}{\overline{u}_n - c_n} \cdot \frac{d^2\overline{u}_n}{dy_n^2}\right) + \alpha_n^2\right]V_n \tag{198}$$

Equation (198) is Rayleigh equation and if the function of velocity profile $\overline{U}_n$ os determined the following boundary condition for fluctuation is assumed, the eigenfunction V_n which satisfies Eq. (199) is given for any α_n corresponding to the eigen value c_n.

$$V_n = 0 \quad \text{at } y_n = \pm\infty \tag{199}$$

This V_n and is Eqs. (195)–(197) give the values of U_n and P_n, while U_n, P_n and Eqs. (192)–(194) give the complex solutions of $\hat{U}_n$, $\hat{V}_n$, $\hat{P}_n$. Real solutions for fluctuation are given by $\tilde{U}_n = (1/2)(\hat{u}_n + \hat{u}_n^*)$ and so on because the complex conjugates $\hat{u}_n^*$, $\hat{v}_n^*$ and $\hat{p}_n^*$ of $\hat{u}_n$, $\hat{v}_n$ and $\hat{p}_n$ are also the solutions group for fluctuation. Now, as a result of expressing the velocity distribution of the main flow by the following equation and examining the above eigenvalue problem, it was found that Eq. (198) has a complex eigenvalue c_n for the complex value of α_n.

$$u_n = \text{sech}^2 y_n \tag{200}$$

That is, for $\alpha_n = \alpha_{nr} + i\alpha_{ni}$, the following equation is obtained.

$$c_n = c_{nr} + ic_{ni} \tag{201}$$

As can be seen from Eqs. (192)–(194), solutions such as $\hat{v}_n$ take the form of spatial growth or decay according to the positive or negative value of $-\alpha_{ni}$, and growth or decay with time according to the positive or negative value of the imaginary part

$[\mathrm{Im}(\alpha_n c_n) = \alpha_{nr} c_{ni} + \alpha_{ni} c_{nr}]$ of $\alpha_n c_n$. However, in the following discussion, we will ignore the passage of time in which the amplitude of the disturbance gradually increases, and focus on the steady oscillation, and will limit our discussion to case $\mathrm{Im}(\alpha_n c_n) = 0$. That is, if it is expressed as

$$\alpha_n c_n (\text{real number}) = \omega_n = 2\pi f_n = 2\pi (1/T_n) \tag{202}$$

where, ω_n, f_n and T_n represent angular frequency, frequency, and period of disturbance (all dimensionless), respectively.

Note that there are two cases in which $V_n(y_n)$ is an even function of y_n and an odd function of y_n in the real solution of the disturbance, but if x_n is viewed at a constant position, the oscillation phenomenon in which the jet oscillates in the y_n coordinate direction as a whole can be seen. In the present case, we will consider only the case of even functions. Now, Fig. 223a, b shows the result of solving the aforementioned boundary value problem under the constraint $\alpha_n c_n = $ real number. The solution method is numerical integration using a computer. In the calculation, Eq. 199 was approximated as $V_n = 0$ at $y_n = \pm 5$, but this had little effect on the results. In addition, when integrating, $V_n = \mathrm{A}(1 + \mathrm{i})$ and $dV_n/dy_n = 0$ were set to initial values at $y_n = 0$. Here, A is an arbitrary constant. Figure 223a shows the real and imaginary parts of α_n and c_n with ω_n as the horizontal axis. From Fig. 223a, we know that in the range $\alpha_{nr} < 2$ (in the figure, a small circle is marked at position $\alpha_{nr} = 2$), the disturbance is spatially amplified (spatially unstable), but in the range of $\alpha_{nr} > 2$, it becomes $c_{ni} = 0$, $\alpha_{ni} = 0$, and α_n, c_n are real numbers. Next, Fig. 223b shows the calculation results of the eigenfunction when the above arbitrary constant A is set to A $= 1$, with y_n as the horizontal axis and ω_n as the parameter.

The left half of Fig. 223b shows only the range of $y_n < 0$ with real part V_{nr} of $V_n(y_n)$ (solid line). $V_n(y_n)$ in the range of $y_n > 0$, is symmetrical to that on the left half with respect to the vertical axis. In Fig. 223, the range of $\alpha_n > 2$ corresponds to the range of $\omega_n > 3/4$, and in this range $V_n(y_n)$ also becomes a real function of y_n, and the corresponding curve directly represents the velocity distribution shape of $\tilde{v}_n$. Note that the right half of Fig. 223b is an example of the imaginary part V_{ni} of

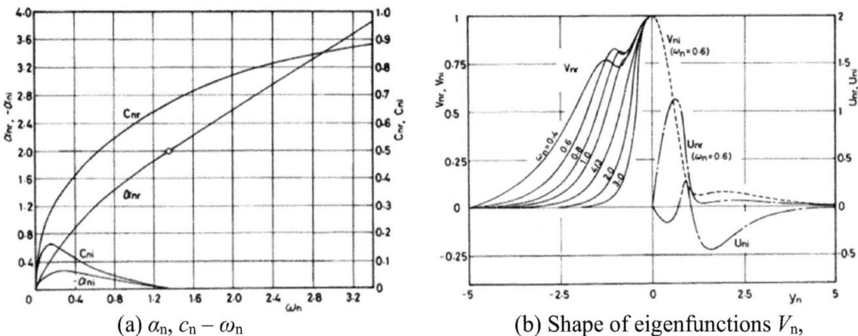

(a) α_n, $c_n - \omega_n$ (b) Shape of eigenfunctions V_n,

Fig. 223   Solution of boundary value problem under $\alpha_n c_n = $ real number

$V_n(y_n)$ (broken line, only the right half is shown). The left half is symmetrical about the vertical axis, and is an example of calculating the real part, imaginary part U_{nr}, and U_{ni} of $U_n(y_n)$ (dashed line, the left half is symmetrical about the origin).

(b.2.2) Behavior of a jet with disturbance and mechanism determining oscillatory frequency

As can be seen from Fig. 223a, $c_{nr} > 0$ and the disturbance represented by Eqs. (192)–(194) are waves that propagate in the downstream direction of the jet. Here, the flow field of the jet including this disturbance was analyzed analytically using the approximations described below. As a result, it is shown that the behavior of the jet flow during oscillation can be qualitatively explained well as the characteristics shown by the above-mentioned flow field. This examination also includes speculation regarding the theoretical prediction of the oscillatory frequency, which is an unresolved and important issue. First, let's focus on the streamlines of the flow field.

The streamline is obtained by considering Eqs. (189) and (190) and integrating the following differential equation.

$$\frac{dx_n}{u_n}\left[=\frac{d_{xn}}{\bar{u}_n+\tilde{u}_n}\right]=\frac{dy_n}{v_n}\left[=\frac{dy_n}{\tilde{v}_n}\right] \tag{203}$$

where, $u_n = u/U_0$, $v_n = v/U_0$

u_n and v_n are generally not obtained in the form of analytical solutions, and the streamline equation cannot be obtained in a closed form. However, the following approximation allows analytical consideration to some extent. In other words, considering the assumption that $\tilde{u}_n$ is the smallest quantity compared to $\bar{u}_n$, and the fact that $-\alpha_{ni} \lesssim \alpha_{nr}$ and $c_{ni} \lesssim c_{nr}$ hold true except for the extremely small range of ω_n from Fig. 223a, in addition, if we ignore $\tilde{u}_n$, $-\alpha_{ni}$, and c_{ni} with respect to $\bar{\bar{u}}_n$, α_{nr}, and c_{nr}, respectively, and consider only the vicinity of the jet center, $\tilde{v}_n$ can be expressed in the following form.

$$\tilde{v}_n = A_n \cos\alpha_n(x_n - c_n t_n) \tag{204}$$

where, A_n: amplitude, $c_n \fallingdotseq c_{nr}$, $\alpha_n \fallingdotseq \alpha_{nr}$

Next, when Eq. (204) is used in Eq. (203) to find the streamline passing through the center ($x_n = 0$, $y_n = o$) of the nozzle exit, the following equation is obtained.

$$y_n = \frac{2A_n}{\alpha_n} sin\frac{\alpha_n x_n}{2} cos\left(\frac{\alpha_n x_n}{2} - \alpha_n c_n t_n\right) \tag{205}$$

Equation (205) shows that the centerline of the jet oscillates sinusoidally with respect to time. Also, from Eq. (203), it can be seen that the streamline above always passes through a fixed point x_{nk} ($k = 1, 2, 3, \ldots$, hereafter, it is called node) on the x-axis, regardless of time t_n.

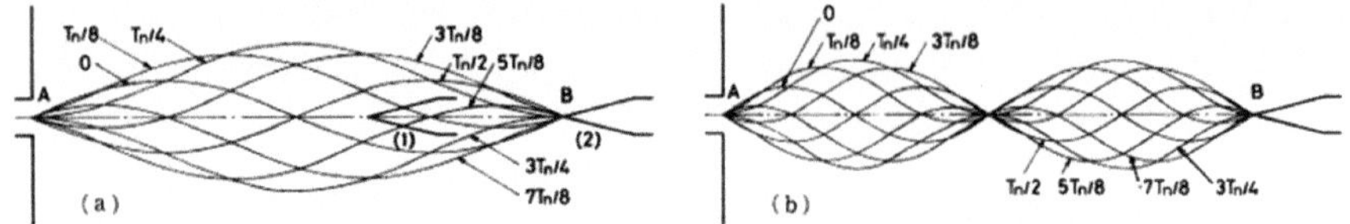

Fig. 224 Behavior of a streamline passing through the center of the nozzle exit

$$x_{nk} = k\frac{2\pi}{\alpha_n} = k\lambda_n \tag{206}$$

where, $\lambda_n = \lambda/\delta$, λ: wave length

This leads to the following important inferences regarding the oscillation frequency.

(1) Figure 224a shows the shape of the streamline for each $T_n/8$ ($T_n = U_0 T/\delta$, T: period). For example, if an edge exists at position (1) in the figure, the oscillation of the wavelength (i.e., frequency) corresponding to the temporal change of the streamline as shown in the figure will be disturbed by the edge and will perform a steady periodic motion. That would be impossible. On the other hand, an edge located immediately after a node, at position (2) in the figure, can be considered to be a boundary condition that does not impede periodic oscillations between positions A and B.

That is, when the edge distance is h_n ($= h/\delta$), disturbance $\lambda_n = h_n$ [$k = 1$ in Eq. (206)] is selected as the fundamental mode (mode I), and it is considered that this is the dominant mode in which the jet oscillates.

(2) Similarly, as shown in Fig. 224b, disturbances of wavelengths that satisfy $h_n = k\lambda_n$ can also occur, indicating the possibility of oscillation in higher-order modes.

(3) If the edge distance h_n is fixed, the corresponding λ_n (therefore, α_n, c_n) is determined from the above, and the frequency f is determined from Eq. (202) as follows.

$$f = \frac{U_0}{\delta} \cdot f_n = \frac{c_n \alpha_n}{2\pi\delta} U_0 \tag{207}$$

Equation (207) shows that f is proportional to U_0, which is in good agreement with the experimental results (Fig. 222).

Next, for comparison with the visualization experiment, we will examine the characteristics of the streak line passing through the center of the nozzle exit.

The equation for this streak line is determined by Eq. (209) using the equation for the streak line found by integrating the following differential Eq. (208) for the path line under various approximations.

$$\frac{dx_n}{u_n}\left[=\frac{d_n}{\bar{u}_n+\tilde{u}_n}\right]=\frac{dy_n}{v_n}\left[=\frac{dy_n}{\tilde{v}_n}\right]=dt_n \tag{208}$$

$$y_n=\frac{2A_n}{\alpha_n(1-c_n)}\cdot\sin\frac{\alpha_n x_n(1-c_n)}{2}\cdot\cos\left(\frac{\alpha_n x_n(1+c_n)}{2}-\alpha_n c_n t_n\right) \tag{209}$$

Equation (209) represents that the streak line, unlike the stream line that becomes a node at $x_n = k\lambda_n$, oscillates with a finite amplitude at that position. Note that the shape of the streak line determined from Eq. (209) and its behavior over time were qualitatively consistent with the experimental results (Figs. 219 and 220). However, this comparison yields to the case of the theoretical model described in the next section.

(b.2.3) Theoretical model of jet oscillation

In the previous section, we approximated the base flow as a parallel jet and investigated the nature of the disturbance that occurs there. As a result, we showed that various characteristics of the jet oscillation phenomena in the jet-edge system can be qualitatively explained as the behavior of the flow field including these disturbances. Here, we will describe a more accurate theoretical model of jet behavior based on the results of the previous section and constructed by taking into account the effect of gradual changes in jet velocity distribution in the flow direction due to diffusion.

First, the basic flow is expressed by the following velocity distribution equation by Goertler, which has been well verified experimentally.

$$\bar{u}=U_0\cdot\text{sech}^2\frac{\sigma y}{x+x_0}=U_o\cdot\text{sech}^2\eta \tag{210}$$

where,

$$U_0=[3J\sigma/\{4\rho(x+x_0)\}]^{1/2},\quad x_0=\sigma b_0/3 \tag{211}$$

J: jet momentum, σ: spreading parameter ($=7.67$, Reichardt's experimental value for two-dimensional jet, x_0: hypothetical origin distance

We treat Euler's equation of motion and continuity equation as a group of basic equations in the same way as in the previous section. At that time, when making various quantities dimensionless, if we take the value $[\delta=(x+x_0)/\sigma]$ of y where the similar variable $\eta\left[=\sigma y/(x+x_0)\right]$ in Eq. (211) is 1 as δ, the equation describing the eigenfunction V_n of the disturbance in the y direction becomes Eq. (198). obtained in a similar form. Therefore, the results shown in Fig. 223a, b can be applied to this case as well. Next, it is assumed that the dimensional quantities α, c, etc. in the equation corresponding to the various quantities α_n, c_n, etc. gradually change with x as a parameter. This means that for any specified value of x, each minute section in the vicinity is treated as a parallel jet, and in that sense it can be called a quasi-parallel jet. Next, using this model, a streamline passing through the center of the nozzle exit is determined, and this is compared with the analysis result (Eq. (205)) in the previous section. The streamlines are calculated as follows. Assuming that the nozzle width b_0 is known, first, the values of the frequency f and the jet flow rate Q are appropriately selected. Assuming the velocity distribution form of the jet flow at the nozzle exit, the value of J in Eq. (211) is determined, and the flow field of the basic flow is determined from Eqs. (210) and (211). Note that the velocity distribution at the nozzle exit was assumed to be uniform, assuming that the Reynolds number was relatively large. Next, the equations for this basic flow and disturbances $\tilde{u}$ and $\tilde{v}$ (represented in a dimensional form) were used in Eq. (203) (a dimensional equation with the subscript n removed), and it was numerically integrated to determine the behavior of the streamline passing through the center of the nozzle exit. At this time, the following equation was used for $\tilde{v}$.

$$\tilde{v} = A_0 \cdot e - \int_0^x \alpha_i d_x Re\{V(y)e^{i(\alpha_r x - \omega t)}\} \tag{212}$$

In addition, the corresponding equation [determined from Eqs. (195)–(197)] was used for $\tilde{u}$. Note that $\alpha_i = \alpha_{ni}/\delta$, $\alpha_r = \alpha_{nr}/\delta$ and $V(y) = U_0 V_n(y_n)$ are determined from the relationship shown in Fig. 223. Here, A_0 is a dimensionless number specifying the amplitude of the disturbance at $x = 0$, and $Re(\Phi)$ is the real part of the function Φ in parentheses. An example of calculating streamlines is shown in Fig. 225 Although the nodes are not as clear as in the case of Fig. 224, there are positions where the streamlines almost converge, and these correspond to the nodes of $k = 1, 2, 3 \ldots$.

(c)   Verification of the theoretical model based on experimental results

(c.1)   oscillatory frequency

Figure 226 shows the relation between oscillatory frequency fb_0/U_m and edge distance h/b_0 and the comparison of experimental results and theoretical results shown in the previous chapter. The white and black circles in the figure correspond to the measured values in Figs. 221 and 222. Furthermore, the half-white circles in the figure are measured values obtained by conducting an experiment similar to that in Fig. 221 in the range of $Re \fallingdotseq 7.0 \times 10^2 \text{\sim} - 5.5 \times 10^3$. The figure also includes part of the experimental results (triangle mark, $Re \fallingdotseq 5.0 \times 10^2 \text{\sim} - 2.2 \times 10^3$) of Hayashi et al., who used air as the working fluid using a nozzle with a relatively long parallel section, similar to the present study. We also wrote some of his Rockwell experimental results (square mark, $Re \fallingdotseq 7.0 \times 10^2 \text{\sim} - 7.5 \times 10^3$, working fluid: water) in the case where there is no parallel part and the nozzle has a strong three-dimensionality with an aspect ratio of unity. To create a theoretical line, first perform the streamline calculation described at the end of the previous chapter for various sets of specified values of f and u_m, and then find the coordinate value x_k ($k = 1, 2, 3 \ldots$) of the node corresponding to each value of f and u_m. For h ($= x_k$), the relationship between St ($= fb_0/u_m$) and h/b_0 was determined assuming that when the average flow velocity is u_m, oscillation in mode k of frequency f occurs.

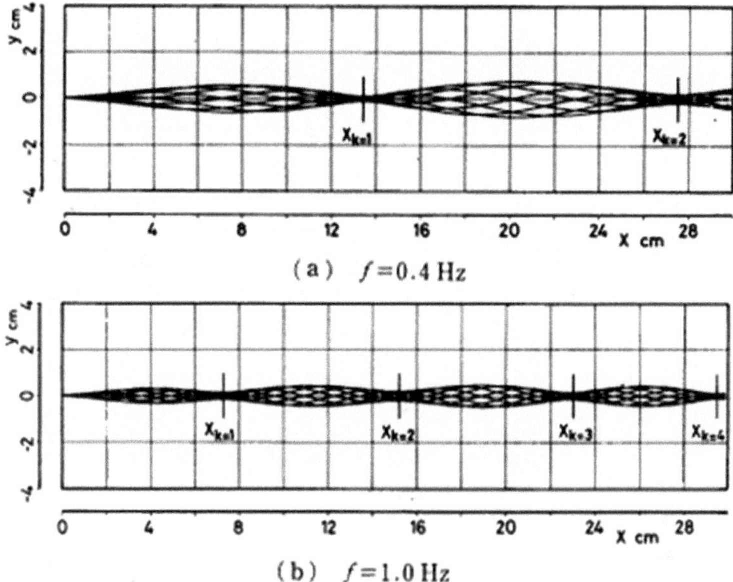

Fig. 225 Behavior of a streamline passing through the center of the nozzle exit by quasi-parallel jet model ($u_\mathrm{m} = 20$ cm/s, $A_0 = 0.05$)

From Fig. 226, it can be seen that the theoretical results give a good approximation to the experimental values, except when the Rynolds number is small (black circles in the figure). It was confirmed through a flow visualization experiment that the jet oscillated regularly even when the edge was excluded, but the frequency of the oscillation was completely different from the result shown in Fig. 226.

(c.2) Behavior of jet flow

Here, we examine the validity of the model from the perspective of jet behavior.

Figure 227 shows the streak line of mode I calculated theoretically using the various values of u_m, b_0 and f corresponding to the experiment shown in Fig. 219. However, the value of the dimensionless number A_0, which determines the amplitude of the disturbance, was appropriately selected. Furthermore, the distance between each point forming the streak line corresponds to the density of the dye in the visualization experiment. The time intervals in each of Fig. 227a–e correspond to those of Fig. 219a–e.

In addition, Fig. 228 shows similar calculation results for the streak line of mode II corresponding to the experiment in Fig. 220. However, the streak lines are drawn using a disturbance equation that was derived without considering changes in the base flow due to the presence of edges. Therefore, comparison of the shape with experimental results is meaningful only in the section from the nozzle exit to the edge tip. In both Figs. 227 and 228, the calculated results show good qualitative

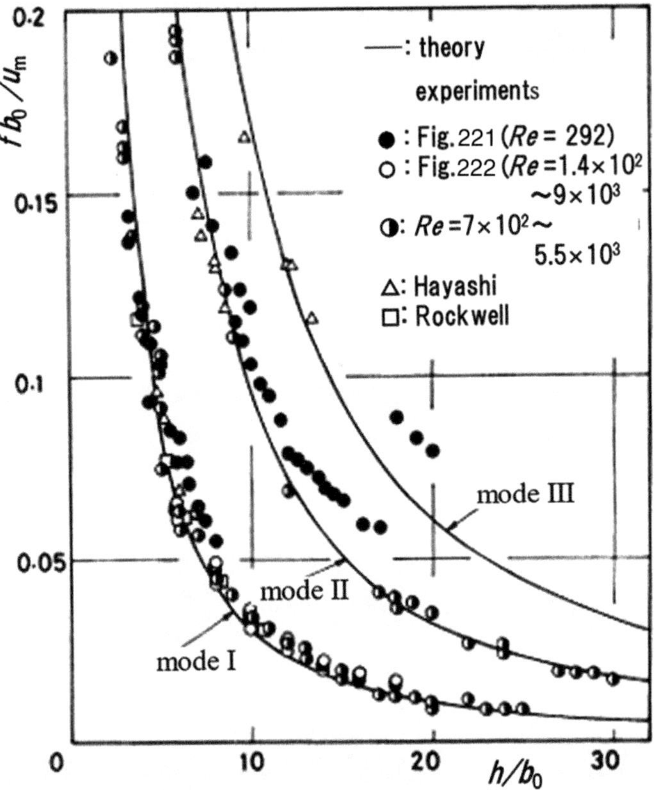

Fig. 226 Comparison between experiments and theoretical results of oscillatory frequency

agreement with the experimental results. This result, along with the results in the previous section, can be said to support the validity of the model.

The theoretical results of the streak lines were obtained as follows.

First, for $\bar{u}$, $\tilde{u}$ and $\tilde{v}$ in Eq. (208) (which has dimensions), use the equation for $\bar{u}$, $\tilde{u}$ and $\tilde{v}$ used in the calculation of the streamline in Section (b-2.3). Next, the position at $t = t_s = $ const. of the fluid particles that momentarily pass through the center of the nozzle exit and the positions on both sides of the nozzle exit is determined by numerical integration.

Here, theoretical and experimental examination on oscillatory phenomena of a jet in a jet-edge system, edge tone oscillation phenomena, were made, and the theoretical model explaining the jet behavior and oscillatory frequency was proposed. However, this does not include consideration of the contribution of edges to the oscillation mechanism of the system. The oscillation mechanism of edge tone is thought to be due to the alternating pressure fluctuations at the edge position caused by the jet impinging on the edge, which together with the stability of the jet itself described in this model contribute to the oscillation phenomenon.

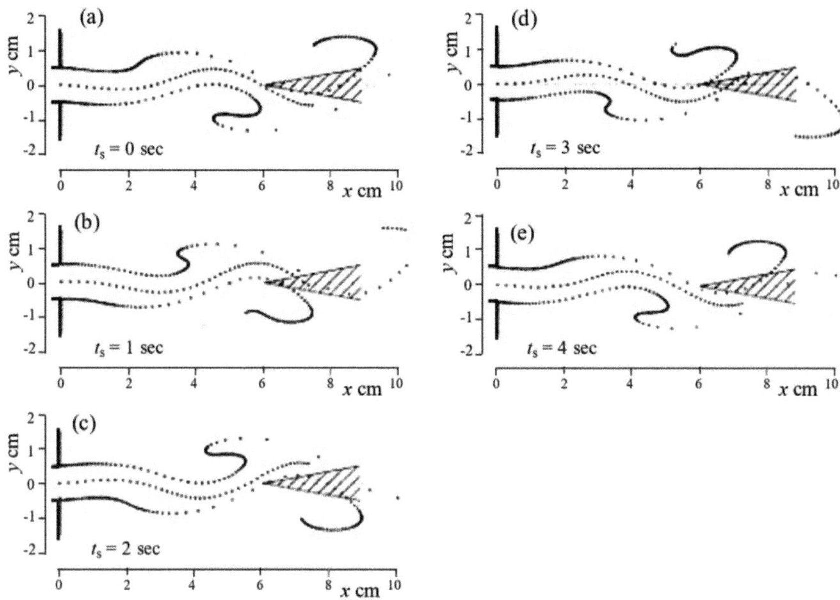

Fig. 227 Calculation results of vein lines (mode I) using the quasi-parallel jet model ($b_0 = 10.0$ mm, $u_m = 2.5$ cm/s, $T = 4.5$ s ($f = 0.22$ Hz), $A_0 = 0.05$)

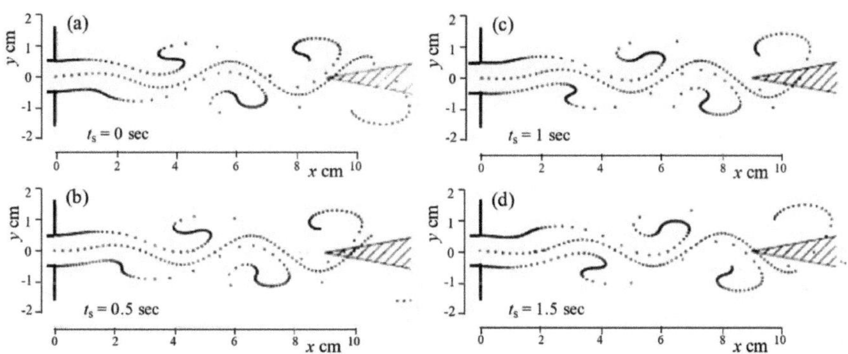

Fig. 228 Calculation results of vein lines (mode II) using the quasi-parallel jet model ($b_0 = 10.0$ mm, $u_m = 2.5$ cm/s, $T = 1$ s ($f = 0.32$ Hz), $A_0 = 0.05$)

7.2 Fluidic Oscillation

* Cavity oscillation (Interaction of external flow and cavities)

Rockwell and Naudascher (1978) classified the oscillating phenomena of flow, flow-induced vibration, which arise as a result of the interaction of external flow and cavities of various shapes, as shown in Fig. 229. The cavity oscillation was classified

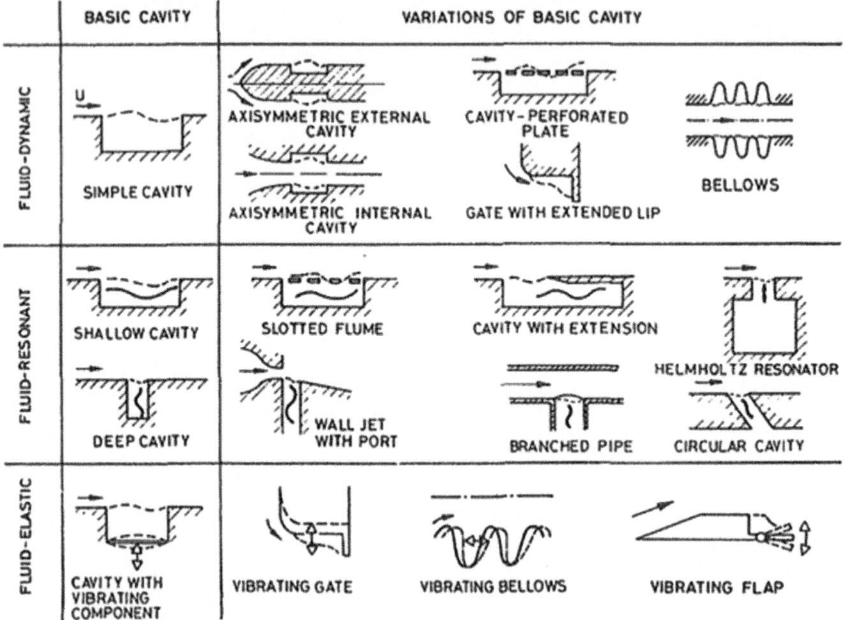

Fig. 229 Classification of cavity oscillation (Rockwell and Naudascher 1978)

into three types of fluid-dynamic, fluid resonant and fluid elastic groups. Where, the shape of cavity of three types of simple cavity, shallow or deep cavity, and cavity with vibrating component.

* Cavity oscillation (Interaction of internal flow and cavities, Cavity type fluidic oscillation)

Oscillation phenomena also occur in confined jets depending on the shape of the flow passage. As shown in Fig. 230a, when the jet impinges on an opposing wall placed at a certain distance, the jet bifurcates and oscillates from side to side while surrounding the vortex in the opposite direction, and exits from the exit. Similar ones include (b), (c), and so on.

In addition to the above, there are various oscillation phenomena that occur in confined jets, and they have been studied as fluidic oscillations. As an example, feedback type fluidic oscillator is shown in Figs. 231 and 249. The jet from the nozzle attaches to one of the side walls and flows, and part of its output passes through the feedback loop to the base of the attached jet, increasing the pressure there. As a result, the attachment of the jet is switched. After the same process, the attachment of the jet stream is switched, and an oscillation phenomenon occurs in the jet stream.

Some of others are described below.

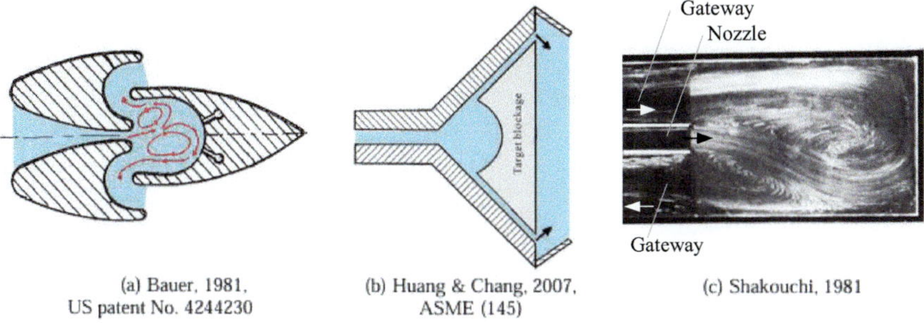

(a) Bauer, 1981,
US patent No. 4244230

(b) Huang & Chang, 2007,
ASME (145)

(c) Shakouchi, 1981

Fig. 230 Cavity jet fluidic oscillators (Ghanami and Farhadi 2019)

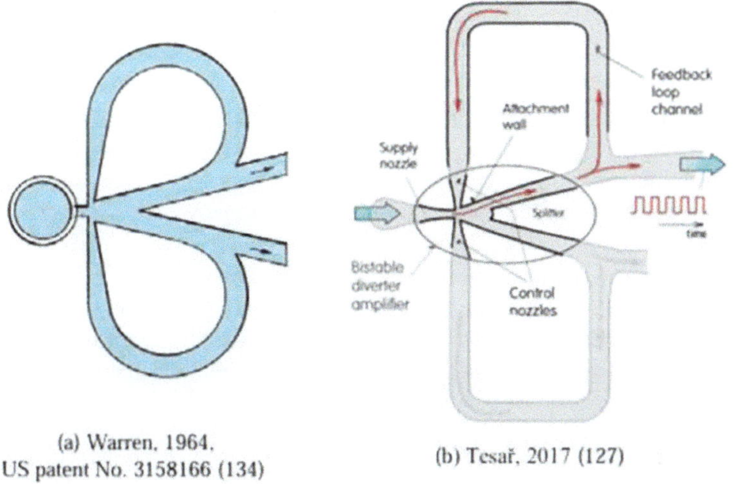

(a) Warren, 1964,
US patent No. 3158166 (134)

(b) Tesař, 2017 (127)

Fig. 231 Feedback type fluidic oscillator (Ghanami and Farhadi 2019) **a** Feedback fluidic oscillator, **b** Schematic of switching mechanism

7.2.1 New Fluidic Oscillator Without Control Port and Feedback Loop

Recently, flowmeters using various oscillatory phenomena of fluid flow, hydraulic oscillator type meters are of three types. Vortex precession flowmeters use the precession of the axis of a rotating flow in an expanded pipe to measure flow. Fluidic flowmeters use fluidic oscillators to effect measurement. Karman vortex flowmeters use the phenomenon of the Karman vortex arising in wake of a body. These hydraulic oscillator type flowmeters have advantages of high range ability, using digital output of the oscillatory frequency proportional to volume flow rate, high stability, etc.

Fluidic flowmeter is of three types. These are those which use feedback oscillation, relaxation oscillation (two control ports are connected together), and edge-tone oscillation (Fig. 232a–c). Previously, the configuration of the nozzle in fluidics was

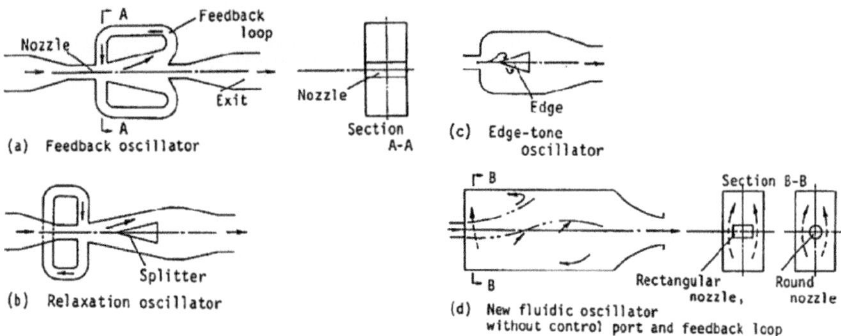

Fig. 232 Fluidic oscillators (Shakouchi 1991a, b)

almost always the rectangular, Two-dimensional type in the sense that the height, a, of the nozzle equals the height of the flow passage, H (Fig. 232a). Shakouchi (1991a, b) pointed out the fact that the regular oscillation of a three-dimensional, rectangular jet which occurs when the jet squirts into a suddenly enlarged flow passage from rectangular nozzle which is set smaller than the flow passage ($a/H < 1$, Fig. 232), depends on the shapes of the nozzle, flow passage, and jet velocity. He also inferred that a simpler fluidic oscillator, flowmeter, without control port and feedback loop could be constructed by using the above-mentioned phenomenon.

Here, the conditions which causes the oscillation and relations between the oscillatory frequency and the various conditions such as the dimensions of nozzle and flow passage or jet velocity are made clear by experimental results and dimensional analysis. The mechanism of oscillation and the range of oscillatory frequency are explained qualitatively from the observation of an oscillatory flow pattern.

(a) **Test section and procedure**

Figure 233 shows a test section, new fluidic oscillator without control port and feedback loop. The test section consists of a nozzle, side walls, an exit contraction, and upper and lower end plates, which are transparent acrylic plate. The width of the nozzle is $W = 5.0$ mm, and its height can be varied from $a = 5.0$mm to 30.0 mm by setting the spacers at various height, h, at the upper and lower parts of the nozzle. The flow passage is 30.0 mm high and 150.0 mm long. The width of the flow passage can be varied from $B = 30.0$ mm to 100.0 mm. The oscillatory frequency of the jet was obtained from the pressure measurements using the total pressure tube with outer and inner diameters of 1.0 mm and 0.6 mm, respectively, which is set at the half depth of the flow passage, the point p in the figure. The pressure transducer is semi-conductor type (JTECT, PD-104K-0.1FW). As example of measurements is shown in Fig. 234. In this case, working fluid is air.

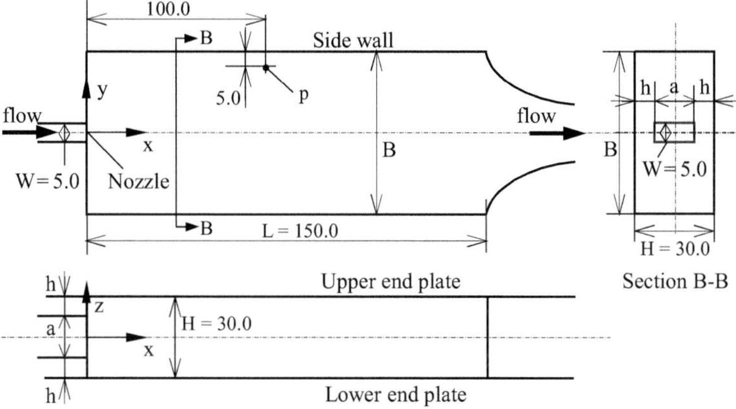

Fig. 233 Test section

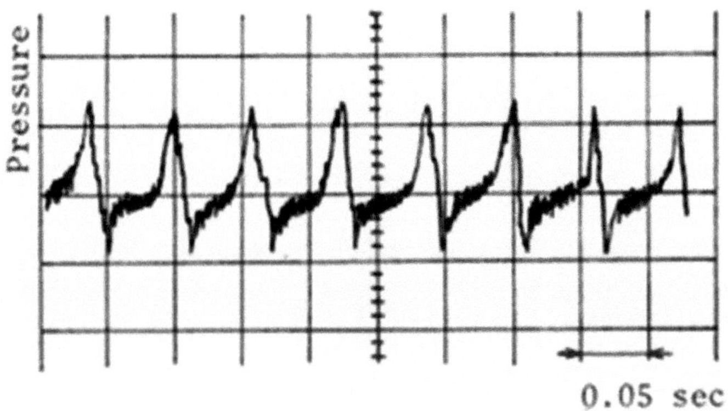

Fig. 234 Pressure fluctuation ($a/H = 1/3$, $B/W = 12$, $Re = 1.68 \times 10^4$, $f = 16.0$ Hz, working fluid: air)

(b) **Mechanism of jet oscillation and flow pattern**

Figure 235 shows sketches of a series of periodic oscillatory flow patterns near the plane at the middle height ($z = 0$) of the flow passage at every 1/6 period based on the visual observation of the oscillatory water flow. In this case, the height of nozzle is $a/H = 1/2$, the width of flow passage is $B/W = 12$, and Reynolds number is $Re = 8.93 \times 10^3$, the oscillatory frequency is $f = 0.38$ Hz. The solid lines in the figure roughly show the regions of vortex, and the thin and thick arrows indicate the flow direction and the jet direction, respectively. As shown in Fig. 235a, the jet from a rectangular nozzle with $a/H < 1$ attaches to the side wall I forming the vortex A (a low-pressure region). Vortex B is also formed between the side wall I and II. Next, the flow [the dotted arrow lines in Fig. 235a, b] from the unattached side comes into vortex A through the throat near nozzle exit between the upper (or lower) edge of the

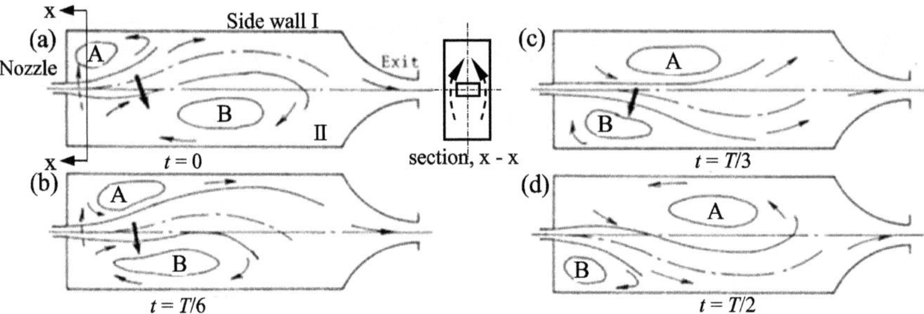

Fig. 235 Sketches of flow pattern in oscillation ($a/H = 1/2$, $B/W = 12$, $Re = 8.94 \times 10^3$, $f = 0.38$ Hz, working fluid: water)

jet and upper (or lower) end plate. The pressure within vortices A and B increases and decreases, respectively. As a result, attachment of the jet switches from the side wall I to the side wall II. Vortex A drifts downstream and vortex B drifts upstream (Fig. 235b–d). The observed circumstances are repeated and continuous oscillation occurs. This oscillatory phenomenon is called relaxation oscillation.

(c) **Condition of oscillation**

Figure 236 shows the flow pattern map under $Re = (0.5 \sim - 4) \times 10^4$. Domains (a), (b) and (c) in the figure show the regions which feature the stable, unstable and no oscillations, respectively. It is inferred that a certain dimension of the distance, $h[= (H - a)/2]$, from the upper (or lower) edge of the nozzle to the upper (or lower) end plate and the width of flow passage, H, with reference to the nozzle is required in order to cause the regular jet oscillation.

(d) **Oscillatory frequency**

Oscillatory frequency, f, is considered to have a considerable dependence on the configuration of the nozzle, flow passage, and jet velocity, respectively.

Figure 237 shows the relation between f and mean velocity, u, and flow rate, Q, at the nozzle exit Reynolds number, Re. The nozzle height is $a/H = 1/2$. f varies linearly with Q, and increases with decreasing B/W values. This is caused by the fact that the mass of fluid which must move to oscillate decreases as the width of the flow passage decrease. Using above results, oscillatory frequency is expected as follows.

$$f = k_1 Q + k_2 \tag{213}$$

where, k_1 and k_2 are derived from experimental results.

Consider the oscillatory frequency using dimensional analysis. Adopting the width of the nozzle, W, mean velocity, u, at the nozzle exit and kinematic viscosity, v, as the basic quantities, and applying dimensional analysis to this oscillatory phenomenon, we obtain

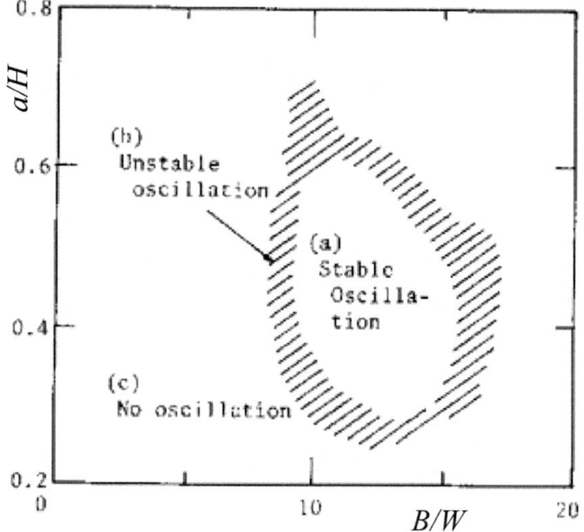

Fig. 236 Flow pattern map $[Re = (0.5 \sim - 4) \times 10^4]$

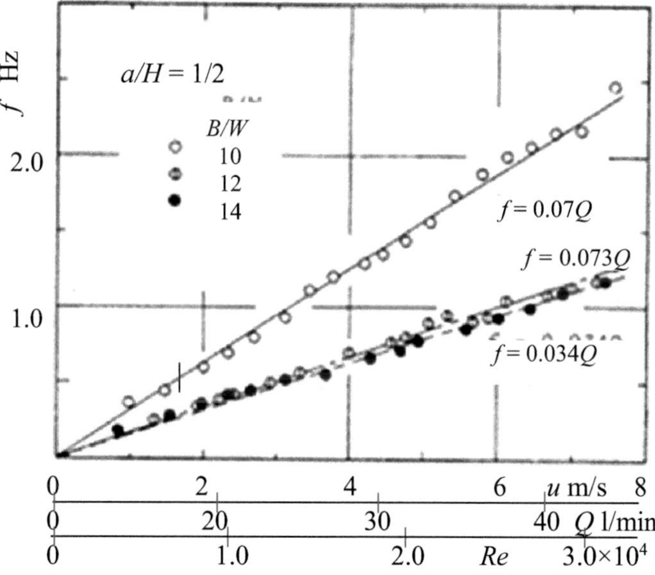

Fig. 237 Oscillatory frequency, $a/H = 1/2$

$$F\left(\frac{fW}{u}, \frac{uW}{v}, \frac{a}{W}, \frac{H}{W}, \frac{h}{W}, \frac{B}{W}, AR\right) = 0 \tag{214}$$

where, F is a functional symbol.

Multiplying the first variable in Eq. (214) by the second variable gives

$$F\left(\frac{fW^2}{u}, \frac{uW}{v}, \frac{a}{W}, \frac{H}{W}, \frac{h}{W}, \frac{B}{W}, AR\right) = 0 \tag{215}$$

From Eq. (10.3),

$$\frac{fW}{u} = S_t = F\left(R_e, \frac{a}{W}, \frac{H}{W}, \frac{h}{W}, \frac{B}{W}, AR\right) = 0 \tag{216}$$

where, St is Strouhal number.

Since $u = Q/(aW)$, from Eq. (216)

$$f = St\frac{Q}{aW^2} = St\frac{Q \cdot AR^2}{a^2} \tag{217}$$

From this equation, we know that f is in proportion to Q when St is constant. From Eq. (215)

$$\frac{f}{v} = \frac{1}{W^2}F\left(R_e, \frac{a}{W}, \frac{H}{W}, \frac{h}{W}, \frac{B}{W}, AR\right) \tag{218}$$

From this equation, we know that the ratio, (f/v), of oscillatory frequency to kinematic viscosity has the same value even if a different working fluid is used when the similarity of the configuration of the test section is maintained and $F(Re)$ is constant.

Figure 238 shows the relationship between the Strouhal number, St, which is calculated from the results in Fig. 237 and Reynolds number, Re. St takes the constant value in the region of $Re \gtrsim 1.0 \times 10^4$.

The relationship between St and Re is expressed from Eq. (213) as follows.

$$St = (k_1Q + k_2)\frac{W}{v} = c_1 + \frac{c_2}{Re} \tag{219}$$

where,

$$c_1 = AR \cdot W^3 k_1, \quad c_2 = \left(W^2/v\right)k_2 \tag{220}$$

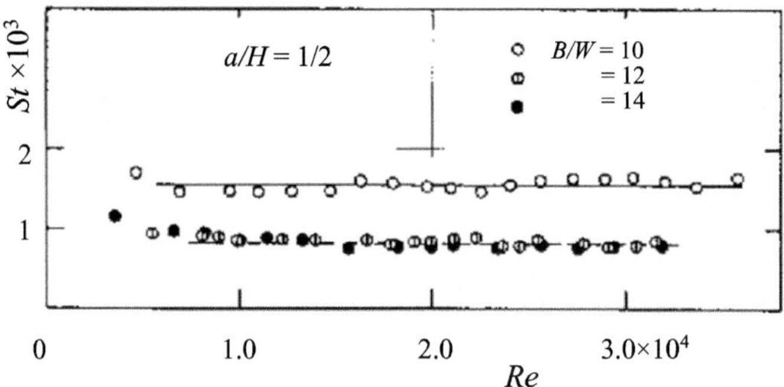

Fig. 238 *St–Re (a/H = 1/2)*

It is known from Eq. (219) that St approaches c_1 as *Re* increases. As an example, the value of c_1 calculated from the value of k_1 for $a/H = 1/3$ and $B/W = 10$ is represented by the dotted line in Fig. 235 This result is in good agreement with experimental results. From the forgoing analysis, it is known that Eq. (216) obtained from dimensional analysis is expressed by Eq. (219) for the case of regular jet oscillation.

Figure 239 shows the relation between (f/v) obtained from the measurements of oscillatory frequency for air or water used as a working fluid and *Re*. In this case, $a/H = 1/2$, $1/3$ and $B/W = 10$. It is recognized from Fig. 239 that the next relation that can be obtained using Eq. (218) of the dimensional analysis is given by

$$\left(\frac{f}{v}\right)_{water} = \left(\frac{f}{v}\right)_{air} \tag{221}$$

Here, a new oscillatory phenomenon of a jet which is issued into a suddenly enlarged flow passage from rectangular nozzle with a smaller height than the flow passage was investigated.

Oscillatory frequency, f, was shown to increase linearly with an increase of Reynolds number (flow rate) for the regular oscillation. f decreases with an increase of nozzle height, a/H, and width, B/W, of flow passage. This is caused by the fact that the mass of fluid which must move to oscillate decreases as a/H and B/W decrease. The ratio, f/v, of f to kinematic viscosity has a constant value under the same Reynolds number even though different working fluid are used.

7.2.2 Oscillatory Phenomenon of Gas–liquid, Two-Phase Jet Flow

Previously, it was shown that when a jet is ejected from a rectangular nozzle that is smaller than the height of the channel into a suddenly expanding channel with a rectangular cross-section, a regular oscillation phenomenon occurs in the jet, but the same is true when the nozzle is circular rather than rectangular (Fig. 240). Here, we examine that a regular oscillation phenomenon occurs in the jet, similar to the above,

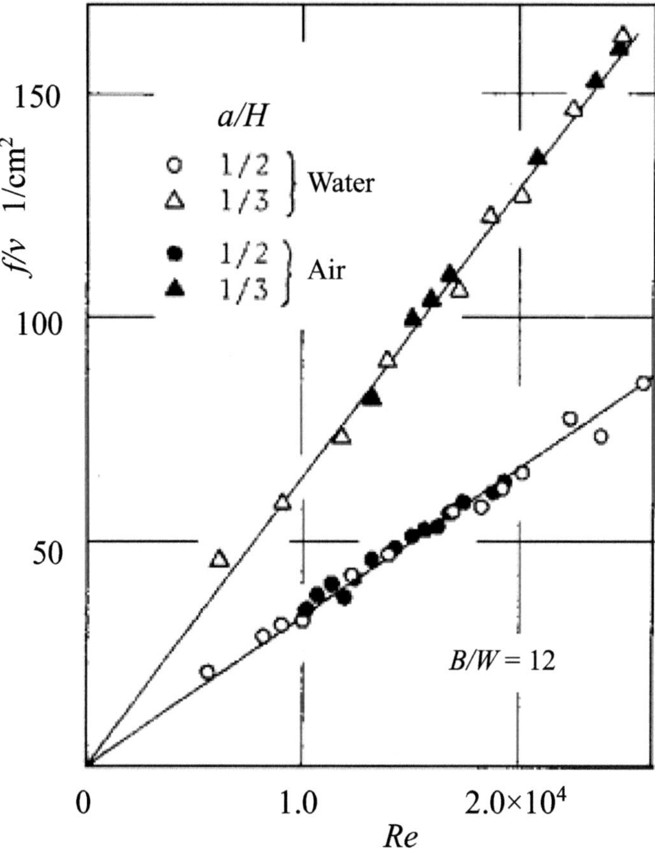

Fig. 239 *f/v–Re (B/W = 12)*

when the working fluid is a two-phase gas–liquid flow. Furthermore, we show that this oscillation phenomenon can be used to measure the flow rate of a two-phase gas–liquid flow. In other words, because the flow resistance and oscillation frequency of the suddenly expanding channel (element) change according to the volume fraction of gas (void fraction), the average volume fraction of gas and liquid in a two-phase gas–liquid flow can be measured.

(a)   **Test section (Fluidic oscillator with a circular nozzle)**

A bubble flow with any void fraction is created by mixing bubbles with a mean bubble diameter $d_a = 2.5$ mm into the water flow, and the flow passage through a nozzle with an inner diameter $d = 15.0$ mm and reaches a measurement section installed vertically. Volumetric flow rate ratio, void fraction α, of air to water can be changed arbitrarily. The measurement section is a rectangular flow passage made of transparent acrylic resin with a width $B = 90.0$ mm, depth $H = 45.0$ mm, and length $L = 1080$ mm (Fig. 240).

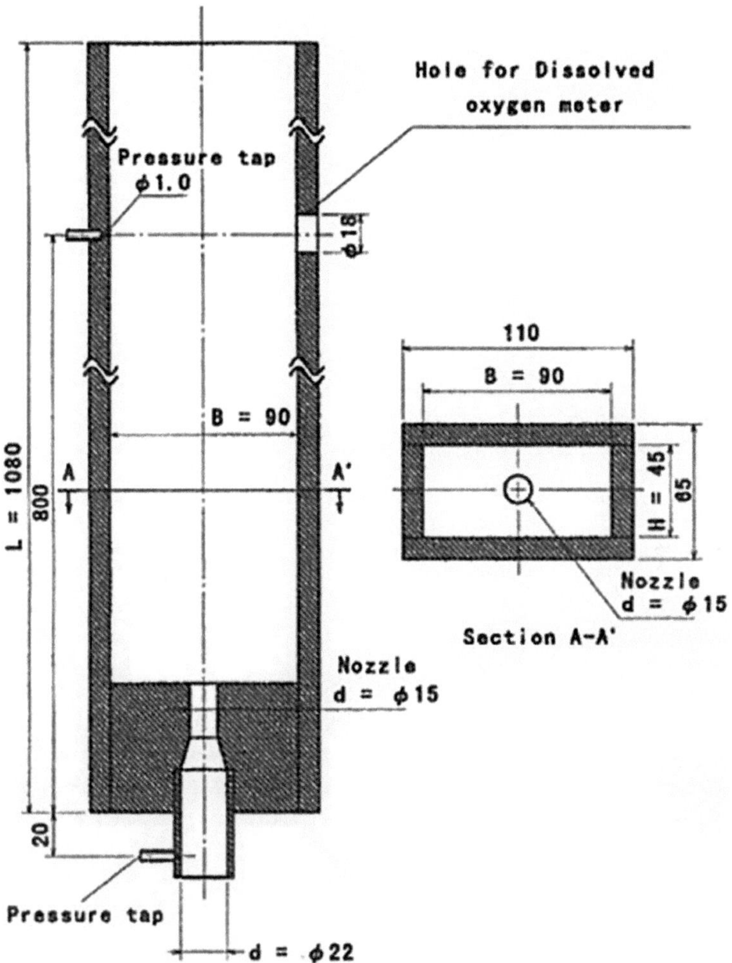

Fig. 240 Test section (Fluidic oscillator with a circular nozzle)

(b) **Flow pattern of oscillation jet**

Figures 241a shows the visualized flow pattern of a half pattern of oscillating jet. The white colored area in the figure shows the bubble region. Figure 241b, c are velocity vector obtained by a high-speed video camara and Particle Image Velocimetry (PIV, Flow-vec 32) and sketches of the oscillating jet, respectively. The jet from the nozzle attaches to one of the side walls (in this case, side wall I), and at that time a flow $Q_{\mathrm{II}\text{-}1}$ is generated near the nozzle exit toward the constant-pressure vortex region A. As a result, the pressure in vortex region A increases, and the jet attachment switches from side wall I to wall II (Fig. 241a, (b)① \sim − ④). This situation is repeated, causing the jet to periodically oscillate.

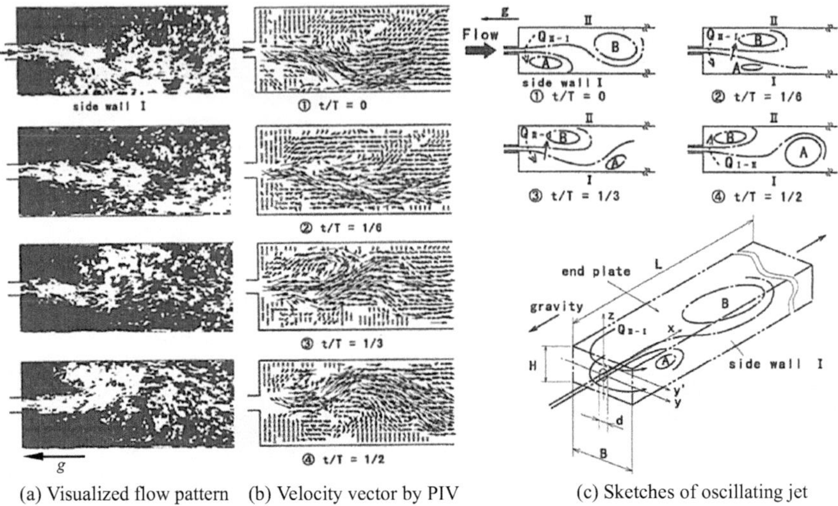

Fig. 241 Oscillating jet

(c) **Oscillatory frequency**

Figure 242 shows the relationship between oscillatory frequency f of jet and Re. f increases with increasing Reynolds number Re ($= u_w\, d/v$, u_w: velocity of water, v: kinematic viscosity of water) and void fraction α.

Figure 243 shows the f and α. f is hardly affected by α when Re is small, but it increases as α increases with increasing Re. The rate of increase increases with Re.

Fig. 242 Oscillatory frequency f versus Re

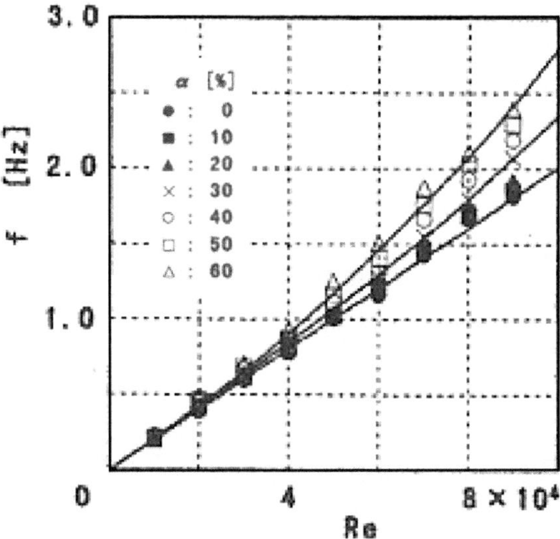

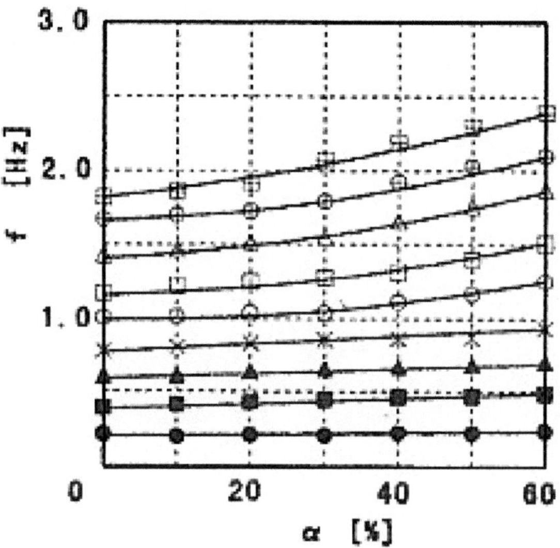

Fig. 243 Oscillatory frequency f versus α

(d) **Pressure loss**

Figure 244 shows the relationship between the pressure loss Δp of the test section and Re. Δp was measured using the pressure taps (Fig. 240) installed 20 mm before and 800 mm after the inlet to the test section.

This oscillatory phenomenon of a single phase (e.g., air or water only) jet flow can be used as a flow meter using the relationship between f and Re. In the case of two-phase flow, for example, air–water bubble flow, we can measure the volumetric

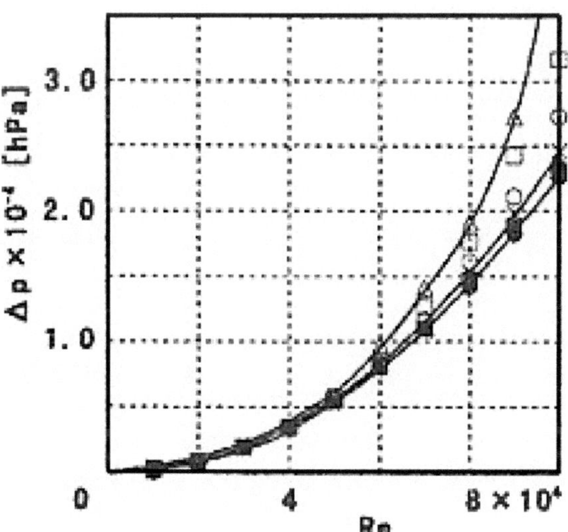

Fig. 244 Pressure loss Δp versus Re

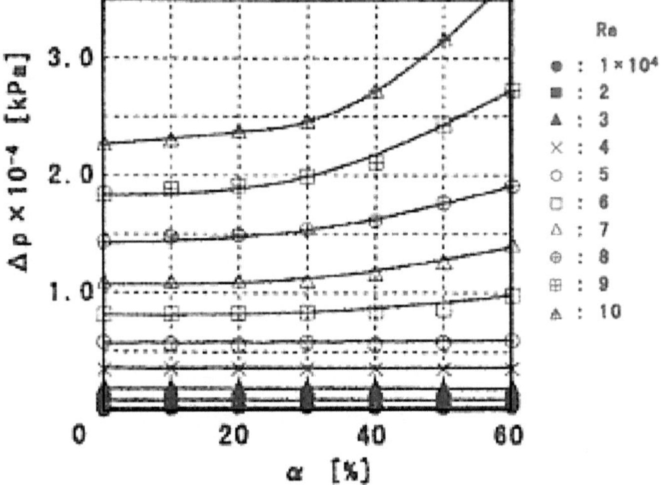

Fig. 245 Pressure loss Δp versus α

flow rate of air and water using the relationship between f and Re for various α, and Δp and Re for various α. Furthermore, this type of oscillation phenomenon can also be used as a fluid mixing device.

Figure 245 shows the Δp and α for various Re. Δp increases as increasing α, and as increasing Re the increasing rate of Δp increases as α increases.

The fluidic oscillator proposed here exhibits regular oscillation phenomena not only in water single-phase flows but also in gas (air)/liquid (water) two-phase bubbly flows. The oscillation frequency f and the flow loss ΔP of the element increase with increasing flow velocity u_w and void fraction α. A calibration curve created using this information can be used to determine the measured values of f and Δp for a bubbly flow with unknown volumetric flow rates of gas (air) and liquid (water). Where, the calibration curve is created by measuring f and ΔP for bubbly flows with various u_w and α, and then creating graphs of f versus α and Δp versus α, and plotting them on the same graph.

7.2.3   Feedback Type Fluidic Oscillator (Gas Absorption)

The gas diffusion, mixing or absorption into liquid is very important operation in the fields of nature and industry. For example, water organism absorbs dissolved oxygen to live and it is a very important to adjust or enhance the dissolved oxygen artificially by aeration in order to make an artificial raising of a water organism efficiently. The aeration is used in many industry fields, for ex ample, to purify waste or sewage water, and to diffuse or absorb a gas into a liquid by bubble column at various operations in chemical industry (JSME 1989, Hoefken et al. 1991, Ramanathan and Arndt 1996).

Now, it is desired to improve the mixing, diffusion and absorption rates of gas into liquid and to make a more com pact device.

Here, first the performance of fluidic oscillator operated by gas–liquid two-phase flow is made clear experimentally. For example, the relations among oscillatory frequency, mean nozzle exit velocity of water flow, flow resistance of oscillator and void fraction. Next, the aeration performance of the oscillator, namely the diffusion or absorption rate of air, dissolved oxygen, into water under various kind of operation conditions are examined experimentally (Shakouchi 2003).

(a) **Test section**

Figure 246 shows the details of the test section, feedback type fluidic oscillator. The oscillator is consisted with nozzle and exit parts, two feedback loops and side walls and upper and lower end plates. The cross section of nozzle is rectangular and the width and height are $d_f = 7.5$ mm and $H_f = 20.0$ mm, respectively. The aspect ratio is $AR = 2.7$. The oscillator is made by transparent acrylic resin and has feedback loop, and it worked well when the working fluid is a single-phase flow, air or water flow, or a gas–liquid two phase flow. There is an observatory box made by trans parent acrylic resin and filled with water just before and after the oscillator in order to observe the flow patterns in a pipe without the effect of pipe's curvature. The pressure loss, flow resistance, Δp was measured by pressure tapings of diameter 1.0 mm which were set at just before and after the oscillator and mercury column manometer. The air–water mixing part is made by coaxial double pipe, and the air from a compressor is flowed into the water f low through an inner pipe wall of sintered porous metal and then the air bubble flows into the water flow. The volumetric flow rate of air is normalized at the pressure and temperature just before the oscillator. The mean size of air bubble is about 2.2 mm which distributes from $d_a = 0.5$ to 4 mm. The oscillatory frequency f was measured by pressure tapping of diameter 1.0 mm set on the side wall, semi-conductor type pressure transducer and digital storage type oscilloscope and was obtained as the mean value of 3 \sim − 5 times measurements of periodic pressure change for 5 \sim − 8 cycles. The observation of periodic oscillatory flow visualized by air bubbles was also used to measure the oscillatory frequency. The absorption of air, dissolved oxygen, into water was measured by sodium sulfite process as follows. First, sodium sulfite is put into the water in the reservoir and then the dissolved oxygen was consumed to be zero by the following chemical reaction.

$$Na_2SO_3 \cdot 7H_2O + (1/2)O_2 = Na_2SO_4 + 7H_2O$$

In the chemical reaction, $CoCL_2\ 6H_2O$ was used as a catalyst. Next, the experiment starts and the dissolved oxygen after the oscillator by dissolved oxygen meter. In this experiment, the water passes through the oscillator does not back to reservoir and is thrown away to the outside. The dissolved oxygen for only the pipe flow without the oscillator was also measured.

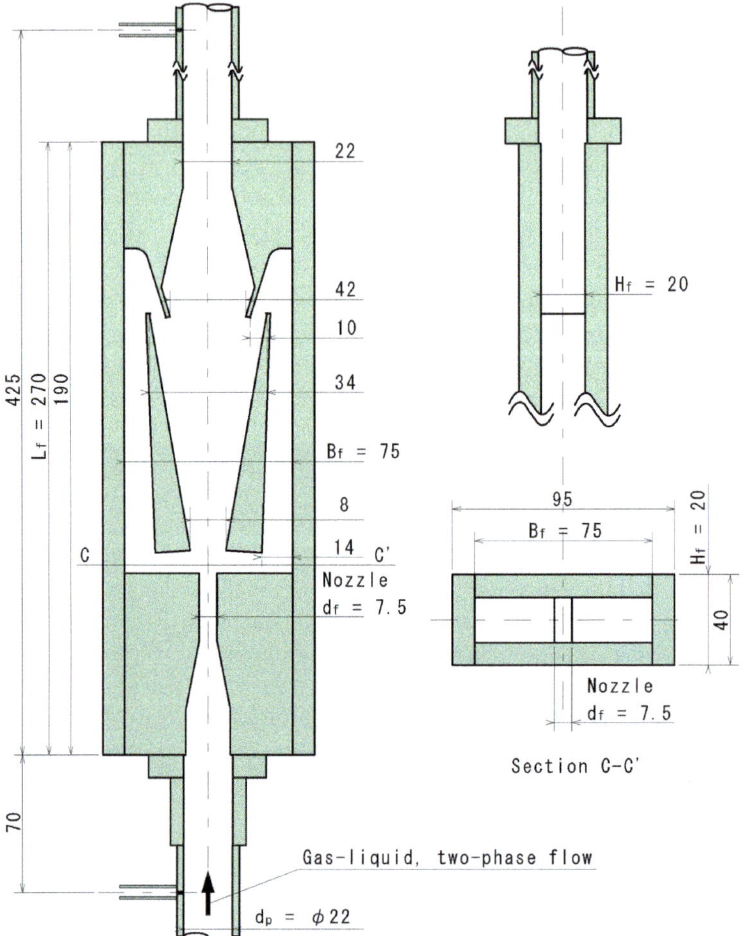

Fig. 246 Test section of feedback type fluidic oscillator

(b) **Bubble size distribution**

Figure 247 shows a few examples of bubble size distribution which were obtained from the photograph of visualized f low. The measuring area is just after the oscillator. The Reynolds number Re_p was defined by $(u\, d_p/v)$. The bubble size d_a distributes from under 1–2 mm, and the mean size increases with increasing α.

Figure 248 is the relation between the mean bubble size da and Re_p. The d_a decreases rapidly with increasing Re_p. The d_a also increases with increasing α. It may depend on that the chance of collision and coalescence of bubble increases with increasing α.

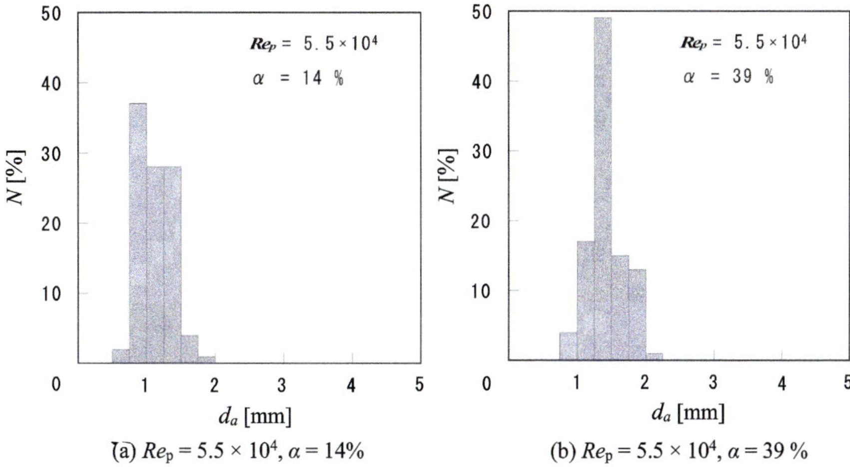

(a) $Re_p = 5.5 \times 10^4$, $\alpha = 14\%$ (b) $Re_p = 5.5 \times 10^4$, $\alpha = 39\%$

Fig. 247 Bubble diameter distribution (Feedback type oscillator)

Fig. 248 Mean bubble diameter

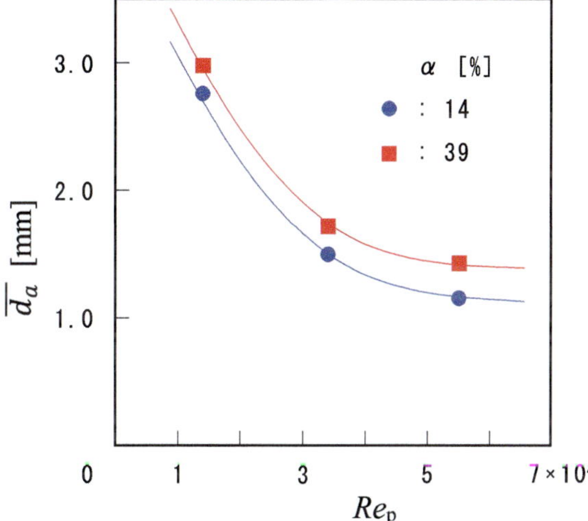

(c) **Flow pattern in oscillation**

Figure 249 shows the flow patterns of the oscillatory jet, and Fig. 249a, b, c ① to ⑤ are the flow patterns for almost half period at every one sixth of period. Figure 249a–c are visualized flow pattern, velocity vector and streamline, respectively. In Fig. 249a white colored parts are air bubbles. The velocity vector and streamline were obtained using the high-speed video camera, particle image velocimeter, PIV, and the volume cross-correlation method. Where, Reynolds number Re ($= u d_f / v$, u: mean velocity of water flow at the nozzle exit, d_f: nozzle width, v: kinematic viscosity) $= 1.5 \times 10^4$, $\alpha = 30\%$ and oscillatory frequency $f = 0.66$ Hz. At Fig. 249 ①, the jet which is

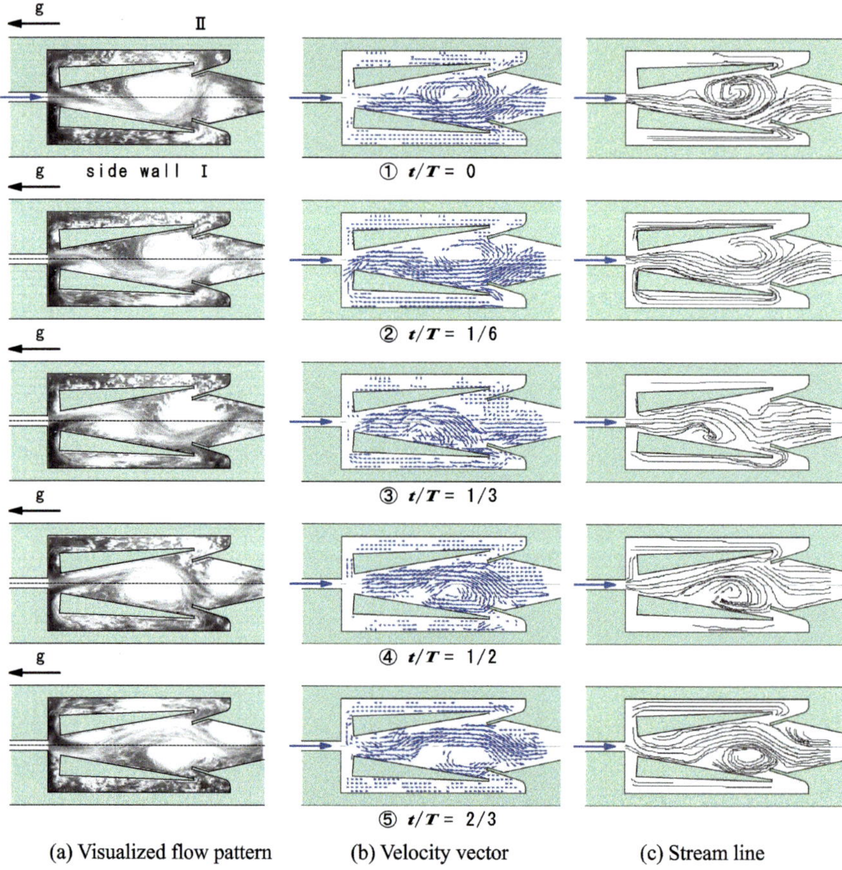

Fig. 249 Flow pattern (x–y plane, $Re_f = 1.5 \times 10^4$, $\alpha = 30\%$, $f = 0.66$ Hz)

issued from the nozzle deflects to the side wall I mostly and at the downstream the anti-clockwise vortex is formed between the jet and the side wall II. At Fig. 249 ②, the jet detaches from the side wall I by increased pressure which is caused by the flow to the upstream passing through the feedback loop I forming the clockwise vortex between the jet and the side wall I but at the downstream the jet attaches to the side wall I yet. At Fig. 249 ③ to ⑤, the clockwise vortex is flowed to the downstream and then the attachment of the jet switches to the side wall II entirely. Next, the contrary process is occurred and then the attachment of the jet switches from the side wall II to the I. This is one cycle of the jet oscillation, and it is repeated and then the jet oscillation is continued. The flow pattern in the pipe just before the oscillator is plug- or slug-flow, but in the oscillator it is changed to bubbly flow with relatively small and uniform size of air bubbles. It depends on that sudden acceleration and deceleration of a gas–liquid two-phase flow by passing through the nozzle contraction and sudden enlargement of the oscillator, respectively, make air bubble to smaller and to bubbly flow.

This fluidic oscillator may be useful for making gas liquid flow to bubbly flow and for mixer or mass transfer enhancement device between gas and liquid phases.

(d) **Dimensional analysis**

Since this oscillatory phenomenon is three-dimensional, unsteady and gas–liquid two-phase flow and then the flow condition is very complex, it is very difficult to examine it by theoretical and numerical analyses. In this research it is examined by dimensional analysis.

Defining nozzle width d_f, volumetric flow rate of two-phase flow Q_m and density of fluid ρ as fundamental quantities and making dimensional analysis,

$$F\left(fd_f^2/v,\, u_m d_f/v,\, B_f/d_f,\, H_f/d_f,\, L/d_f\right) = 0 \tag{222}$$

where, F: functional symbol, u_m: mean velocity of two-phase flow at the nozzle exit, B_f, L: width and length of flow passage.

From Eq. (222),

$$fd_f/u_m = St = F(Q_m\rho/\mu d_f,\, B/d_f,\, H_f/d_f,\, L/d_f) \tag{223}$$

where, St: Strouhal number $(= fd_f/u_m)$, μ: viscosity

Since there is an almost linear relationship between the oscillatory frequency f and volumetric flow rate of working fluid Q_m from the experimental result which will be shown later,

$$f = kQ_m + k' = kQ_w/(1 - \alpha) + k' \tag{224}$$

where, $k,\, k'$: experimental constant

Using $\alpha'\ (= Q_a/Q_w)$, Eq. (224) is reduced to

$$f = kQ_w\alpha' + \left(kQ_w + k'\right) \tag{225}$$

From Eqs. (222) and (223),

$$St = fd_f/u_m = k\pi d_f^3 + k'd_f/u_m \tag{226}$$

St approaches to a constant $(= k\,\pi d_f{}^3)$ as in creasing u_m.
From Eq. (224),

$$fd_f^2/v = F(Q_m\rho/\mu d_f,\, B_f/d_f,\, H_f/d_f,\, L/d_f) \tag{227}$$

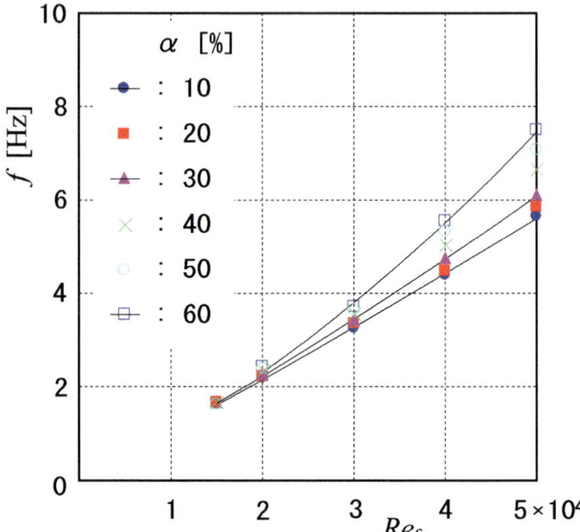

Fig. 250 f–Re_f (Feedback oscillator)

From Eq. (227), it results in that if the geometry of the oscillator is similar and $F(Q_m \rho /\mu d_f)$ is the same the ratio of the oscillatory frequency to kinetic viscosity $(f d_f^2/v)$ is the same even if the working fluid is different which has different kinetic viscosity.

(e) **Oscillatory frequency**

Figure 250 is the relation between the oscillatory frequency f and Re_f for $\alpha = 0$ to 60%. The f increases almost linearly with increasing Re_f.

$$f = kQ_w + k' \tag{228}$$

But, exactly the relation can be expressed as follows.

$$f = a_1 (Re_f)^{b_1} \tag{229}$$

where, $a_1 \times 10^5 = -0.154\,\alpha + 9.28$, $b_1 = 0.00529\,\alpha + 0.955$, ($\alpha$: %)

Figure 251 shows a few rearranged examples of the relation between f and α which were derived from Fig. 250. As increasing Re_f, f increases almost linearly with increasing α, and the increasing rate increases with increasing Re_f.

But, exactly the relation can be expressed as follows.

$$f = a_2\alpha + b_2 \tag{230}$$

where, $a_2 = -0.00738\,Re_f + 0.012$, $b_2 = 1.14\,Re_f - 0.082$

Fig. 251 f–α

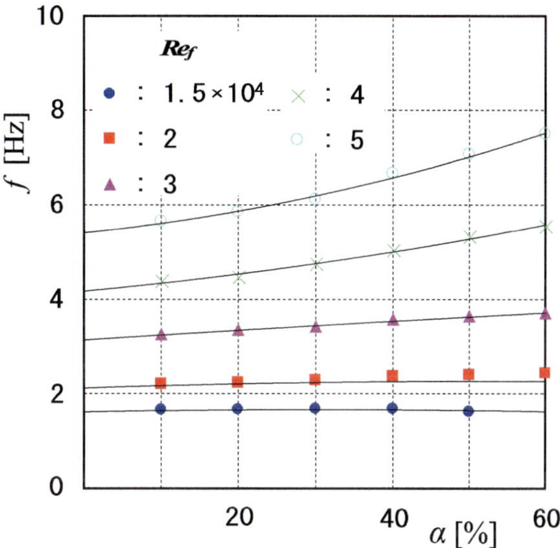

The f increases when α increases. This may depend on that the mass flow rate which have to move in the oscillation decreases when α increases.

The f increases when α increases. This may depend on that the mass flow rate which have to move in the oscillation decreases when α increases.

(f) **Pressure loss of oscillator**

Figure 252 shows the pressure loss, flow resistance, Δp of the oscillator. The Δp increases abruptly with increasing Re_f.

$$\Delta p = c_1 (Re_f)^{d_1} \tag{231}$$

where, $c_1 = -1.71 \times 10^{-8}\, \alpha + 8.92$, $d_1 = 0.00774\, \alpha + 1.97$

Figure 253 shows a few rearranged examples of the relation between Δp and α when Re_f is constant which were derived from Fig. 252. As increasing Re_f, Δp increases with increasing α, and the increasing rate increases with increasing Re_f. The relation between Δp and α is

$$\Delta p = c_2 \alpha + d_2 \tag{232}$$

where, $c_2 = -1.42\, Re_f^{\,2} - 9.89\, Re_f + 10.724$, $d_2 = 759\, Re_f + 1125$

(g) **Dissolved oxygen**

In this section, the performance of fluidic oscillator as a gas-absorber, aerator, will be shown.

Fig. 252 $\Delta p - Re_\mathrm{f}$

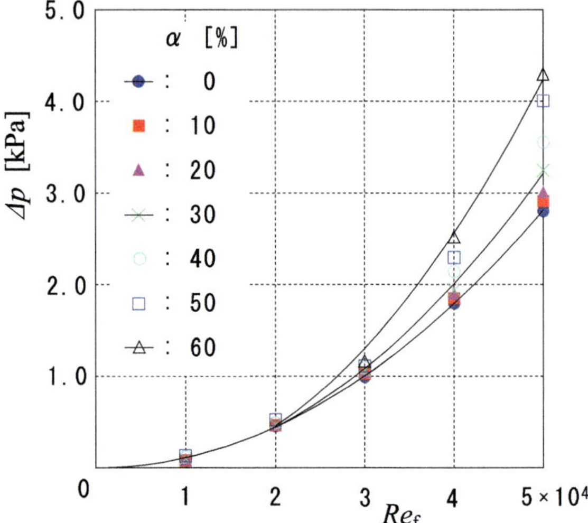

Fig. 253 $\Delta p - \alpha$

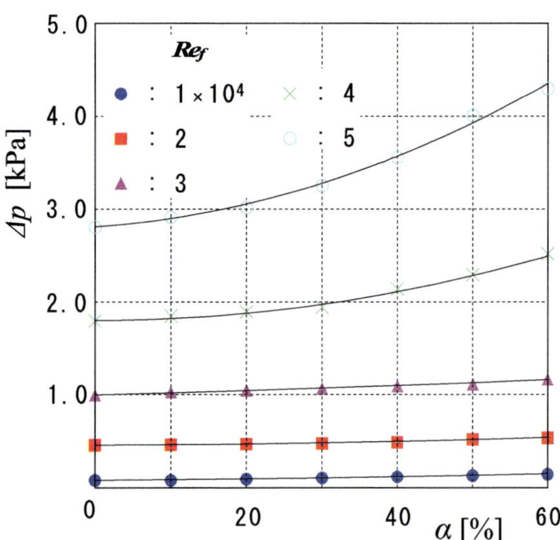

Figure 254 shows the dissolved oxygen C. The C increases rapidly with increasing Re_p. In the figure, the results for the pipe flow without the oscillator are also shown. For the pipe, C increases almost linearly with increasing Re_p.

The relation between C and Re_p can be expressed as follows.

For the oscillator,

$$C = A_1 Re_\mathrm{p} + B_1 \quad (\alpha \leq 20\%) \tag{233}$$

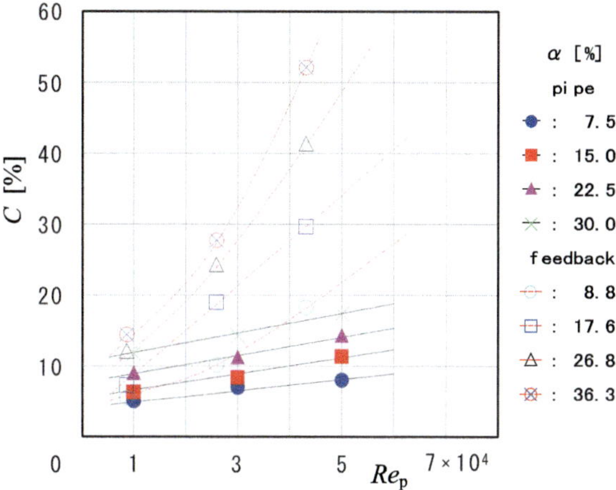

Fig. 254 C–Re_p (pipe and oscillation)

$$C = A_2 \exp(B_2) \quad (\alpha \gtrsim 20\%) \tag{234}$$

where,

	$\alpha = 8.8\%$	17.6	26.8	36.3
A_1	3.7×10^{-4}	6.29×10^{-4}		
B_1	1.1	2.6	–	–
A_2	–	–	9.1	10.4
B_2	–	–	$0.36 Re_\mathrm{p}$	$0.38 Re_\mathrm{p}$

For the pipe,

$$C = A_1 Re_\mathrm{p} + B_1 \tag{235}$$

where,
$A_1 = 1.38 \times 10^{-4}$

	$\alpha = 7.5\%$	15.0	22.5	30.0
B_1	3.0	4.8	7.4	10.4

The effect of oscillation of two-phase flow on the absorption of air, oxygen, into water is remarkably distinguished. For example, when $\alpha \doteqdot 30\%$ and $Re_\mathrm{p} = 4.0 \times 10^4$, the C for the oscillator is almost 2.6 times of that for the pipe.

Figure 255 is the relation between C and α. The C is the ratio to the saturated quantity at the same temperature and pressure. The C increases linearly with increasing α, and it can be expressed by the following equations.

$$C = C_1 \alpha + D_1$$

Fig. 255 C–α (pipe and oscillation)

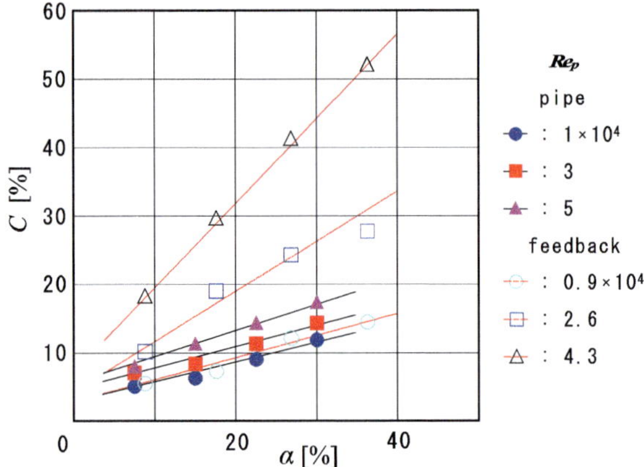

For the oscillator,

	$Re_\mathrm{p} = 0.9 \times 10^4$	2.6	4.3
C_1	0.32	0.76	1.24
D_1	3.0	4.5	7.1

For the pipe,
$C_1 = 75$.

	$Re_\mathrm{p} = 1.0 \times 10^4$	3.0	5.0
D_1	0.8	3.0	6.0

Using the fluidic oscillator, a very compact type aerator with high performance will be able to construct.

Here, we examined the performance of fluidic oscillator which operated by gas–liquid two-phase flow and its application to the compact type aerator.

(1) Even if the working fluid is a gas–liquid two-phase flow fluidic oscillator works normally and shows a periodic oscillation.
(2) The oscillatory frequency f increases linearly with increasing Reynolds number Re_f and void fraction α.
(3) The pressure loss, flow resistance, Δp of the oscillator increases rapidly with increasing Re_f and linearly increasing α.
(4) Dissolved oxygen in water can be increased considerably by fluidic oscillator operated by air–water two-phase f low, and it was well known that a new compact type aerator can be constructed by a fluidic oscillator.

7.2.4 Measurements of Volumetric Flow Rates of Gas (Air) and Liquid (Water) Two-Phase Flow

The flow rates of gas–liquid (air–water) bubbly flow passing through a pipe can be measured using a fluidic oscillator as shown in the above (Shakouchi 1990, 1991a, 1991b, 1995).

The method is as follows.

Fist, we obtained the relationship between oscillatory frequency f and the flow rate of water Q_w for various void fractions α of air by experiments. The relation could be expressed by

$$f = K_1 Q_w + K_2 \tag{236}$$

where, f: [Hz], Q_w: [L/min], K_1, K_2: experimental constants

Using these results, we obtained the relationship between f and α under the condition that $Q_w = $ const. This was

$$f = kQ_w\alpha + \left(kQ_w + k'\right) \tag{237}$$

Next, we obtained the relationship between the pressure loss Δp of the oscillator and Q_w for various α. The relation can be expressed by

$$\Delta p = C_1(Q_w)^{C_2} \tag{238}$$

where, Δp: [kPa], C_1, C_2: experimental const.

Using these results, the relationship between Δp and α under the condition that $Q_w = $ const. could be expressed by the following equation.

$$\Delta p = C_3\alpha + C_4 \tag{239}$$

where, C_3, C_4: experimental const.

Using the above results, the relationship between $Q_w - \alpha$, $f - \alpha$, and $\Delta p - \alpha$ can be plotted on a single graph putting Q_w, f, and Δp on the vertical axis and α on the horizontal axis to obtain the calibration curve for the flow rate of the oscillator.

Now, for a two-phase gas–liquid bubbly flow with an unknown flow rate inside a pipe, introduce it into the fluidic oscillator described above and measure the jet oscillation frequency f and the pressure loss Δp of the oscillator. Then, find the intersection point between these on the calibration curve and read the vertical and horizontal axes to find Q_w and α, i.e. flow rate of air Q_a. In other words, it is possible to measure the flow rates (time averaged values) of the gas and liquid phases of the two-phase gas–liquid bubbly flow, Q_w and Q_a.

7.2.5   Cavity Oscillation and Cavity Oscillator

A two-dimensional jet injected into a cavity which is a hexahedron in shape, shows periodic oscillation under certain conditions as pointed out by Molloy et al. They present the experimental results mainly for the relation between the oscillatory frequency of the jet and the configuration of a cavity or flow rate. However, the mechanism of oscillation of a jet in a cavity has not been clarified yet. Here, the flow patterns of the oscillatory jet are clarified by visual observation and a mechanism of the oscillation is explained using a numerical analysis of the oscillatory jet in a cavity and the flow study by the conformal mapping method.

It is found that a pair of vortices induced on both sides of the jet plays an important role in the oscillatory phenomenon. The oscillatory frequency is examined by the dimensional analysis and the relations between the oscillatory frequency and the configuration of the cavity or flow rate are investigated experimentally (Shakouchi et al. 1982).

(a) **Test section and procedure**

Figure 256 shows the configuration of the cavity. The cavity consists of nozzle ①, side wall ②, end wall ③, upper and lower plates ④, ⑤. The configuration of the cavity is axisymmetric with respect to the nozzle axis OO'. The geometry of the cavity is fixed by three non-dimensional quantities of b/a, c/a and d/a. a, b, c and d are the width of the nozzle, width of the cavity, distance from the exit of the cavity to the nozzle tip, nozzle length, and the distance from the nozzle tip to the lower plate. The flow resistance at the exit flow passage can be varied by the sliding resistance plate ⑥. The cavity is set underwater in a large tank with a constant water level, and water from a pump is injected through a nozzle into the cavity.

(b) **Flow pattern of oscillatory jet flow**

Figure 257 shows a series of photographs of the oscillatory jet for $c/a = 3.33$, $d/a = 16$, $Re = 2400$, and the periodic time is $T = 21$ s. The broken lines with arrows in the figure indicate approximately the positions and directions of rotation of vortices. The arrows in the figure show the direction of the flow at the exit of the cavity. There are prominent counter-current vortex regions on either side of the jet. These vortices wax and wane as the jet oscillates. Figure 257a shows a jet most deflected to the right, and the left vortex is situated downstream of the nozzle and the right vortex near the entrance of the right exit flow passage. In this case, there is an out flow from the left exit but no flow from the right one. Figure 257b–d represent the process of the jet singing from the right to the left. Figure 257e shows a jet most deflected to the left. Figure 257a–e correspond to a half cycle of the jet oscillation, and the jet reverses the movement during the remaining half cycle.

(c) **Conditions of oscillation and frequency**

Figure 258 shows the relations between the occurrence of the oscillation and the configuration of the cavity. Numbers (I), (II) and (III) in the figure represent the

Fig. 256 Configuration of
cavity

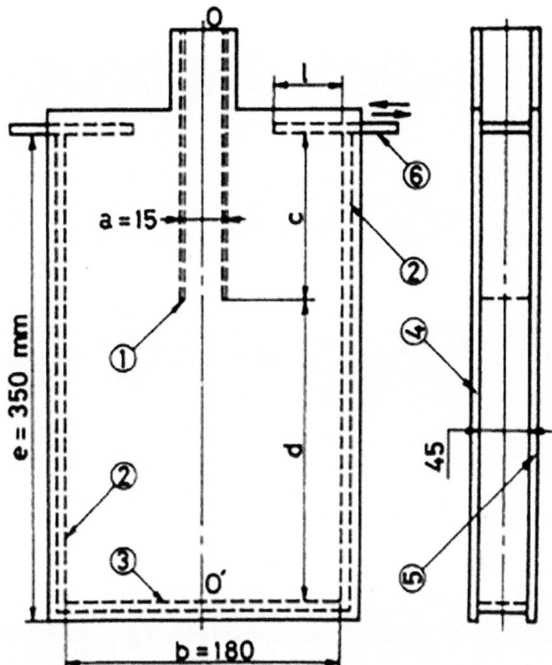

stable-, unstable- and no-oscillation regions, respectively, and the solid lines show
the approximate bounds. In order to get a stable-oscillation d mast exceed a certain
lower limit when c is constant and the limit increases as the flow rate increases.
Figure 258b shows the effect of the width b of cavity. A larger value of d is needed as
b becomes larger for the occurrence of the oscillation when the flow rate and nozzle
width, a, are constant.

Figures 259 and 260 show the relation between the oscillatory frequency f and the
flow rate or Re under the stable oscillation. The frequency of oscillation is proportional to the flow rate and the constant of proportion depends on the configuration of
the cavity, and that an increase in c and resistance at the exit flow passage, namely
increase of l in Fig. 256, causes a decrease in the frequency of oscillation.

Dimensional analysis for this oscillatory phenomenon under the condition of
geometrical similarity gives a functional relation

$$\varphi(St, Re) = 0 \tag{240}$$

where, St is Strouhal number $(= f \cdot a/u_{\mathrm{m}})$

Therefore, $f = (U_m/a)\overline{\psi}(Re)$ and this shows that f is proportional to U_{m} when
$\overline{\psi}(Re)$ is constant. (a is constant in Figs. 259 and 260. Furthermore, Eq. (236) can
be also expressed as

$$\varphi_1(StRe, Re) = 0 \tag{241}$$

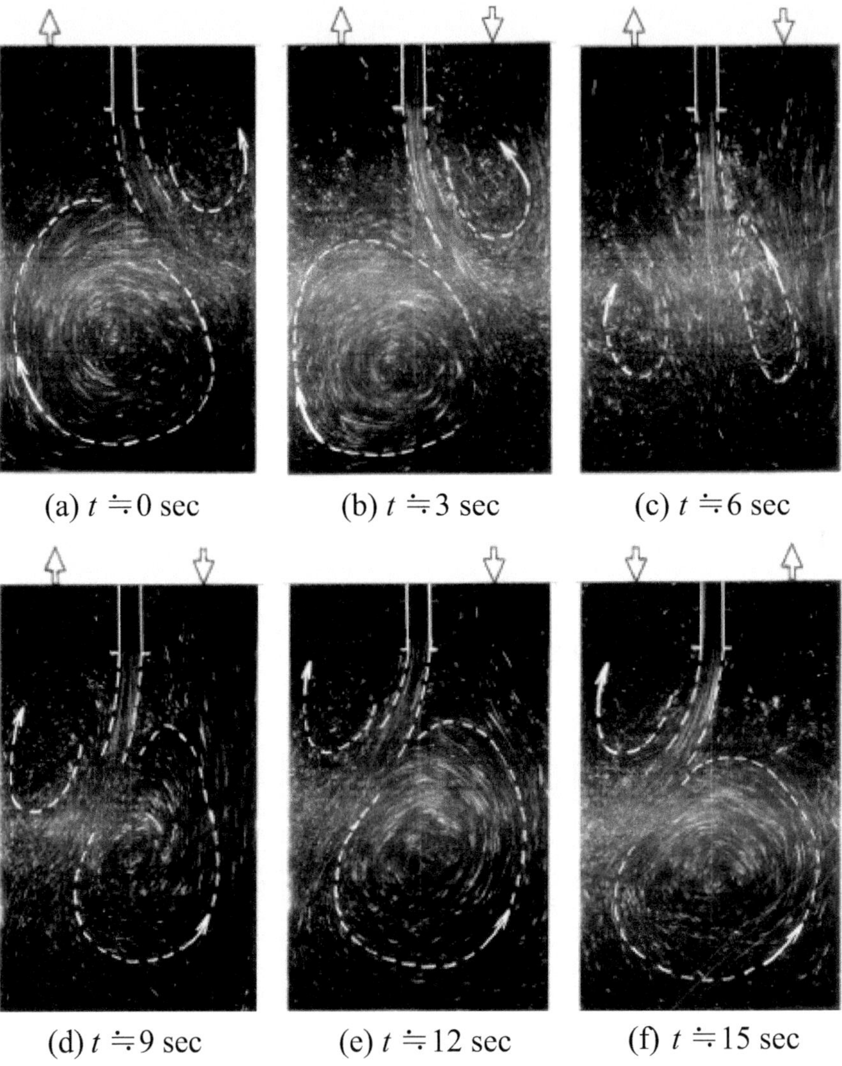

(a) $t \doteq 0$ sec   (b) $t \doteq 3$ sec   (c) $t \doteq 6$ sec

(d) $t \doteq 9$ sec   (e) $t \doteq 12$ sec   (f) $t \doteq 15$ sec

Fig. 257 Flow pattern in oscillation ($c/a = 3.33$, $d/a = 16$, $Re = 2400$, $T = 21$ s)

From this we can obtain $f/v = \left(1/a^2\right)\overline{\psi}_1(Re)$. This shows that f/v is constant for different working fluids within a range of Renolds numbers in which $\overline{\psi}_1(Re)$ is constant.

(d) Analysis of the flow in a cavity

(d.1) Numerical analysis

Visual observation of an oscillatory jet as shown in Fig. 257s shows that there exist two distinguishable vortices on both sides of the jet and the deflection of the jet

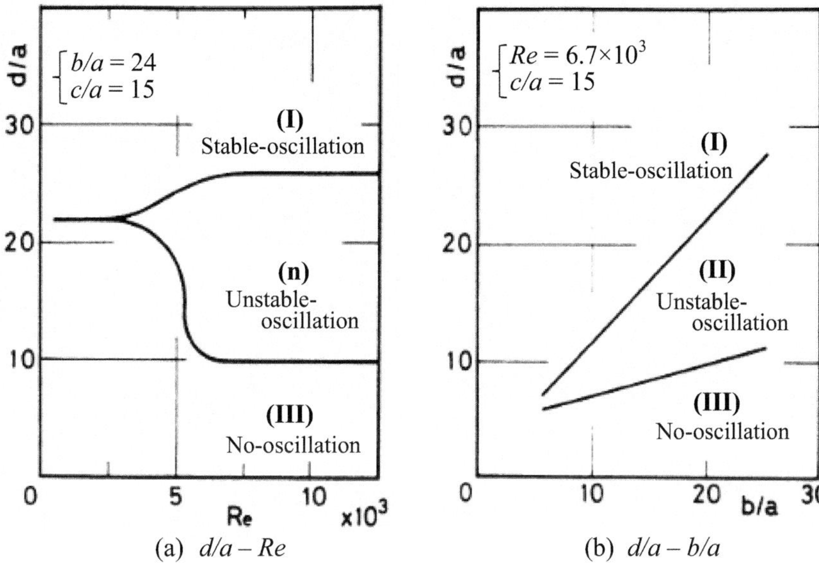

Fig. 258 Flow pattern map ($a = 8.0$ mm, $b = 192.0$ mm)

Fig. 259 f–Re ($c/a = 0, 33$)

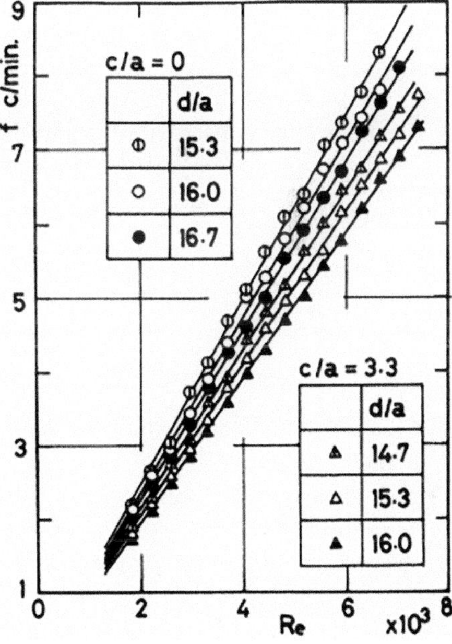

Fig. 260 *f–Re (d/a = 16)*

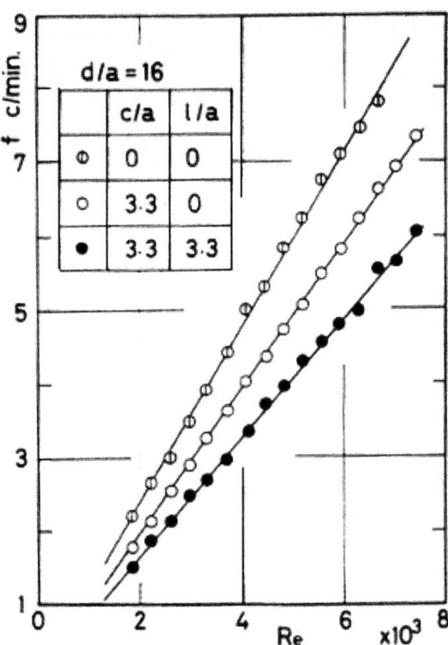

is closely related with the positions of the vortices, and the relations between the positions of vortices and the deflection of the jet are also examined analytically. The flow passage for numerical analysis is shown in Fig. 261.

The basic equations derived from Navier–Stokes equations and equation of continuity by stream function ψ and vorticity ζ are

$$\zeta = \frac{\partial v}{\partial x} - \frac{\partial u}{\partial y} = -\left(\frac{\partial^2 \psi}{\partial x^2} + \frac{\partial^2 \psi}{\partial y^2}\right) \tag{242}$$

$$\frac{\partial \zeta}{\partial t} + \frac{\partial \psi}{\partial y}\frac{\partial \zeta}{\partial x} - \frac{\partial \psi}{\partial x}\frac{\partial \zeta}{\partial y} = v\left(\frac{\partial^2 \zeta}{\partial x^2} + \frac{\partial^2 \zeta}{\partial y^2}\right) \tag{243}$$

where, u and v are velocity components in x and y directions, respectively.

Boundary conditions:

The velocity distribution at far upstream boundary, EF, of the nozzle has a parabolic profile and the value of ψ and ζ the downstream boundaries, AF and ED, are obtained by the application of Hung's extrapolation. Equations (241) and (242) are converted to non-dimensional forms using the following dimensionless quantities.

$$\Psi = \varphi/(aU_m), \Theta = a\varsigma/U_m, H = h/a,$$
$$K = k/a, T = tU_m/a, \mathrm{Re} = aU_m/v$$

Fig. 261 Calculation region

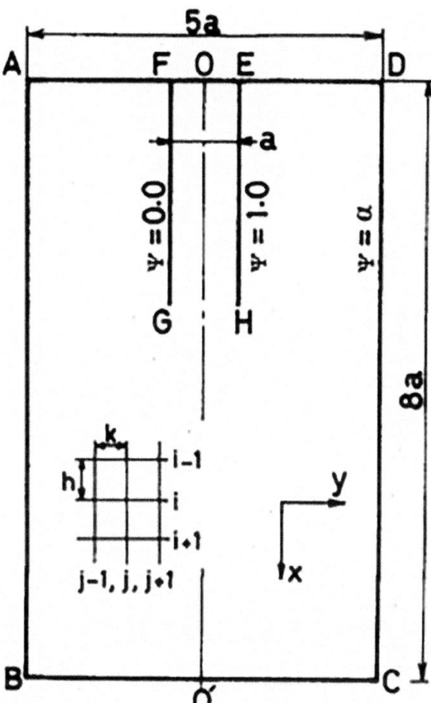

where, h and k: lengths of the sides of square cells (Fig. 261), t: time.

Figure 262a, b show the calculated results for $Re = 100$, $\alpha = 0.5$ and 0.7, respectively. These results are solutions for the steady floe and do not directly correspond to the flow pattens in oscillation as shown in Fig. 257. Nevertheless, these results reveal the characteristics of the flow pattern; i.e., the presence of two distinguishable vortices on both sides of the jet and the movement of a vortex center depending on the deflection of the jet.

(d.2) Expression of the flow by potential model

*Conformal mapping of a flow passage and complex cavity potential

The interior of a cavity in the physical plane (Z-plane) a shown in Fig. 263a is mapped on the lower half in the transformed θ-plane by Schwarz-Christoffel transformation. The both side walls and the nozzle are supposed to extend infinity in the nozzle upstream direction. That is, the position B, D and F are assumed to locate at infinity in the Z-plane. If the points O, A, B, C and D in the Z-plane correspond to the O', A', B', C' and D' on the real axis in the θ-plane, a mapping function by which the interior of a cavity is mapped on the lower half of the θ-plane is obtained by integrating the equation with $1 > m > n > 0$.

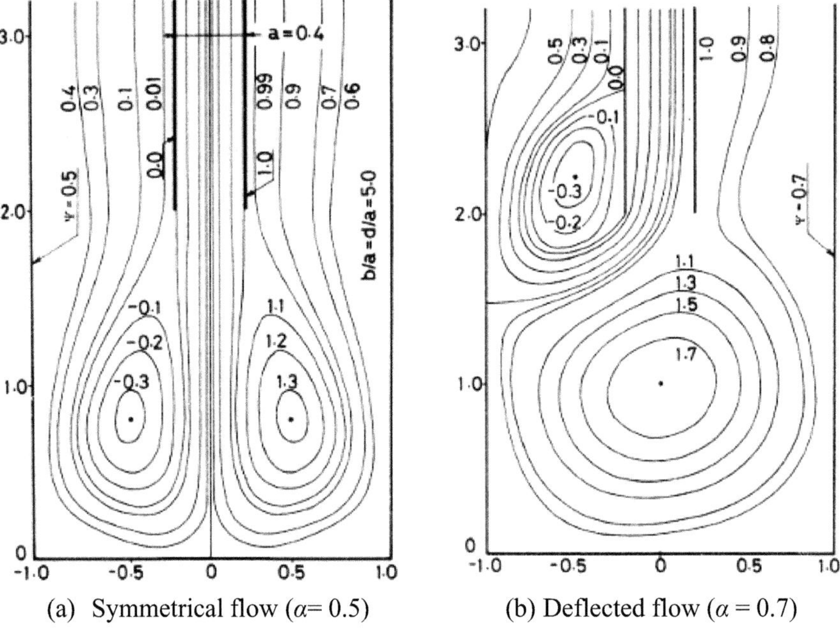

(a) Symmetrical flow (α= 0.5)   (b) Deflected flow (α = 0.7)

Fig. 262   Stream line

$$\frac{dz}{d\theta} = \frac{-iL(\theta^2 - n^2)}{\theta\sqrt{1-\theta^2}(\theta^2 - m^2)} \tag{244}$$

$$z = \frac{iL}{2}\left(\frac{n^2}{m^2}\log\frac{\sqrt{1-\theta^2}+1}{\sqrt{1-\theta^2}-1} + \frac{m^2-n^2}{m^2\sqrt{1-m^2}}\log\frac{\sqrt{1-\theta^2}+\sqrt{1-m^2}}{\sqrt{1-\theta^2}-\sqrt{1-m^2}}\right) \tag{245}$$

where,

$$L = \frac{b}{\pi}\frac{m^2\sqrt{1-m^2}}{n^2\sqrt{1-m^2}+(m^2-n^2)} \tag{246}$$

$$\frac{a}{b} = \frac{n^2\sqrt{1-n^2}}{n^2\sqrt{1-m^2}+(m^2-n^2)} \tag{247}$$

$$\frac{2d}{b} = \frac{1}{\pi}\frac{a}{b}\log\frac{1+\sqrt{1-n^2}}{1-\sqrt{1-n^2}} + \frac{m^2-n^2}{n^2\sqrt{1-m^2}+(m^2-n^2)}\log\frac{\sqrt{1-n^2}+\sqrt{1-m^2}}{\sqrt{1-n^2}-\sqrt{1-m^2}} \tag{248}$$

Next, let the flow in the z-plane be represented by a potential flow with a source of strength μ, two sinks of $\alpha\mu$ and $\beta\mu$ ($\alpha + \beta = 1$; i.e., α and β are functions of flow rate) and also with two vortex filaments rotating clockwise and counterclockwise.

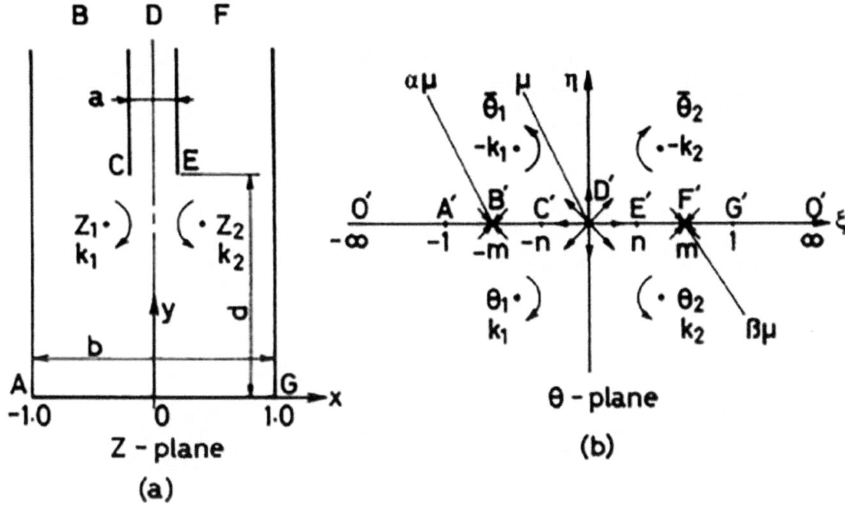

Fig. 263 Mapping of flow passage

Strengths of vortices are K_1 and K_2 which are positive when they rotate clockwise. The source and sinks are placed at the points D, B and F, respectively. Vortices are at the points Z_1 an Z_2 near the points C and E, respectively. This flow in the Z-plane is obtained from the mapping relation Eq. (244), when the flow in the θ-plane is represented with a source of strength μ at D', sink of strength $\alpha\mu$ and $\beta\mu$ at B' and F' and also with vortex filaments of K_1 and K_2 at the points θ_1 and θ_2 (Fig. 263b). The flow field on the lower half of the θ-plane in which the real axis is a stream line can be obtained by placing reflected images of the two vortex filaments on the upper half plane as shown in Fig. 263b. The complex velocity potential, $W(\theta)$, of the flow in the θ-plane is given by

$$W(\theta) = iK_1\{\log(\theta - \theta_1) - \log(\theta - \bar{\theta}_1)\} + iK_2\{\log(\theta - \theta_2) - \log(\theta - \bar{\theta}_2)\}$$
$$+ \mu\{\log\theta - \alpha\log(\theta + m) - \beta\log(\theta - m)\} \tag{249}$$

where, the bar on θ_1 and θ_2 indicates a complex conjugate.

From the above equation the velocity field is given by

$$u - iv = \frac{dW(z)}{dz} = \frac{dW(\theta)}{d\theta}\frac{d\theta}{dx}$$
$$= \frac{\theta(\theta^2 - m^2)\sqrt{1 - \theta^2}}{-iL(\theta^2 - n^2)}\left[iK_1\left\{\frac{1}{\theta - \theta_1} - \frac{1}{\theta - \bar{\theta}_1}\right\} + iK_2\left\{\frac{1}{\theta - \theta_2} - \frac{1}{\theta - \bar{\theta}_2}\right\}\right.$$
$$\left. + \mu\left\{\frac{1}{\theta} - \frac{\alpha}{\theta + m} - \frac{\beta}{\theta - m}\right\}\right] \tag{250}$$

(d.3) **Expression of flow in a cavity**

Figure 264a, b are examples of the calculated flow field in the Z-plane for $b/a = d/a = 5.0$, using the above-mentioned relations, The solid lines in Fig. 264a show the symmetric stream lines with respect to the center line, ($\alpha = \beta = 0.5$). In this case, it is assumed that $\mu = 1$, $K_1 = -K_2$ ($K_1 > 0$), $Z_1 = (-0.5, 1.7i)$ and $Z_2 = (0.5, 1.7i)$. The strengths of the vortices, K_1 and K_2, are determined by the condition that the nozzle tips C and E are the branch points. Mathematically this is given by

$$(dW/d\theta)_{\theta=\pm n} = 0 \tag{251}$$

μK_1 and μ_2 are determined from Eqs. (249) and (251). When $\mu = 1$, $K_1 = |K_2| = 2.74$.

Figure 265b shows the stream lines for $\alpha = 0.7$ ($\mu = 1$). In this case, it is assumed that $Z_1 = (-0.5, 2.2i)$ and $Z_2 = (0.0, 1.0i)$. These positions are selected in order to approximate the positions of the vortices in Fig. 263e. The strength, K_1, of the vortex situated at Z_1 is determined from the condition that the nozzle tip C is a branch point and strength, $|K_2|$, of vortex at Z_2 equals to K_1. The stream line pattern shown

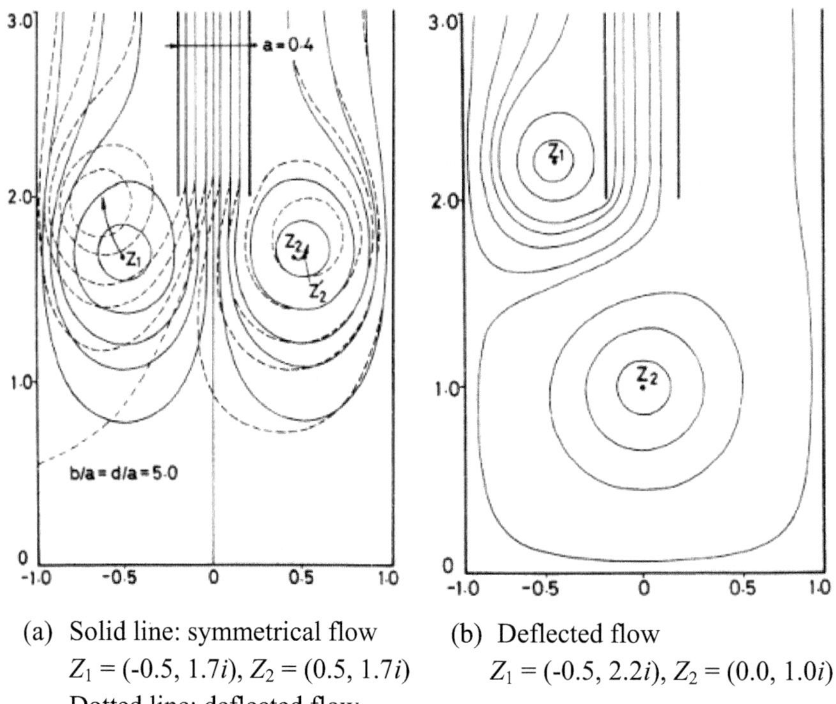

(a) Solid line: symmetrical flow
$Z_1 = (-0.5, 1.7i)$, $Z_2 = (0.5, 1.7i)$
Dotted line: deflected flow
$Z_1 = (-0.5, 1.7i)$, $Z_2 = (0.45, 1.7i)$

(b) Deflected flow
$Z_1 = (-0.5, 2.2i)$, $Z_2 = (0.0, 1.0i)$

Fig. 264 Stream line (Potential flow model)

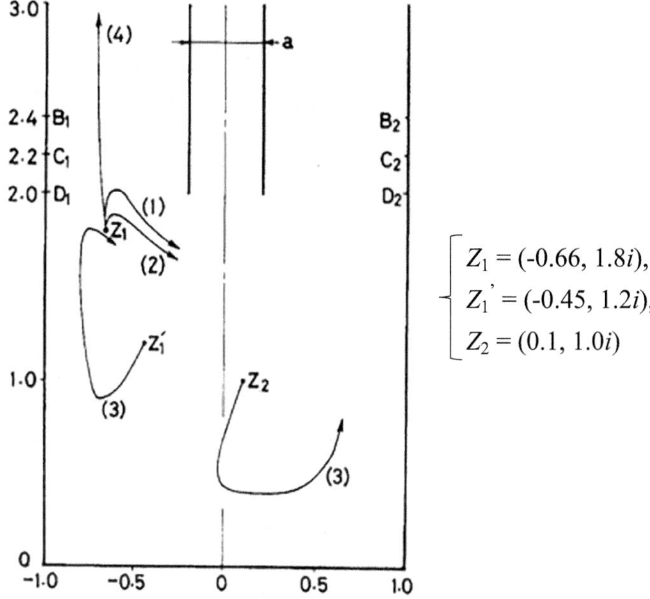

Fig. 265 Behavior of vortex ($b/a = d/a = 0.5$, $|K| = 2.74$)

in Fig. 265b is qualitatively equal to one obtained from the condition that the both nozzle tips C and E are branch points. By comparison of Fig. 264 with 262, it can be recognized that the numerical solutions based on Navier–Stokes equation are similar to those based on the potential flow concerning the relative positions of the vortices and the deflection of the jet. Circulating flow region around Z_1 in spite of $K_1 = |K_2|$ as shown in Fig. 264b. This suggests that an increase of circulating flow region as shown in Fig. 257e, for example, does not always imply an increase of strength of the vortices.

(d.4) Mechanism of oscillation

*Stability of a jet in a cavity

Drift velocities of the two vortex filaments, Z_1 and Z_2 in Fig. 262a, are calculated by Routh's theorem as follows:

$$(u - iv)_{zj} = \left[\frac{d}{dz}\left\{ W(z) - ik_j\log(z - z_j) \right\} \right]_{zj} = \left[\frac{d}{dz}\left\{ \tilde{W}(\theta) + ik_j\log\frac{\theta - \theta_j}{z - z_j} \right\} \right]_{zj}$$

$$= \left[\left\{ \frac{d}{d\theta}\tilde{W}(\theta) \right\} \frac{d\theta}{dz} + \frac{ik_j}{2}\frac{d^2\theta/dz^2}{d\theta/dz} \right]_{zj} \tag{252}$$

where, $j = 1$ or 2.

$\tilde{W}(\theta)$ is obtained from $W(\theta)$ subtracting the complex velocity potential of a vortex filament situated at θ_j;

$$\tilde{W}(\theta) = W(\theta) - ik_j\log(\theta - \theta_j) \tag{253}$$

Therefore, the condition that the vortices stop at certain positions having zero drift velocities is obtained putting the right-hand side of Eq. (252) equal to zero. After some calculations using Eqs. (249) and (252) and (253), the following condition that the vortex stands still is obtained:

$$\theta_1^2\left[\theta_1^6 - \theta_1^4(2 + 3m^2 - n^2 + 2i\sigma m^2) + \theta_1^2\{(4m^2 + m^2n^2) + 2i\sigma m^2(1 + n^2)\}\right.$$
$$\left.-2m^2n^2(1 + i\sigma)\right]$$
$$+\bar{\theta}_1^2\left[3\theta_1^6 - \theta_1^4(2 + m^2 + 5n^2 - 2i\sigma m^2) + \theta_1^2\{(3m^2n^2 + 4n^2) + 2i\sigma m^2(1 + n^2)\}\right.$$
$$\left.-2m^2n^2(1 - i\sigma)\right] = 0$$
$$\tag{254}$$

From Eq. (254), it can be proved that there are no values of θ_1 which satisfy the above equation for any combination of m, n and σ $(= \mu/K_1)$ in the lower half of Fig. 263b. This result implies that the symmetrical flow field with a pair of vortices on both sides of the jet which is represented by the solid lines in Fig. 264a is not maintained stable regardless of the configuration of a cavity and flow rate.

The solid lines with arrows in Fig. 264a represent the calculated path lines of the two vortices by Eq. (251) when the position of one of the symmetrical vortices which are located initially at $Z_1 = (-0.5, 1.7i)$ is shifted from Z_2 to $Z_2{}'$ in a small disturbance. The dotted lines in the figure indicate the stream lines when the centers of vortices come at the points indicate by the arrows. The vortices situated at Z_1 and $Z_2{}'$ move along the arrows. This shows that the un-symmetry of the flow is enhanced and the deflection of the jet increases, as the symmetry of vortices is lost.

(d.5) Mechanism of switching of a jet

The mechanism of the switching phenomenon may be closely related the motion of a pair of vortices which are formed by the effect of viscosity or Reynolds stress and their growth and decay. However, an analytical estimation of growth and decay of the vortex is very difficult without uncertain assumption. Here, it is shown that the mechanism of the switching of the jet can be explained by the potential flow model without consideration of their growth and decay.

Both exits, B and F in Fig. 263, of a cavity are located at finite distance from the nozzle tips, but it is assumed that they are placed at infinity in the potential flow model. Since the vortex approaches considerably an exit before the switching of the jet (Fig. 257) it is appropriate to taken into account the effects of the exit tips (openings) for the examination of the mechanism of the switching. It is assumed that the pressure at the two reference points on the side walls (for example, B_1 and B_2 or C_1 and C_2 corresponding to the both exits) are equal, because the pressure at the exit is nearly equal to the pressure at the outside. Assuming that the switching proceeds quasi-statically, Bernoulli's equation is given by

$$\frac{\rho V_{B1}^2}{2} + P_{B1} = \frac{\rho V_{B2}^2}{2} + P_{B2} \tag{255}$$

where, subscripts B1 and B2 imply the quantities at each station.

After all, the condition at the exit is given by

$$V_{B1}^2 = V_{B1}^2 \tag{256}$$

Figure 265 shows some path lines which are calculated by Eqs. (252) and (256). The strength of the vortices, $|K_j|$, is constant. The curves (1) and (2) in the figure show the path lines of the vortex Z_1 near the exit where the reference points are B_1 and B_2 or C_1 and C_2 and the vortex Z_2 is neglected. The path lines show that the vortex turns near the exit and the jet switches swinging direction from the left to the right. The curve (3) shows the path lines when existence of the vortex Z_2 is considered. The qualitative behavior of the vortex Z_1 (it is denoted Z_1' in the figure) does not change appreciably with neglect of the vortex Z_2. When the vortex approaches the exit, the pressure around it decreases and reversed flow from the outside of the cavity is induced. This turns the switching direction of a jet. The het switching can be recognized from the evolution of the stream lines which are drawn corresponding to the positions of the vortices.

Here, the following were shown.

(1)   The behavior of the jet depends on a pair of vortices which are formed by the effect of viscosity.
(2)   A symmetric flow field with a pair of vortices on both sides of a jet is not maintained and the jet begin to move.
(3)   One vortex approaches an exit when the vortex decreases and a reversed flow is induced. An oscillation of the jet in a cavity is induced, only when the above conditions are satisfied

In this study on the mechanism of oscillation of a jet, effects of the strength of the vortices, their growth and decay on the oscillation are not take into account. Nevertheless, it seems that this study gives an effective model for the mechanism of oscillation as a first approximation.

8   Compressible Flows

8.1   Thermodynamic Relationship

First, the physical quantities necessary to consider compressible flow are defined simply.

8.1.1   Gas Equation of State

The equation of state of perfect gas, or ideal gas is expressed by the following equation.

$$p = \rho RT \tag{257}$$

where, p: pressure [Pa], T: absolute temperature [K], ρ: density [kg/m^3], R: Gas constant [J/(kg K]

This equation of state gives a good approximation for normal pressure and temperature ranges.

8.1.2   Internal Energy and Constant Volume Specific Heat

Now, consider a closed system with volume V and mass m shown in Fig. 266.

If the heat of dQ is added to the system from the surroundings, the internal energy increases by dE and the work of the surroundings is dW, from the first law of thermodynamics,

$$dQ = dE + dW \tag{258}$$

If the work is only due to the expansion of the gas,

$$dW = pdV \tag{259}$$

In terms of unit volume,

$$dq = de + pd\overline{v} = de + pd\left(\frac{1}{\rho}\right) \tag{260}$$

The constant pressure specific heat C_p [J/(kg•K] is

$$C_\mathrm{p} = \left(\frac{dq}{dT}\right) = \frac{de}{dT} \tag{261}$$

Fig. 266 Control volume,
First law of thermodynamics

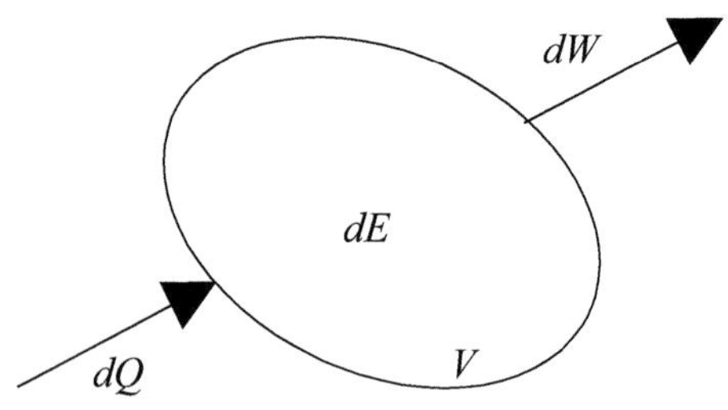

When C_v is constant, the internal energy is

$$e = C_v T \tag{262}$$

8.1.3 Enthalpy and Specific Heat at Constant Pressure

Enthalpy is considered to be the energy possessed by the system under the condition of constant pressure, and is defined by the following equation.

$$h = e + \frac{p}{\rho} \tag{263}$$

Rewriting the Eq. (260) in terms of enthalpy, we get

$$dq = dh + \frac{1}{\rho}dp \tag{264}$$

The specific heat at constant pressure C_p and the enthalpy h are expressed by the following equations.

$$C_p = \left(\frac{dq}{dT}\right) = \frac{dh}{dT} \tag{265}$$

$$C_p T = h \tag{266}$$

8.1.4 Specific Heat

For perfect gas, Eqs. (257), (260), and (262) give the following equation.

$$RdT = d\left(\frac{p}{\rho}\right) = \frac{1}{\rho}dp + pd\left(\frac{1}{\rho}\right) = \frac{1}{\rho}dp + dq - de$$

$$= \frac{1}{\rho}dp + dq - C_p dT \tag{267}$$

$$\left(\frac{dq}{dT}\right) = C_p = R + C_v \tag{268}$$

Defining the ratio of specific heats, $\kappa = C_p/C_v$, for the perfect gas,

$$C_p = \frac{\kappa R}{\kappa - 1}, \; C_v = \frac{R}{\kappa - 1} \tag{269}$$

where, C_p: specific heat at constant volume

8.1.5 Isentropic Change

Entropy is a quantity that expresses the statistical mechanical randomness of a system, and its change is expressed by the following equation.

$$dq = Tds \tag{270}$$

where, from Eq. (260),

$$ds = C_\mathrm{v}\frac{dT}{T} - R\frac{d\rho}{\rho} = C_\mathrm{v}\frac{dp}{p} - C_\mathrm{p}\frac{d\rho}{\rho} \tag{271}$$

Assuming an isentropic change,

$$C_\mathrm{v}\frac{dT}{T} = R\frac{d\rho}{\rho}, \quad C_\mathrm{v}\frac{dp}{p} = C_\mathrm{p}\frac{d\rho}{\rho} \tag{272}$$

By integrating,

$$\frac{\rho}{T^{1/(\kappa-1)}} = \text{const.}, \quad \frac{p}{\rho^\kappa} = \text{const.}, \quad \frac{\rho}{T^{\kappa/(\kappa-1)}} = \text{const.} \tag{273}$$

8.1.6 Sound Velocity and Mach Number

In compressible flow, sound velocity and Mach number are important indices.

***Sound velocity**

Now, consider the case where a piston is placed in a tube. The flow in the tube is stationary, and if the piston starts to move at a very small velocity u from a certain moment, the pressure in front of the piston and the mist increase by dp and $d\rho$, respectively. This change in pressure travels at velocity a, i.e., the speed of sound, which is the propagation velocity of pressure waves. Now, if we move at the propagation velocity of the pressure wave, the velocity at the position where the pressure wave reaches zero, the velocity is—a on the right side, and the velocity is $(u–a)$ on the left side. Mass must also be conserved here, so,

$$-a\rho = (\rho + d\rho)(u - a) \tag{274}$$

Neglecting the square of the infinitesimal amount in this equation and considering the conservation of momentum,

$$dp = \rho a[a - (a - u)] \tag{275}$$

Then,

$$a = \sqrt{\frac{dp}{d\rho}} \tag{276}$$

Assuming an isentropic change, since the sound wave is a rapidly decreasing change that prevails in minute changes,

$$a = \sqrt{\frac{dp}{d\rho}} = \sqrt{\kappa \frac{p}{\rho}} = \sqrt{\kappa R T} \tag{277}$$

Mach number is defined by

$$M = \frac{u}{a} \tag{278}$$

8.2 Supersonic Jet

8.2.1 Nozzle and Supply Pressure

A nozzle is used to increase the ejection velocity of the jet. A monotonically shrinking nozzle shown in Fig. 267 is called convergent nozzle in general or sonic nozzle for a high-speed jet. Now, if the supply pressure P_0 is made somewhat larger (constant) than the ambient pressure P_a and P_a is reduced, the flow Mach number at the nozzle outlet will be $M = 1.0$ when the pressure ratio $P_a/P_0 = 0.528$. After that, even if P_a is decreased and the pressure difference with P_0 is increased, the Mach number at the nozzle exit remains $M = 1.0$ and the flow rate does not increase. This is called choked flow.

Conversely, if P_a is kept constant (for example, atmospheric pressure) and P_0 is increased, a similar phenomenon occurs. However, when P_0 is increased, the Mach number at the nozzle exit remains $M = 1.0$, but the nozzle exit pressure P_e increases, and as a result, the mass flow rate also increases.

In the subsonic flow P_e equals P_a, but in the supersonic flow $P_e \neq P_a$ and $P_e > P_a$, and then expansion waves and shock waves occur after the nozzle exit, and the pressure of the jet decreases to P_a.

A jet ejected under the condition of $P_e > P_a$ is called an under-expansion jet, and expands after the nozzle exit.

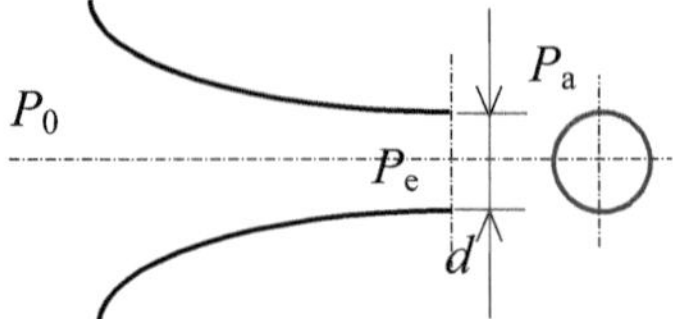

Fig. 267 Convergent nozzle, and convergent and divergent nozzle

(a) Convergent nozzle or
Sonic nozzle for high speed jet

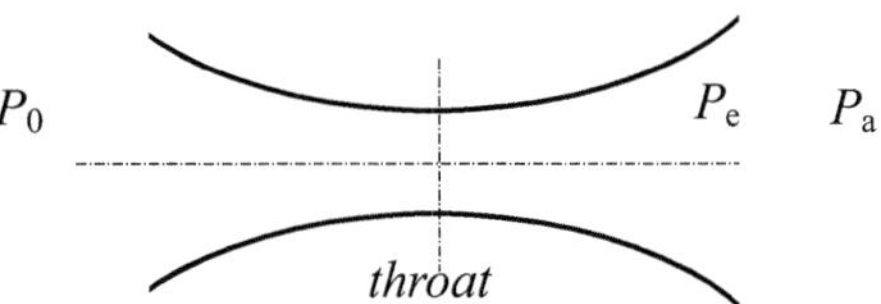

(b) Convergent and divergent nozzle or
Supersonic nozzle for supersonic jet

A nozzle that smoothly shrinks to the downstream as shown in Fig. 267 is called a supersonic nozzle. In supersonic nozzles, the flow Mach number is $M = 1.0$ at the throat and then expands to supersonic flow.

A nozzle that smoothly contracts and expands as shown in Fig. 267 is called a supersonic nozzle. In supersonic nozzles, the flow Mach number is again $M = 1.0$ at the throat and then expands to supersonic flow.

The flow pattern after the convergent nozzle exit is classified as follows depending on the pressure ratio of P_0/P_e.

(a)   Flow pattern in the pressure ratio of $1.0 < P_0/P_a < 1.89$ [shown in Fig. 268a];

For a pressure ratio of $1.0 < P_0/P_a < 1.89$, the pressure P_e at the nozzle exit is equal to the ambient pressure P_a. In addition, the nozzle exit velocity u_e is constant in the core region of the length x_c of the jet center. The diameter of the core region decreases linearly downstream to zero, around which there is a region of mixing with the surrounding fluid. The outer edge of the jet is a free boundary. For example, the position y where the velocity u becomes $u/u_m = 0.1$ of the maximum velocity u_m is generally defined as the jet boundary. In addition, the width of the jet and the half width $b_{1/2}$, that is, the position of y at half $u_{m/2}$ of the maximum flow velocity are used for the spread of the jet.

After the core region, the transition region leads to the development region where the velocity distribution is self-similar. The length of the core region is $x_c \fallingdotseq 5.0d$, and $x_c \fallingdotseq 6.0b_0$ (b_0: nozzle width) for the two-dimensional jet.

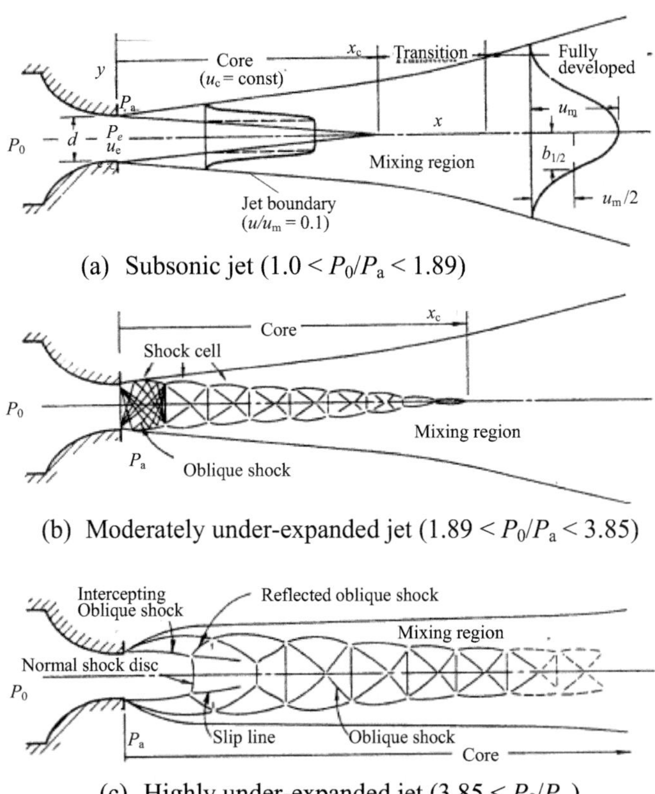

(a) Subsonic jet ($1.0 < P_0/P_a < 1.89$)

(b) Moderately under-expanded jet ($1.89 < P_0/P_a < 3.85$)

(c) Highly under-expanded jet ($3.85 < P_0/P_a$)

Fig. 268 Three variations in jet flow from a sonic nozzle (Donaldson and Snedeker 1971a, b)

(b) $1.89 < P_0/P_a < 3.85$ 1.89 [shown in Fig. 268b]

In moderately under-expanded jets with a pressure ratio of $1.0 < P_0/P_a < 1.89$, when the nozzle exit pressure is $1.0 < P_e/P_a < 2.0$, the jet expands to the surrounding pressure P_a and produces a series of oblique shock waves from the nozzle edges and diamond shock waves, shock cells, downstream. The jet mixes with the surrounding at the outside of the shock cells.

(c) $3.85 < P_0/P_a$ [shown in Fig. 268c];

A Mach disk occurs when the pressure ratio is $85 < P_0/P_a$. As P_0 increases, the flow after the nozzle exit rapidly expands and the pressure decreases, and rapid recompression occurs. As a result, the oblique shock waves converge to the end of the first shock cell to produce a vertical shock wave, a Mach disk. Velocity upstream of the Mach disk is supersonic and downstream of it is subsonic. In this case, the nozzle exit pressure is $P_e/P_a > 2.0$, and when it becomes $P_e/P_a > 4.0$, the subsonic flow downstream of the Mach disk is rapidly accelerated and a second shock disk forms.

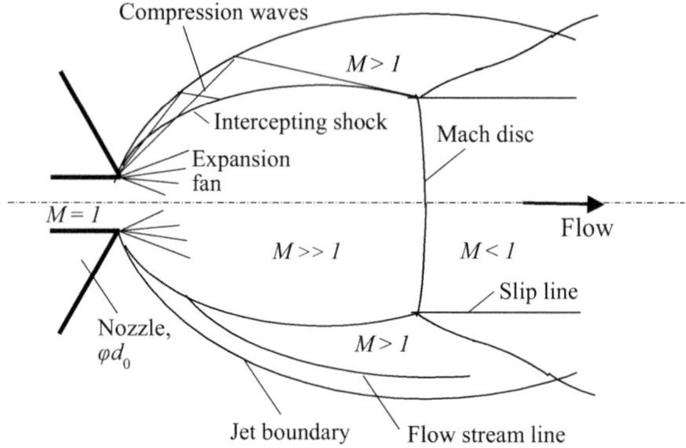

Fig. 269 Schematic diagram of first shock cell of a highly under-expanded jet (Crist et al. 1966)

The details of the construction of the first shock cell of a highly under-expanded jet are shown in Fig. 269. The jet from the nozzle expands rapidly to form a high-density jet boundary, where expansion waves generated from the nozzle corners are reflected and converge to form a high-density Mach disk, normal shock disc.

8.2.2 Static Pressure and Velocity Distributions in the First Shock Cell of Under-Expanded Jet

Figure 270 shows the sketch of static pressure and velocity distributions in the first shock cell by Kojima and Matsuoka (1986). The Mach number is in the range of 1.1 $\lesssim Ma \lesssim 1.3$. They explained it as follows. The static pressure distribution becomes a convex distribution immediately after the nozzle exit and a W-shaped distribution slightly downstream. The reason why the pressure is high in the central part is that the pressure just before the nozzle exit is higher than the atmospheric pressure due to the choking phenomenon, and it is less susceptible to rapid expansion than the jet end. Judging from the Schlieren photograph, the positions of the two minimum values of the W-type are generated near the expansion wave indicated by the dashed line ($y/d_0 \fallingdotseq \pm 0.25$), and are not caused by the entrainment phenomenon. This W-shaped central peak gradually decreases due to expansion, and becomes a concave distribution near the maximum expansion cross-section.

Next, in the compression region, there are two pressure maxima near the compression wave indicated by the solid line, resulting in an M-shaped distribution. Thereafter, the same distribution is repeated downstream. In addition, at the edge of the jet where $y/d_0 > \pm 0.5$ in any cross section, the pressure change is monotonous and gradually recovers to the atmospheric pressure. Considering that all pressure distributions are convex distributions, the shape of the velocity distribution is likely to be affected by the shape of the static pressure distribution as shown in the figure. Therefore, the shape of the velocity distribution corresponding to the convex or W-shaped static pressure distribution near the expansion region and the front shock wave is a saddle shape distribution with a slightly depressed center as shown in Fig. 270a, d.

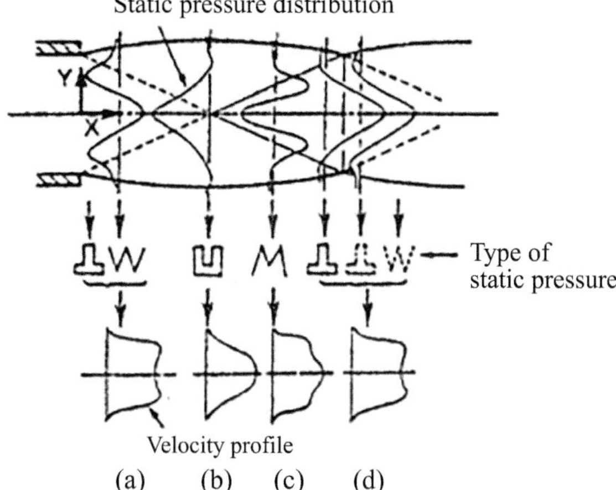

Fig. 270 Static pressure and velocity distributions in the first shock cell of under-expanded jet (Kojima and Matsuoka 1986)

The velocity distribution corresponding to the concave static pressure distribution near the maximum expansion cross section becomes the convex distribution shown in Fig. 270b. Also, the velocity distribution corresponding to the M-shaped static pressure distribution in the compression region is a papillary distribution shown in Fig. 270c.

Figure 271 shows the flow pattern of a jet issuing from a long pipe nozzle with a diameter of $d_0 = 4.0$ mm toward atmospheric pressure P_a at various supply pressures P_0. They were taken by Schlieren photography. The dotted line shows the approximate jet boundary on the Schlieren photo. Mach number at the nozzle exit is $M \fallingdotseq$ 0.45, 1.14, 1.52 for $P_0 = 0.115, 0.225, 0.38$ MPa, respectively.

Figure 271a at $P_0/P_a = 0.2$ shows a subsonic jet. The supply pressure is $P_0 = 0.02$ [MPa]. The jet spreads almost linearly to the downstream. The flow structure is that appeared in Fig. 268a.

Figure 271b at $P_0/P_a = 1.15$ shows before the moderately under-expanded jet appeared in Fig. 268b. The dark and bright colored areas are expansion and compression regions, respectively. In the figure something like a shock cell can be seen. The jet expands slightly just after the nozzle exit and keeps a constant width, call it shock cell width, to the downstream and the interval of the shock cell is smaller than that in Fig. 268b.

Figure 271c at $P_0/P_a = 2.25$ shows the same flow pattern as the moderately under-expanded jet in Fig. 268b. The shock cell appeared clearly, and when the pressure increases further and reaches $P_0/P_a = 8$, the flow pattern becomes moderately under expanded jet appeared in Figs. 271d and 268b. The jet expands rapidly at the nozzle exit, and the shock cell width becomes about $1.8d_0$ and it remains almost constant. However, the jet width w_j increases to the downstream following

$$w_j/d_0 = 1.2802 \cdot \ln(x/d_0) + 2.4862 \quad \left(\text{for } x/d_0 \gtrsim 2 \right) \tag{279}$$

Fig. 271 Visualized flow pattern of pipe jet by Schlieren method ($d_0 =$ 4.0 mm, $P_0/P_a = 0.2 \sim 3.8$) (Shakouchi et al. 2019)

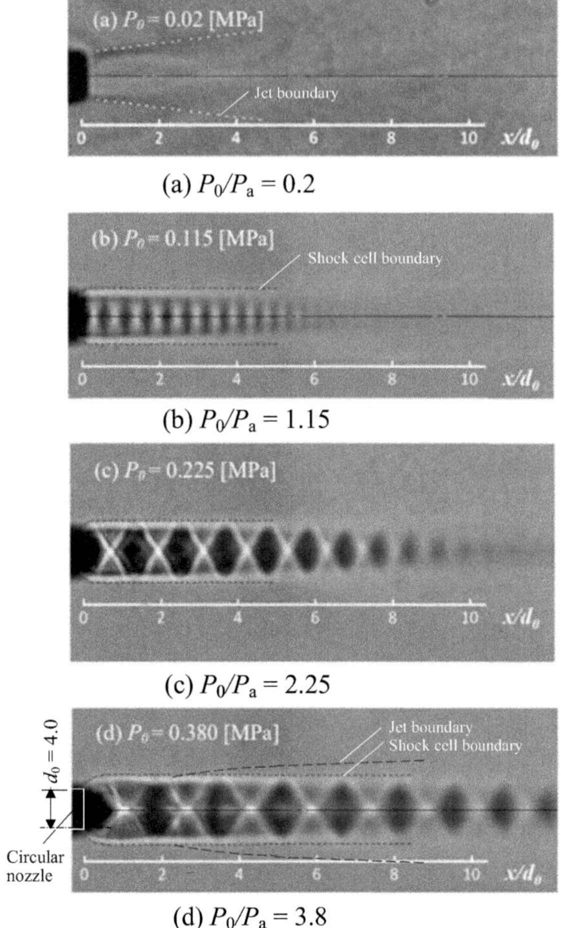

(a) $P_0/P_a = 0.2$

(b) $P_0/P_a = 1.15$

(c) $P_0/P_a = 2.25$

(d) $P_0/P_a = 3.8$

As increasing P_0/P_a, the size, width and length, of the shock cell becomes larger. The size of the shock cell becomes smaller in the downstream direction. This is in the transition state from moderately to highly under-expanded jet. In the case of $P_0/P_a > 3.85$ the flow pattern is highly under-expanded jet shown in Fig. 268c, and the details of the first shock cell is appeared in Fig. 269.

In order to examine the change of shock cell length L_c to the downstream, in Fig. 272a L_c of each shock cell at $P_0 = 0.38$ MPa is shown, and the 1st one has a maximum value and repeats alternately increases and decreases toward the downstream, however decreases almost linearly overall.

Figure 272b shows the length of the 1st shock cell, L_{c1}, and it increases with increasing P_0 following the relation of

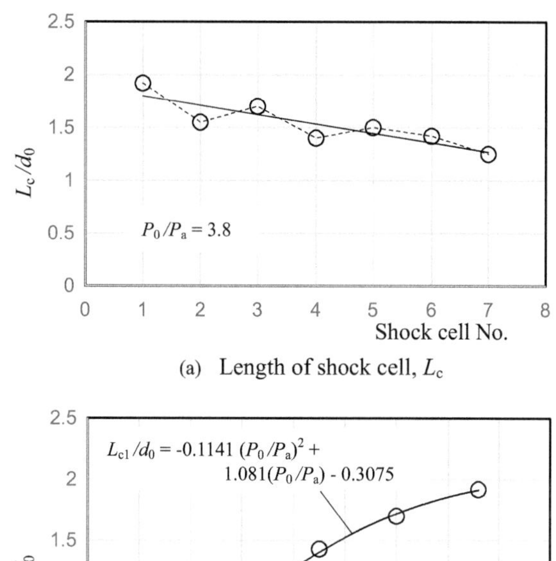

Fig. 272 Length of shock cell

(a) Length of shock cell, L_c

(b) Length of 1st shock cell, L_c and P_0

$$L_{c1}/d_0 = -0.1141(P_0/P_a)^2 + 1.081(P_0/P_a) - 0.3075 \quad (\text{for } 0.115 < P_\circ/P_a < 3.8) \tag{280}$$

8.2.3 Velocity Distribution

(a) **Jet centerline velocity**

The jet center line velocity u_c of supersonic jet has a different profile from a subsonic jet as shown in Fig. 273. u_c of the jet after it leaves the nozzle exit at pressure $P_0 = 0.38$ MPa ($P_0/P_a = 3.8$) is large as a result of the rapid expansion, and then decreases and increases repeatedly as the flow compresses and expands. u_c is shown on the right side of the figure along with a visualization photograph of $P_0 = 0.38$ MPa. It can be seen that the minimum and maximum values of u_c correspond to the compression and expansion areas of the flow. The maximum velocity of u_c is about 470 m/s at $x/d_0 \backsimeq 9$, and downstream of $x/d_0 \gtrsim 11$, u_c decreases proportionally to $(x/d_0)^{-0.329}$. u_c at $P_0 = 0.18$ and 0.115 MPa decrease proportionality to $(x/d_0)^{-0.923}$ for $x/d_0 \gtrsim$

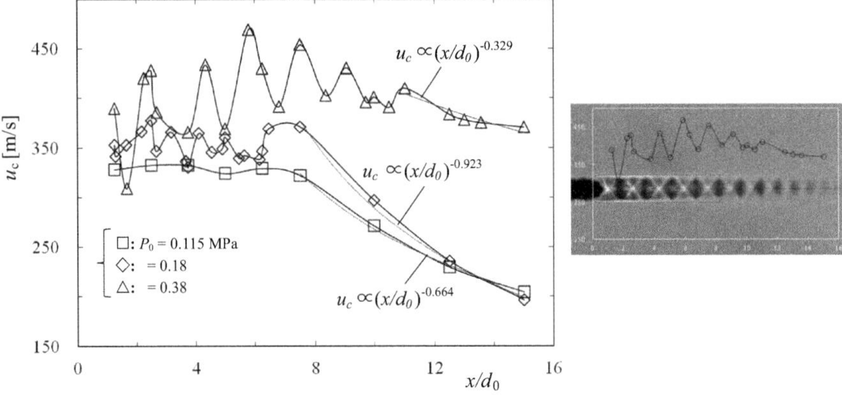

Fig. 273 Jet center line velocity ($P0 = 038$ MPa)

7.5 and $(x/d_0)^{-0.664}$ for $x/d_0 \gtrsim 7.5$, respectively. They are different from a subsonic circular jet which is $u_c \propto (x/d_0)^{-1}$ for $x/d_0 \gtrsim 5$ through a constant velocity ($u_c =$ const.) region, core region, of $x/d_0 \lesssim 5$.

(b) **Velocity distribution at cross section**

Figure 274 shows the velocity distributions at various cross sections of $x/d_0 = 2.1 \sim$ 8.3. The positions of r where velocities become $u_c/2$ and $u_c/10$ are shown in the figure. Velocity profile is axisymmetric at $x/d_0 = 2.1$ in near the end of expansion area has a minimum at the jet center and two maxima. This depends on rapid jet expansion The profile is different from a subsonic jet which has axisymmetric profile with maximum value at the center of the jet. The local minimum value at the center of $x/d_0 = 2.1$ recovers at $x/d_0 = 4.2$.

(c) **Spreading of jet**

Figure 275 shows the jet width w_j and half width $w_{1/2}$. w_j and $w_{1/2}$ are defined as the jet boundary is the radius position of $u = 0.1u_{max}$ ant the position of r at $u_{max}/2$, respectively. The jet expands quickly at the nozzle exit, and the jet width is almost constant until $x/d_0 \fallingdotseq 2$ and then it increases to the downstream according to the following equation. In the case of under expanded supersonic jet,

$$w_j = 1.2802 \ln(x/d_0) + 2.4862 \quad \left(\text{for } 2 \lesssim x/d_0 \leq 8.5 \right) \tag{281}$$

The half width increases slightly almost linearly to the downstream according to the following equation. The jet spreading of under expanded supersonic jet is much smaller than that of subsonic jet which is $w_{1/2} \propto (x/d_0)^{1/2}$.

$$w_{1/2} = 0.0557(x/d_0) + 2.7214 \quad \left(\text{for } 2 \lesssim x/d_0 \leq 8.5 \right) \tag{282}$$

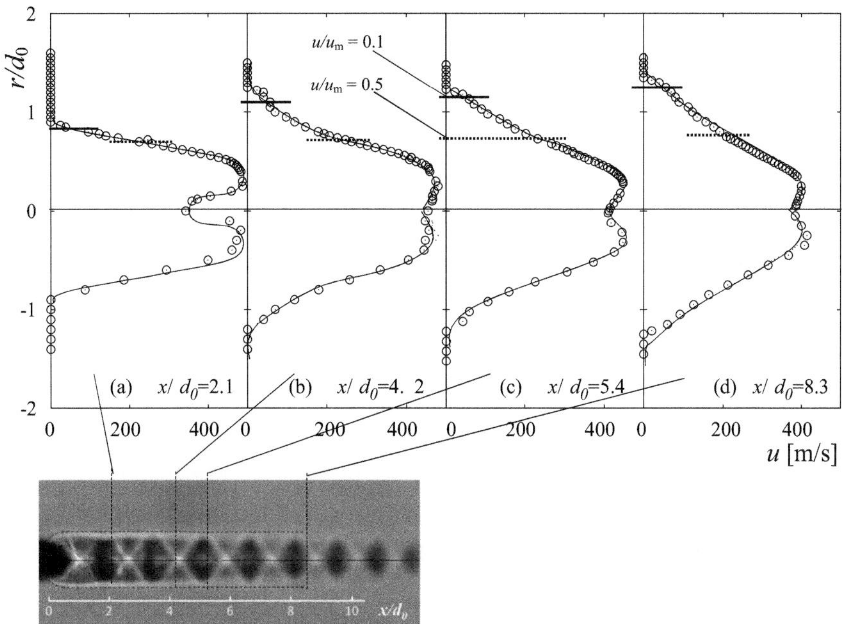

Fig. 274 Velocity distribution by Pitot tube ($d_0 = 4.0$ mm, $P_0/P_\mathrm{a} = 0.38$)

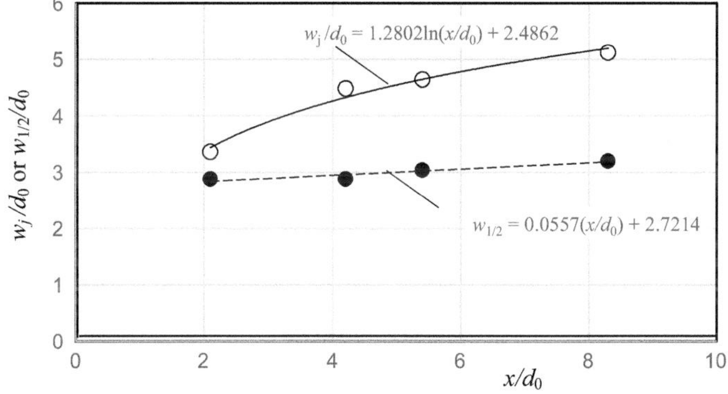

Fig. 275 Jet width and half width ($P_0/P_\mathrm{a} = 0.38$)

(d) **Flow rate**

The volumetric flow rate Q was calculated by the following equation and shows in Fig. 276. The flow rate is almost constant up to $x \fallingdotseq 2$ and does not involve the surrounding fluid, but increases rapidly from $x \fallingdotseq 3 \sim 4$ and then becomes almost constant. This is different from subsonic jet which increases in proportion to x. The jet appears to be flowing through a stationary fluid.

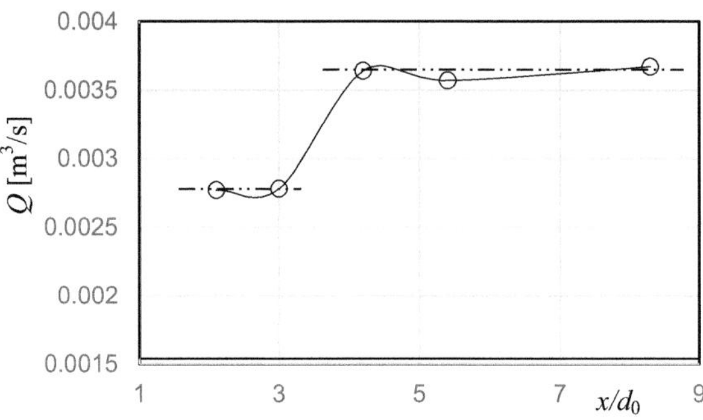

Fig. 276 Volumetric flow rate ($P_0/P_a = 0.38$)

$$Q = \int\limits_{-\infty}^{\infty} 2\pi r u \, dr \tag{283}$$

8.2.4   Measurement of Velocity of Supersonic Jet Flow by Pitot Tube

The velocity measurement of compressible flow by Pitot tube is shown here.

The flow velocity u in an incompressible fluid can be obtained by the following equation if the stagnation pressure, or total pressure P_0 and static pressure P are measured with a Pitot tube.

$$P_0 - P = \frac{1}{2}\rho u^2 \tag{284}$$

However, for compressible flow, P_0 and P give the Mach number, M. If M and static temperature T at anywhere in the flow field are known, u can be obtained from the following equation, but it is generally difficult to obtain T. Where, κ is specific heat ratio.

$$u = M\sqrt{\kappa R T} \tag{285}$$

If the flow is adiabatic, then the relationship of Eq. (285) between the total temperature T_0 of the reservoir and the local static temperature T is determined, and u is determined from the measurements of T_0 and M.

Now, the sound velocity a_0 at the stagnation point is given by the following equation from the isentropic relation of perfect fluid.

$$a_0 = \sqrt{\frac{\kappa p}{\rho}} = \sqrt{\kappa R T_0} \tag{286}$$

Also, from the energy equation,

$$\frac{1}{2}u^2 + \frac{a^2}{\kappa - 1} = \frac{a_0^2}{\kappa - 1}.$$

(287)

Now, since $M = u/a$,

$$\left(\frac{a_0}{a}\right)^2 = \frac{T_0}{T} = 1 + \frac{\kappa - 1}{2}M^2$$

(288)

$$\therefore u = Ma = M\sqrt{\kappa RT_0}\left(1 + \frac{\kappa - 1}{a}M^2\right)^{1/2}$$

(289)

8.3 *Effects of Nozzle Shape on Spreading of Supersonic Jet*

8.3.1 Spread of Jet

As in the case of subsonic jets shown in Section x, the nozzle shape influences also the flow characteristics of supersonic jets greatly. Similarly for case supersonic jet, Zaman (1996) experimentally investigated the effects of nozzle geometry using circular, oval, rectangular and generic lobed nozzles which are shown in Fig. 277. The diameter of circular nozzle is $d_0 = 1.27$ cm, and the equivalent diameter (D) is the same for all the asymmetric nozzles and 1.47 cm. The asymmetric nozzle used are 3:1 elliptic, 3:1 rectangular and a generic 4-loved nozzles. The circular nozzle with four equally small delta tabs having a triangular shape attached on the nozzle wall and an apex leaning angle at 45°. The area blockages are about 8 and 12% for the circular and rectangular cases, respectively.

Figure 278 shows the iso-surface of Mach number $M/M_{\max} = 0.3$. Mach number at the nozzle exit is $M_j = 1.63$ (M_j: jet Mach number assuming fully expanded flow

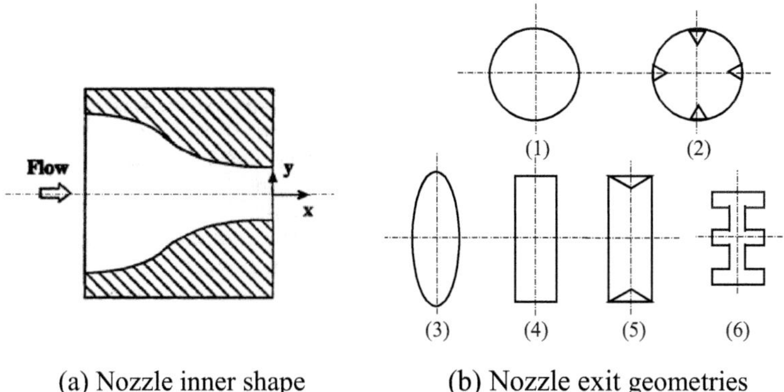

(a) Nozzle inner shape   (b) Nozzle exit geometries

Fig. 277 Schematics of four nozzles with and without tabs (Zaman 1996)

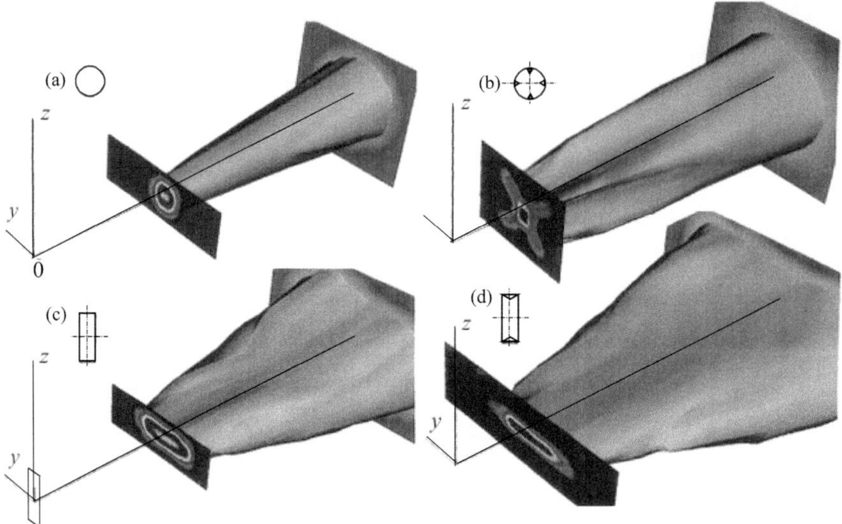

Fig. 278 For $M_j = 1.63$, iso surfaces of Mach number, $M/M_{max} = 0.3$, M_{max}: maximum Mach number at a given x, Data cover x/D range of 10 \sim –30 in (**a**) and (**c**), and 5–30 in (**b**) and (**d**) (Zaman 1996)

for a given P_T/P_a, P_T: plenum chamber pressure, P_a: ambient pressure). The circular nozzles with and without tabs are shown in Fig. 277b-(1) and -(2), respectively. In the case of circular jet, the iso-surface of velocity expands into a cone shape downstream. The jet in case Fig. 278b, which has four small tabs inside the circular nozzle, expands significantly in the downstream direction. In the case of the rectangular nozzle shown in Fig. 278c, a so-called switching phenomenon occurs downstream of the nozzle exit, in which the major and minor axes of the jet are switched, and the jet then expands into an elliptical cone shape.

8.3.2   Jet Switching

The jet switching is explained as follows. In the corners of a non-circular nozzle, such as the rectangular nozzle shown in Figs. 278c and 279, longitudinal vortices that rotate in opposite directions are generated. When these in $z > 0$ are ejected from the nozzle, they move in the -z-direction (the large arrow in Fig. 279) due to the induced velocity generated between the vortex pairs. The induced velocity u_i can be expressed as $u_i = \Gamma/(4\pi a)$, where Γ is the vortex strength, circulation, and $2a$ is the distance between the vortex pairs. When the two vortex pairs collide, the jet expands (in $\pm y$ directions) in in such a way that its major and minor axes are swapped.

Fig. 279 Vortex pairs from non-circular nozzle, rectangular nozzle (Vortex pair and induced velocity, Zaman 1995, 1996)

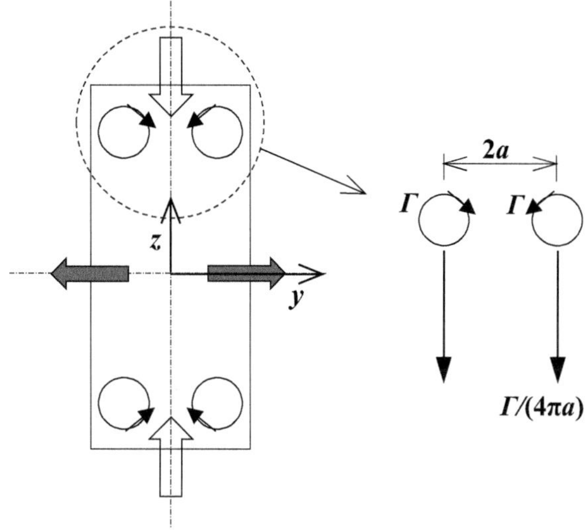

8.3.3 Maximum Mach Number

The maximum Mach number, M_{max}, for $M_j = 1.63$ decreases to the downstream in every case as shown in Fig. 280, and the value for the circular jet is larger than that for other jets without the hook-shaped jet.

Fig. 280 Maximum Mach number ($M_j = 1.63$) (Zaman 1996)

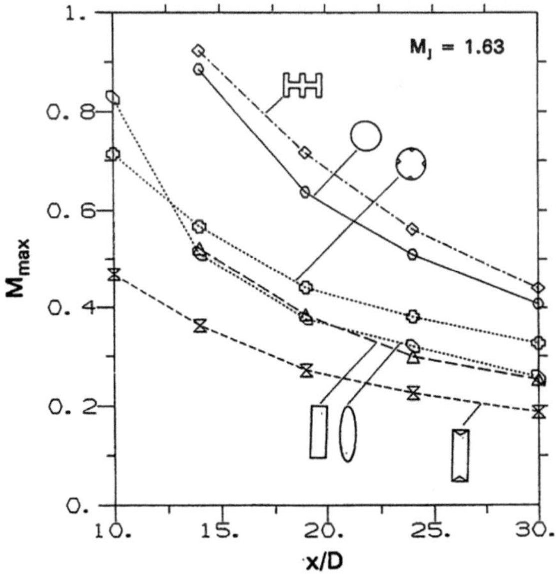

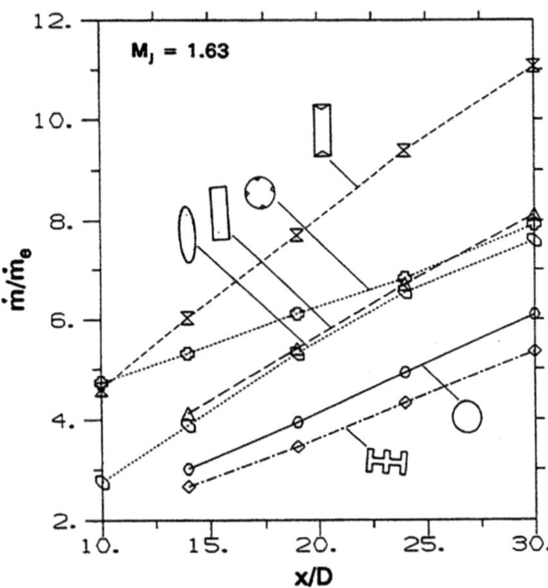

Fig. 281 Streamwise normalized mass flux, $M_\mathrm{j} = 1.63$, (Zaman 1996)

8.3.4   Mass Flux

The mass flux, m/m_e (m_e: mass flow rate at the nozzle exit), of the circular nozzle increases with increasing x/d_0 almost linearly and the slop is about 0.18 (Fig. 281). This was much smaller than that ($= 0.27$) of the subsonic case at $M_\mathrm{j} = 0$. The mass fluxes for both the rectangular and elliptic jets are significantly larger than that of the circular jet. The circular jet with four tabs shows a much larger mass flux, and the increase rates are approximately 1.7 and 1.35 times at $x/d_0 = 15.0$ and 25.0, respectively, compared to the circular jet. Also, in the case of the rectangular nozzle with two-small tabs the increase rates are approximately 2.0 and 1.85 times at $x/d_0 = 15.0$ and 25.0, respectively, compared to the circular jet.

8.3.5   Far Field Noise Spectra

Figure 282 shows the far field noise spectra. The microphone was located approximately $70d_0$ away from the jet axis and about $110°$ from the downstream axis. In the case of circular jet, the dominant screech component of 9 kHz appeared clearly, but this is disappeared when the tabs applied and the sound pressure level, SPL, is reduced in the region $f < 4.0$ kHz. In both cases of rectangular and elliptic jets the screech noises are also appeared, but the SPL are smaller considerably than that of the circular jet.

Here, we examined the flow characteristics of supersonic jet issued from circular and rectangular nozzle with and without small tabs, and the followings were obtained.

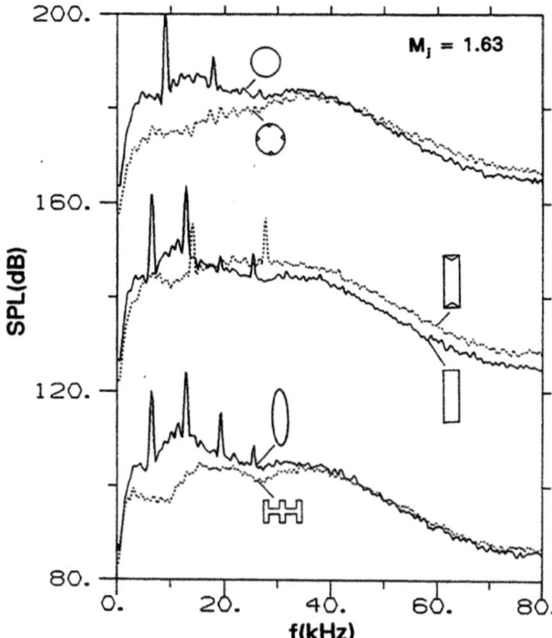

Fig. 282 Far field noise spectra ($M_j = 1.63$) (Zaman 1996)

(1) The circular jet with four tabs shows a much larger mass flux, and the increase rates are approximately 1.7 and 1.35 times at $x/d_0 = 15.0$ and 25.0, respectively, compared to the circular jet. Also, in the case of the rectangular nozzle with two-small tabs the increase rates are approximately 2.0 and 1.85 times at $x/d_0 = 15.0$ and 25.0, respectively, compared to the circular jet.

(2) Switching phenomenon occurs downstream of the nozzle exit, in which the major and minor axes of the jet are switched, and the jet then expands into an elliptical cone shape.

(3) In both cases of rectangular and elliptic jets the screech noises are also appeared, but the SPL are smaller considerably than that of the circular jet.

8.4 Supersonic Reattached Jet

If there is a wall near the two-dimensional jet, the jet will reattach to the wall and flow due to the Coanda effect. The jet entrains surrounding fluid from the outer edge of the jet on the wall surface side, and as a result, the pressure between the outer edge of the jet flow and the wall surface decreases, and the jet flow is deflected toward the wall surface. This also occurs in supersonic jets. The same is true for three-dimensional jets, but in this case the pressure drop is small and the degree of jet reattachment is weak (e.g., the reattaching distance is large).

Here, first, supersonic free jets will be briefly discussed, and then supersonic reattached jets will be examined (Ohmura et al. 2021; Shakouchi et al. 2021).

8.4.1   Two-Dimensional Free Jet

(a)   Test section

In order to examine the flow characteristics of a two-dimensional supersonic free and reattached jets, the test section which consists of the rectangular nozzle of the width $b = 2.5$ mm and aspect ratio $AR = 4.0$, and upper and lower transparent glass end plates. The nozzle is a contraction nozzle with a parallel section of length $4b$.

(b)   Numerical analysis

The CFD program, SOLIDWORKS Flow Simulation, was used for the numerical analysis. The governing equations obtained using the k–ε turbulence model and the Favre-averaged Navier–Stokes equation were analyzed based on the finite sedimentation and participation methods.

(c)   Visualized flow pattern

Figure 283 shows the visualized flow pattern of the two-dimensional supersonic free jet which were taken by Schlieren method. The supply pressure is $P_0 = 0.2$–0.5 MPa. Only the approximate upper outline of the shock cell is shown with a dotted line since the flow was axisymmetric. The jet from the two-dimensional parallel nozzle shows the under-expanded supersonic jet flow with barrel-shaped shock cells. The jet from the nozzle repeatedly expands and then compresses, forming a shock cell. The shock cell disappears in the downstream and the length increases with increasing P_0. In the case of $P_0 = 0.4$ MPa, contour map of density due to numerical calculation is also shown, and it can be expressed well the experiments by Schlieren method.

Figure 284a, b show the contour map of velocity and pressure distributions at $P_0 = 0.4$ MPa, respectively. The velocity u and the pressure in the expansion area (read colored area) of shock cell are large and small, respectively, and in the subsequent compression region they are the opposite, respectively.

8.4.2   Two-Dimensional Supersonic Reattached Jet

When there is a wall near the jet, the jet attaches to and flow round nearby the wall. This Coanda-reattached jet flow is used for controlling the flow in a wall attached type fluidic device, preventing flow separation, and increasing the wing lift force. The flow characteristics of incompressible two-dimensional (2D) Coanda-reattached jet flows have been elucidated considerably owing to the development of fluidics. For example, Bourque et al. studied the reattachment distance and pressure within the separation bubble of a 2D reattached jet by a theoretical analysis using momentum theory, and compared them with experiments in terms of the offset distance and

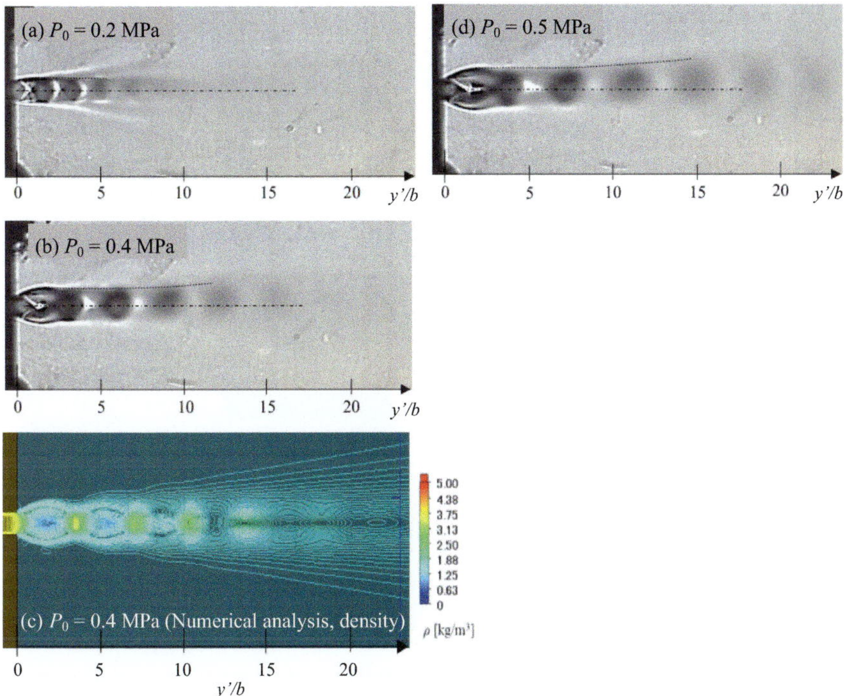

Fig. 283 Visualized flow pattern of two-dimensional supersonic free jet

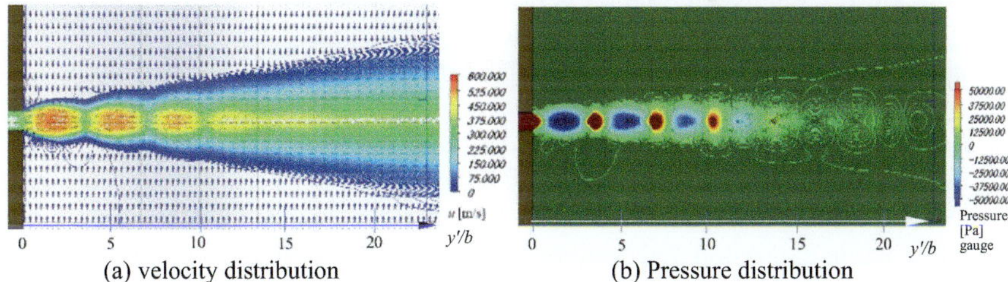

Fig. 284 Contour map of velocity and pressure distributions, $P_0 = 0.4$ MPa

inclined angle of the side wall. This has been introduced precisely in the previous Sect. 5. Here, we will examine the flow characteristics of a supersonic reattached jet flow by a numerical and experimental analyses.

(a) Flow model and coordinate system

Figure 285a shows the flow model and an illustration of how the 2D air jet issued from the nozzle of width b reattaches to the side wall with an offset distance D at the reattaching distance x_R while surrounding a separation bubble region of pressure P_b.

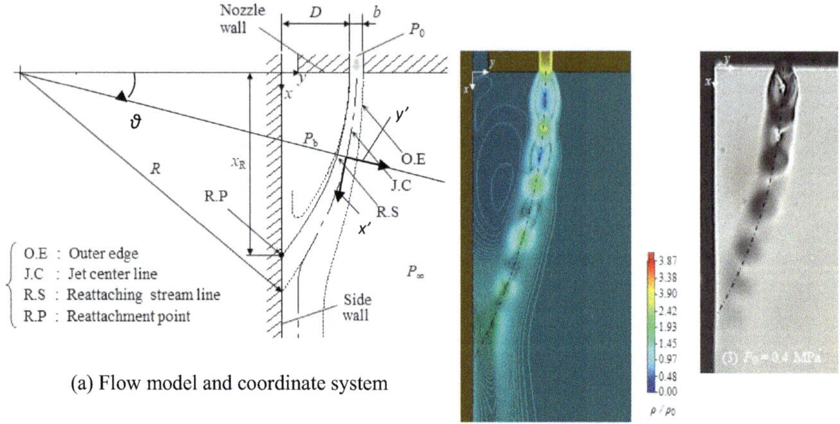

(a) Flow model and coordinate system

(b) Numerical flow visualization by   (c) Schlieren photo.
contour map of density

Fig. 285 Flow model and coordinate system of Two-dimensional supersonic Coanda reattached jet flow [(**b**), (**c**): $D/b = 5.0$, $P_0 = 0.4$ MPa]

In Fig. 285b, the visualized flow pattern by a contour map of density obtained by a numerical analysis. The offset distance and supply pressure are offset distance of the side wall $D/b = 5.0$ and $P_0 = 0.4$ MPa, respectively. Figure 285c shows Schlieren photo. In the case of a supersonic jet, the jet also reattaches to the side wall due to the Coanda effect. Shock cells are also seen in the jet, and it flows as an under-expanded jet. After exiting the nozzle, the jet continues to expand and compress, forming a shock cell. In the case of a 2D supersonic wall-reattached jet, the jet centerline is also approximately represented as a circular arc of radius R (Fig. 285).

(b) **Shock cell size**

The jet issued from the nozzle formed a shock cell after expansion and compression; thereafter, while expansion and compression continued, it curved and reattached to the side wall and flowed downstream.

Figure 286 shows the size, length C_L, and maximum width C_w of the first shock cell. C_L was obtained as the distance from the nozzle exit to the position where the measured jet centerline velocity is initially minimal. C_w was obtained from the visualized flow pattern and the density distribution. At that time, the jet boundary was set at the position where the shade suddenly changed in the jet width direction. Both C_L and C_w increase almost linearly with supply pressure P_0. The numerical analysis is consistent with the experimental results. In addition, the sizes of the second and third shock cells were similar to the first one.

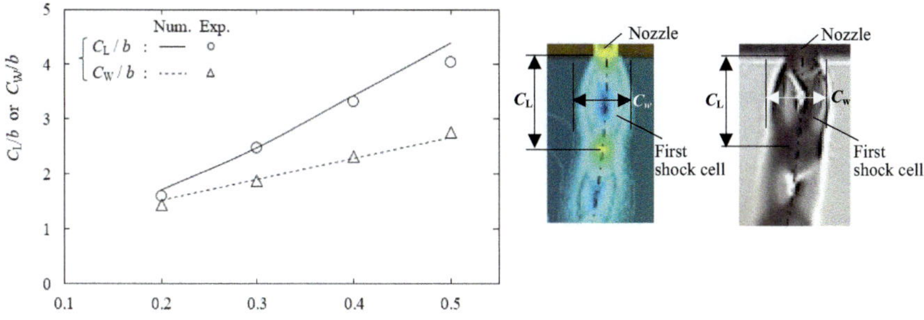

Fig. 286 Length C_L and width C_w of 1st shock cell

(c) **Radius of curved reattached jet centerline and reattached distance**

As mentioned above, in the case of a 2D supersonic wall-reattached jet, the jet centerline is also approximately represented as a circular arc of radius R.

Figure 287 shows the radius R along with the numerical analysis when $P_0 = 0.4$ MPa. The centerlines for the numerical analysis and the experimental result were obtained using the least-squares method by specifying the position of the maximum velocity from the velocity distribution of each jet cross-section. R increases with increasing P_0 and is almost constant at $R/b = 34.6$ after reaching the maximum at $P_0 = 0.4$ MPa. Furthermore, the numerical analysis of R and reattachment distance x_R reflects the experimental results accurately for $P_0 \geq 0.4$ MPa. x_R/b, shown in Fig. 287 increases with P_0 and then remains constant at 16.1. However, the numerical calculations for $P_0 = 0.2, 0.3$ don't reflect the experimental results well.

(d) **Pressure difference on both sides of jet and jet reattachment to side wall**

Bourque and Newman (1960) reported that when a 2D incompressible jet flow reattaches to a side wall which is set parallel to the nozzle axis at an offset distance, the jet centerline can be approximated to a circular arc of radius R (Figs. 149 and 285).

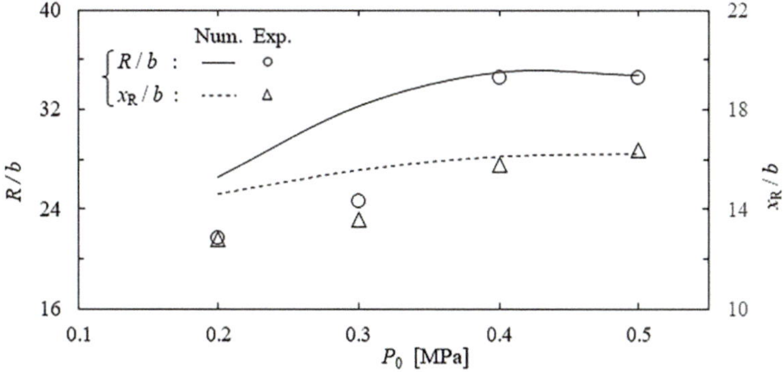

Fig. 287 Radius of curvature and reattachment distance of reattached jet centerline

The relationship between the pressure difference ΔP on both sides of the reattached jet and the centrifugal force exerted on the jet can be expressed as follows. Where the inside pressure of the reattached jet is the minimum pressure of the separation bubble which is approximated by the minimum pressure on the side wall.

$$\Delta P = J/R \tag{290}$$

where, J is the jet momentum.

Equation (290) can be rewritten as

$$\Delta P = \frac{\rho_0 Q^2}{bH^2 R} \tag{291}$$

where, Q and ρ_0 are the flow rate and density of air at standard conditions, respectively.

This relationship was examined for a supersonic under-expanded reattached jet of $P_0 = 0.4$ MPa, assuming that the jet momentum was conserved, as in Bourque and Newman. Here, the minimum pressure on the side wall, flow rate, and arc radius were obtained from the experiments $P_{wmin} = 1.58 \times 10^4$ [Pa], $Q = 58.0$ [Nm^3/h], and $R = 86.5$ mm, respectively. Inserting the jet momentum derived from Q and R into Eq. (290) yields $\Delta P = 1.55 \times 10^4$ [Pa]. This is almost the same as P_{wmin}, which is reflected in the case of the supersonic reattached jet. Therefore, the relationship shown in Eq. (290) holds well.

(e)   **Pressure distribution on side wall**

Figure 288 shows the pressure distribution P_w/P_0 on the side wall for $P_0 = 0.2 \sim - 0.5$ MPa. The jet entrains the surrounding fluid due to a large velocity gradient and the fluid viscosity. When a solid wall exists near the jet, it entrains and removes the fluid from the finite area between the jet and the wall. Consequently, the pressure decreases, and the jet deflects and reattaches enclosing a circulating separation bubble to the side wall. In this case, the jet centerline can be approximated by a circular arc of radius R. The pressure P_w/P_0 on the side wall has a negative value corresponding to the separation bubble. It decreases with increasing x/b and reaches a minimum and approaches the ambient pressure.

The minimum of the negative value and maximum of P_w/P_0 increase with P_0, and are almost the same at $P_0 \geq 0.4$ MPa. This is because increasing the velocity owing to the increase in P_0 hinders the jet from entraining the surrounding fluid. The numerical analysis reflected the experimental well, particularly when $P_0 \geq 0.3$ MPa.

Figure 289 shows the minimum pressure P_{wmin}/P_0 on the side wall. P_{wmin}/P_0 decreases with increasing P_0 and approaches a constant at $P_0 \geq 0.4$ MPa. The numerical analysis reflects the experiment effectively. In addition, according to Bourque et al., the position x of the minimum pressure on the side wall of a 2D incompressible reattached jet corresponds to the position x of the minimum bubble pressure. This is

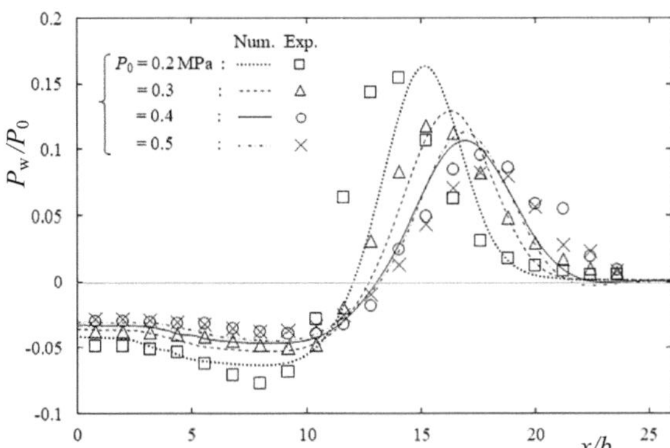

Fig. 288 Pressure distribution P_w/P_0 on side wall

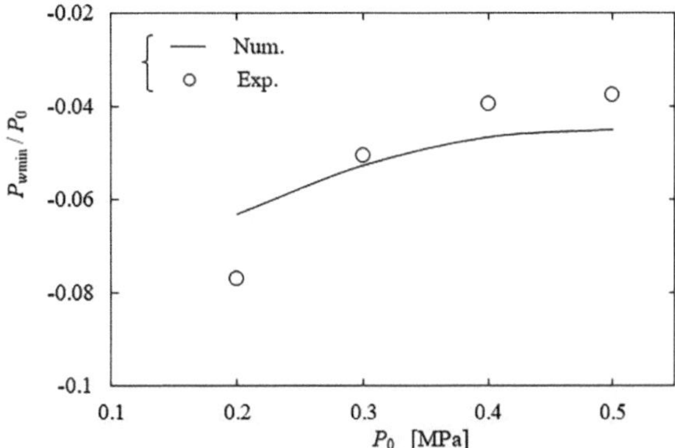

Fig. 289 Minimum pressure P_{wmin}/P_0 on side wal

reflected in the numerical analysis of the supersonic under-expanded reattached jet investigated in this study.

(f) **Velocity and pressure distributions**

Figure 290a, b show respectively an example of the contour map of velocity and pressure distributions for the case $D/b = 5.0$ and $P_0 = 0.4$ MPa. The contour map of the density and visualized flow pattern by Schlieren method were already shown in Fig. 285b, c. The jet from the nozzle expands and the velocity increases (red colored area), and then compresses and the velocity decreases. The pressure behaves inversely to velocity. The jet reattaches to the side wall, and recirculation flow is in the separation bubble. The jet flow decreases the velocity to the downstream, while expansion and compression is repeated.

Figure 291 shows the experimental results of the velocity distribution u at the cross sections of (a): $\theta = 3.25°$ in the middle of the compression region; (b): $\theta = 6.5°$ in the maximum compressed region of the first shock cell; and (c), (d): $\theta = 12.5°$ and $18.25°$ in the second and third most compressed regions (Fig. 290a), respectively.

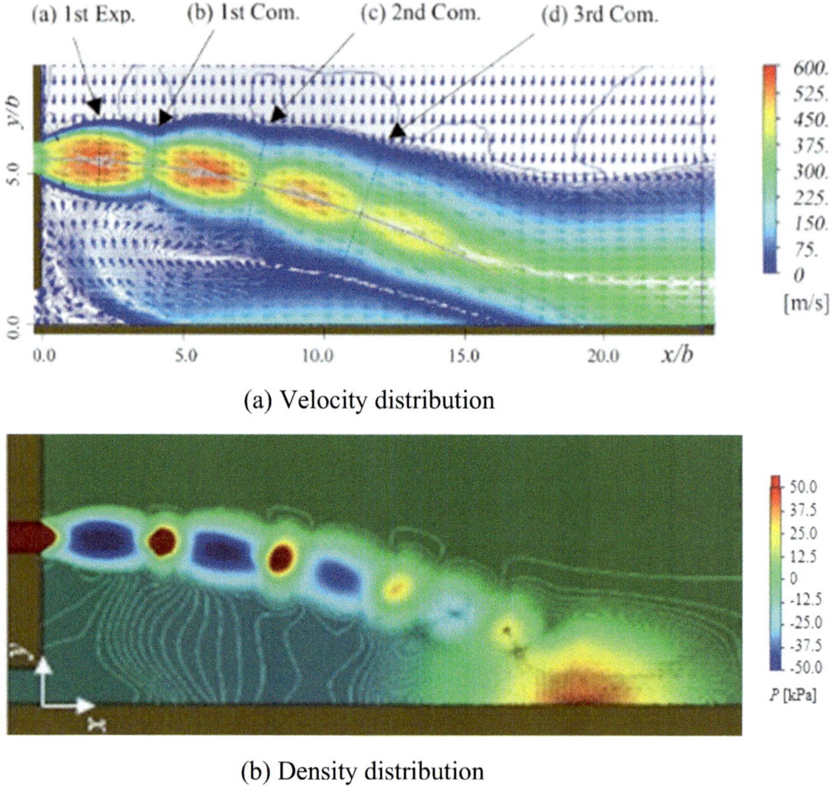

(a) Velocity distribution

(b) Density distribution

Fig. 290   Contour maps of velocity and pressure ($D/b = 5.0$, $P_0 = 0.4$ MPa)

The y' on the vertical axis in Fig. 285a is the distance in the radius direction from the centerline of the reattached jet.

The cross-sectional velocity profile of a supersonic under-expanded jet differs depending on the expansion or compression regions. As mentioned at the Sect. 8, Compressible flow (Fig. 270), Kojima, T. et al. (1986) showed that the velocity profile in the shock cell of a supersonic under-expanded free jet was concave initially, convex in the middle, and nipple-shaped at the end. The velocity profile at $\theta = 25°$ was concave and asymmetric, with a minimum at the jet center, which differs from Kojima's classification of convex. This could because this jet was a reattached jet, and the pressure on the separation bubble side was negative. The profiles at $\theta = 6.5$, 12.5, and 18.25° are convex at the most compressed the first shock cell is convex, that at $\theta = 6.5°$ ("1st Com.") is concave, and those at $\theta° = 12.5, 18.25°$ (2nd and 3rd Com.) are convex. As shown by the jet centerline velocity in point of the local minimum. Additionally, numerical calculations are shown in Fig. 292. The jet width and range of the numerical results are smaller than those of the experimental results, however, the numerical calculation reflects the approximate distribution of the experiment.

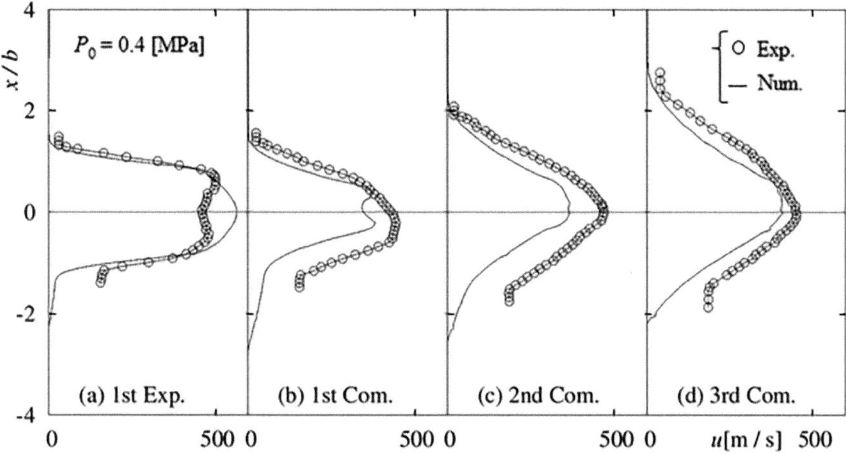

Fig. 291 Velocity distribution u at cross section, $D/b = 5.0$, $P_0 = 0.4$ MPa

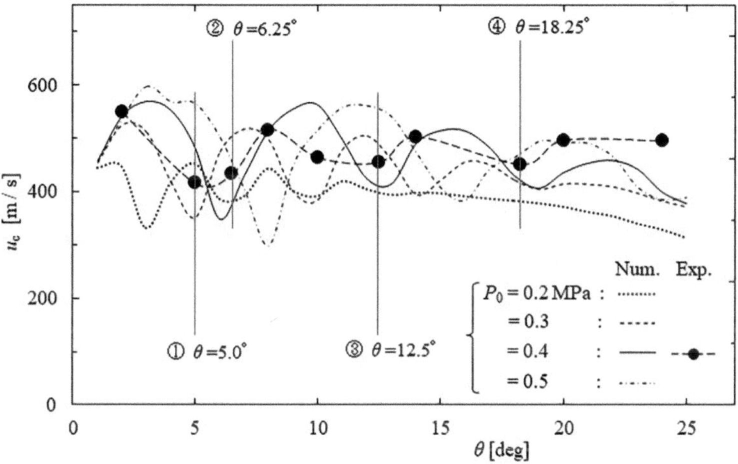

Fig. 292 Centerline velocity u_c of reattached jet

(g) **Jet centerline velocity of supersonic reattached jet**

Figure 292 shows the numerical calculations of the centerline velocity u_c of the supersonic reattached jet. The abscissa is the angle θ [angle in the circumferential direction with nozzle exit set at $\theta = 0°$, Fig. 285a] when the jet centerline is approximated by a circular arc. The experimental results for $P_0 = 0.4$ MPa are also shown in Fig. 292. According to the numerical results for $P_0 = 0.4$ MPa, the jet issues from the nozzle initially expanding and the u_c reaches a maximum at $\theta \fallingdotseq 25°$ and then decreases by compression to a minimum at $\theta \fallingdotseq 6.0°$. It is assumed that the region from $\theta = 0 \sim -6.0°$ corresponds to the first shock cell. u_c decays gradually while repeating expansion and compression, and the amplitude of the increase/decrease in u_c reduces. The positions θ of the second and third shock cells are almost the same based on numerical analysis and experiment. However, the behavior of velocity u_c in the shock cells differs, and the values from the numerical analysis are larger and smaller than those of the experiments in the back and front halves of the shock cell, respectively. The numerical analysis for $P_0 = 0.4$ MPa expresses the experimental results qualitatively. However, it does not represent the reality of the experiments. In addition, the maximum of u_c and the length of the shock cell increased with P_0.

(h) **Effects of offset distance**

Figure 293 shows the static pressure distribution P_w/P_0 on the side wall surface when $P_0 = 0.4$ MPa and $D/b = 0 \sim 7.0$. The position corresponding to the vortex region of the flow field shows negative pressure, and the position where the jet reattaches to the side wall shows positive pressure, which asymptotically approaches atmospheric pressure as it flows down. The negative pressure region increases as D/b increases, and the jet reattaches the side wall farther from the nozzle exit. In addition, the numerical analysis results represent the experimental results fairly well.

Here, the flow characteristics, such as shock cell formations, moving trajectory of reattached centerline approximated by a circular arc, reattachment distance, pressure distribution on a side wall, jet centerline velocity, and velocity distribution at the cross-section of a 2D supersonic under-expanded jet reattached were studied through experiment and numerical analyses.

(1) A supersonic under-expanded jet reattached to the side wall with an offset distance, surrounding the separation bubble region, as in a 2D incompressible reattached jet. Additionally, expansion and compression shock waves and shock cells were formed in this jet.

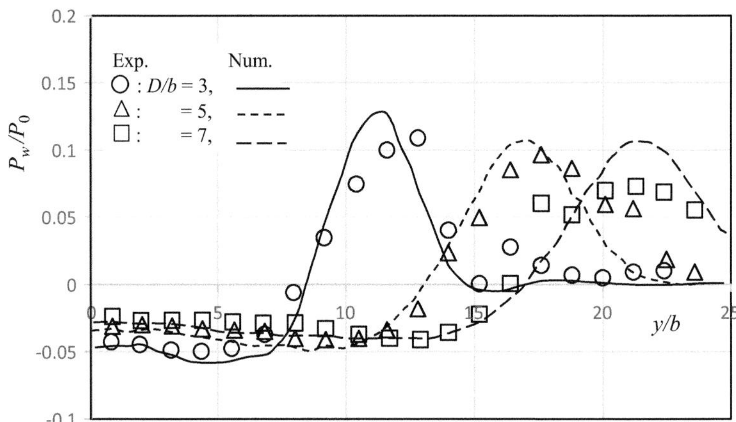

Fig. 293 Static pressure distribution on the side wall (Coanda reattached jets, $P_0 = 0.4$ MPa, $D/b = 0 \sim 7.0$)

(2) The centerline of a 2D supersonic under-expanded reattached jet can be approximated by a circular arc of radius R, in which R and the reattachment distance x_R increased with supply pressure P_0. Furthermore, the numerical calculations for supersonic reattached jet represent the experimental results fairly well. The relationship $\Delta P = J/R$, obtained by assuming that the force owing to the pressure difference ΔP applied to both sides of the jet and the centrifugal force applied to the jet element were balanced, was applicable.

(3) The length and width of the shock cell of the supersonic reattached jet increased with P_0, and the minimum pressure on the side wall decreased with increasing P_0. The numerical calculations represented the experimental outcomes well.

(4) The jet centerline velocity fluctuated with its maximum and minimum values according to the expansion and compression of the jet, and the values increased with P_0. The numerical calculations for the supersonic reattached jet represented the experimental outcomes fairly well. The experimental and numerical results of the cross-sectional velocity distribution and jet width differed a little, however, the numerical calculations represented the experimental results fairly well.

(5) The above results for the 2D reattached jet will also be useful in understanding the flow characteristics from a gas-atomization nozzle approximately. Conflicts of Interest The authors declare no conflicts of interest regarding the publication of this paper.

9 Aerospace Applications

9.1 Thrust of Jet and Rocket Engines

In jet engines and rocket engines, high-temperature, high-pressure combustion gas is ejected from a nozzle to obtain propulsion. At that time, the shape of the nozzle is devised to efficiently expand the high-pressure gas to generate a supersonic or hypersonic jet to obtain a large thrust.

9.2 Helicopter Without Tail Rotor

Usually, a helicopter has a main rotor and a tail rotor. When the main rotor rotates, the recoil force causes the aircraft to rotate in the opposite direction to the direction of rotation of the main rotor. Therefore, in order to stop the rotation of the aircraft and obtain a straight-flying attitude, a small rotor, tail rotor, suppresses the rotation of the aircraft (Fig. 294). The tail rotor is located in the tail boom and driven via the drive shaft and gearbox. This drive system is mechanically complex and can cause failure.

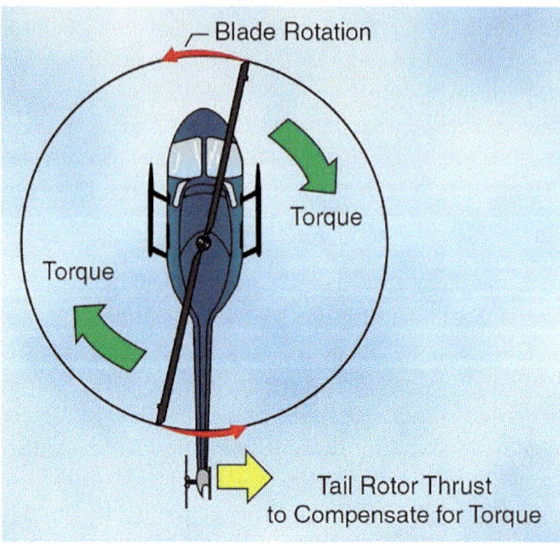

Fig. 294 Generation of recoil torque due to rotation of the main rotor (by Wikipedia)

Figure 295 shows the control of flow to create a recoil force that counteracts the torque generated by the main rotor of a helicopter without a tail rotor, NOTAR. A variable-pitch fan ② housed in the rear fuselage just before the tail boom ③ is driven by the main rotor transmission (MD 520 Helicopter, Technical Description: https:// support.mdhelicopters.com/files/Models/MD520N_Tech_Desc.pdf).

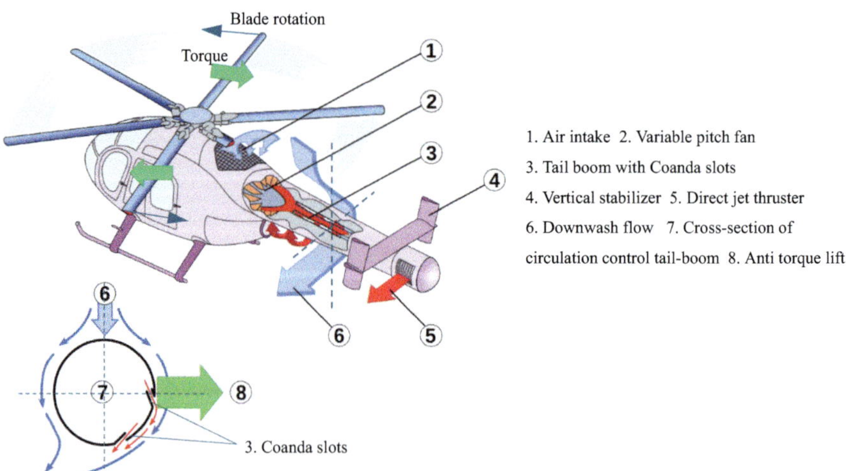

Fig. 295 Flows in and around tail boom of helicopter without tail rotor (by Wikipedia)

Fig. 296 Flows in and around tail boom of helicopter without tail rotor (by Wikipedia)

The low-pressure air generated by the fan is ejected from the two slots on the right side of the tail boom as shown in Fig. 295, allowing the downwash ⑥ of the main rotor to flow along the tail boom by the Coanda effect and generate lift ⑧. The jet flow from the Coanda slot accelerates the flow of downwash flowing down the wall of the tail boom. It flows along the wall surface by the Coanda effect (7. Cross-section of circulation control tail-boom in Fig. 295). At that time, the pressure decreases because the flow is accelerated. It generates a counter-torque force ⑧ proportional to the air flow rate of the rotor downwash to cancel the clockwise torque of the main rotor, which rotates counterclockwise (as seen from above). In addition, the tail (including the vertical stabilizer ④), which does not have a moving part at the rear end of the tail boom, compensates for the reaction torque force, and controls the direction by direct jet ⑤.

Figure 296 shows the flight status of the helicopter without a tail rotor (MD Helicopters 520N NOTAR).

9.3 Boundary Layer Control on Wing Surface in Aircraft

9.3.1 Vortex Generator

The Vortex Generator is a device that is installed on the leading edge of an aircraft wing and intentionally creates turbulence at the place of air flow at a constant flow velocity during flight, suppressing boundary layer separation and reducing air resistance.

Figure 297 shows the vortex generators mounted on the leading edge of the wings of the Hawker Siddeley Harrier (vertical/short take-off and landing aircraft, V/STOL aircraft).

Fig. 297 Vortex generators mounted on the leading edge of the wings of the Hawker Siddeley Harrier (by Wikipedia)

9.3.2   Blowing Out and Suction

When the attack angle of the wing is too large, the airflow on the upper surface of the wing loses kinetic energy due to the viscosity of the air and peels off without being able to follow the wing (boundary layer separation), and the pressure on the upper surface of the wing does not decrease and lift cannot be generated. In order to solve this problem, mechanical mechanisms such as leading-edge flaps, trailing edge flaps, and slats are used to prevent separation of the airflow on the upper surface of the wing. However, the maximum lift coefficient when using it is limited to about 36, and as the speed of the airplane increases, the wing loading and takeoff and landing speed increase, and the runway distance increases. Therefore, the boundary layer control, BLC, device is a device that prevents the separation of the airflow and dramatically increases the maximum lift coefficient by artificially applying kinetic energy to the boundary layer.

In addition to boundary layer control by blowing out, there is a boundary layer control by a so-called suction wing that sucks in the airflow that has lost kinetic energy on the upper surface of the wing and detached from the suction hole (slot) installed on the upper surface of the main wing.

Figure 298 shows the boundary layer control of the flow around the main wing and tail of the US-2, a rescue flying boat manufactured by ShinMaywa Industries, Ltd. Japan. The US-2 has a small gas turbine dedicated to blowing out on its back, and controls the boundary layer on the main wing and tail surfaces. As a result, it has a maximum lift coefficient of 7 or more, about twice that of a normal airplane, and can cruise at extremely low speeds of about 90 km/h and take off and land on the water within a very short distance.

9.3.3   Plasma Actuator

The Dielectric-Barrier-Discharge Plasma Actuator, DBD-PA, has two electrodes separated a dielectric subject, as shown in Fig. 299. Applying a high voltage alternating current between two electrodes causes the air near the exposed electrode to be

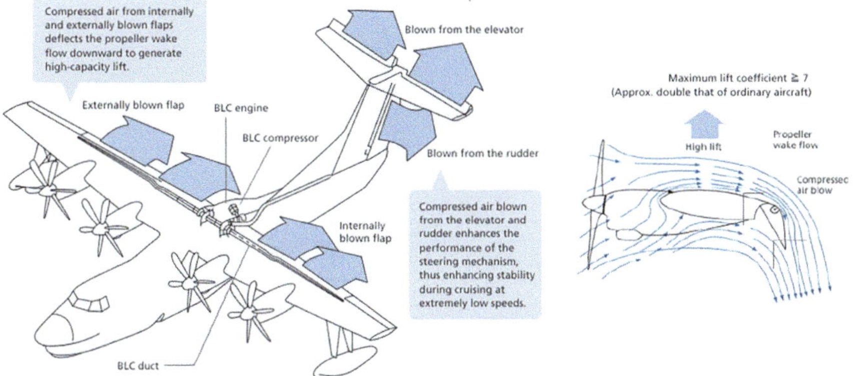

Fig. 298 Boundary layer control of the flow around the main wing and tail of the ShinMaywa US-2, Japan https://www.shinmaywa.co.jp/english/products/aircraft/amphibian/us2/capability.html

Fig. 299
Dielectric-Barrier-Discharge Plasma Actuator, DBD-PA

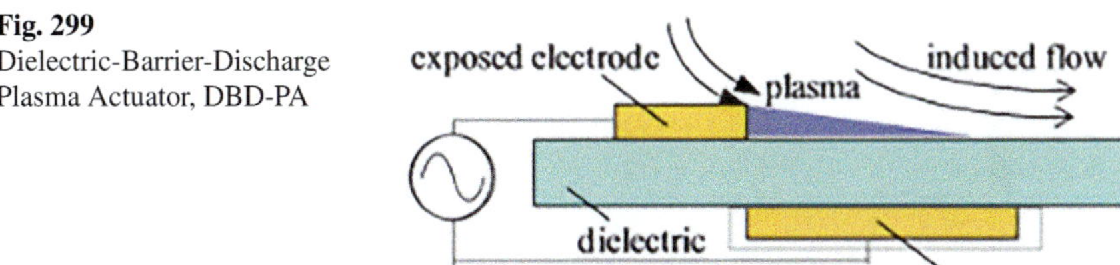

ionized, and the ions accelerate the surrounding air, and induce a flow. The induced flow can be used to control the main flow. The induced flow by the plasma actuator can be used as a blowout flow to suppress the separation of the boundary layer on the wing surface.

9.4 Winglet

During flight, an airplane generates a vortex of air called a wingtip vortex from its wingtips. This is due to causing a flow from the underside of the high-pressure wing through the wingtip to the upper surface of the high-pressure wing (Fig. 300). This creates a force that pulls the wingtip backward, increasing air resistance (induced drag). The winglet (Fig. 301) reduces the wingtip vortex by rectifying the air flow near the wing tip, or moves the generation direction upward to reduce air resistance.

Figure 302 shows the flight of a sea eagle during gliding. The sea eagle spreads its wings wide, and flies with its wings separated one by one at the tip. This, including the fact that feathers are also elastic bodies, seems to be effective in suppressing the generation of wingtip vortices.

Fig. 300 Wingtip vortices

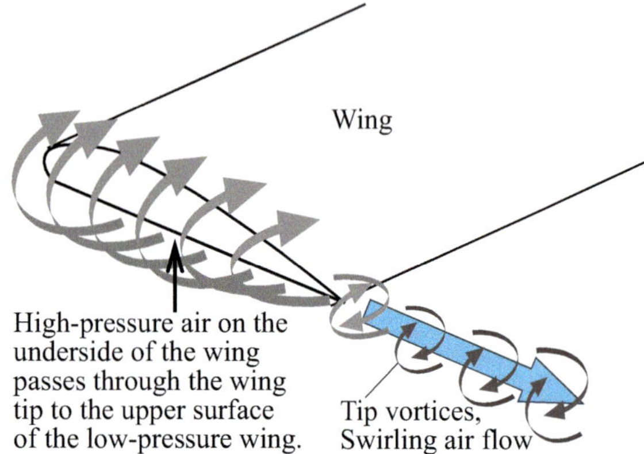

Fig. 301 Conventional
wingtip and winglet

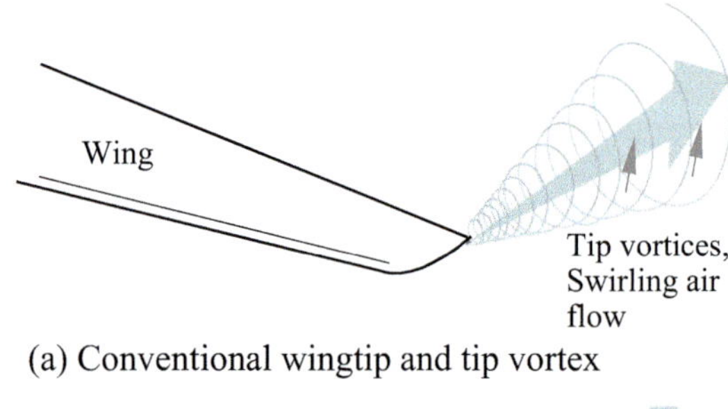

(a) Conventional wingtip and tip vortex

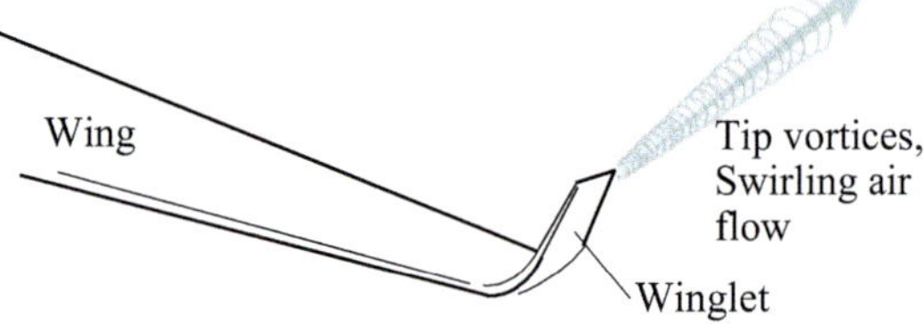

(b) Winglet and reduced tip vortices

Fig. 302 Gliding flight of
sea eagle (by Wikipedia)

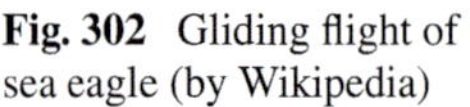

In order to suppress the generation of wingtip vortices, in addition to winglets, there are similar ones such as wingtip fence, drooped wingtip, sharklet, raked wingtip, and so on.

9.5 Chevron Nozzle and Noise Reduction of High-Speed Jet

We can see the wave-shaped exit of the jet engine, as shown in Fig. 303. This is called "Chevron nozzle", and is used to reduce the jet noise (Gutmark et al. 2006, Munday et al. 2013). Gutmark et al. (2006) examined the effect of the chevron nozzle on noise reduction.

9.5.1 Test Section

Figure 304 shows the separate flow exhaust model with primary and secondary exit areas of 23.23   223 and 80.65 cm^2, respectively. One conical and three chevron nozzles with 8 and 12 lobe chevrons were used. The chevron nozzle of which the level of penetration of the chevrons into the flow was equal to the boundary layer thickness and twice was used.

9.5.2 Visualized Flow Pattern

Figure 305a, b show the instantaneous shadowgraph images of jet flows from the baseline nozzle and chevron nozzle with 8 lobes, respectively. The Mach number is $M_i = 1.56$. The shock wave in the jet flow from baseline nozzle can be seen and the

Fig. 303 Boeing 787, jet engine with chevron nozzle https://www.quora.com/Why-does-the-Boeing-787-have-a-curved-engine

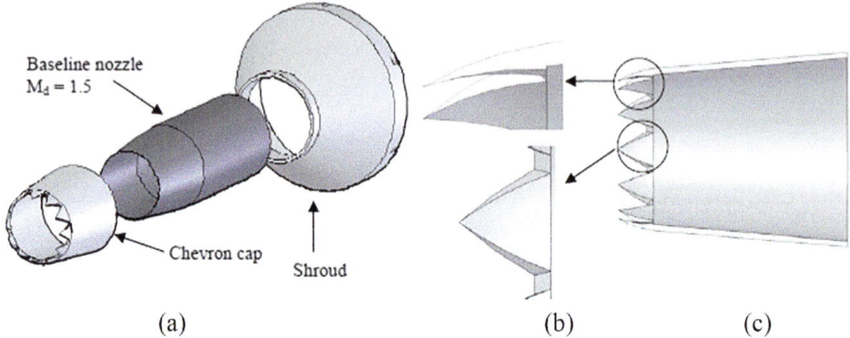

Fig. 304 Test section of chevron nozzle arrangement (Gutmark et al. 2006) **a** Schematics, **b** Details of chevron, **c** Cross section of cap

jet flows almost without diffusion. In comparison, in the jet from the chevron nozzle, there is no clear shock wave from the nozzle outlet, and the mixing in the jet stream progresses and the jet flow spreads largely.

Figure 306 shows average photograph of 100 instantaneous shadowgraph images superimposed. The clear shock waves can be seen in the jet from the baseline nozzle, but the shock waves from the uneven part of the nozzle exit interfere with that from the chevron nozzle, and the shock wave is weakened.

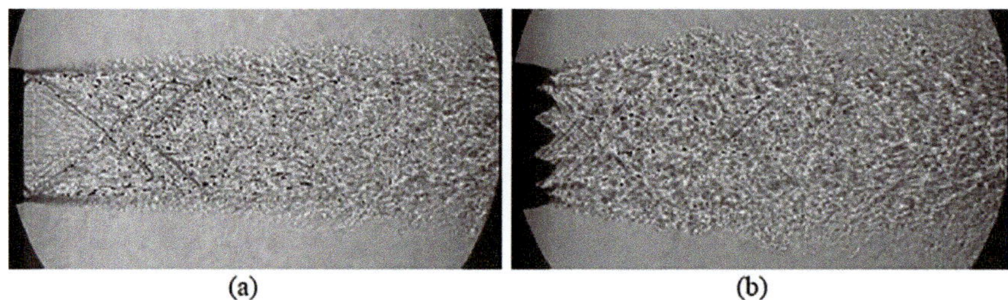

Fig. 305 Instantaneous shadowgraph image, $M_\mathrm{j} = 1.56$ **a** Baseline, **b** Chevron nozzle with 8 lobes

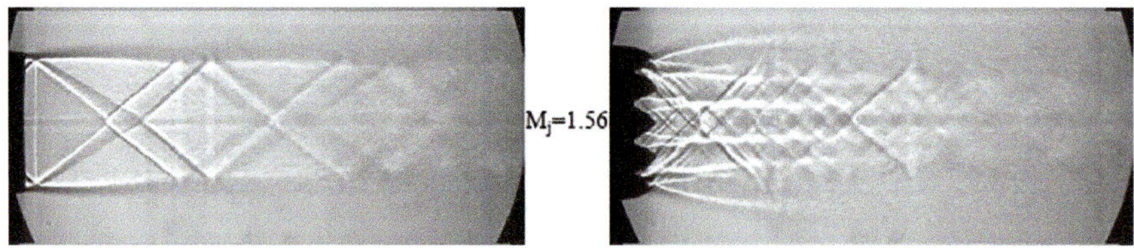

Fig. 306 Averaged shadowgraph image (100 images averaged)

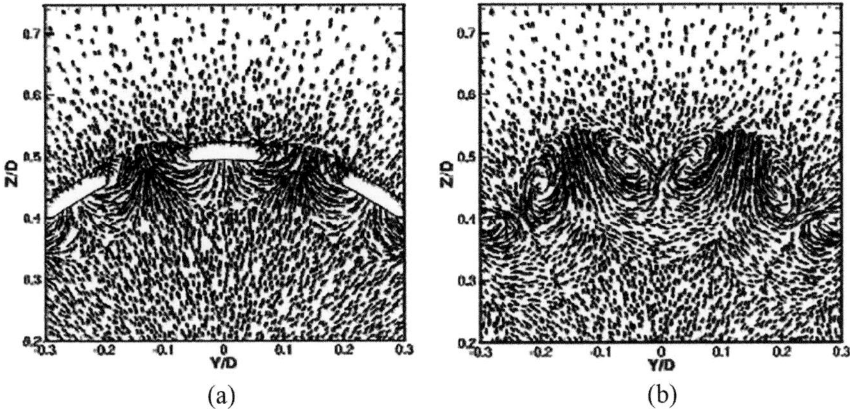

Fig. 307 Velocity vectors by LES, **a** in a plane half way between the chevron tips and roots, **b** in a plane containing the chevron tips

9.5.3 Velocity Vectors in Cross Sectional Planes by LES

Figure 307 shows the numerical results of velocity vectors in a plane half way between the chevron nozzle tips and roots, and in a plane containing the chevron tips by large eddy simulation, LES.

9.5.4 Acoustic Characteristics and Noise Reduction

Figure 308 shows the far-field acoustic measurements of the supersonic jet flow from baseline nozzle and chevron nozzle. The sound pressure level, SPL, decreases with decreasing Mach number, but the dominant frequency increases. In addition, the chevron nozzle reduces SPL in the vicinity of the predominant frequency. That is, the jet noise is reduced.

Their acoustic studies identified that nozzle with a higher penetration provides a superior low frequency benefit. However, nozzle with a nominal penetration provides an inferior high frequency. The chevron penetration controls the strength of the mixing, and stronger mixing makes the jet potential core length shorter, which leads to an additional noise reduction of lower frequency. However, this strong mixing generates an additional high frequency noise. To get an optimum benefit, it is important to tailor this design parameter.

Here, we examined the flow characteristics of the jet floe issued from chevron nozzle and the followings were made clear.

(1) The clear shock waves can be seen in the jet from the baseline nozzle, but the shock waves from the uneven part of the nozzle exit interfere with that from the chevron nozzle, and the shock wave is weakened.

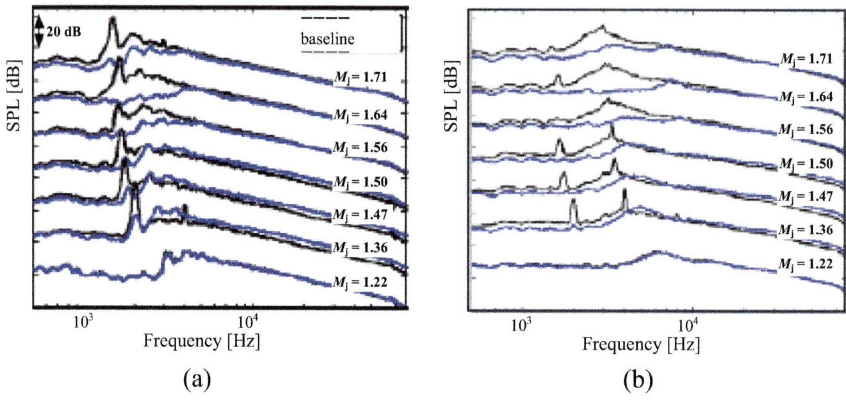

Fig. 308 far-field acoustic measurements of $M_d = 1.50$ nozzle with and without chevrons at a distance of 47 exit diameters from the nozzle exit, **a** $\Psi = 35°$ (Ψ: angle from the centerline with nozzle exit as the origin), **b** $\Psi = 90°$

(2) There are many other studies on the chevron nozzle, and they concluded that the SPL, can becontrolled and reduced by the chevron nozzle. Furthermore, the chevron nozzle is used practically to reduce the jet noise from jet engine (Fig. 303).

9.6 Thrust Vector Control of Supersonic Jet

Supersonic jets are widely used in the industrial field, in applications such as jet and rocket engines, ejectors, gas-atomization, and many other devices. There have been significant researches on the flow characteristics of supersonic under-expanded jet flows (SSU-jets).

Donaldson and Snedeker (1971a, b) and Kojima and Matsuoka (1986) showed the flow structure of the SSU-Jet by measuring the pressure and velocity distributions, and Nourl and Whitelaw (1966), Zapryaggaez et al. (2015a, 2015b), and Shakouchi et al. (2019), Shakouchi (2020) investigated the flow characteristics of the SSU-Jet by visualizing its flow pattern with the Schlieren method and measuring the pressure and velocity distributions to show the applications of this device. Shadow et al. (1990) investigated the spreading rate of the SSU-jet, and Gutmark et al. (2006) studied the aeroacoustics of turbulent jets and noise reduction by the chevron nozzle. The vector control of jet thrust through propulsion is an important research subject. Thrust vector control of the SSU-jet is mainly applied to the altitude control of aircrafts by a jet engine or rocket engine. There are two approaches in general, namely mechanical moving systems and fluidic thrust vector control systems without moving parts, with mechanical moving systems being the most common. Recent typical examples of such systems are the mechanical thrust vector control system of the jet engine in V/STOL aircrafts and rocket engines. However, this is very complicated system, and simpler methods and devices are desired. Some fluidic systems without moving parts have been proposed.

Here a new and simple method for the vector control of supersonic jets is proposed (Shakouchi and Fukushima 2022). This method involves a fluidic Coanda nozzle (FC-nozzle) and uses the entrainment of the surrounding fluid and the Coanda effect. The FC-nozzle consists of a pipe nozzle (Pi-nozzle), spacer, and linearly expanded Coanda nozzle (Co-nozzle) with eight suction pipes (Su-pipes) installed to surround the jet from the Pi-nozzle. The jet from the Pi-nozzle flows straight with entrainment of the surrounding fluid. When some suction pipes are closed, the pressure between the jet and Co-nozzle wall decreases, and subsequently, the jet deflects to the closed side of the Su-pipe and attaches to the wall by the Coanda effect. The flow characteristics and deflection characteristics of the supersonic jet from the FC-nozzle were investigated by the visualized flow pattern using the Schlieren method and measurements of the velocity distribution. We show that by changing the number of Su-pipes and the locations at which they close, the deflection angle and circumferential position of the jet can be controlled. Furthermore, when there is no fluid to be entrained by the jet, that is, when the surroundings are in a vacuum state, the same type of control can be performed by blowing small amounts of fluid and air from the Su-pipe. This new fluidic thrust vector control system of the supersonic jet is extremely simple compared with other conventional methods that require control with additional devices, such as secondary jets.

9.6.1 Fluidic Coanda Nozzle

Figure 309 shows a schematic diagram of the fluidic Coanda nozzle (FC-nozzle) with no moving parts. The FC-nozzle is an extremely simple system that consists of a pipe nozzle (Pi-nozzle), spacer, and linearly expanded Coanda nozzle (Co-nozzle) with eight suction pipes (Su-pipes) installed to surround the jet from the Pi-nozzle. The diameter of the Pi-nozzle is $d_0 = 4.0$ mm, and the inlet diameter, length, and expansion angle of the Co-nozzle are $d_{ic} = 5.0$ mm, $L_c/d_{is} = 2.4$, and $\alpha = 20°$, respectively. On the circumference of the Co-nozzle, eight Su-pipes ($n = 8$) with diameters of $d_{is} = 2.4$ mm are installed. The thickness of the spacer is 2.0 mm, and the inlet and outlet diameters are 4.0 and 4.7 mm, circumference of the Co-nozzle, eight Su-pipes ($n = 8$) with diameters of $d_{is} = 2.4$ mm are installed. The thickness of the spacer is 2.0 mm, and the inlet and outlet diameters are 4.0 and 4.7 mm, when all 8 Su-pipes are opened, the jet entrains the surrounding fluid from the Su-pipes and around the jet, and flows downstream along the nozzle axis, as shown in Fig. 309a. On the other hand, when some Su pipes are closed, the jet cannot entrain surrounding fluid from their sides, and subsequently, the pressure between the outer edge of the jet and the wall of the Co-nozzle decreases, and then, the jet deflects to the closed side of the Su-pipe by the Coanda effect, as shown in Fig. 309b. Therefore, arbitrary circumferential vector control of the jet from the Pi-nozzle can be performed by changing the number of Su-pipes and the locations at which they are closed. If these Su-pipes are opened and closed using a solenoid valve and a computer control system,

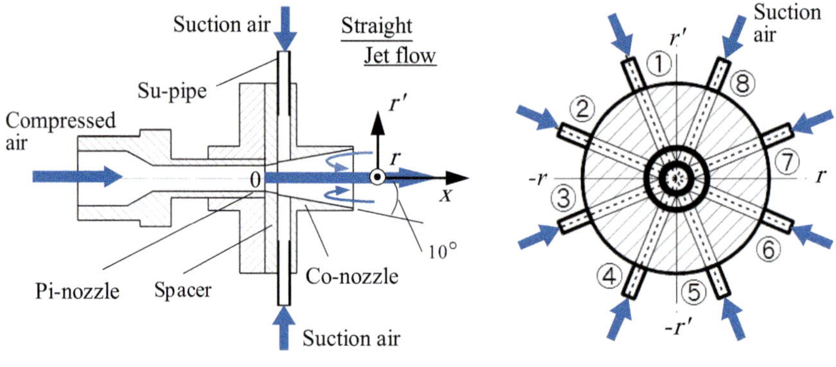

(a) Straight jet flow (all Su-pipes: open)

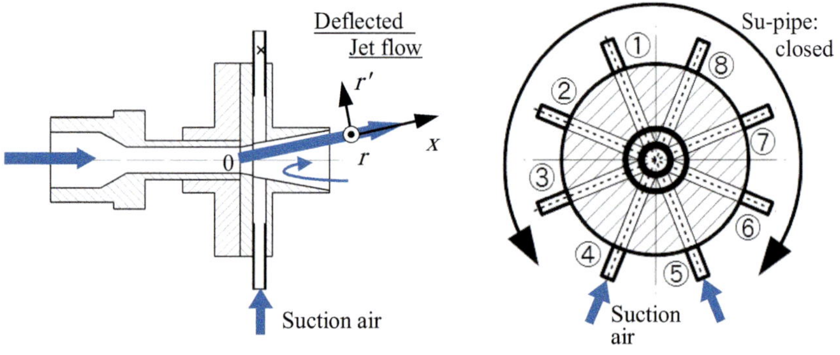

(b) Deflected jet flow (Su-pipe ①-③ and ⑥-⑧: closed)

Fig. 309 FC-nozzle (Fluidic thrust vector control of the Pi-jet by the FC nozzle with 8 SU-pipes): When all Su-pipes are opened, the jet flows downstream along the nozzle axis, as shown in Fig. 309a. When some Su-pipes are closed, the pressure between the outer edge of the jet and the wall of the Co-nozzle decreases and then the jet deflects to the closed side of Su-pipe by the Coanda effect as shown in Fig. 309b

dynamic thrust vector control becomes possible. There does not necessarily have to be eight Su-pipes. All that is necessary is to block part of the circumference of the jet from the Pi-nozzle.

9.6.2 Jet Flow from the Pipe Nozzle (Pi-Nozzle)

(a) Visualized flow pattern of free jet

The visualized flow pattern of the under-expanded supersonic free jet from the pipe nozzle (Pi-nozzle) (Fig. 309) produced by the Schlieren method is shown in Fig. 310. The supply pressure is $P_0 = 0.38$ MPa. The black and white colored areas in the

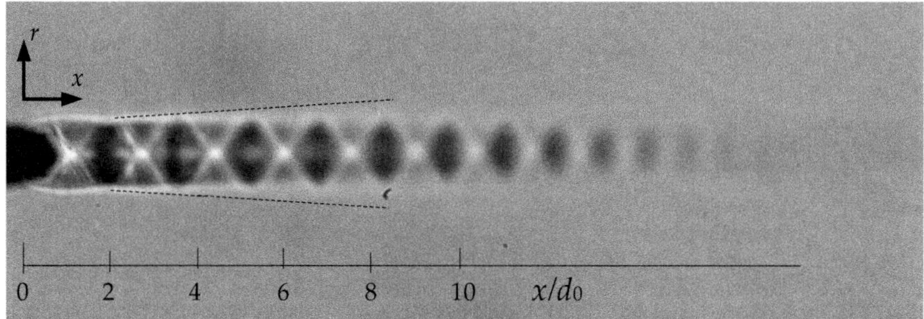

Fig. 310 Visualized flow pattern of the under-expanded supersonic free issued from the Pi-nozzle produced by the Schlieren method, $P_0 = 0.380$ MPa: The dotted lines indicate the approximate jet boundary with the position of r at $u/u_c = 0.1$

jet are compression and expansion regions, respectively, and white colored x shaped pseudo-shock waves can be seen in the shock cell. The dotted lines in the figure indicate the approximate jet boundary derived from the measurements of the cross-sectional velocity distributions at $x/d_0 = 2.1, 4.2, 5.4, 8.3$. The jet boundary is at the position of r at $u/u_c = 0.1$. The jet expands at the nozzle exit and the jet width increases almost linearly in the range of $x/d_0 > 3.0$.

(b) Centerline velocity of free jet

The jet centerline velocity u_c is indicated by a white circle mark in Fig. 317. This was obtained from the measurements of the total and static pressure distributions. The jet repeatedly expands and compresses and reaches the maximum value at $x/d_0 = 5.8$. In the downstream area of $x/d_0 > 12.5$, u_c decay occurs slowly.

(c) Jet flow from fluidic Coanda nozzle (FC-nozzle)

Next, the flow characteristics and vector control of the jet issued from the fluidic Coanda nozzle (FC-nozzle) with the Co-nozzle were investigated using the visualized flow pattern and measurements of the velocity distribution.

(d) Visualized flow pattern

Figure 311a shows the visualized flow pattern of the FC-jet issued from the FC-nozzle when the all Su-pipes are opened, 0-close. The supplied pressure is $P_0 = 0.38$ MPa. The jet flows straight to the downstream entraining the surrounding fluid from the Su-pipes and around the jet (Fig. 309a). The shock waves can be seen in the jet, and the white and black colored areas from the nozzle exit represent the expansion and compression shock waves, respectively.

However, the FC-jet with 0-close decays faster than the Pi-jet shown in Fig. 309, because the entrainment of surrounding fluid is suppressed by the Co-nozzle.

Figure 311d shows the case in which Su-pipes ① \sim – ③ and ⑥ \sim – ⑧ are closed (6-closed). The entrainment from those Su-pipes is prevented, and the pressure between the outer edge of the jet from the Pi-nozzle and the wall of the Co-nozzle

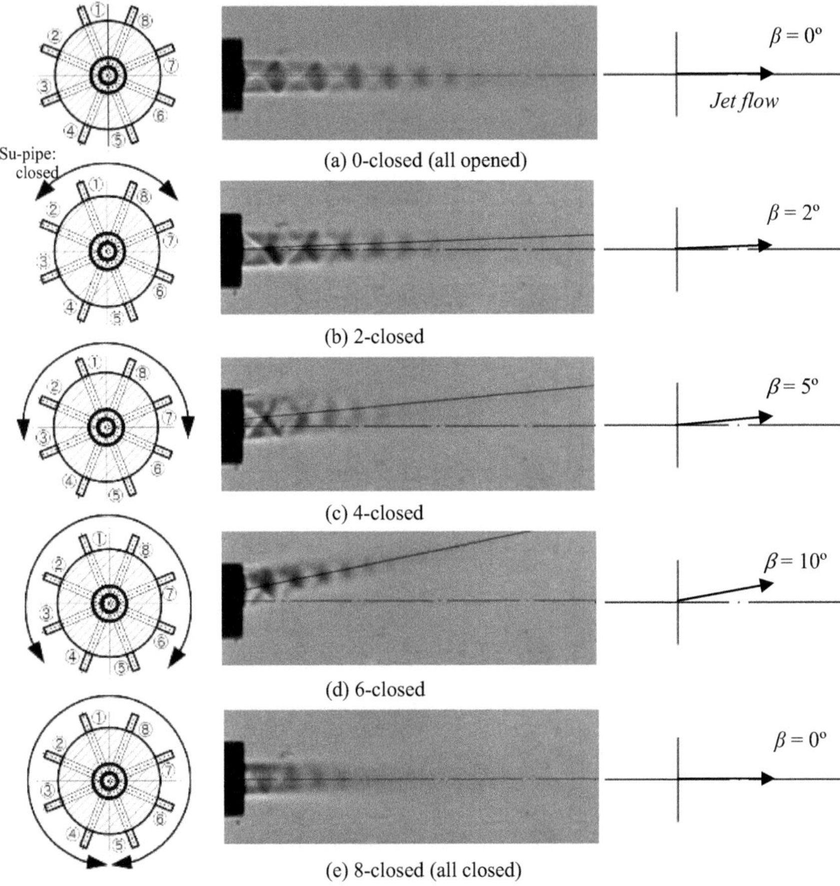

(a) 0-closed (all opened)

(b) 2-closed

(c) 4-closed

(d) 6-closed

(e) 8-closed (all closed)

Fig. 311 Visualized flow pattern of the jet issued from the FC-nozzle by Schlieren method, $P_0 =$ 0.38 MPa (Effects of suction flow): The deflection angle of the jet was measured using visualized flow pattern

decreases. Consequently, the jet deflects and reattaches to the wall of the Co-nozzle by the Coanda effect. The deflection angle is $\beta = 10°$, which is the same as the opening angle of the Co-nozzle.

Figure 311b, c show the 2- and 4-closed cases, respectively. Figure 311e shows the 8-closed case, where the jet flows straight to the downstream area: however, since the jet cannot entrain the surrounding fluid from the Su-pipe and the flow resistance of the nozzle increases, the jet velocity is slower than those of the others.

(e) **Jet deflection angle**

Figure 312 shows the deflection angle, β, and the number of closed Su-pipes, n_c. β increases as n_c increases and has a maximum value of $\beta = 10°$ at $n_c = 6$. In the figure, the results for the greater supply pressure, $P_0 = 0.46$ MPa, are also shown

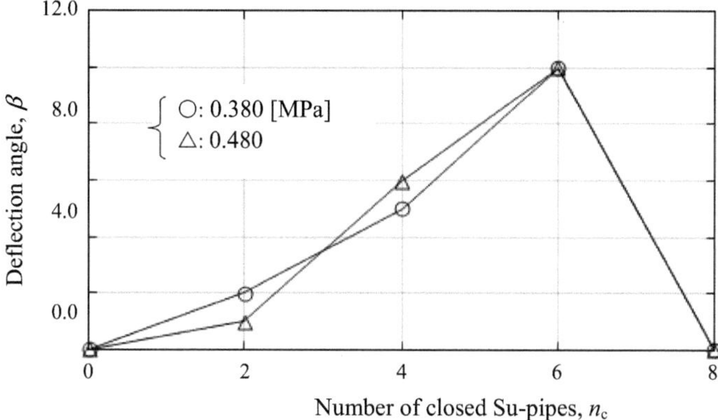

Fig. 312 Deflection angle, β ($P_0 = 0.380, 0.48$ MPa): The deflection angle has a maximum value of $\beta = 10°$ at $n_c = 6$

for reference, and both values are approximately the same. By changing the number of Su-pipes and the locations at which the Su-pipes close, the deflection angle and circumferential position of the jet can be controlled.

(f) Practical operation

In practical applications, it is conceivable to use a computer-controlled solenoid valve to change the number of Su-pipes and the locations at which they close.

(g) Velocity distribution

In order to clarify the flow characteristics of the straight and deflected jets, the jet center line and cross-sectional velocity distributions were measured.

(g.1) Cross-Sectional Pressure and Velocity Distributions

* Total Pressure Distribution

Figure 313a shows the total pressure distributions, P_t/P_0, in the cross-section at $x/d_0 = 4.2$ ($s/d_0 = 0.7$) for the 0-closed condition as an example. The jet of the 0-closed case flow straight to the downstream area, and the profile is axisymmetric and dented at the center. In comparison, the 6-closed case has a small maximum total pressure because of the large flow resistance of the Co-nozzle and the asymmetric profile.

Figure 313b, c are the results at $x/d_0 = 5.4, 8.3$, respectively. The profile at $x/d_0 = 8.3$ for the 0 closed case is axisymmetric, and the maximum value is approximately twice that of the 6-closed case.

* Static pressure distribution

Figure 314a shows the static pressure distribution, P_s/P_0, in the cross section at $x/d_0 = 5.4$ ($s/d_0 = 1.9$) for the 0-closed case. The profile is dented at the center by the expansion shock wave.

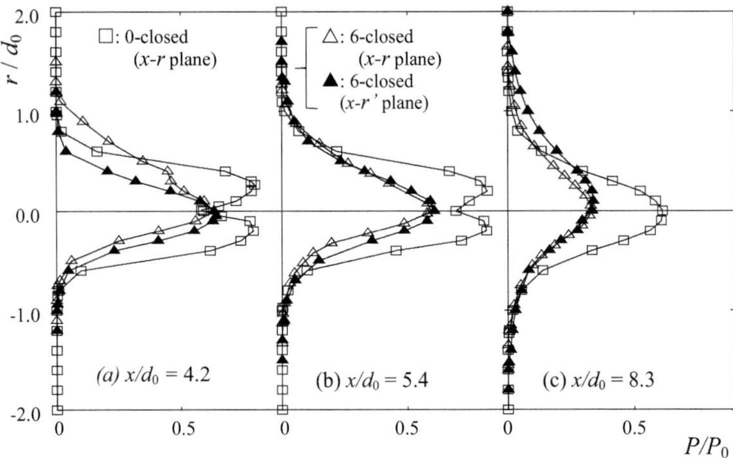

Fig. 313 Total pressure distribution, P/P_0, in the cross section (0-, 6-closed jets, $P_0 = 0.380$ MPa)

Figure 314b shows the results for the 6-closed case and Fig. 314c is a three-dimensional drawing of Fig. 314b. The profile is asymmetric, because the jet deflects and reattaches to the wall of the Co-nozzle, and consequently, the shape of the jet deforms. The jet reattaches to the wall of the Co-nozzle at the $+r'$-axis side and entrains the surrounding fluid from the $-r'$-axis side because of negative pressure.

* Sectional velocity distribution

Figure 315 shows the velocity distribution obtained with the total and static pressure measurements under the consideration of the compressibility of the fluid in the cross-section at $x/d_0 = 5.4$. The velocity profile of the 6–closed case is smaller than that of the 0-closed case because of a larger flow resistance of the Co-nozzle.

Figure 316 shows the velocity distribution in the cross section at $x/d_0 = 4.2, 5.4,$ and 8.3 for the 0-, 6-, and 8-closed cases. In Fig. 316a, the results for the Pi-jet are also shown for reference. The velocity was obtained by taking the static pressure, P_s, as the ambient pressure, i.e., the atmospheric pressure, P_a. The velocity distribution of the 0-closed case at $x/d_0 = 5.4$ and the results shown in Fig. 315a are approximately the same, because the static pressure is much smaller than the total pressure, as shown in Figs. 315 and 316. In this case, it seems that the velocity may be obtained by using P_s as P_a. The maximum velocity of the 0-closed case is approximately equal to that of the Pi-jet and larger than that of the 8-closed case, because the flow resistance of the Co-nozzle for the 8-closed case is greater than those of the others.

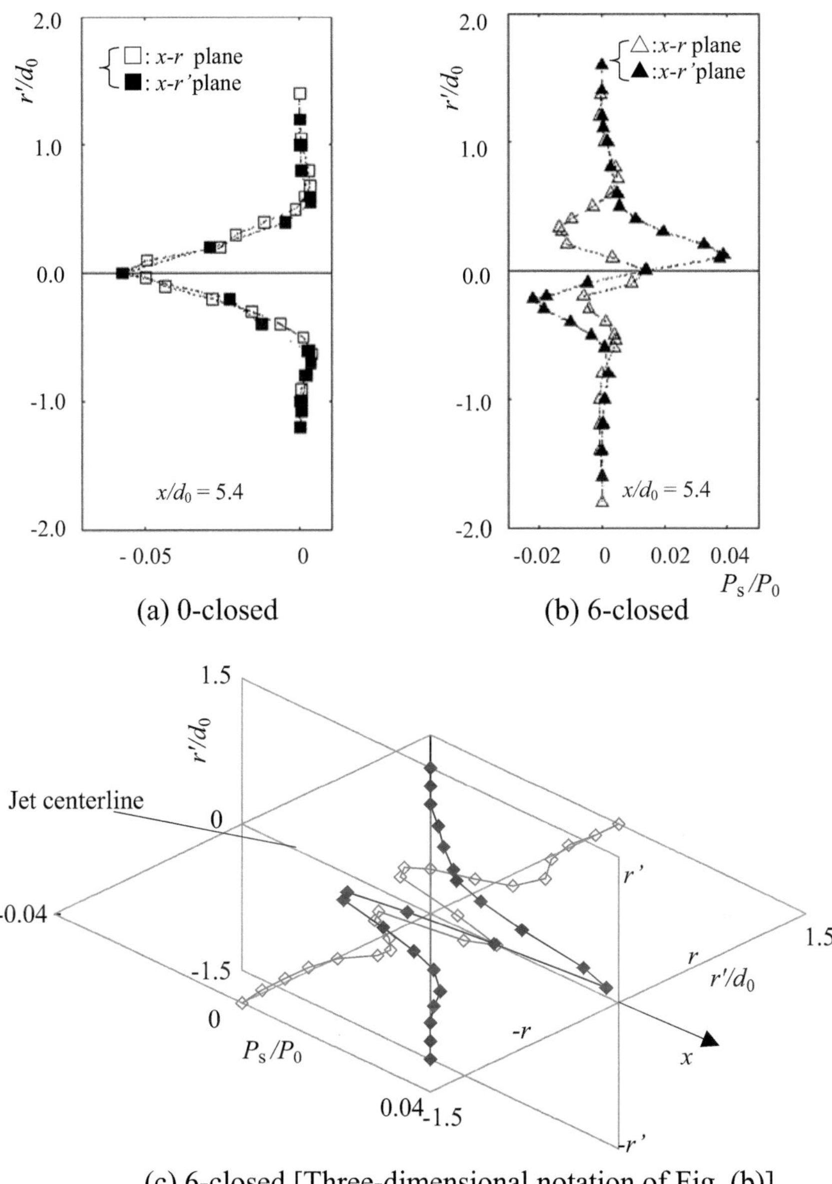

(c) 6-closed [Three-dimensional notation of Fig. (b)]

Fig. 314 Static pressure distribution, P_s/P_0, in the cross section at $x/d_0 = 5.4$ ($P_0 = 0.380$ MPa): The jet reattaches to the wall of the Co-nozzle at the $+r'$-axis side and entrains the surrounding fluid from the $-r'$-axis side of the negative pressure

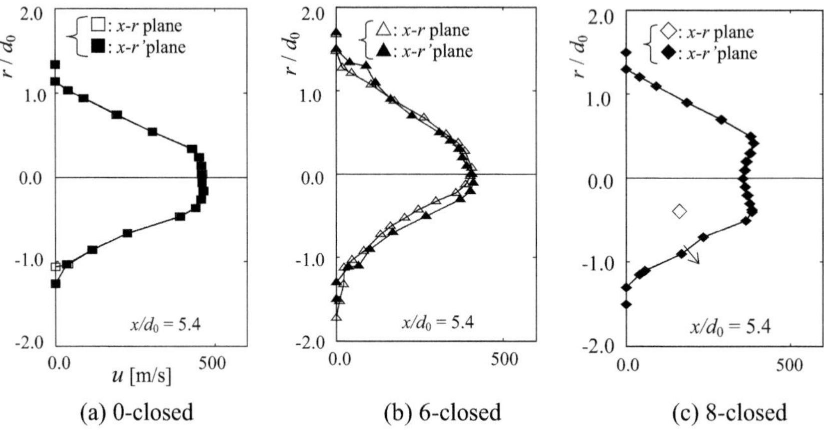

(a) 0-closed   (b) 6-closed   (c) 8-closed

Fig. 315 Cross sectional velocity distribution at $x/d_0 = 5.4$ ($P_0 = 0.380$ MPa): The velocity profile of the 6-closed nozzle is asymmetric because the jet deflects and reattaches to the side wall of the Co-nozzle

The flow characteristics of the straight and deflected flows were made clear by the velocity measurements.

(g.2) Jet centerline velocity distribution

Figure 317 shows the centerline velocity distribution of the jets u_c at $P_0 = 0.38$ MPa. The velocity was obtained by measurements of the total and static pressure distributions with consideration of the compressibility of the fluid, air. In Fig. 317, the position of the nozzle exit of the FC-nozzle is $x/d_0 = 3.5$ (the distance in the x-direction from the exit of the FC-nozzle: $s/d_0 = 0.0$). The Pi-jet flows downstream with a large fluctuation of u_c, corresponding to the expansion and compression shock waves of the flow, and u_c decreases monotonously in the region of $x/d_0 > 11$. In the 0-closed case, u_c is smaller than in the Pi-jet due to the large flow resistance of the Co-nozzle. Moreover, as the number of closed Su-pipes increases, the entrainment of the surrounding fluid from the Su-pipe decreases. The u_c of the 8-closed condition decreases monotonously, and it approximately equals that of the 6-closed case in the region of $x/d_0 > 8$.

(h) Thrust and approximate estimation

The deflected jet has an asymmetric velocity profile, as shown in Fig. 316b. The thrust, F, of the jet was calculated by the following equation under the assumption that the cross-section of the jet is approximately circular. At this time, the mean values of the x–r and x–r' planes were taken for the asymmetric velocity distribution. The density, ρ, was estimated approximately using the equation of the state of gas.

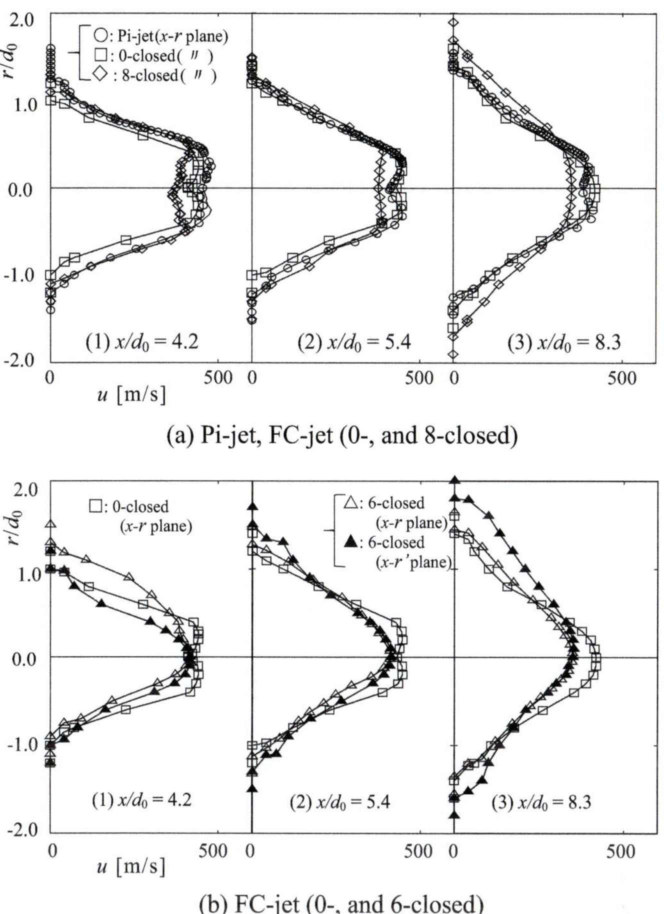

(a) Pi-jet, FC-jet (0-, and 8-closed)

(b) FC-jet (0-, and 6-closed)

Fig. 316 Cross sectional velocity distribution at $x/d_0 = 4.2, 5.4, 8.3$ ($P_0 = 0.380$ MPa, $P_s = P_a$): The maximum velocity of the 0-closed case is approximately equal to that of the Pi-jet and is larger than that of the 8-closed case

Fig. 317 Jet centerline velocity distribution, u_c (0-, 6-, 8-closed jets and Pi-jet, $P_0 = 0.380$ MPa): The u_c of the 0-closed nozzle is smaller than that of the Pi-jet owing to the large flow resistance of the Co-nozzle

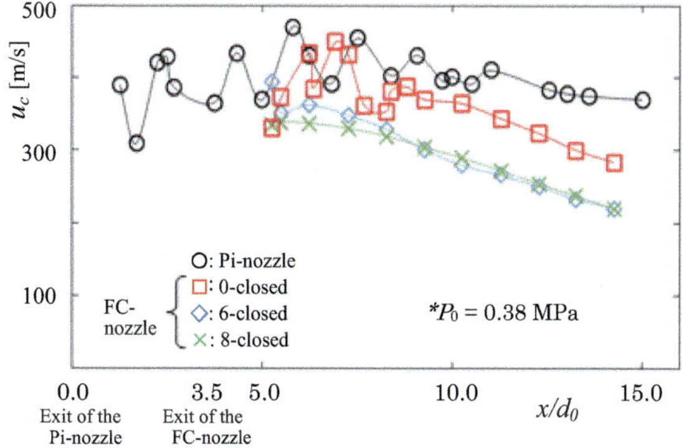

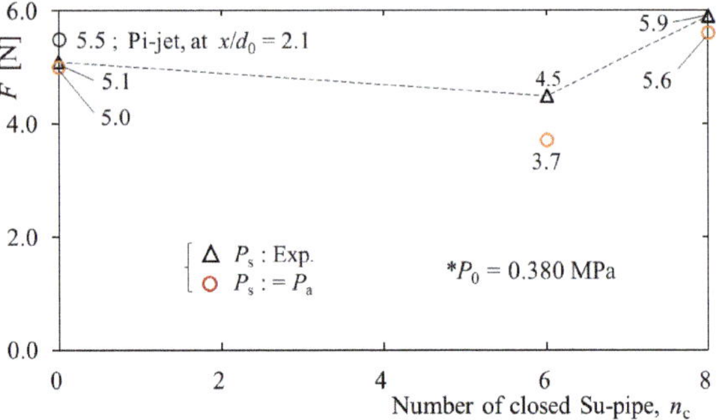

Fig. 318 Thrust, F, of the supersonic jet from the FC-nozzle: The thrust of the 8-closed case is greater than that of the 0-closed case, and this seems to be due to the jet approaching proper expansion

Figure 318 shows the thrust, F, of the FC-jet calculated at $x/d_0 = 5.4$ ($s/d_0 = 1.9$) for $P_0 = 0.38$ MPa. The F of the 6-closed case is smaller than that of the 0-closed case because the entrainment from the surroundings of the 6-closed case is restricted. On the other hand, the F of the 8-closed case is larger than that of the 0-closed case, and this seems to be due to the jet approaching proper expansion. The differences in the F value when the static pressure is atmospheric pressure, and when the measurements were used, the value was $\pm 6\%$ for the 0- and 8-closed cases and $\pm 29\%$ for the 6-closed case, respectively. The large difference for the 6-closed case is probably due to the deformation and asymmetry of the jet's shape. We showed that by changing the number of the Su-pipes and the locations at which they close, the deflection angle and circumferential position of the jet can be controlled, and consequently, the thrust of the supersonic jet can be controlled by a new simple fluidic vector control system.

9.6.3   Thrust Vectoring by Sub Jets

When there is no fluid to be entrained by the jet, that is, when the surrounding area is in a vacuum state, the same control can be performed by blowing a small amount of fluid from the Su-pipe. Using the same nozzle, the FC-nozzle, to blow sub-jets from the two Su-pipes (2-blow) at a small portion of the total flow rate of $Q_c = 10$ [L/min], the jet supplied with $P_0 = 0.38$ MPa is deflected, similar to the result shown in Fig. 311d, at an angle of $\beta = 10°$ to the side that does not blow out, because the pressure on the blowing side increases. $Q_c = 10$ [L/min] is about 1.8% of the main jet flow estimated under the standard state.

Figure 319 shows the visualized flow pattern, and the number of Su-pipes blowing is $n_b = 2$ from ④ and ⑤ in Fig. 309 with all remaining Su-pipes are closed.

(a)   Effects of blowing flow rate on jet deflection angle

Figure 320 shows the relationship between β and Q_c when air is blown from Su-pipes 2, 4, and 6. β increases as Q_c increases, and β also increases as the number of blowing nozzles, n_b, decreases, because by increasing n_b, the blowing flow rate per Su-pipe decreases. The deflection angle has a maximum value of $\beta = 10°$ at $n_b = 2$ and $Q_c = 8$ [L/min].

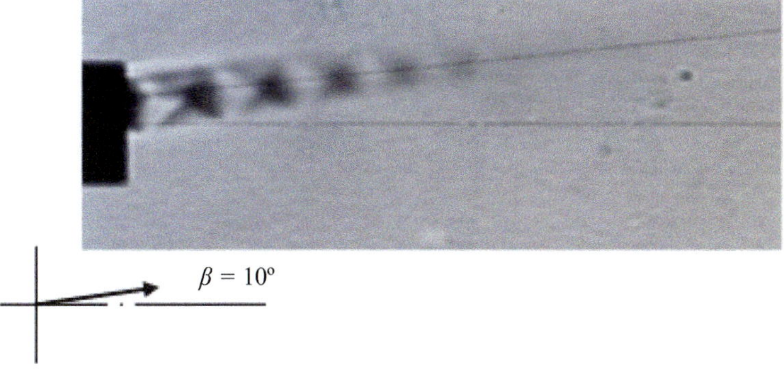

Fig. 319 Deflection angle, β, of the 2-blow condition ($P_0 = 0.38$ MPa, $Q_c = 10$ [L/min])

Fig. 320 Deflection angle, β and blowing flow rate, Q_c ($P_0 = 0.38$ MPa): The deflection angle increases as Q_c in-creases, and β also increases as the number of blowing nozzle n_b decreases

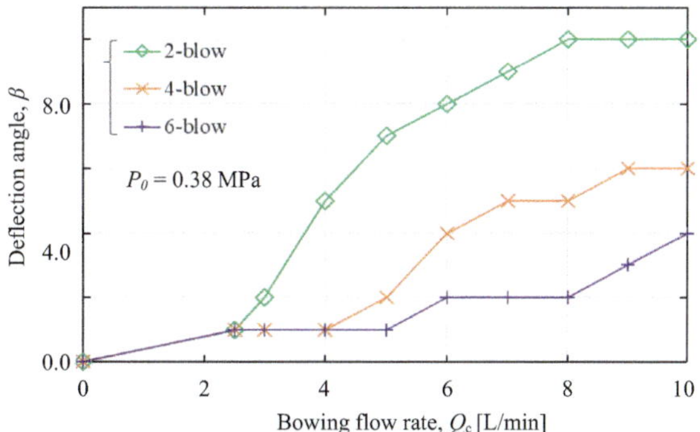

(b) Effects of number of blowing nozzles on jet deflection angle

Figure 321 indicates the relation between β and n_b. The deflection angle has a maximum value of $\beta = 10°$ at $n_b = 2$, and in the 4-, 6-, and 8-blow cases, the deflected angles are 6, 4, and 0°, respectively. It is considered that the deflection angle β can be increased by increasing the opening angle α of the FC-nozzle: however, investigation of the characteristics of the shape of the FC-nozzle, such as the optimal length and α, is a subject for further study.

Here, thrust vector control of the supersonic under-expanded jet by a non-moving system was examined, and a new and simple vector control method for supersonic jets was proposed. The method uses a fluidic Coanda nozzle (FC-nozzle) entrain the surrounding fluid and the Coanda effect. The FC nozzle consists of a pipe nozzle (Pi-nozzle), spacer, and linearly expanded Coanda nozzle (Co-nozzle) with eight suction pipes (Su-pipes). The jet from the Pi-nozzle flows straight with entrainment of the surrounding fluid from the Su-pipes and around the jet. When some Su-pipes are closed, the pressure between the jet and Co-nozzle wall decreases, and subsequently, the jet deflects to the closed side of the Su-pipe and attaches to the wall of the Co-nozzle by the Coanda effect. The flow characteristics and deflection characteristics

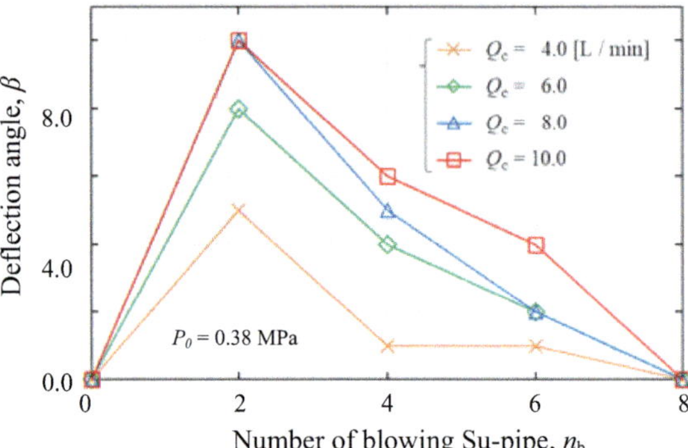

Fig. 321 Deflection angle, β, and the number of Su-pipes blowing, n_b ($P_0 = 0.38$ MPa, $Q_c = 10$ [L/min]): The deflection angle has a maximum value of $\beta = 10°$ at $n = 2$

of the supersonic jet issued from the FC-nozzle were assessed by the visualized flow pattern using the Schlieren method and measurements of the velocity distribution.

By changing the number of the Su-pipes and the locations at which the Su-pipes close, the deflection angle and circumferential position of the jet can be controlled. That is, vector control of the thrust of the supersonic jet can be performed with this new and simple method and apparatus.

In practical applications, it is conceivable that a computer-controlled solenoid valve could be used to change the number of Su-pipes and the locations at which they close.

References

Abramovich GN (1963) The theory of turbulent jets. MIT Press, Cambridge, Mass., Chap. 1. Fundamentals of Turbulent Jets, Microsoft Word—CH1-Fundamentals of Turbulent Jet-EV

Aihara Y, Koyama H (1981) Secondary instability of Goertler vortices-formation dimensional coherent structure. Trans Jan Soc Aero Space Sci 24:78–94

Albertson ML, Dai YB, Jensen RA, Rouse H (1950) Diffusion of submerged jets. ASCE Trans 115:630–697. https://doi.org/10.1061/TACEAT.0006302

Aldaş K, Yapici R (2014) Investigation of effects of scale and surface roughness on efficiency of water jet pumps using CFD. Eng Appl Comput Fluid Mech 8(1):14–25. https://doi.org/10.1080/19942060.2014.11015494

Allen D, Smith BL (2009) Axisymmetric coanda-assisted vectoring. Exp Fluids 46:55–64. https://doi.org/10.1007/s00348-008-0536-y

André B, Castelain T, Bailly C (2014) Experimental exploration of underexpanded supersonic jets. Shock Waves 24:21–32. https://doi.org/10.1007/s00193-013-0457-4

Antonia RA, Zhao Q (2001) Effect of initial conditions on a circular jet. Exp Fluids 31(3): 319–323 ISSN 0723-4864

Bakke P (1957) Experimental investigation of a wall jet. J Fluid Mech 2(5):467–472

Barlow RS, Johnston JP (1988) Structure of a turbulent boundary layer on a concave surface. J Fluid Mech 191:137–176. https://doi.org/10.1017/S0022112088001545

Behrouzi P, Feng T, McGuirk JJ (2008) Active flow control of jet mixing using steady and pulsed fluid tabs. In: Proceedings of IMechE 2008, pp 222 (Part I); J Syst Control Eng 381–392. https://doi.org/10.1243/09596518JSCE543

Belovich VM, Samimy M (1997) Mixing process in a coaxial geometry with a central lobed mixing nozzle. AIAA J 35(5):838–841

Blevins RD (1984) Applied fluid dynamics handbook. Van Nostrand Reinhold Co. ISBN 0-442-21296-8

Boguslawski L, Popiel CZ (1979) Flow structure of the free round turbulent jet in the initial region. J Fluid Mech 90:531–539

Bourque C, Newman BG (1960) Reattachment of a two-dimensional, incompressible jet to an adjacent flat plate. Aeronaut Q 11:201–232. https://doi.org/10.1017/S0001925900001797

Brown FT (1964) A combined analytical and experimental approach to the development of fluid-jet amplifiers. J Basic Eng 86(2):175–182. https://doi.org/10.1115/1.3653023

Brown GL, Roshko A (1974) On density effects and large structure in turbulent mixing layer. J Fluid Mech 64(4):775–816. https://doi.org/10.1017/S002211207400190X

Calletti MJ, Rogers CB, Parekh DE (1996) Parametric study of jet mixing enhancement by vortex generator, tabs, and deflector plates. In: Proceedings of fluids division conference FED-237, ASME, pp 303–312

Carletti MJ, Rogers CB, Parekh DE (1995) Use of streamwise vorticity to increase mass entrainment in a cylindrical ejector. AIAA J 33(9):1641–1645. https://doi.org/10.2514/12704

Chanpagne FH, Wygnanski IJ (1971) An experimental investigation of coaxial turbulent jets. Int J Heat Mass Transfer 14:1445–1464. https://doi.org/10.1016/0017-9310(71)90191-8

Clauser FH (1954) Turbulent boundary layers in adverse pressure gradients. J Aeron Sci 21:91–108. https://doi.org/10.2514/8.2938

Corrsin S, Uberoi MS (1951) Spectra and diffusion in a round turbulent jet. NACA-TR-1040:1–21. ID: 19930092091

Crighton DG, The jet edge-tone feedback cycle; linear theory for the operating stages. J Fluid Mech 234:361–391. https://doi.org/10.1017/S002211209200082X

Crow CT, Sharma MP, Stock DE (1977) The particle-source-in-cell Model for gas-droplet flows. Trans ASME J Fluid Eng 99(2):325–332. https://doi.org/10.1115/1.3448756

Curle N (1953) The mechanism of edge-tones. Proc R Soc Lond A216:412–424. https://doi.org/10.1098/rspa.1950030

Davis MR, Winarto H (1980) Jet diffusion from a circular nozzle above a solid surface. J Fluid Mech 101(1):201–221. https://doi.org/10.1017/S0022112080001607

Deere KA (2003) Summary of fluidic thrust vectoring research conducted at NASA langley research center. In: Proceedings of the 21st AIAA, applied aerodynamics conference, AIAA-2003-3800, pp 1–18. https://doi.org/10.2514/6.2003-3800

Donaldson CD, Snedeker RS (1971a) A study of free jet impingement, Part 1, Mean properties of free and impinging jets. J Fluid Mech 45:81–319. https://doi.org/10.1017/S0022112071000053

Donaldson SD, Snedeker RS (1971b) Abst: under-expanded jet: http://www.kbwiki.ercoftac.org/w/index.pkp/Abstr:Under-expandedjet

Downs SJ, James EH (1987) Jet impingement heat transfer–a literature survey. ASME National Heat Transfer Conf 87-H-35

Drobniak S, Elsner JW, El-Kassem E-SA (1998) Relationship between coherent structures and heat transfer processs in the initial region of a round jet. Exp Fluids 24:225–237

Dumitrache A, Frunzulica F, Preotu O, Ionescu T (2018) Thrust and jet directional control using the Coanda effect. INCAS Bull 10(2):199–205. (P) ISSN 2066-8201, (E) ISSN 2247-4528

Echigo K, Tsujimoto K, Shakouchi T, Ando T (2020) Flow and heat transfer characteristics of blooming jets impinging upon wall using DNS. J Fluid Sci Technol 15(2):1–10. https://doi.org/10.1299/jfst.2020jfst0010

El-Taher RM (1983) Experimental investigation of a curvature effects on ventilated wall jet. AIAA J 21(11):1505–1511. https://doi.org/10.2514/8281

Fiedler HE (1998) Control of free turbulent shear flows. Lect Note Phys 53:335–429. https://doi.org/10.1007/3-540-69672-5_6

Foerthman E (1936a) Turbulent jet expansion, NACA TM-789. (original paper in Germany, 1934: Ing. Archiv., 5)

Foerthman E (1936b) Turbulent jet expansion. NACA, TW-789. ID-19930094628

Franquet E, Perrier V, Gibout S, Bruel P (2015) Free under-expanded jets in a quiescent medium: a review. Prog Aerosp Sci 77:25–5 https://doi.org/10.1016/j.paerosci.2015.06.006

Fujita S, Osaka H, Ueno G (1984) Three-dimensional turbulent jet issuing from the cruciform nozzle–1st Rep., Mean flow properties-, Trans JSME 50(458 B), 2586–2591. https://doi.org/10.1299/kikaib.50.2586

Fujisawa N, Kobayashi R, Shirai H (1985) Centrifugal instability of turbulent wall jets along strongly concave surfaces. J Appl Mech ASME 52(2):492–49 https://doi.org/10.1115/1.3169080

Gauntner JW, Livingood NB, Hrycak P (1970) Survey of literature on flow characteristics of a single turbulent jet impinging on a flat plate. NASA Technical Note D-5652.

Ghanami S, Farhadi (2019) Fluidic oscillators' applications, structures and mechanism—a review. Trans Phenom Nano Micro Scales 7(1):9–27. https://doi.org/10.22111/tpnms.2018.25051.1153

Glauert MB (1956) The wall jet. J Fluid Mech 1(6):625–643. https://doi.org/10.1017/S0022112056000041X

Goertler H (1942) Berechnung von Auhgaben der freien Turbulenz auf Grundeines neunen Naeherungsatzes. ZAMM 22:224–254

Gohil TB, Saha AK, Muralidhar K (2015) Simulation of the blooming phenomenon in forced circular jets. J Fluid Mech 783(215):567–674. https://doi.org/10.1017/jfm.2015.571

Görtler H (1942) Berechnung von Aufgaben der freien Turbulenz auf Frund eines neuen Naherungsansatzes. Ziet F Angrew Math Mech 22(5):240–254

Guitton DE, Newman BG (1977) Self-preserving turbulent wall jets over convex surface. J Fluid Mech 81:155–185. https://doi.org/10.1017/S0022112077001967

Gutmark EJ, Callender BW, Martens S (2006) Aeroacoustics of turbulent jets: flow structure, noise sources, and control. JSME Int J Ser Fluid Therm Eng 49:1078–1085. https://doi.org/10.1299/jsmeb.49.1078

Guyot D, Bobusch B, Paschereit CO, Raghu S (2008) Active combustion control using a fluidic oscillator for asymmetric fuel flow modulation. In: The 44th AIAA/ASME/SAE/ASEE joint propulsion conference and exhibit AIAA 2008-4956, pp 1–19. https://doi.org/10.2514/6.2008-4956

Heo JY, Too KH, Lee Y, Sung HG, Cho SH, Jeon YJ (2009) Fluidic thrust vector control of supersonic jet using co-flow injection. In: Proceedings of the 45th AIAA/ASME/SAE/ASEE joint propulsion conference exhibit AIAA 2009-517, pp 1–2

Heskestadt G (1965) Hot-wire measurements in a plane turbulent jet. Trans ASME J Appl Mech 32(4):721–734. https://doi.org/10.1115/1.3627309

Hilberg D (1996) Repeatedly deflected jets and flames, unpublished, Hermann-Foettinger-Institute, TU-Berlin

Hofken M, Bishof F, Durst F (1991) Novel hyperboloid stirring and aeration system for biological and chemical reactors, FED-132. Ind Appl Fluid Technol ASME, pp 47–56

Holger DK, Wilson TA, Beavers (1977) Fluid mechanics of the edge tone. J Acoust Soc Am 662:1116–1128. https://doi.org/10.1121/1.381645

Homann F (1936) Einfluß großer Zähigkeit bei Strömung um Zylinder. Forschg Ing-Wes 7:1–10

Hortensius R, Dutton JC, Elliot GS (2017) Near field of an axisymmetric under-expanded jet and an adjacent parallel surface. AIAA J 55:2489–2501. https://doi.org/10.2514/1.J055515

Hrycak P (1981) Heat transfer from impinging jets: a literature review. AWAL-TR-81-3054

Hu H, Kobayashi T, Saga T, Segawa S, Taniguchi N (2000) Particle image velocimetry and planar laser-induced fluorescence measurements on lobed jet mixing flows. Exp Fluids (Suppl.):S141–S157

Hussain AKMF (1986) Coherent structures and turbulence. J Fluid Mech 173(1986):303–356. https://doi.org/10.1017/S0022112086001192

Husain HS, Hussain AKMF (1983) Controlled excitation of elliptic jets. Phys Fluids 26(10):2763–2766. https://doi.org/10.1063/1.864062

Hussain F, Husain HS (1989) Elliptic jets, Part 1. Characteristics of unexcited and excited jets. J Fluid Mech 208:257–320. https://doi.org/10.1017/S0022112089002843

Hussain AKMF, Zaman KBMQ (1980) Vortex pairing in a circular jet under controlled excitation—Part 2, coherent structure dynamics. J Fluid Mech 101(3):493–544. https://doi.org/10.1017/S0022112080001772

Ichimiya K, Yamada Y (2003) Three dimensional heat transfer of a confined circular impinging jet with buoyancy effects. ASME J Heat Transfer 125:250–256; JSME ed., JSME Date Handbook: Heat Transfer (1979):123

Ikui T, Matsuo K (2001) Compressible fluid dynamics. Rikogakusha Pub. Co., Ltd, Japan

Iriyama S (2012) Flow characteristics of submerged free jet flow from petal-shaped nozzle (Mixing and Diffusion Characteristics, and Effects of Sub-Nozzle) (In Japanese). Mie University Master's Thesis

Ishihara T, Tsuchida A, Jet flow. J JSME 66(537):1333–1340

Iwagami S, Kobayashi T, Takahashi K, Hattori Y (2019) Reproducibility of mode transition of edge tone with DNS and LES. In: Proceedings 23rd international congress on acoustics, pp 5573–5579

JSME (1989) Hand book of gas-liquid two-phase flow technology. Corona Pub. Co.

Kataoka K (1990) Impingement heat transfer augmentation due to large scale eddies. In: Proceedings of 9th international heat transfer conference, pp 255–273

Katanoda H, Miyazato Y, Masuda M (1995) Numerical calculation of under-expanded axisymmetric supersonic jet—Effects of nozzle divergence angle and Mach number on shock cell length. Trans JSME Ser B 61:3236–3242. https://doi.org/10.1299/kikaib.61.3236

Katanoda H, Miyazato Y, Masuda Y, Matsuo K (2000) Pitot pressures of correctly-expanded and under-expanded free jets from axisymmetric supersonic nozzles. Shock Waves 10:95–101. https://doi.org/10.1007/s001930050182

Kataoka K, Suguro M, Degawa H, Maruo K, Mihata (1987) The effect of surface renewal due to large-scale eddies on jet impingement heat transfer. Int J Heat Mass Transfer 30(3):559–567

Kaushik M, Rathakrishnan E (2006) Studies on the effect of notches on circular sonic jet mixing. J Propul Power 22(1):211–214. https://doi.org/10.2514/1.8424

Kirshner JM, Katz S (1975) Design theory of fluidic components. Academic Press, Cambridge. ISBN 0-12-410250-6

Kito M, Shakouchi T, Tsujimoto K, Ando T (2008a) Flow and heat transfer characteristics of cone orifice free and impinging jrts—effects of cone angle. 74(737 B):1593–1599. http://id.crossref.org/issn/03875016

Kito M, Shakouchi T, Nuntadusit C, Tsujimoto K, Ando T (2008b) Flow control of orifice free jet by nozzle plate thickness. Trans JSME 74(745 B):1935–1941. https://doi.org/10.1299/kikaib.74.1935

Kojima T, Matsuoka Y (1986) Study on underexpanded sonic jet—1st Rep., On the mechanism of pseudo-shock waves and pressure and velocity in jets. Trans JSME Ser. B 52:2021–2027. https://doi.org/10.1299/kikaib.52.2021

Kojima T, Matsuoka Y (1987) Study of an under-expanded sonic jet—1st Rep., On the mechanism of pseudo-shock waves and pressure and velocity in jets. Bull JSME 30(259):190–191. CRID 1573950401911582592

Koklu M, Owens LR (2017) Comparison of sweeping jet actuators with different flow-control techniques for flow-separation control. AIAA J 55(3):848–860. https://doi.org/10.2514/1.J055286

Korschelt D (1980) Experimantelle Untersuchung zum Waerme und Stofftransport im turbulenten ebenen Freistrahl mit periodisher Anregung Duesenaustritt, Dissertation, TU-Berlin

Kudirka AA, Gluntz DM (1974) Development of jet pumps for boiling water reactor recirculation systems. Trans ASME J Eng Power 96:7–12. https://doi.org/10.1115/1.3445755

Kumar MA, Kurian J (1995) Studies on free jets from radially lobed nozzles. Exp Fluids 19:95–102

Kurimoto N, Suzuki Y, Kasagi N (2004) Active control of coaxial jet mixing with arrayed micro actuators. Trans JSME 70(694 B):1417–1424. https://doi.org/10.1299/kikaib.70.1417

Launder BE, Rodi W (1981) The turbulent wall jet. Prog Aerosp Sci 19:81–128. https://doi.org/10.1016/0376-0421(79)90002-2Getrightsandcontent

Lavernia EJ, Srivatsan TS (2010) The rapid solidification processing of material: science, principles, technology, advances, and applications. J Mater Sci 45:287–325. https://doi.org/10.1007/s10 853-009-3995-5

Lee J, Lee SJ (2000a) The effect of nozzle configuration on stagnation region heat transfer enhancement of axisymmetric jet impingement. Int J Heat Mass Transfer 43:3497–3509

Lee J, Lee SJ (2000b) The effect of nozzle aspect ratio on stagnation region heat transfer characteristics of elliptic impinging jet. Int J Heat Mass Transfer 43:555–575

Lee M, Reynolds WC (1985) Bifurcating and blooming jets. In: The 5th symposium on turbulent shear flows. Springer, pp 1.7–1.12

Liu JF, Luo ZB, Deng X, Zhao ZJ, Li SQ, Liu Q, Zhu YX (2022) Dual Synthetic jets actuator and its applications—Part II: novel fluidic thrust-vectoring method based on dual synthetic jets actuator. Actuators 11(8) 209:1–12. https://doi.org/10.3390/act11080209

Livingood JNB, Hrycak P (1973) Impingement heat transfer from turbulent air stream jets to flat plate, a literature survey. NASA TM X-2778

Lytle D, Webb BW (1991) Secondary heat transfer maxima for air jet impingement at low nozzle-plate spacing. In: Experimental heat transfer mechanics and thermodynamics. Elsevier, pp 776–783

Maita M (1986) Flow visualization around propulsion system on Asuka's development. J Flow vis 6(21):106–11 https://doi.org/10.3154/jvs1981.6.106

Maki H, Ogawa N, Nishimura, H (1993) Large-scale fluctuations existing in flow fields of two-dimensional impinging jet. Trans JSME 59–568B:3756–3761

Martin H (1977) Heat and mass transfer between impinging gas jets and solid surface. In: Advances in heat transfer, vol 1 Academic, pp 1–60. ISBN 0-12 020013-9

Martinelli M, Viktorov V (2011) A mini fluidic oscillating flowmeter. Flow Meas Instrum 22(6):537–54 https://doi.org/10.1016/j.flowmeasinst.2011.10.004

Marukami E, Papamoschou D (2000) Mixing layer characteristics of coaxial supersonic jets. Proceedings of AIAA/CEAS Aeronautics Conference, pp 1–16. https://doi.org/10.2514/6.2000-2060

Mataoui A, Schiestel R, Salem A (2001) Flow regimes of interaction of turbulent plane jet into a rectangular cavity: experimental approach and numerical modelling. Flow Turbul Combust 67:267–304

Matsuo K (2011) Fundamentals of compressible fluid dynamics. Jupiter Pub. Com., Japan. ISBN-10: 4990748336, ISBN-13: 978-4990748333

Matsuo K (2020) Compressible fluid dynamics—theory and analysis in internal flow. Ohmusha Ltd., Japan. ISBN-10: 4274225860, ISBN-13: 978-427422586

Matsuda T, Sakakibara J (2005) On the vortical structure in a round jet. Phys Fluids 17(2), 025106:1–11. https://doi.org/10.1063/1.1840869

Matsuo S, Setoguchi T, Kudo T, Yu S (1998) Study on the characteristics of supersonic Coanda jet. J Thermal Sci 7:165–175. https://doi.org/10.1007/s11630-998-0012-2

MD 520N Helicopter: https://support.mdhelicopters.com/files/Models/MD520N_Tech_Desc.pdf

Meslem A, Nastase I, Meaim KA (2008) Experimental investigation of the mixing performance of a lobed jet flow. J Eng Phys Thermo-Phys 81(1):106–111

Mi J, Nathan GJ, Nobes DS (2001) Mixing characteristics of axisymmetric free jets from contoured nozzle, an orifice plate and a pipe. J Fluids Eng Trans ASME 123(4):878–88 ISSN 0098-2202

Michalke A, Wehrmann O (1964) Akustoshe Beeinflussing von Freistrahlgrenzshichten. In: Proceedings of international council aeronautical science 3rd congress, pp 773–785

Miller R, Yagle PJ, Hamstra JW (1999) Fluidic throat skewing for thrust vectoring in fixed-geometry nozzles. In: Proceedings of the 37th aerospace sciences meeting and exhibit, AIAA-99-0365. https://doi.org/10.2514/6.1999-365

Miyakoshi K, Haniu H (2003) Coherent structure in a two-dimensional jet passably controlled by wake of a cylinder installed in nozzle contraction. Trans JSME 69(678):338–345. https://doi.org/10.1299/kikaib.69.338

Molloy NA, Tayler PL (1969) Oscillatory flow of a jet into a blind cavity. Nature 224:1192–1194

Moody LF (1944) Friction factors for pipe flow. Trans ASME 66:671–684. https://doi.org/10.1115/1.4018140

Munday D, Heeb N, Gutmark E, Liu J, Kailasanath K (2012) Acoustic effect of chevrons on jets existing conical convergent-divergent nozzles. AIAA J 50(11):2336–2350. https://doi.org/10.2514/1.J051337

Myers GE, Schauer JJ, Eustis RH (1963) The plane turbulent wall jet flow development and friction factor. J Basic Eng ASME Int 85(1):47–5 https://doi.org/10.1115/1.3656534

Myers GE, Schauer JJ, Eustis RH (1964) The plane turbulent wall jet. Part 2—Heat transfer. Technical Report. 2, Eng. Dept., Stanford University

Nakayama Y (1989) Fluid mechanics (in Japanese). Yokendo Ltd. Pub. Japan. ISBN4-8425-8906-x

Nakamura H, Shakouchi T (2004) A new surface fusing, globular forming method of fine particle. Trans JSME 70(699 B):255–262

Nakamura H, Shakouchi T, Kusuda A (2003) High temperature air jet flow and its application to globular forming of fine particles. Trans Kagaku Kogaku Ronbunshu 6:36–841

Narabayashi T, Yamazaki Y, Kobayashi H, Shakouchi T (2006) Flow analysis for single and multi-nozzle jet pump. JSME Int J Seri B 49(4):933–940. https://doi.org/10.1299/jsmeb.49.933

Nastase I, Bode F (2018) Impinging jets—a short review on strategies for heat transfer enhancement. E3S Web Conf 32:0101 https://doi.org/10.1051/e3sconf/20183201013

Nei H, Narabayashi T (1994) Characteristics of jet pumps for nuclear reactors 1st Rep. Basic tests under high temperature. Transf JSME 60(569 B):194–200. https://doi.org/10.1299/kikaib.60.194

Ninomiya N, Kasagi N (1993) Turbulence statistics in the self-preserving region of an axisymmetric free jet. Trans JSME 59(561):1532–1538. https://doi.org/10.1299/kikaib.59.1532

Nourl JM, Whitelaw JH (1996) Flow characteristics of an under expanded jet and its application to the study of droplet breakup. Exp Fluids 21:243–247. https://doi.org/10.1007/BF00190673

Nyborg WL (1954) Self-maintained oscillations of the jet in a jet-edge system. J Acoust Soc Am 26(2):174–182

Ohmura T, Shakouchi T, Fukushima S, Tsujimoto K (2021) Flow characteristics of two-dimensional supersonic under-expanded Coanda-reattached jet. J Flow Control Meas Vis 9(1):1–14. https://doi.org/10.4236/jfcmv.2021.91001

Olsson M, Fuchs L (1998) Large-eddy simulations of a forced semi-confined circular impinging jet. Phys Fluids 10(2):476–486

Oshima R (1988) Studies on the optimum throat length of the jet pumps. Trans JSME 54(497 B):125–130. https://doi.org/10.1299/kikaib.54.125

Ostermann F, Woszidlo R, Nayeri CN, Paschereit CO (2018) Properties of a sweeping jet emitted from a fluidic oscillator. J Fluid Mech 857:216–238

Pascoa L, Dumas A, Trancossi M, Stewart P, Vucinic D (2013) A review of thrust vectoring in support of V/STOL non-moving mechanical propulsion system. Cent Eur J Phys 3:374–388. https://doi.org/10.2478/s13531-013-0114-9

Phan LQ, Johnstone AD, Kosasih B, Renshaw W (2020) Vortex dynamics and fluctuations of impinging planer jet. ISIJ Int J 60(5):1030–1039. https://doi.org/10.2355/isijinternational.ISIJINT-2019-566

Poreh M, Tsuei YG, Cermark JE (1967) Investigation of a turbulent radial wall jet. J Appl Mech 34(2):457–463 https://doi.org/10.1115/1.3607705

Pourhashem H, Kumar S, Kalkhoran IM (2019) Flow field characteristics of a supersonic jet influenced by downstream micro jet fluidic injection. Aerosp Sci Technol AESCTE 2019 93:1–15. https://doi.org/10.1016/j.ast.2019.07.014

Powell A (1961) On the edgetone. J Acoust Soc Am 33(4):395–409. https://doi.org/10.1121/1.1908677

Quinn WR (2006) Upstream nozzle shaping effects on near field flow in round turbulent free jets. Euro J Mech B/fluid 25(3):279–301. https://doi.org/10.1016/j.euromechflu.2005.10.002

Raghu S (2013) Fluidic oscillators for flow control. Exp Fluids 54(2):1455. https://doi.org/10.1007/s00348-012-1455-5

Rajaratnam N (1976) Turbulent jets. Elsevier. ISBN O-444-41372-3

Rajaratnam N, Subramanya K (1967) Diffusion od rectangular wall jets in wider channel. J Hydraulic Res 5:281–294. https://doi.org/10.1080/00221686709500212

Ramanathan V, Arndt REA (1996) Fluid mechanical considerations in the design of improved aerator. Trans ASME J Fluids Eng 118:736–742

Reichardt H (1942) Gasetzmaessigkeiten der freien Turbulenz. VDI-Fotshungsheft 414

Rembold B, AdamsNA KL (2002) Direct numerical simulation of transitional rectangular jet. Int J Heat Fluid Flow 23:547–553

Resta E, Malsilio R, Ferlauto M (2021) Thrust vectoring of a fixed axisymmetric supersonic nozzle using the shock-vector control method. Fluids 6(12) 441:1–17. https://doi.org/10.3390/fluids 6120441

Reynolds WC, Parekh DE, Juvet PJD, Lee MJD (2003) Bifurcating and blooming jets. Ann Rev Fluid Mech 35:295–315. https://doi.org/10.1146/annurev.fluid.35.101101.161128

Rockwell D, Naudascher E (1979) Self sustained oscillations of impinging free shear layer. Ann Rev Fluid Mech 11:69–74. https://doi.org/10.1146/annurev.fl.11.010179.000435

Saga T, Hu H, Kobayashi T (1999) Mixing process in the jet flow of lobed nozzle. In: Proceedings of '99 Korea-Japan joint seminar on particle image velocimetry, pp 39–50

Sanger NL (1970) An experimental investigation of several low-area-ratio water jet pumps. Trans ASME J Basic Eng 92:11–20. https://doi.org/10.1115/1.3424917

Schlichting H (1933) Laminare Strhlenausbreitung. ZAMM 13:260–263

Schlichting H (1968) Boundary-layer theory, 6th edn. McGraw-Hill Book Co., Inc

Schlichiting H, Gersten K (2000) Boundary-layer theory, 8th edn. Springer

Schwarz WH, Cosart WP (1961) The two-dimensional turbulent wall jet. J Fluid Mech 10(4):481–495. https://doi.org/10.1017/S0022112061000299

Seo IL W, Lyu S, Park YS (2002) Experimental investigation of the velocity measurements and turbulent structures of turbulent jets using PIV. In: Advances in fluid modeling and turbulence measurements, pp 185–192. https://doi.org/10.1142/9789812777591_0022

Sfeir AA (1979) Investigation of three-dimensional turbulent rectangular jets. AIAA J 17(10):1055–1060

Shadow KC, Gutmark E, Wilson KJ (1990a) Compressible spreading rates of supersonic coaxial jets. Exp Fluids 10:161–167. https://doi.org/10.1007/BF00215025

Shakouchi T (1990) A new fluidic oscillator type flowmeter using a three-dimensional nozzle—effect of nozzle shape. Trans JSME 56(524 B):975–982. https://doi.org/10.1299/kikaib.56.975

Shakouchi T (1991a) A Study on the flow measurement of gas-liquid two-phase flow by a new hydraulic oscillator-type flowmeter—bubble flow in a vertical pipe. Trans JSME 57(543 B):3647–3652. https://doi.org/10.1299/kikaib.57.3647

Shakouchi T (1991b) A new fluidic oscillator, flowmeter, without control port and feedback loop. J Dyn Syst Meas Control Trans ASME 111(3):535–539. https://doi.org/10.1115/1.3153085

Shakouchi T (1995) Flow measurement of gas-liquid two-phase flow in a horizontal pipe by a new hydraulic oscillator. In: Advanced in multiphase flow. Elsevier, pp 793–802. ISBN 0-444-81811-1

Shakouchi T (2003b) Gas absorption, aeration, by fluidic oscillator operated by gas-liquid two-phase flow. In: Proceedings of ASME FEDSM'3, pp 1–8

Shakouchi T (2004a) Jet flow engineering—fundamentals and application. Morikita Pub. Co. Ltd., Japan. ISBN 978-4-627-67201-7

Shakouchi T (2004b) Jet flow engineering (in Japanese). Morikita Pub. Co. Ltd., Japan. ISBN 4-627-67201-2

Shakouchi T (2020) Passive flow control of jets. J Mech Eng Rev JSME 17:1–18. https://doi.org/10.1299/mer.19-00386

Shakouchi T, Fukushima S (2022) Fluidic thrust, propulsion, vector control of supersonic jets by flow entrainment and the Coanda effect. Energies 15(22) 8513:1–1. https://doi.org/10.3390/en1 5228513

Shakouchi T, Onohara Y (1989) Analysis of a two-dimensional, turbulent wall jet along a circular cylinder–2nd Rep., effects of nozzle width. Trans JSME 55(511 B):662–668. https://doi.org/10.1299/kikaib.55.662

Shakouchi T, Onohara Y (1990) Analysis of a two-dimensional, turbulent wall jet along a circular cylinder–3rd Rep.,—effects of nozzle exit velocity profile and injection angle. Trans JSME 56(532 B):3650–3657. https://doi.org/10.1299/kikaib.56.3650

Shakouchi T, Aoki T, Uesugi M (1993) Study on the three-dimensional, round-wall jet along a concave surface. Trans JSME 59(563 B):2157–2164. https://doi.org/10.1299/kikaib.59.2157

Shakouchi T, Ito T, Suematsu Y, Shia Y (1985) Oscillatory phenomena of a jet in jet-edge system—1st Rep., Behavior of jet and oscillatory frequency. Trans JSME 51-(469 B):2896–2905. https://doi.org/10.1299/kikaib.51.2897

Shakouchi T, Ito T, Suematsu Y (1986) Oscillatory phenomena of a jet in jet-edge system—2nd Rep., Behavior of jet and oscillatory frequency. Trans JSME B 52(480 B):2872–2880. https://doi.org/10.1299/kikaib.52.2872

Shakouchi T, Iriyama S, Kawashima Y, Tsujinoto K, Ando T (2014) Flow characteristics of submerged free jet flow from petal-shaped nozzle. J Fluid Sci Tech JSME 9(3):1–7. https://doi.org/10.1299/jfst.2014jfst0037

Shakouchi T, Kito M, Sakamoto T, Tsujimoto K, Ando T (2008b) Flow analysis of orifice free jet (Effects of contraction ratio). Trans JSME 74(737 B):42–48. https://doi.org/10.1299/kikaib.74.42

Shakouchi T, Kito M, Tsuda M, Tsujimoto K, Ando Y (2011) Flow and heat transfer of impinging jet from notched-orifice nozzle. Trans JSME 6(4):453–464. https://doi.org/10.1299/jfst.6.453

Shakouchi T, Matsumoto A, Watanabe A (2008c) Heat and fluid flow properties of circular impinging jet with a low nozzle to plate spacing—improvement by notched nozzle. Trans JSME 66(650 B):2655–2660. https://doi.org/10.1299/kikaib.66.650_2655

Shakouchi T, Nakamura H, Morimoto H, Okumoto M (2003) Flow and heat transfer of high temperature circular jet and its application. Proc JSME 03–30:283–284

Shakouchi T, Onohara Y, Kato S (1988) Analysis of a two-dimensional, turbulent wall jet along a circular cylinder—1st Rep., velocity and pressure distributions. Trans Jpn Soc Mech Eng JSME 54(500 B):783–790. https://doi.org/10.1299/kikaib.54.783

Shakouchi T, Suematsu Y, Ito T (1982) A study on oscillatory jet in a cavity–1st Rep., Mechanism of oscillation-, Bull JSME 25(206), 1258–11265

Shakouchi T, Sugimoto H, Ando T, Tsujimoto K, Tanabe T (2003a) Flow and heat transfer of impinging jet with low nozzle-plate spacing and their control—effects of nozzle wall thickness and geometry. Trans JSME 69(686):2305–23212. https://doi.org/10.1299/kikaib.69.2305

Shakouchi T, Tanoue T, Tsujimoto K, Ando T (2019) Flow characteristics of sub- and supersonic jets issued from orifice and notched orifice nozzles, and application to ejector. J Fluid Sci Technol JSME 14:1–12. https://doi.org/10.1299/jfst.2019jfst0009

Shakouchi T, Tsujimoto K, Maeda T (2008a) Fluid mechanics (in Japanese). Yokendo Ltd. Pub., Japan. ISBN: 9784842504377

Shakouchi T, Uesugi M, Kato S (1992a) A study on the two-dimensional, turbulent wall jet along a concave surface of a circular cylinder. J Jpn Soc Aeronaut Space Sci 38(442):34–41. https://doi.org/10.2322/jjsass1969.38.600

Shakouchi T, Wakamatsu T, Ando T, Yasuda M (1996) Flow and forced convective heat transfer of round impinging jet—control by coaxial annular subjet. Trans JSME 62(593 B):248–255. https://doi.org/10.1299/kikaib.62.248

Shakouchi T, Yoshida Y, Kato S (1992b) A study on the three-dimensional, round wall jet along a circular cylinder—effects of ejection angle. Trans. JSME 58(552 B):2374–2379. https://doi.org/10.1299/kikaib.58.2374

ShinMaywa US-2 https://www.shinmaywa.co.jp/english/products/aircraft/amphibian/us2/capability.html. Wikipedia: https://www.wikipedia.org/

Shirahama Y, Toyada K (1989) Large-scale structure of a turbulent pipe jet. Trans JSME 55-513B:1372–1376

Sigalla A (1958) Measurement of skin friction in a plane turbulent wall jet. J R Aeronaut Soc 62:837–877. https://doi.org/10.1017/S0368393100070231

Smith BL, Glezer A (1998) The formation and evolution of synthetic jets. Phys Fluids 10(9):2281–2297. https://doi.org/10.1063/1.869828

Sridhar K, Tu PKC (1969) Experimental investigation of curvature effects on turbulent wall jets. J R Aero Sci 73:977–981

Sugawara S, Nakano S, Miyazato Y, Ishino Y (2020) Three-dimensional reconstruction of a microjet with a Mach disc by Mach-Zehnder interferometer. J Fluid Mech 893:A25. https://doi.org/10.1017/jfm.2020.217

Suzuki H, Kasagi N, Suzuki Y (1999) Active control of axisymmetric jet with an array of micro electro-magnetic flap actuators. Trans JSME 65(639 B):3644–3651

Takahashi H (1986) Flow visualization in the STOL research aircraft named Asuka. J Flow vis 6(21):97–105. https://doi.org/10.3154/jvs1981.6.97

Takahashi M, Fukui R, Tsujimoto K, Ando T, Shakouchi T (2023) Helical structures in a temporally developing round jet in the developed state. Flow Turbul Combust 111:59–79. https://doi.org/10.1007/s10494-023-00423-4

Takezawa S, Hayakawa M, Iida S (1988) Near field measurements of circular pipe jets with a fullyturbulent initial condition. Trans JSME 54–502B:1313–1318

Tani I (1962) Production of longitudinal boundary vortices in the layer along a concave wall. J Geophys Res 67:3075–3080. https://doi.org/10.1029/JZ067i008p03075

Tani I, Komatsu Y (1966) Impingement of a round jet on a flat surface. In: Proceedings of 11th international congress of applied mechanics. Springer, pp 672–676

Tesař V (2012) "Master and Slave" fluidic amplifier cascade. EPJ Web Conf, EDP Sci 25:0109 https://doi.org/10.1051/epjconf/20122501093

The Japan Society of Mechanical Engineers, ed., Flow resistance of pipe and duct flows (1979):53

Tollmien W (1926) Berechnung turbulenter Ausbreitungsvorgaenge. ZAMM 6:468–478

Toyoda K (2005) Vortices in jets. Nagare JSFM 24(2):11–160. https://doi.org/10.11426/nagare1982.24.151

Toyoda K, Hussain F (1989) A study of the vortical structures of noncircular jets. Trans JSME 55(514 B):1542–1545. https://doi.org/10.1299/kikaib.55.1542

Toyoda K, Shirahama Y, Kotani K (1992) Manipulation of vortical structures in noncircular jets. Trans JSME 58(545):6–1. https://doi.org/10.1299/kikaib.58.7

Trancossi M (2011) An overview of scientific and technical literature on Coanda effect applied to nozzles. SAE Technical Paper, p 2591. https://doi.org/10.4271/2011-01-2591

Trancossi M, Dumas A (2011a) ACHEON aerial Coanda high efficiency orienting-jet nozzle; No. 2011-01-2737. SAE International. https://doi.org/10.4271/2011-01-2737

Trancossi M, Dumas A (2011b) Coanda synthetic jet deflection apparatus and control; No. 2011-01-2590. SAE International, pp 1–1 https://doi.org/10.4271/2011-01-2590

Trancossi M, Madonia M, Dumas A, Angeli D, Bingham C, Das SS, Grimaccia F, Marques JP, Porreca E, Smith T, Stewart P, Subhash M, Sunol A, Vucinic D (2016a) A new aircraft architecture based on the ACHEON Coanda nozzle: flight model and energy evaluation. Eur Transp Res Rev 8(11):1–21. https://doi.org/10.1007/s12544-016-0198-4

Trancossi M, Stewart J, Maharshi S, Angeli D (2016b) Mathematical model of a constructal Coanda effect nozzle. J Appl Fluid Mech 9(6):2813–2822. https://doi.org/10.29252/jafm.09.06.23508

Trentacoste N, Sforza PM (1967) Further experimental results for three-dimensional free jets. AIAA J 5(5):885–891

Truepel T (1915) Uber die Einwirkung eines Luftstrahles auf die umgebende Luft. Zeitschlift fuer das gesammte Turbinenwesen 5(6) (see Abramovich 1963)

Tsuchiya Y, Horikoshi C (1986) On the spread of rectangular jets. Exp Fluids 4:197–204. https://doi.org/10.1007/BF00717815

Tu GV, Hooper JD, Wood DH (1992) Wall pressure and shear stress measurements for normal jet impingement. In: Proceedings of 11th Australasian fluid mechanics conference, 9A-6, pp 1109–1112

Tyliszczak A (2015) Multi-armed jets: a subset of the blooming jet. Phys Fluids 27(4):04170. https://doi.org/10.1063/1.4917179

Uddin N, Neumann SO, Weigand B (2013) LES simulations of an impinging jet: On the origin of the second peak in the Nusselt number distribution. Int J Heat Mass Transfer 57(1):356–368

Uddin N, Newmann SO, Weigant B, Younis BA (2009) Large-eddy simulation and heat-flux modeling in a turbulent impinging jet. Numer Heat Transfer Part A 55:906–930

Üral A (1990) Production of rapidly aluminum alloy powders by gas atomization and their applications. Powder Metall 33:53–64

Verhoff A (1963) The two-dimensional turbulent wall jet with and without an external stream, Rep. 626, Princeton University

Viskanta R (1993) Heat transfer to impinging isothermal gas and flame jets. Exp Thermal Fluid Sci 6:111–134

Wienand J, Riedelsheimer A, Wiegand B (2018) Numerical study of s turbulent impinging jet for different jet-to-plate distances using two-equation turbulence models. In: Proceedings of international symposium of transport phenomena and dynamics of rotating machinery, pp 1–9. https://hal.science/hal-01891323v1

Wilson DJ, Goldstein RJ (1976) Turbulent wall jets with cylindrical streamwise surface curvature. J Fluids Eng Trans ASME 98(3):550–557. https://doi.org/10.1115/1.3448393

Wooly JP, Karamcheti K (1973) The role of jet stability in edgetone generation. AIAA J 12(11):1457–1458. https://doi.org/10.2514/49525

Woszidlo R, Ostermann F, Schmidt H-J (2019) Fundamental properties of fluidic oscillators for flow control applications. AIAA J 57(3):978–992. https://doi.org/10.2514/1.J056775

Wu K, Kim HD, Jin Y (2018) Fluidic thrust vector control based on counter-flow concept. Proc Inst Mech Eng Part G J Aerosp Eng 233(4):1412–1422. https://doi.org/10.1177/0954410017752580

Yamada H, Yahata S, Osaka H, Kageyama Y (1980) Turbulent properties of two-dimensional turbulent wall jet. Memoirs Faculty Eng Yamaguchi Univ 30(2):381–388

Yamazaki Y, Yamazaki A, Narabayashi T, Suzuki J, Shakouchi T (2005) Studies on mixing process and performance improvement of jet pumps—effect of nozzle and throat shapes. J Fluid Sci Tech JSME 2(1):237–247. https://doi.org/10.1299/jfst.2.238

Yamazaki Y, Nakayama T, Narabayashi T, Kobayashi H, Shakouchi T (2006) Effect of surface roughness on jet pump performance. JSME Int J 49(4 B):928–932. https://doi.org/10.1299/jsmeb.49.928

Yan D, Wei Z, Xie K, Wang N (2020) Simulation of thrust control by fluidic injection and pintle in a solid rocket motor. Aerosp Sci Technol AESCTE 99:1–10. https://doi.org/10.1016/j.ast.2020.105711

Yokobori S, Kasagi N, Hirata, M, Nishiwaki N (1978) Role of large-scale eddy structure on enhancement of heat transfer in stagnation region of two-dimensional, submerged, impinging jet. In: Proceedings of 6th international heat transfer conference, pp 305–310

Zaman KBMQ (1995) Axis switching and spreading of an asymmetric jet—Role of vorticity dynamics-, NASA Technical Momo. 106385, AIAA-95-0889, 1–18

Zaman KBMQ (1996) Spreading characteristics and thrust of jets from asymmetric nozzles, NASA Technical Memo. 107132, AIAA-96-0200, 1–17

Zaman KBMQ (1999) Spreading characteristics compressible jets from nozzles of various geometries. J Fluid Mech 383:197–228. https://doi.org/10.1017/S0022112099003833

Zapryagaev VI, Kiselev NP, Pivovarov AA (2015a) Gasdynamic structure of an axisymmetric supersonic underexpanded jet. Fluid Dyn 50:87–97. https://doi.org/10.1134/S001546281501010X

Zapryagaez VI, Kiselev NP, Pivovarov AA (2015b) Gas dynamic structure of an axisymmetric supersonic under-expanded jet. Fluid Dyn 50:95–107. https://doi.org/10.1134/S001546281501010X

Zhou DW, Lee SJ, Ma CF, Bergles AE (2006) Optimization of mesh screen for enhancing jet impingement heat transfer. J Heat Mass Transfer 42(6)6:501–510. ISSN 0947-7411

Chapter 4
Wakes and Separated Flows

The flow around and behind an object placed in the flow (Fig. 1) is related to boundary layers, flow separation, vortex formation, etc. is important in relation to the resistance force received from the flow (flow resistance). As shown in Fig. 1, when a cylinder is placed in a uniform flow with velocity U, the boundary layer that has developed along the surface from the stagnation point in front of the cylinder separates at a certain point (separation point), and a free turbulent region (wake flow) with a defective velocity is formed behind the cylinder. The free turbulence region flows downward while mixing with the surrounding fluid at its boundary, and the velocity deficit region eventually disappears. In this way, the wake flow is a flow with a free turbulent boundary layer, similar to the jet described above. Furthermore, when the Reynolds number is $Re = Ud/v \fallingdotseq 40$ to 200 (d: cylinder diameter), a regular vortex street, the so-called Karman vortex street, occurs in the wake of the object.

1   Deficit Velocity Distribution in Wake

1.1   Velocity Distribution

Here, the deficit velocity distribution in the wake is determined using the turbulent boundary layer equation described earlier.

$$u\frac{\partial u}{\partial x} + v\frac{\partial u}{\partial y} = \varepsilon\frac{\partial^2 u}{\partial y^2} \tag{1}$$

Now, since, $u_d (= U - u) \ll U$, and $v \ll U$

$$u\frac{\partial u}{\partial x} = (U - u)\frac{\partial(U - u_d)}{\partial x} \fallingdotseq -U\frac{\partial u_d}{\partial x}$$

© The Author(s), under exclusive license to Springer Nature Singapore Pte Ltd. 2025
T. Shakouchi, *Jets, Wakes and Separated Flows,*
https://doi.org/10.1007/978-981-96-9439-6_4

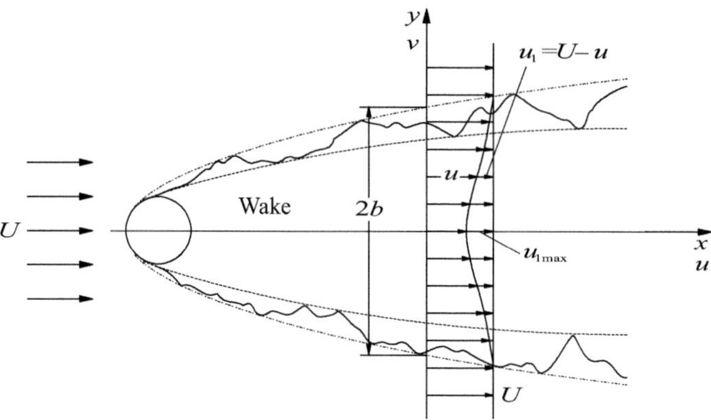

Fig. 1 Wake of two-dimensional cylinder

$$v\frac{\partial u}{\partial y} \dot{=} 0$$

$$\frac{\partial^2 u}{\partial y^2} = \frac{\partial^2 (U - u_d)}{\partial y^2}$$

From these equations, Eq. (1) becomes:

$$U\frac{\partial u_d}{\partial x} = \varepsilon \frac{\partial^2 u_d}{\partial y^2} \tag{2}$$

Assuming that the eddy viscosity ε in this case is $\varepsilon \propto b \cdot u_{dm}$ as in the case of a jet flow, the half width and maximum deficit velocity of the wake are expressed as $b = C_1 \cdot x_{1/2}$ and $u_{dm} = C_2 \cdot x^{-1/2}$, respectively, so the following relationship is obtained.

$$\varepsilon \propto b u_{dm} = C_1 x^{\frac{1}{2}} C_2 x^{-\frac{1}{2}} = C_1 C_2 \tag{3}$$

The defect velocity distribution u_d is also the same as in the jet case.

$$u_d = u_{dm} f(\eta), \; \eta = y/b \tag{4}$$

Then, we obtain the following relationship.

$$u_d = C_2 x^{-\frac{1}{2}} f(\eta) \tag{5}$$

$$\eta = y/\left(C_1 x^{1/2}\right) \tag{6}$$

From Eqs. (2) and (5),

$$\frac{f}{2} + \eta\frac{f'}{2} + \frac{\varepsilon}{UC_1^2}f'' = 0 \tag{7}$$

U and ε in the term $\varepsilon/(UC_1^2)$ in the above equation are constants, and C_1 is a constant arbitrarily determined according to the relationship $b = C_1 \cdot x^{1/2}$. Therefore, if we now set $\varepsilon/(UC_1^2) = 1$, the above equation becomes

$$\frac{f}{2} + \eta\frac{f'}{2} + f'' = 0 \tag{8}$$

Now, when we solve Eq. (8) under boundary conditions, $f = 1$ and $f' = 0$ at $\eta = 0$, and $f = 0$ at $\eta = \infty$, we get

$$u_d = u_{dm}\exp\left(-\eta^2/4\right) \tag{9}$$

Therefore,

$$\frac{u_d}{u_{dm}} = \exp\frac{-\eta^2}{4} = \exp\left\{\frac{-1}{4}\left(\frac{y}{b}\right)^2\right\} \tag{10}$$

Now, if the half width of the wake is $b_{1/2}$, then

$$\frac{u_d}{u_{dm}} = 0.5 = \exp\left\{\frac{-1}{4}\left(\frac{b_{1/2}}{b}\right)^2\right\} \tag{11}$$

Therefore,

$$\log_e 0.5 = -0.693 = -\frac{1}{4}\left(\frac{b_{1/2}}{b}\right)^2 - \frac{1}{4}\left(\frac{1}{b}\right)^2 = -\frac{0.693}{b_{1/2}^2} \tag{12}$$

From the above equation and Eq. (10), the deficit velocity distribution u_d in the wake of the two-dimensional cylinder is expressed by the following equation.

$$\frac{u_d}{u_{dm}} = \exp\left\{-0.693\left(\frac{y}{b_{1/2}}\right)^2\right\} \tag{13}$$

*** Other solution of Eq. (8) by Prandtl's mixing length hypothesis;**

Schlichting investigated first the deficit velocity distribution using Prandtl's mixing length hypothesis and had the following equation.

$$\frac{u_d}{U} = \frac{\sqrt{10}}{18\beta}\left(\frac{x}{c_d d}\right)^{-1/2}\left[1 - \left(\frac{y}{b}\right)^{1.5}\right]^2 \tag{14}$$

where, c_d: drag coefficient of circular cylinder, $\beta := l/b$ (l: mixing length).

2   Drag (Flow Resistance) and Lift Forces of Object

When an object moves in a stationary fluid or a stationary object exists in a flow, the object receives a force (flow resistance) from the fluid. Of this force, the components that act parallel and perpendicular to the mainstream are called drag or resistance and lift, respectively. Note that resistance includes the following.

Flow resistance that an object experiences from a fluid:

- Frictional drag
- Pressure (shape) resistance (pressure- or form-drag),
- Induced drag
- Wave drag, etc.

2.1   Drag Force

2.1.1   Frictional Drag (Resistance)

Frictional resistance is the drag force that occurs due to the viscosity of fluid when it flows over the surface of an object, and the integral of its mainstream direction component over the surface of the object is frictional resistance.

2.1.2   Pressure (Shape) Resistance

When an object is placed in a flow, in the case of the ideal fluid, no flow separation occurs and no resistance occurs (d'Alembert's paradox). However, in real fluids, as a result of the action of viscosity, flow separation occurs, creating a wake with a velocity deficit behind the object (see Sect. 4.1). In particular, when the Reynolds number is $Re \fallingdotseq 40 \sim 200$, a regular vortex street, the so-called Karman vortex street, occurs in the wake of the object. For example, the vortex generation frequency f from a cylinder of diameter d at a uniform velocity U is experimentally given by the following equation.

Vortex shedding frequency of Karman vortex street:

$$f = 0.198\frac{U}{d}\left(1 - \frac{19.7}{Re}\right) \tag{15}$$

The pressure in the wake is low, and the object receives force (resistance) from the fluid.

2.1.3 Induced Drag

Example is the flow resistance caused by free vortices emitted from the tips of airplane wings.

2.1.4 Wave Drag

For example, resistance is caused by waves created on the water surface as the ship moves, or by shock waves created by supersonic flow. Note that these resistance forces Sects. 2.1.1–2.1.4 do not act separately, but act simultaneously, although there are differences in magnitude depending on the case.

2.2 Drag Coefficient

Newton believed that the resistance force D exerted on an object of projected area A at a uniform velocity U is proportional to the momentum $\rho A U^2$ of the fluid impinging with the object per unit time, based on the law of momentum, assuming that the fluid consists of particles. Letting the constant of proportionality be C_d',

$$D = C_d'\left(\rho U^2 A\right) \tag{16}$$

Generally, it is given by the following equation based on the dynamic pressure $\rho U^2/2$.

The frag force that an object receives from the fluid is

$$D = C_d A\left(\rho U^2/2\right) \tag{17}$$

In other words, it shows how much (C_d times) of the dynamic pressure component acting on an object with projected area A, that is, the kinetic energy of the fluid $A(\rho U^2/2)$, becomes drag. This C_d (≤ 1.0) is called the drag coefficient, and is an experimental constant that varies depending on the shape of the object. For aircraft wings, the area projected onto the wing chord is used as A.

Figure 2 shows an example of the drag coefficient of a columnar object, and Fig. 15 shows an example of the drag coefficient of a three-dimensional object.

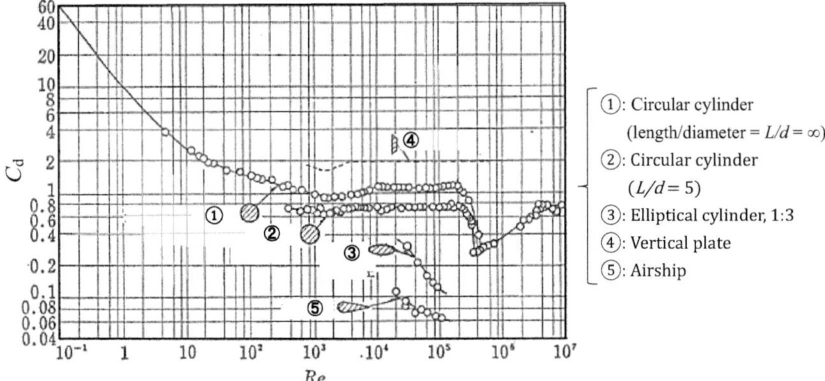

Fig. 2 Drag coefficient of two-dimensional column (Nakayama, 1989)

2.3   Measurement of Drag Force

There are three possible methods for determining the drag of an object. That is, directly measuring the force exerted on an object, measuring the pressure distribution on the object's surface and using it to find the force in the flow direction, or measuring the velocity distribution of the wake and applying the law of momentum, etc. The latter is as follows.

Figure 3 shows an outline of the velocity distribution before and after a two-dimensional object installed in a flow with uniform flow velocity U and static pressure P_s. The total pressure at cross section B-B$'$ near the rear end of the object is:

$$P_{tB} = P_{sB} + (\rho/2)u_b^2 \tag{18}$$

The static pressure P_{sB} is not uniform due to the influence of the object. The static pressure P_{sc} at the cross-section C-C$'$ far downstream is considered to be uniform and approximately equal to P_s. Therefore, the total pressure at cross section C-C$'$ is

$$P_{tc} = P_s + (\rho/2)u_c^2 \tag{19}$$

By the way, the drag force D that an object receives from the flow can be found as the momentum deficit of the cross section C-C$'$, that is,

$$D = \rho \int_{C-C'} u_c(U - u_c)dy_c \tag{20}$$

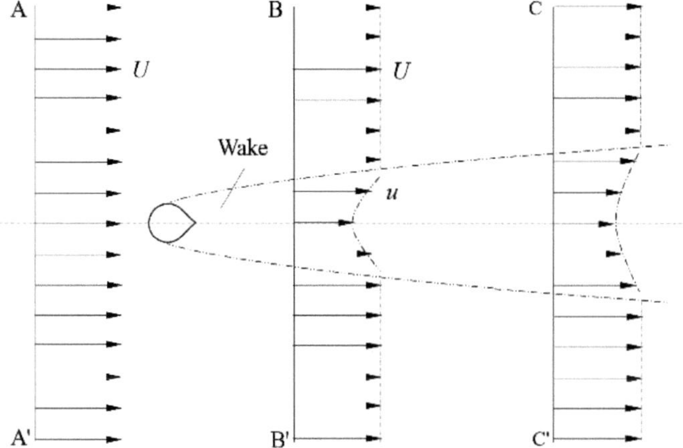

Fig. 3 Flow in front of and behind the object

However, it can actually be found as follows using P_{tB} of the cross section B-B'. Now, since $\rho u_b dy_b = \rho u_c dy_c$ holds true, the above equation is

$$D = \rho \int_{B-B'} u_b(U - u_c)dy_b \tag{21}$$

From Eq. (18),

$$u_b = \sqrt{2(P_{tB} - P_{sB})/\rho} \tag{22}$$

Also, if $P_{tB} = P_{sB}$,

$$u_c = \sqrt{2(P_{tB} - P_s)/\rho} \tag{23}$$

Moreover,

$$P_t = P_s + (\rho/2)U^2 \tag{24}$$

From Eqs. (20)–(24), the drag force is given by

$$D = 2 \int_{B-B'} \sqrt{P_{tB} - P_{sB}}\left(\sqrt{P_t - P_s} - \sqrt{P_{tB} - P_s}\right)dy_b \tag{25}$$

2.4   Drag Force of Circular Cylinder

The drag resistance D per unit length of a two-dimensional cylinder of diameter d placed orthogonally in a uniform flow of velocity U is given by the following equation from Eq. (16).

$$D = C_d A \frac{\rho U^2}{2} = C_d d \frac{\rho U^2}{2} \tag{26}$$

The drag coefficient C_d of the cylinder is a function of the Reynolds number $Re = Ud/v$, and the experimental results are shown in Fig. 15. At $Re = (0.01{\sim}2.0) \times 10^5$, C_d is almost a constant value ($\fallingdotseq 1{\sim}1.2$), but when $Re = 5.0 \times 10^5$, the boundary layer changes from laminar separation to turbulent separation and rapidly decreases to $C_d \fallingdotseq 0.3$. This Reynolds number is called the critical Reynolds number. This sudden decrease in the C_d value can be explained as follows. When $Re > 5.0 \times 10^5$, the position where the laminar boundary layer that develops along the surface from the front stagnation point of the cylinder transitions to the turbulent boundary layer moves forward, and the mixing of fluid particles inside and outside the boundary layer is promoted. Therefore, energy is supplied into the boundary layer, and as a result, the separation point moves backward, the wake region shrinks, and the drag force, or C_d value, decreases.

Note that the frag force D of the cylinder can also be determined from the pressure distribution on the cylinder surface shown in Fig. 4, and the force in the flow direction as described above using the following equation. Note that the drag force D of the cylinder in the flow direction can be determined from the pressure distribution on the cylinder surface shown in Fig. 15 using the following equation.

$$D = \int_A p \cos \theta \, dA \tag{27}$$

where, θ: clockwise angle from forward stagnation point.

Figure 4 shows the results for ideal fluid, laminar- and turbulent-flows of viscous fluid. In the case of an ideal fluid, no drag force acts (d'Alembert's paradox). Additionally, compared to laminar flow where the Reynolds number is smaller than the critical Reynolds number ($Re_c \fallingdotseq 5 \times 10^5$), in turbulent flow with irregularly fluctuating velocity components, energy is replenished from the mainstream to the boundary layer, the flow becomes difficult to separate. For example, in the figure, the separation point in the case of laminar flow is $\theta \fallingdotseq 80°$, but in the case of turbulent flow with a large Reynolds number, it moves much further back to $\theta \fallingdotseq 130°$. In addition, the pressure recovery after separation point is larger in turbulent flow, and as a result, the drag force in turbulent boundary layer is smaller.

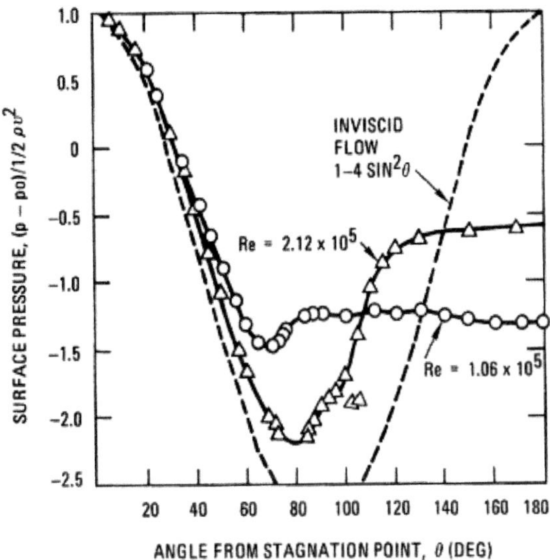

Fig. 4 Pressure distribution on cylinder surface (Blevins 1984)

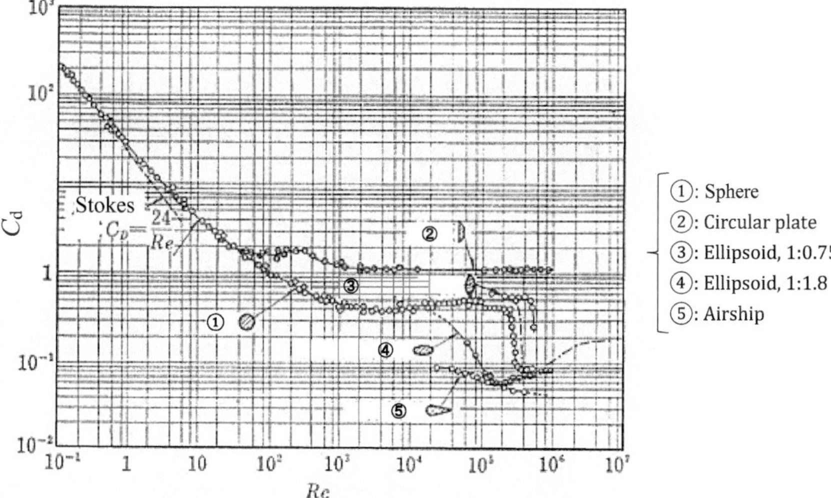

Fig. 5 Drag coefficient of three-dimensional object (Nakayama 1989)

2.5   *Drag Force of Sphere*

The resistance of a sphere with diameter d is obtained from Eq. (26) as $A = (\pi/4)d^2$, and the C_d value is shown in Fig. 5. In particular, flow with a small Reynolds number ($Re < 1$) is known as Stoke's flow, and the drag force and C_d value are

$$D = 3\pi \mu U d$$

$$C_d = 24/Re \tag{28}$$

When $Re > 10^3$, $C_d \fallingdotseq 0.44$, and when $Re \fallingdotseq 3.0 \times 10^5$, the critical Reynolds number is reached, and it rapidly decreases to $C_d \fallingdotseq 0.1$.

3   Lift Force

3.1   *Lift Force of Object*

Now, calculate the lift force L generated when a cylinder of radius r installed in a uniform flow with velocity U rotates at angular velocity ω from the pressure distribution on the cylinder surface. Now, assuming that no separation occurs in the flow and the rotational angular velocity of the cylinder is ω, the flow velocity at any point (θ) on the cylinder surface is the sum of the tangential velocity $2U\sin\theta$ and the circumferential velocity $r\omega (= 2U\sin\theta + r\omega)$.

Letting the pressure at any point on the mainstream and the cylinder surface be P_a and P, respectively, $P_a + \rho U^2/2 = P + \rho(2U\sin\theta + r\omega)^2$ is obtained from Bernoulli's equation, and the following relationship is obtained.

$$\frac{P - P_a}{\rho U^2/2} = 1 - 2\left(\frac{2U\sin\theta + r\omega}{U}\right)^2 \tag{29}$$

Therefore, if Γ is the circulation along the cylinder surface, the lift force per unit width of the cylinder is given by the following equation, the Kutta-Joukowski equation.

Lift force per unit width of cylinder (Kutta-Zukowski equation):

$$L = 2 \int_{-\pi/2}^{\pi/2} -(P - P_a)rd\theta\sin\theta$$

$$= -r\rho U^2 \int_{-\pi/2}^{\pi/2}\left\{1 - 2\left(\frac{2U\sin\theta + r\omega}{U}\right)^2\right\}\sin\theta d\theta$$

$$= 2\pi r^2\omega\rho U = 2\pi ru\rho U = \rho U\Gamma \tag{30}$$

where, Γ: circulation along the cylinder surface (counterclockwise is positive).

When there is clockwise and counterclockwise circulation along a cylinder or an object of arbitrary shape, upward and downward lifting forces $L(= \rho U\Gamma)$ act, respectively. The fact that a ball, such as a baseball, curves when it is rotated is due to this lifting force, and this is called the Magnus effect.

3.2 Lift Force of Wing

3.2.1 Two-Dimensional Wing

When an object is placed in a flow, it receives forces (drag force D, lift force L, and moment M) from the fluid, but blades whose lift force is greater than the drag force are blades used in airplanes, propellers, turbine blades, etc. (wing, airfoil, blade).

Figure 6 shows the cross-sectional shape of the wing and the names of each part.

The cross-sectional shape of a wing is called the wing section, the line connecting the leading edge and trailing edge is called the chord, and its length is called the chord length.

Also, the line connecting the midpoints of the upper and lower surfaces is the camber line, the distance between the wing chord and the camber line is the camber, and the thickness perpendicular to the camber line or chord is the wing thickness. When the span of the wing is B and the maximum projected area of the wing is A, $B^2/A \equiv \lambda$ is called the aspect ratio of the wing. Also, the angle α between the chord and the flow direction U is called the attack angle.

Force acting per unit width of the wing is as follows. If the chord length of the blade is l, then.

(a) Drag force: $D = C_d l(\rho U^2/2)$
(b) Lift force: $L = C_L l(\rho U^2/2)$
(c) Moment:

$$M = C_M l^2 \left(\rho U^2/2\right) \tag{31}$$

Here, C_d, C_L, and C_M are the drag coefficient, lift coefficient, and moment coefficient, respectively, M is a value around the leading edge or around the chord point at l/4 from the leading edge, and is positive in the head-up direction.

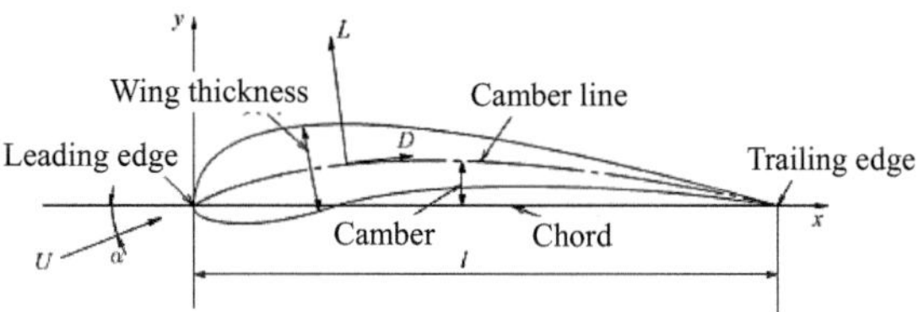

Fig. 6 Cross sectional shape of wing and names of each part

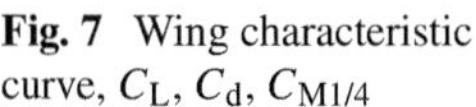

Fig. 7 Wing characteristic curve, C_L, C_d, $C_{M1/4}$

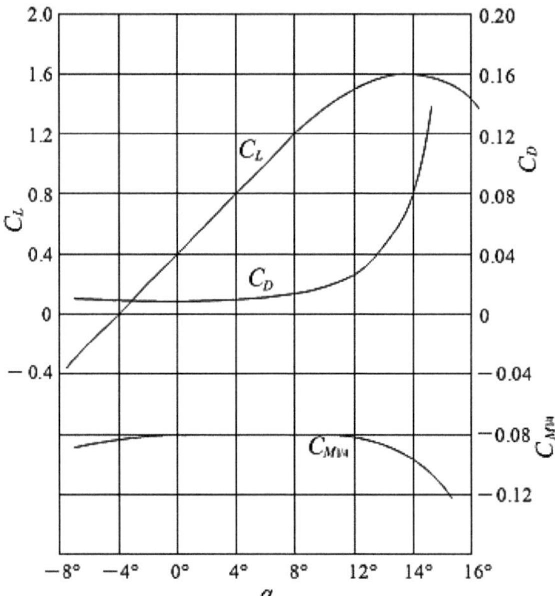

3.2.2   Wing Performance

Figure 7 shows an example of a wing characteristic curve. The lift coefficient C_L increases almost linearly as the angle of attack α increases, then gradually increases, reaches the maximum (maximum lift coefficient), and then rapidly decreases. The reason why C_L suddenly decreases is because α increases and the flow on the wing surface separates. This is called a stall, and its angle of attack α is called the stalling angle. The drag coefficient C_d increases rapidly from a certain α.

Figure 8 shows the relationship between C_L and C_d (lift-drag curve). This allows us to find the angle of attack α that maximizes the lift-drag ratio C_L/C_d.

3.2.3   Three-Dimensional and Finite Width Wing

Cross-sectional (shape) drag D_f and induced drag D_i due to the presence of the wing tip act on the wing of an aircraft during flight. Therefore, the total drag force D_t on the wing is

$$D_t = D_f + D_i = (C_{dt} + C_{di})\frac{\rho U^2 B}{2} \tag{32}$$

where, C_{df}: section drag coefficient, C_{di}: induced drag coefficient, B: wingspan.

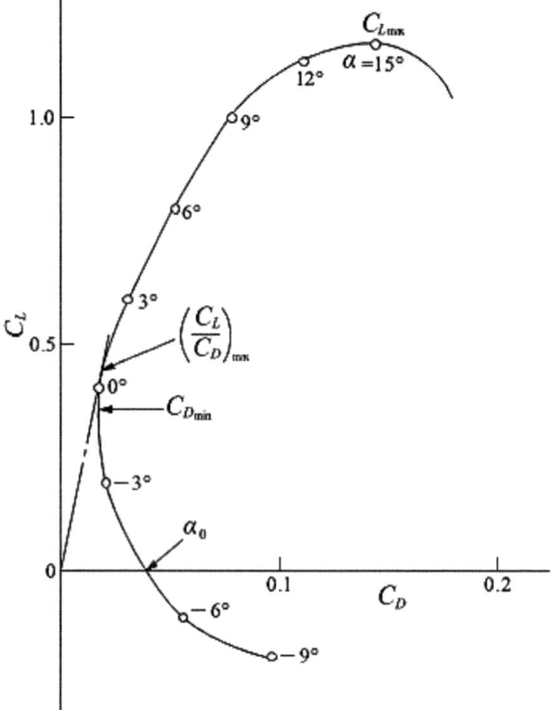

Fig. 8 Wing characteristic curve, $C_L - C_d$

Lift on the wing is generated because the pressure on the lower surface of the wing is higher than that on the upper surface, so a flow from the lower surface to the upper surface occurs at the tip of the wing.

Furthermore, a free vortex called the wing tip vortex is generated at the wing tip, and a bound vortex that circulates around it is generated within the wing itself (Fig. 9).

In reality, vortices are emitted from the entire wing, not just from the tip, but vortices with the same direction of rotation come together downstream to form a pair of free vortices with an induced velocity v. The lift force L of the wing is given by

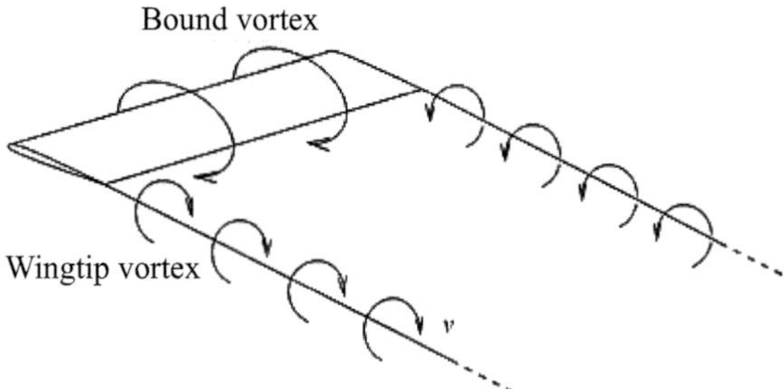

Fig. 9 Three-dimensional, finite width wing (Bound vortex and wing tip vortex)

the law of momentum as $L = mv$, where m is the mass of the fluid pushed downward by the wing. Here, since m is proportional to $\rho B^2 U$,

$$L = mv = k\rho B^2 Uv \tag{33}$$

where, k is a constant of proportionality, which is theoretically determined to be $\pi/4$.

By the way, as shown above, the lift force in the case of a two-dimensional wing is $L = C_L l B \,(\rho U^2/2)$, where B is the wing span, so v is

$$v = C_L U \frac{2}{\pi} \frac{l}{B} \tag{34}$$

If the kinetic energy $mv^2/2$ for the fluid to move downward at velocity v is equal to the work done by the induced drag D_i, then

$$D_i = \pi\rho B^2 v^2/8 \tag{35}$$

Therefore, the induced drag coefficient C_{Di} is

$$C_{Di} = \frac{C_L^2}{\pi}\frac{Bl}{B^2} = \frac{C_L^2}{\pi\lambda} \tag{36}$$

where, $\lambda = B/l$.

By the way, there is no induced velocity v sufficiently ahead or behind the wing, so the velocity near the wing can be approximately given as $v/2$. Now, if the attack angle of a two-dimensional wing is α_t, then the finite width α_f will be smaller by δ, and the lift will be reduced ($\alpha_f = \alpha_t - \delta$). Now, $\delta = (v/2)\,/U = C_L/(\pi\lambda)$, so

$$\alpha_t = \alpha_f + \delta = \alpha_f + C_L/(\pi\lambda) \tag{37}$$

In other words, to obtain the same value of lift coefficient as a two-dimensional wing with a finite width wing, the attack angle must be increased by $\delta = C_L/(\pi\lambda)$ (Fig. 10).

Fig. 10 Angle of attack

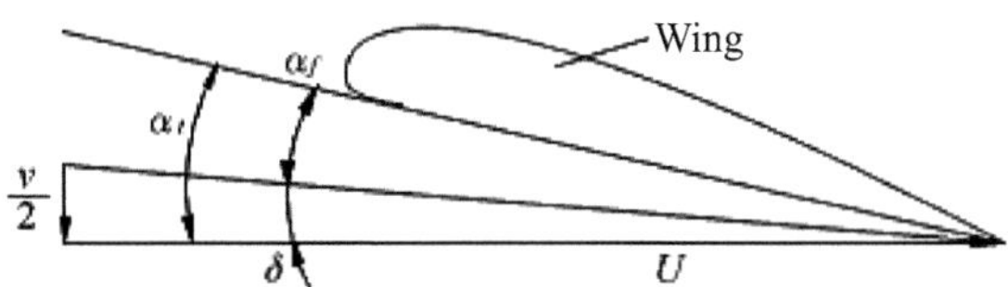

4 Some Examples of Wake Flow

4.1 Karman Vortex Street

4.1.1 Flow Separation from Cylinder Surface

As briefly shown in the Introduction (2. Wake and Separated Flows), among wake flows, Karman vortex is one of the most interesting phenomena in terms of the clarity of the phenomenon, its appearance in the natural world, its adverse effects, and its engineering applications.

As shown in Fig. 11, in the flow of two-dimensional objects, the boundary layer is left out of the wall in the middle of the object, and a vorticity layer is continuously released in the flow. They gather and form a vortex. In the vortices released alternately from the cylinder, the pressure at the center of the vortex decreases in order to balance the centrifugal force due to rotation of the fluid.

The Strouhal number ($St = f\,d/U$, f: vortex shedding frequency, d: reference length, U: free stream velocity $=$ const.), which is a non-dimensional vortex shedding frequency, is also a similar rule of the unsteady flow, and has various values depending on the object cross sectional shape. Also, the value of the Strouhal number changes depending on the Reynolds number ($Re = Ud/v$). In the case of a circular cylinder, it is well known that $St \fallingdotseq 0.2$ (const.) at the $Re = 1.0 \times 10^3{\sim}2.0 \times 10^5$.

4.1.2 Visualized Flow Pattern of Karman Vortex

Figure 12 shows the visualized flow pattern at $Re = 4$ around the circular cylinder which is set in a uniform flow. The velocity is very small, and the flow flows almost along the cylindrical wall.

Figure 13 shows the flow pattern behind the circular cylinder at various Renolds number obtained by Homann. Behind the cylinder, there is a symmetrical vortex in the range of $Re = 5{\sim}45$, and if it exceeds it, the opposite vortex is released alternately.

Fig. 11 Separation of vorticity layer and formation of vorticity (by Kamemoto)

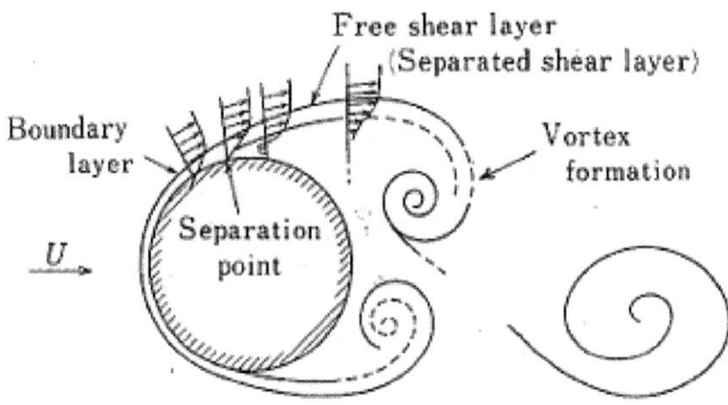

Fig. 12 Flow around circular cylinder at $Re = 4$ in oil flow with uniform velocity (by Homann)

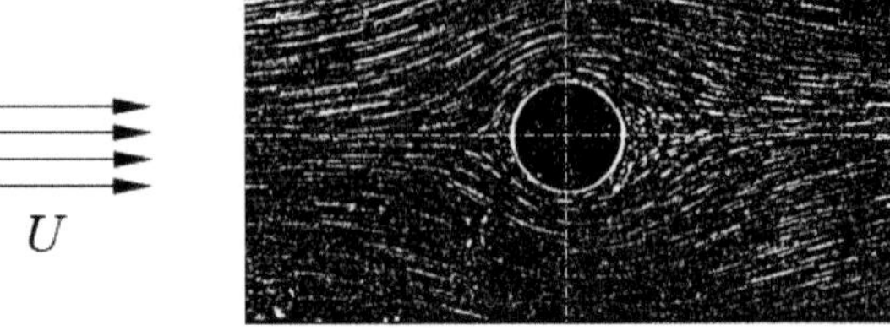

This vortex street is called Karman vortex street. The vortex shedding increases with the increase in Re.

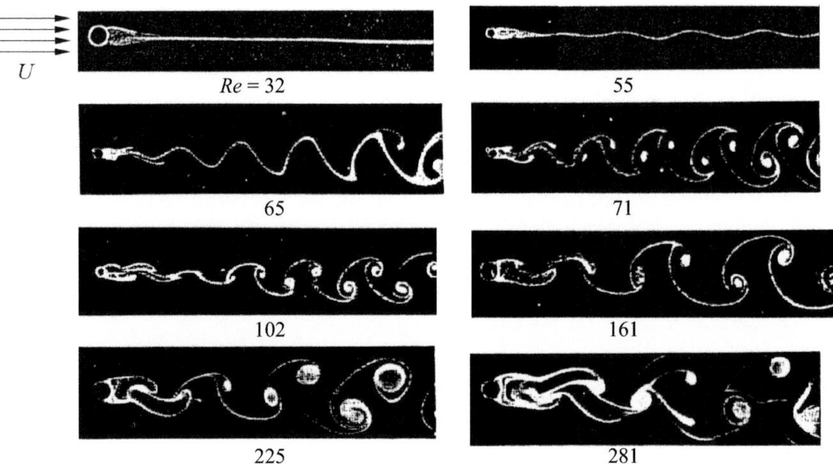

Fig. 13 Flow in wake of circular cylinder in oil flow with uniform velocity (by Homann 1936)

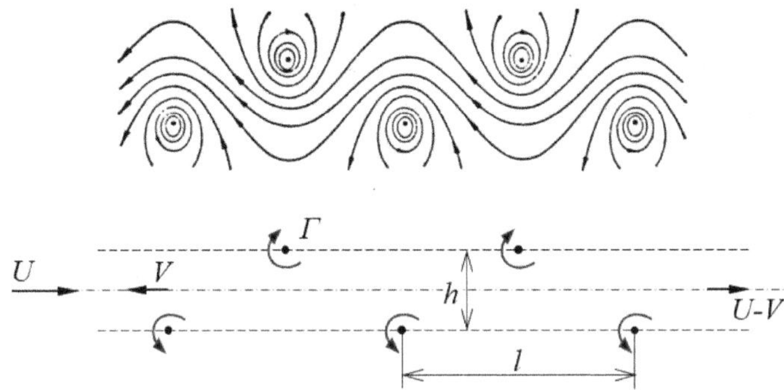

Fig. 14 Karman vortex street

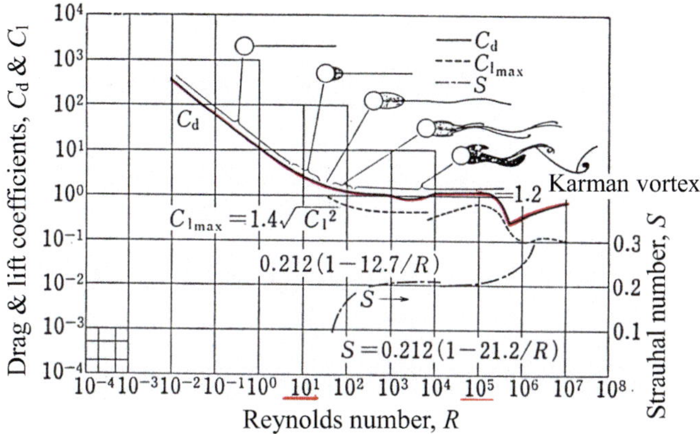

Fig. 15 Drag and lift coefficients, and Karman vortex street

4.1.3 Theoretical Analysis of Stability of Karman Vortex with Perfect Fluids

Karman modeled the Karman vortex street by arranging the vortex street in a perfect fluid as shown below, and investigated its stability. First, it is assumed that the Karman vortex street is composed of vortex filaments of equal strength circulation Γ (clockwise direction is positive) that are lined up in a straight line at equal intervals b and are infinitely long. Next, these two vortex streets are arranged with a spacing of h (Fig. 14) and the vortices in the two vortex streets are just half-period $l/2$, and they are different from each other. The rotation direction of the vortices in each vortex row is reversed. He showed theoretically that these two vortex streets are arranged alternately, and that except in the case of $\cosh(\pi h/l) = \sqrt{2}$, that is, in the special case of $h/l = 0.2806$, the vortex street becomes unstable. In other words, when some kind of disturbance occurs, the arrangement of the vortices collapse. This stable vortex street is called the Karman vortex street. The fact that the Karman vortex street actually has a staggered arrangement, and that it is $h/l \fallingdotseq 0.3$, supports this theory. At this time, the vortex street as a whole moves at the following velocity relative to the stationary fluid.

$$V = \frac{\Gamma}{2l}\tanh\frac{\pi h}{l} \tag{38}$$

4.1.4 Drag of Object Generated by Karman Vortex Street

Now, if we use a large rectangular area surrounding the object and the Karman vortex street as the control volume and consider the changes in the momentum flux there, we can find the drag force acting on the object using the following equation.

$$D = \frac{\rho \Gamma h}{l}(U - 2V) + \frac{\rho \Gamma^2}{2\pi l} \tag{39}$$

Using Eq. (38) and $h/l = 0.2806$ in the above equation, we get

$$D = \rho l U^2 \left\{ 0.7936 \frac{V}{U} - 0.3141 \left(\frac{V}{U} \right)^2 \right\} \tag{40}$$

In this case, the drag is a function of l and V only. In addition, the drag coefficient C_d [$=D/(1/2)\,\rho U^2 d$, where, d: width of object] is

$$C_d = \frac{2l}{d} \left\{ 0.7936 \frac{V}{U} - 0.3141 \left(\frac{V}{U} \right)^2 \right\} \tag{41}$$

Karman and Ruback conducted an experiment at $Re = 2000$~3000 using a cylinder with a diameter of 1.5 cm, and found that $h = 6.4$ cm, $l = 1.8$ cm, and $V/U = 0.14$. Substituting these into Eq. (41), we get $C_d = 0.91$. The measured value by Foeppl is $C_d = 0.9$, which is in good agreement with the theoretical value.

Furthermore, theoretical and experimental comparisons were also made for the case of a flat plate inserted perpendicular to the flow. When the width of the plate is d, $h/d = 5.5$ and $V/U = 0.20$ were obtained from the experiment, and by substituting these into Eq. (41), the theoretical value of the drag coefficient was obtained as $C_d = 1.6$. Flachsbart's experimental value was $C_d = 1.7$, and the error from the theoretical value was about 6%.

In this case, the theory is a good representation of the actual decline.

4.1.5 Drag Coefficient and Renolds Number

Figure 15 shows the relationship between the drag coefficient C_d, lift coefficient C_l, and Reynolds number R of circular cylinder. The sketch of the wake and the generation range of Karman vortex are also shown approximately in the figure. Also, for reference, the vortex shedding frequency f is shown as Strouhal number, $S = fd/v$ (v: kinematic viscosity).

4.2 Global-Scale Karman Vortex Street

Figure 16 shows the distribution of clouds around the western Kyushu region (Japan) as photographed by a Japanese meteorological observation satellite (geostationary satellite) (January 9, 2015, 9:00 a.m.). Clouds hanging on the middle of Mt. Halla (Altitude; 1,950 m) on Jeju Island in Korea visualize the flow caused by the northwest winter monsoon, and it can be seen that Karman vortex streets are being released.

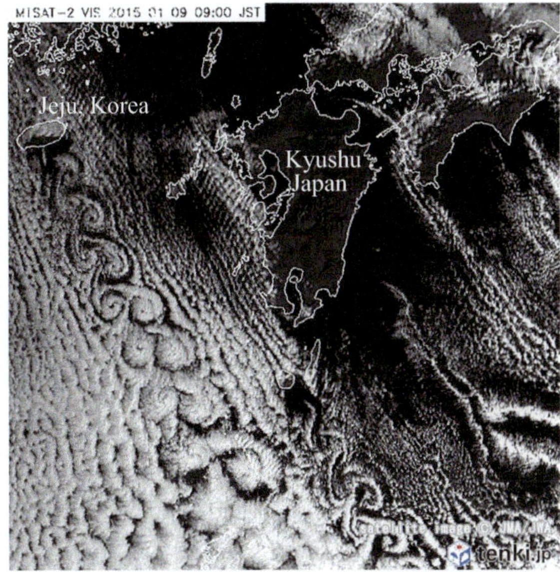

Fig. 16 Global scale Karman vortex street (by Japan Meteorological)

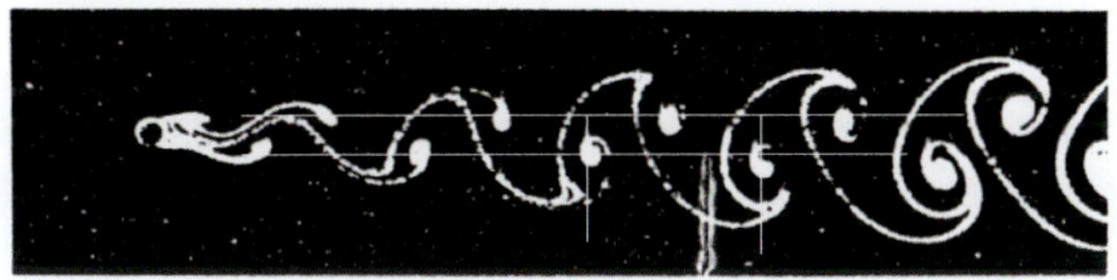

Fig. 17 Karman vortex street (laboratory size, by Homann 1936)

The flow velocity is $U = 10$ m/s, $d = 30$ km, and $Re = 2 \times 10^{10}$. The ratio of the vortex street spacing h to the vortex pitch l is surprisingly almost the same ($h/l \fallingdotseq 0.28$)as the laboratory results shown in Fig. 17.

A vortex street from Yakushima, Japan, is also occurring in the bottom center of the figure. Similar vortex street occurrences have also been observed in several other locations around the globe, such as islands off the coast of the California peninsula.

4.3 Evils of Karman Vortex

4.3.1 Destruction of Tacoma Narrows Bridge by Wind

Each time a vortex is shed from an object placed in a flow, the direction of circulation around the object changes, so according to the Kutta–Joukowski theorem, the object experiences lift forces alternating in the direction perpendicular to the flow. Therefore, objects in the flow or objects moving in the fluid vibrate, which causes sound when the flow is gas. This sound is called the Aeolith sound, named after the Greek god of wind, Aeollith. The sound of the wind on a windy day is the sound produced by the vibrations of air twigs and electric wires due to the Karman vortex.

On November 7, 1940, the Tacoma Narrows Bridge, a 1,600 m long, 853 m suspension bridge spanning the Tacoma Strait in Tacoma, Washington, U.S.A., collapsed. When the wind blowing through the strait reached 19 m/s, the bridge began to vibrate with a node in the center (Fig. 18a), and after one hour of repeated vibrations with a frequency of 0.2 Hz, the bridge girder broke and part of the floor collapsed into the sea (Fig. 18b).

The cause of suspension bridge collapse due to vibration was initially concluded to be a resonance phenomenon between the periodic excitation force generated by the vortices, Karman vortex street, generated behind the bridge girder and the natural frequency of the bridge. However, it was later determined that what was initially a flexural vibration due to vortex-excited vibration became a self-excited vibration called torsional flutter, which was the direct cause of the Tacoma Narrows Bridge collapse (by Wikipedia).

4.3.2 Destruction of Cooling Tower by Wind

In the early days, huge concrete cooling towers installed outdoors at thermal power plants and other facilities were often destroyed by wind force. The outer walls of cooling towers were made of thin concrete, and in areas without major earthquakes, their structures were often simply constructed. As a result, when strong winds blew, Karman vortices were released from the cooling tower, and an alternating force acted on the cooling tower depending on the vortex shedding frequency f. When the natural frequency of the cooling tower matched f, the cooling tower could be destroyed due

| (a) Vibration with a node in the center | (b) Part of the floor collapsed into the sea |

Fig. 18 Takoma Narrows Bridge by Wikipedia (total length: 1600 m, hanging span: 853 m)

Fig. 19 Wind damage to cooling towers

to resonance (Fig. 19). Today, structures that can withstand such winds are being adopted based on this knowledge.

4.3.3 Prevention of Generation of Karman Vortex

One way to prevent damage to structures due to the generation of Karman vortex is to prevent regular vortex shedding from the structure. Possible solutions for this include the installation of spiral or fin-like protrusions or grooves as shown in Fig. 20, or the installation of through holes or grooves in the object to make it transparent. In fact, these solutions have been used to prevent vortex shedding from the structure.

Fig. 20 Several ways to stop shaking caused by Karman vortex

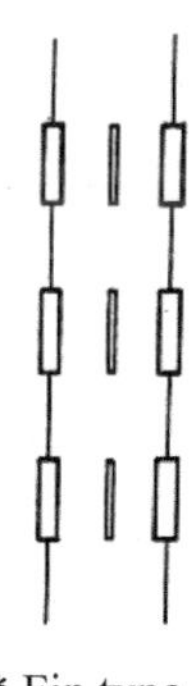

4.4   Use of Karman Vortex

* Karman Vortex Flow Meter

A columnar object is installed inside the pipe perpendicular to the flow to generate
a vortex (Fig. 21). The shape of the vortex generator is other than a circular cylinder
so that the vortex can be generated stably. Since the vortex shedding frequency f of
the Karman vortex from the vortex generator is proportional to the flow velocity in
the pipe, the flow rate in the pipe can be determined by measuring f. In the vortex,
the pressure at the center of the vortex decreases in order to balance the centrifugal
force due to rotation of the fluid. In this way, since the pressure and density at the
vortex center are low, it is possible to detect them and determine the vortex shedding
frequency. The main method of detection is to pick up the vibrations of the vortex
using a piezoelectric element, but there is also a method of detecting the vortex using
ultrasonic waves as shown in Fig. 21.

4.5   Karman Vortex in Gas–Liquid Two-Phase Bubbly Flow

It is well known that the Karman vortex or twin vortex will be shed from an obstacle
when a single-phase flow passes it. It has been proved that Karman vortex is also shed
for a gas–liquid two-phase cross-flow (Inoue et al. 1986; Pettigrew 1994; Yokosawa
et al. 1986). Gas–liquid two-phase flow exists in many industrial components such as
shell-and-tube heat exchangers, chemical machinery and nuclear steam generators.
The geometry of obstacle used for vortex shedding in previous studies was mainly
circular cylinder (Hara et al. 1982a, b; Inoue et al. 1986, Yokosawa et al. 1984). But
gas–liquid two-phase flow around rectangular cylinder has been scarcely studied
apart from some studies (Pettigrew 1994; Roshko 1955). However, since a flow
always separates at the fixed front corners of the rectangular cylinder unlike the flow
past a circular cylinder with variable separation point, the flow might be different for
the case of rectangular cylinder.

Fig. 21 Karman vortex flow
meter (Vortex detection
using ultrasonic waves)

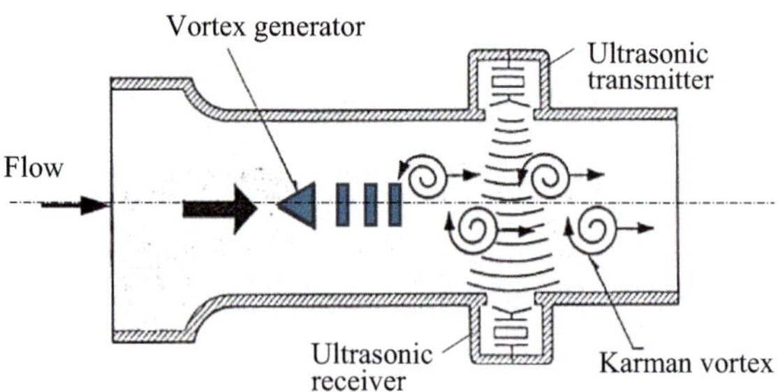

Here, flow characteristics such as the flow pattern, oscillatory frequency and flow resistance of a single and two-tandem rectangular cylinders with blockage ratio of 1/3 and distance between the two cylinders of $s/d = 1$ is presented (Shakouchi et al. 2001, 2002).

4.5.1 Test Set-Up and Procedure

Figure 22 shows the test section of the experimental set-up. The rectangular test section was made of a transparent acrylic resin channel with $L = 1520$mm in length, $D = 90$mm in width and $H = 45$mm in height. The size of rectangular test cylinder is $d = 30$ mm and was mounted at the center of channel perpendicular to the two-phase flow rigidly. The blockage ratio is $d/D = 1/3$.

The gas–liquid two-phase flow was subject to the bubbly flow regime in this research and the void fraction which express the volume ratio of the gas phase in the flow, was varied from $\alpha [= Q_a/(Q_a + Q_w)] = 0 \sim 30\%$ in the experiments. The void fraction was calibrated under the pressure on the sidewall at position $x = 0$ when the test cylinder was not put into the flow. The mean bubble diameter in the region of $x = (-120, -60$ mm$)$ was approximately 2.5 mm at the condition of $Re = U_m d/v = 1.0 \times 10^4$ [U_m: apparent average velocity of water, $= Q_w/(Dh)$] and $\alpha = 10\%$.

4.5.2 Flow Around Single Cylinder

First, results for the case when a single rectangular cylinder is put will be discussed with aspect ratio $AR = 1$. Then, results for the case of tandem cylinder with aspect ratio of cylinder $AR = 1$ and distance ratio $s/d = 1$ will be presented.

(a) **Flow Pattern**

Figure 23a shows the visualized flow pattern of vortex shedding of the two-phase bubbly flow for single rectangular cylinder when $Re = 1.0 \times 10^4$ and $\alpha = 10\%$. The white colored parts denote the bubbles in two-phase flow. For the case of the flow past a square cylinder, the air water bubbly flow separates at the front corners of the cylinder and forms a high local void fraction area on the two sides of it, where the separation area is. There is a dead water region, which contains no bubbles, or few bubbles occasionally just behind the cylinder and a high bubble density region surrounds it. The bubbles swarm in this region, which oscillates and develops into Karman vortex at the downstream and reaches to the sidewalls of the flow passage.

In Fig. 23b, the velocity vector got by PIV is shown for the same flow conditions where the Karman vortex can clearly be seen as well the increase of velocity between the cylinder and the sidewall is caused from the flow area decrease.

*** Numerical analysis**

The two-dimensional unsteady numerical analysis of a gas–liquid two-phase bubbly flow with void fraction $\alpha = 10\%$ and $Re = 1.0 \times 10^4$ through the flow passage with

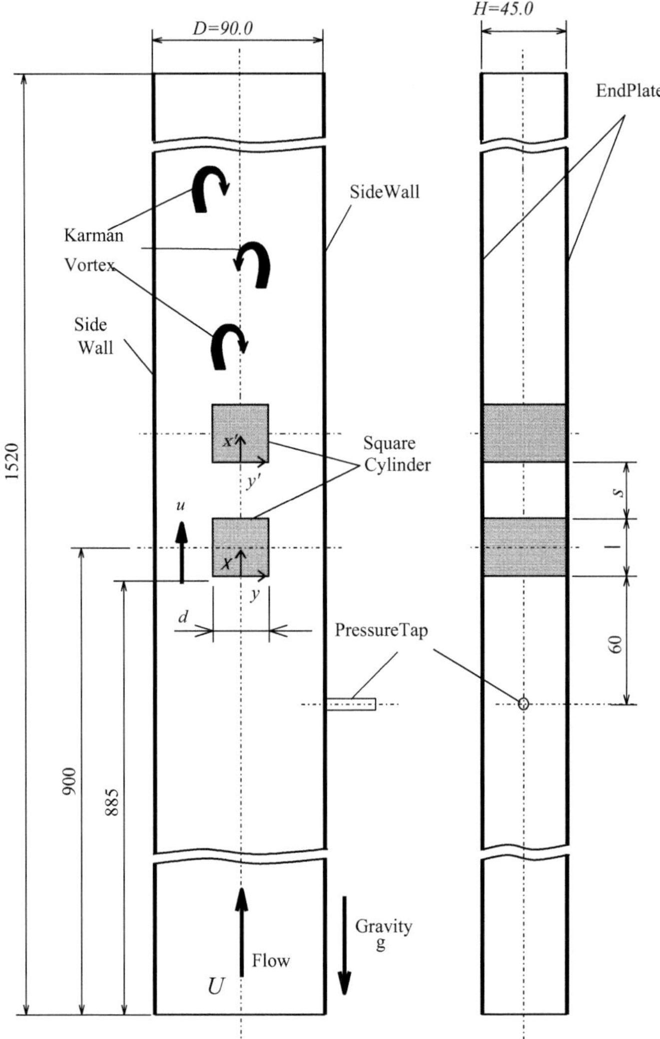

Fig. 22 Test section with two tandem cylinders ($D = 90.0$ mm, $H = 45.0$ mm)

single rectangular cylinder of $\lambda = 1$ in Fig. 22 was conducted. The two-fluid model for bubbly flow and a standard k-ε turbulence model were used.

Figure 24a shows the time histories of pressure coefficient at the position of ($1.2d$, $1.1d$) in Fig. 22. The pressure changes regularly at $f = 2.83$ Hz and then the wake flow oscillates regularly. This agrees well experimental result in Fig. 23. Figure 24b shows the void fraction, the void fraction changes with the change in pressure, i.e., when the pressure is high, the void fraction is low, and when the pressure is low, the void fraction is high.

Fig. 23 Gas–liquid
two-phase Karman Vortex
flow (single cylinder) ($f =$
2.83 Hz, $Re = 1 \times 10^4$, $\alpha =$
10%)

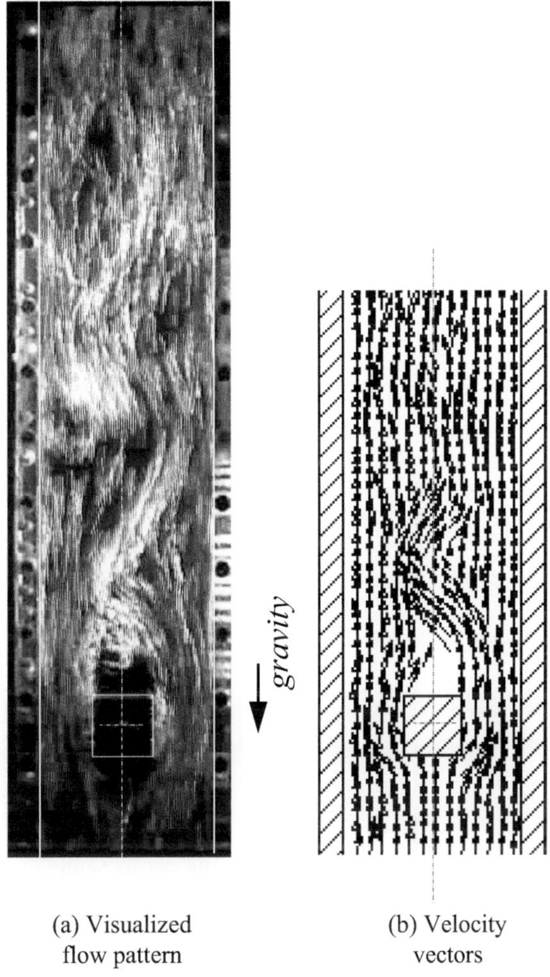

(a) Visualized
flow pattern

(b) Velocity
vectors

Figure 25a–e show the stream lines of air phase at $t = 1.90$ s ~ 2.28 s. A vortex region is formed behind the rectangular cylinder and is released downstream, causing an oscillation phenomenon in the wake.

Figure 26a–e show the time histories of the pressure distribution. The pressure is high in front of the rectangular cylinder and low in the separated vortex region behind it, and the low pressure region meanders along the vortex region released downstream. In areas with dense streamlines, the velocity is large and the pressure is low.

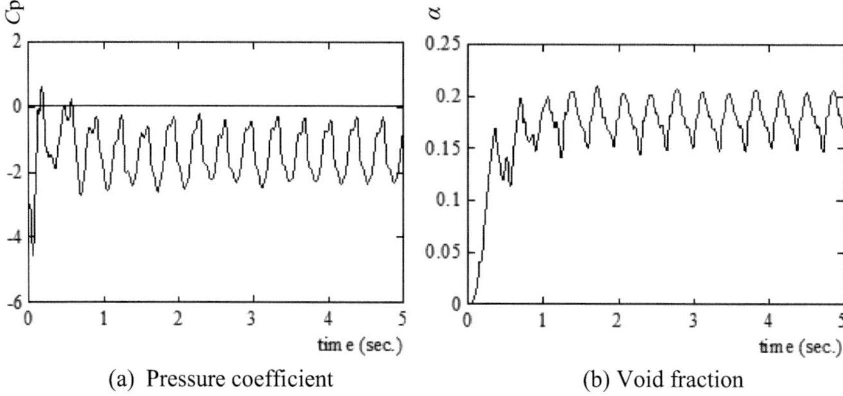

(a) Pressure coefficient   (b) Void fraction

Fig. 24 Time histories of pressure coefficient and void fraction at the position of (1.2*d*, 1.1*d*) (*Re* $= 1.0 \times 10^4$, $\alpha = 10\%$)

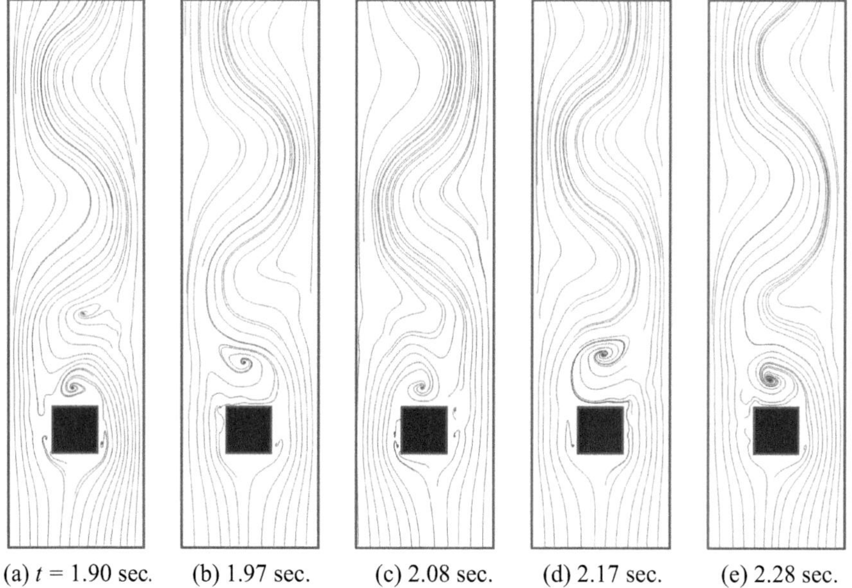

(a) *t* = 1.90 sec.   (b) 1.97 sec.   (c) 2.08 sec.   (d) 2.17 sec.   (e) 2.28 sec.

Fig. 25   Streamlines of air-phase (*Re* $= 1.0 \times 10^4$, $\alpha = 10\%$)

Figure 27 shows the distribution of void fraction. The void fraction is high in front of the rectangular cylinder where the flow impinges, and is large in the low-pressure region downstream. The distribution also meanders downstream of the rectangular cylinder.

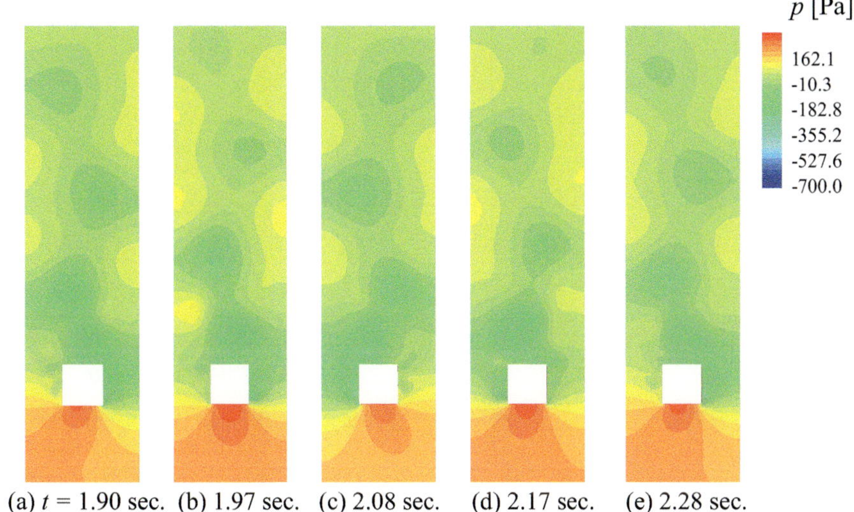

(a) $t = 1.90$ sec. (b) 1.97 sec. (c) 2.08 sec. (d) 2.17 sec. (e) 2.28 sec.

Fig. 26 Pressure ($Re = 1.0 \times 10^4$, $\alpha = 10\%$)

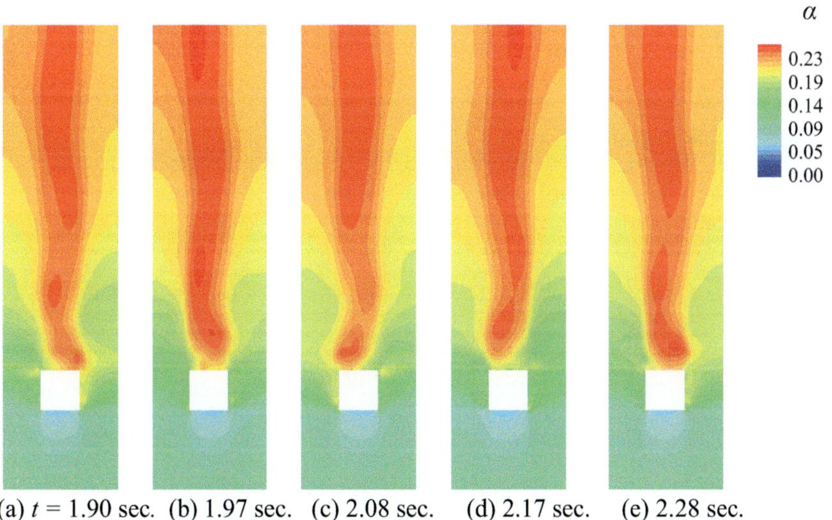

(a) $t = 1.90$ sec. (b) 1.97 sec. (c) 2.08 sec. (d) 2.17 sec. (e) 2.28 sec.

Fig. 27 Void fraction, α ($Re = 1.0 \times 10^4$, $\alpha = 10\%$)

(b) **Oscillatory Frequency**

Figure 28 shows the relationship between Strouhal number $St(= f\,d\,/U_\mathrm{m})$ and void fraction α. St is almost constant and the numerical analysis can be expressed the experiments well.

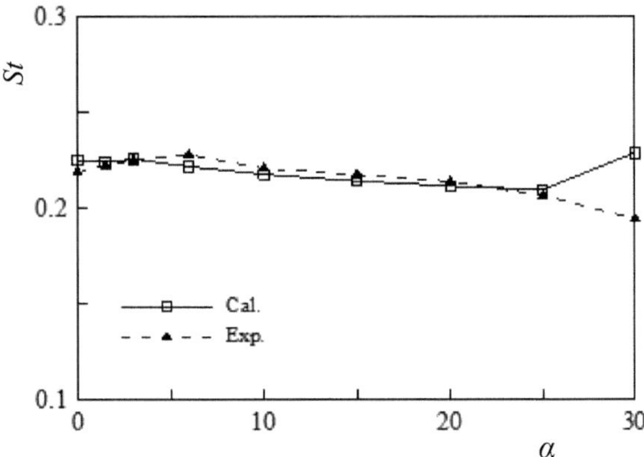

Fig. 28 Simulation and experiment for St ($Re = 1.0 \times 10^4$)

Figure 29 shows the experimental results of the relationship between vortex shedding frequency and Reynolds number for the case of $\alpha = 0$ and 10%. It illustrates that f increases with increasing Re linearly and when bubbly flow is induced there is a slight increase on the values. The straight lines can be expressed generally by the following equation:

$$f = k\,Re + b \tag{42}$$

where, $k = -2.88 \times 10^{-8}\alpha^2 + 8.50 \times 10^{-7}\alpha + 2.38 \times 10^{-4}$, $b = 0.0176\,\alpha + 0.348$ (α: in percentage).

Figure 30, shows the relationship between shedding frequency f and α %. Increasing the void fraction, frequency is raising slightly compared to the effect that Re has.

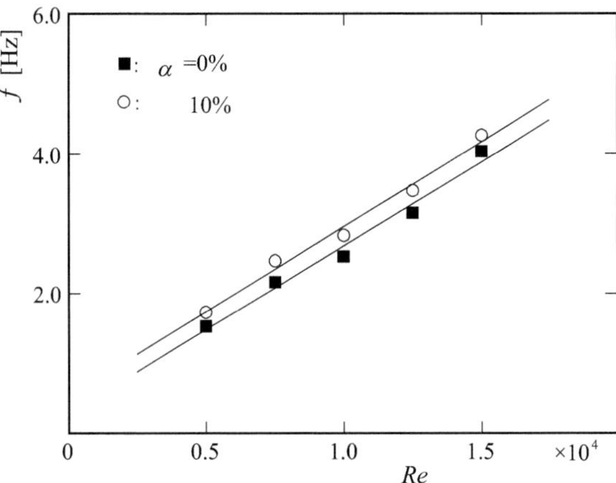

Fig. 29 Vortex shedding frequency, f (Effect of Re)

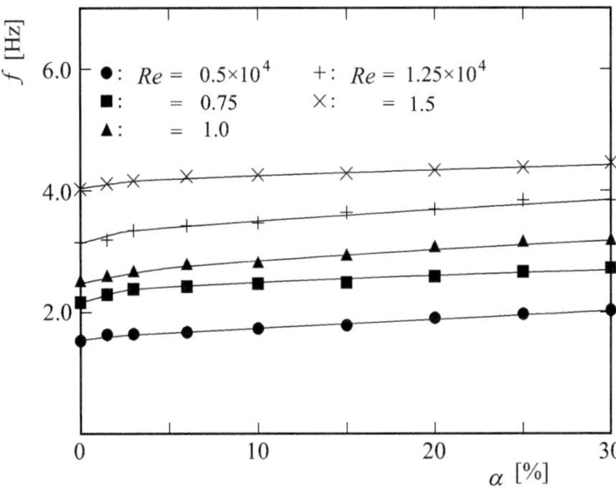

Fig. 30 Vortex shedding frequency, f (Effect of α)

(c) **Flow resistance**

(C.1) **Pressure difference Δp between front and rear stagnation points**

Figure 31a shows the effect of Re on ΔP for $\alpha = 0$ to 10%. It illustrates that ΔP increases with increasing Re in a power function and the experimental results can be expressed by the following equation generally, which has an accuracy of $\pm 10\%$ with the experimental data.

$$\Delta P = aRe^{j} + C \tag{43}$$

where, $a = (9.26 \times 10^{-7})\, e^{-0.066\,\alpha}$, $j = 0.0073\alpha^3 + 3.1$, $C = -0.0054\alpha^3 + 0.31\alpha^2 - 2.77\alpha + 4.3$.

The a, j and c are experimental coefficients which may depend on a certain void fraction and is in percentage in these expressions. Figure 31b, shows the effect of

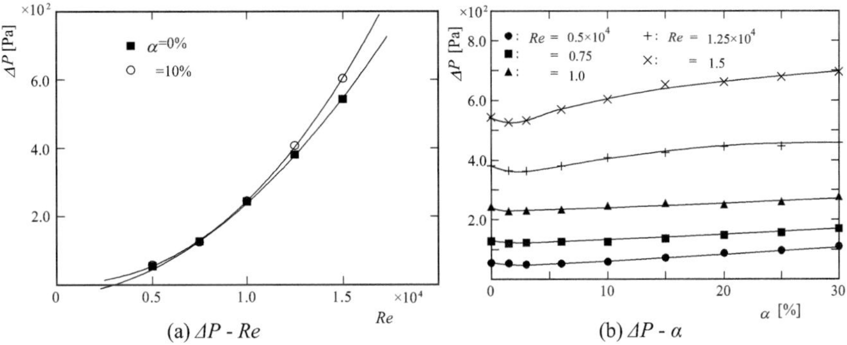

(a) ΔP - Re (b) ΔP - α

Fig. 31 Pressure difference, ΔP

Fig. 32 Pressure distribution on cylinder surface

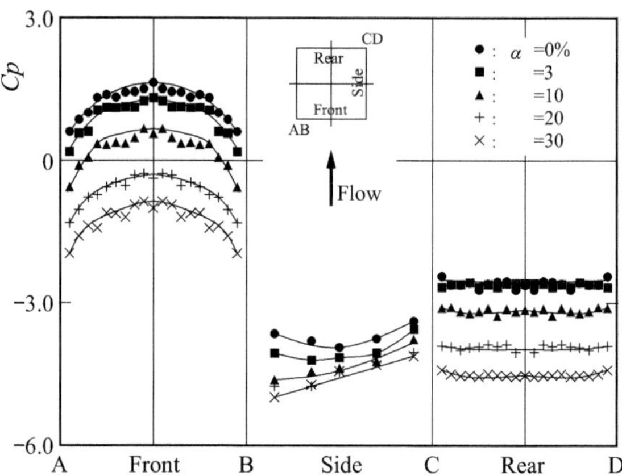

αon ΔP at several Re. It can be noticed that when increasing α, slight increase of ΔP occurs, and it's almost linear for $\alpha > 3\%$, while up to $\alpha = 3\%$ a slight decrease occurs.

(C.2) Drag on cylinder

Figure 32 is an example of pressure distributions on cylinder surface at several void fractions when $Re = 1.0 \times 10^4$. The pressure coefficient C_p is expressed as follows:

$$C_p = 2(P_i - P_{ref}) / (\rho_t U_{mt}^2) \tag{44}$$

where, P_i: pressure at certain position on cylinder surface, P_{ref} : selected at the position $x/d = 2.0$ on the sidewall, U_{mt} : mean free stream velocity of two-phase flow, ρ_t : mean density of two-phase flow.

The pressure has the highest value at the center of the front side of the cylinder and decreases gradually in the direction of both sides. The pressure on the rear side is almost same under certain α, and the relative pressure at the same position decreases with increasing α. According to these results, the drag coefficient C_D can be calculated, and after plotting C_D relatively to the pressure coefficient difference ΔC_p between the front and rear stagnation points as shown in Fig. 33, the following relationship is found:

$$C_D = 0.575\Delta C_p + 1.24 \tag{45}$$

4.5.3   Flow Around Tandem Cylinder

Next, the results on flow characteristics between and after the second cylinder are discussed. When aspect ratio is $AR = 1$ and distance ratio $s/d = 1$ will be presented in

Fig. 33 C_D and ΔC_p

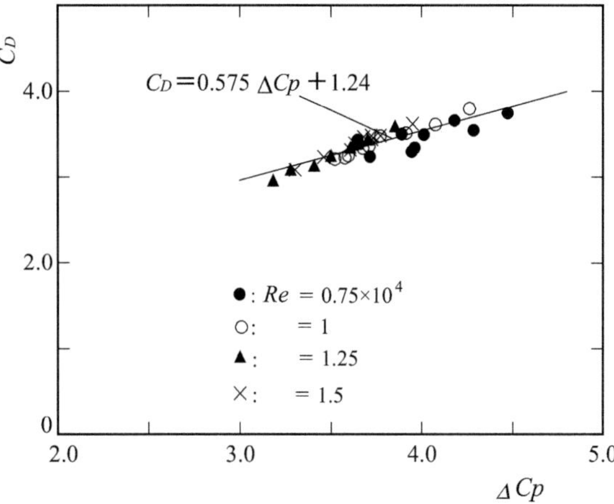

this work while velocity vector, vorticity, vortex shedding frequency and drag were made clear. Also, here a comparison with single-phase flow is conducted for better understanding.

* Flow pattern

Figure 34a, b show the visualized flow patterns of vortex shedding past single rectangular cylinder and two-tandem cylinders with $s/d = 1$ for $Re = 1.0 \times 10^4$, respectively. The flow is from bottom to top and the gravitational direction is opposite. In the case of $s/d = 1$ behind the upstream cylinder, which is the 1st cylinder, twin vortex is produced and after the 2nd cylinder Karman vortex is shed. In Fig. 35, the velocity vector at an instantaneous flow pattern between cylinders measured by PIV, volume cross-correlation method, is plotted. For the case of $s/d = 1$ the twin vortex is asymmetric in respect to the centerline as can be shown. In practice, it fluctuates unstably in $\pm y$ direction.

After examining the case of single-phase flow, two-phase flow pattern with same Re and $\alpha = 10\%$ is shown in Fig. 36, for both cases of single and two tandem rectangular cylinders. The visualized photograph shows that the flow for $s/d = 1$ separates at the front edges of 1st but separates at the rear edges of 2nd cylinder and reattaches at the front edges of 2nd cylinder. For the case of air–water flow, there is a dead water region that contains no bubbles or few bubbles occasionally just behind the cylinder and a bubble rich region surrounding it. Here, it must be noticed that in the case of single flow this region is not clear and this phenomenon result from the existence of void fraction α. The bubble swarms in this region oscillates and develops into Karman vortex in the downstream. Also, same as before, the motion of bubble swarms reaches to the sidewall of the flow passage. Likewise single flow, the vortices between the two cylinders look like twin-vortices but actually fluctuate alternately in $\pm y$ direction.

Fig. 34 Visualized flow pattern of single-phase flow ($Re = 1.0 \times 10^4$)

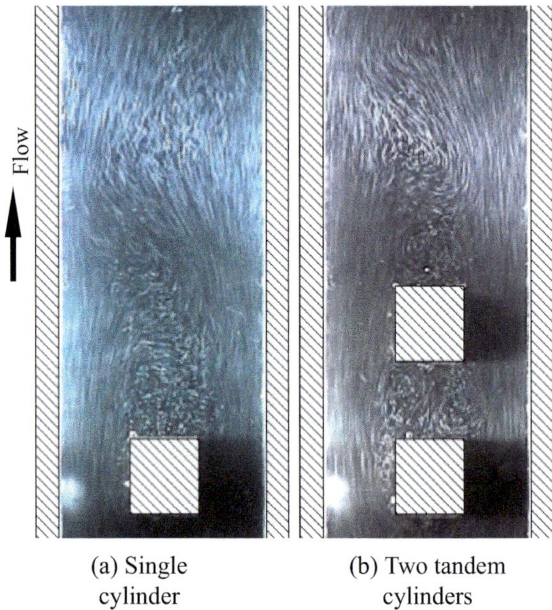

(a) Single cylinder

(b) Two tandem cylinders

Fig. 35 Velocity vectors by PIV (single-phase flow, $Re = 1.0 \times 10^4$)

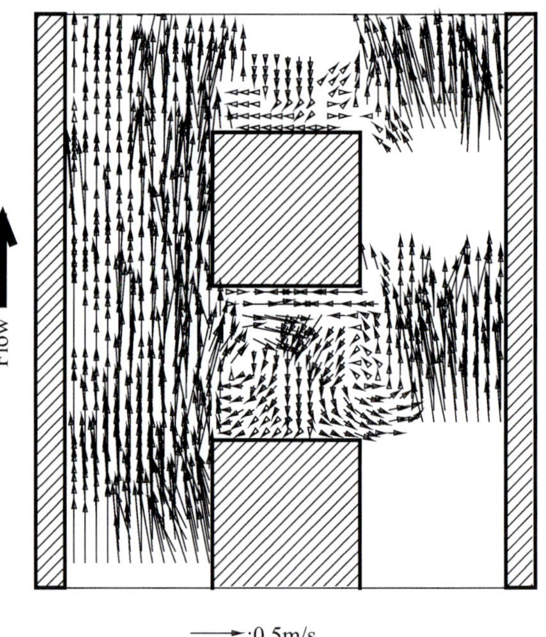

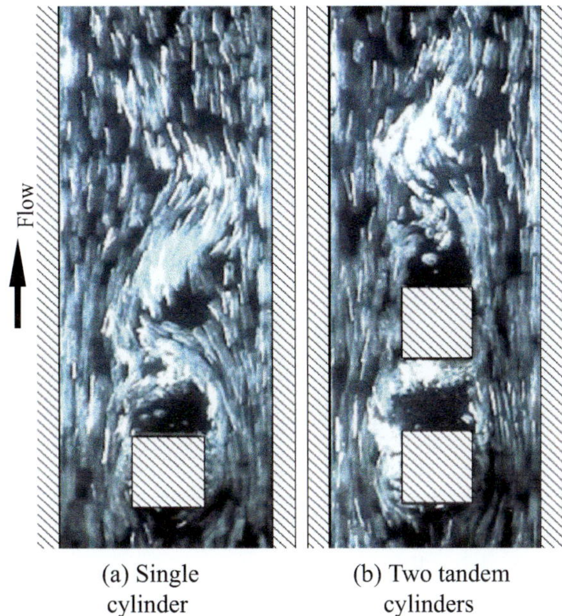

Fig. 36 Visualized flow pattern ($Re = 1.0 \times 10^4$, $\alpha = 10\%$)

4.5.4 Oscillatory Frequency

Karman vortex is shed from the rear sides of the 2nd cylinder. The vortex shedding frequency f was obtained using a high-speed photographic system and measured behind the 2nd cylinder. The measurement position was located at distance $x = 3d$ from the 2nd cylinder.

Figure 37, shows the relation between shedding frequency f and Re with void fraction $\alpha = 0$ and 10%. It is clear that f increases with increasing Re linearly for single cylinder and $s/d = 1$ and can be expressed generally as follows:

$$f = \mathrm{k}Re + b \tag{46}$$

where, k, b: experimental coefficients depending on α.

The f of two-phase flow is at any time always slightly larger than that of single-phase flow.

Figure 38 shows the effect of α on f. The shedding frequency f increases with increasing α slightly at most flow conditions and this depends on the decreased moving mass flow.

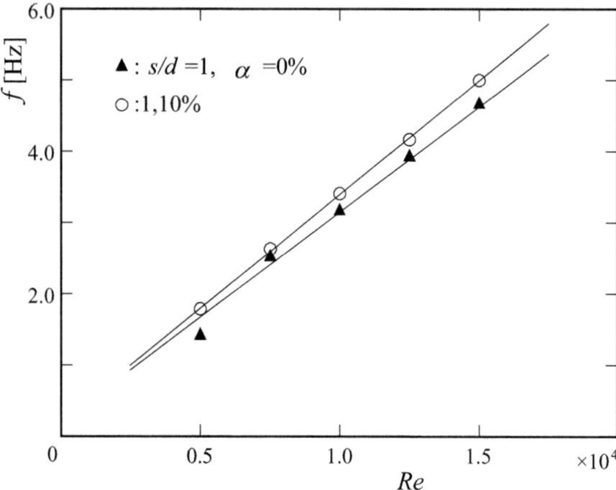

Fig. 37 Vortex shedding frequency (Two cylinders, α = 0, 10%)

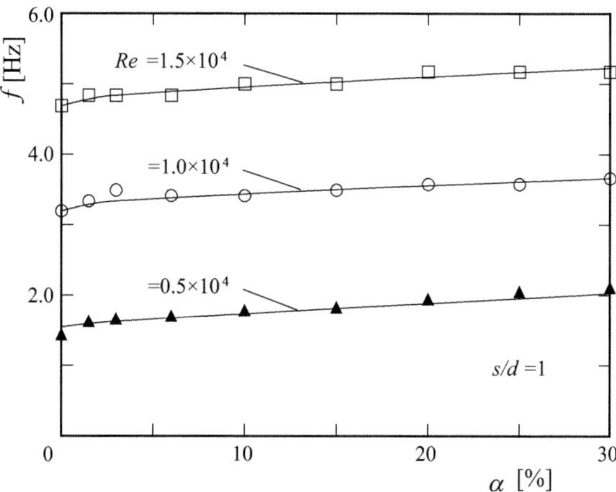

Fig. 38 Vortex shedding frequency [Two cylinders, Re = (0.5 ~ 1.5) × 10^4]

4.5.5   Flow Resistance on First and Second Cylinders

(a)   Pressure difference between front and rear stagnation points

Figure 39, is an example of ΔP at $\alpha = 0\text{~}10\%$ for the 1st and 2nd cylinder. We can see that ΔP increases with increasing Re number while between 0 and 10% void fraction not so much difference can be seen for the 1st cylinder. For the 2nd cylinder a small decrease occurs with decreasing Re number, while the effect of α is greater. For the results taken, a general form of ΔP is shown as follows.

$$\Delta P = A_1 \times Re^x + A_2 \tag{47}$$

where, A_1, A_2: experimental coefficients.

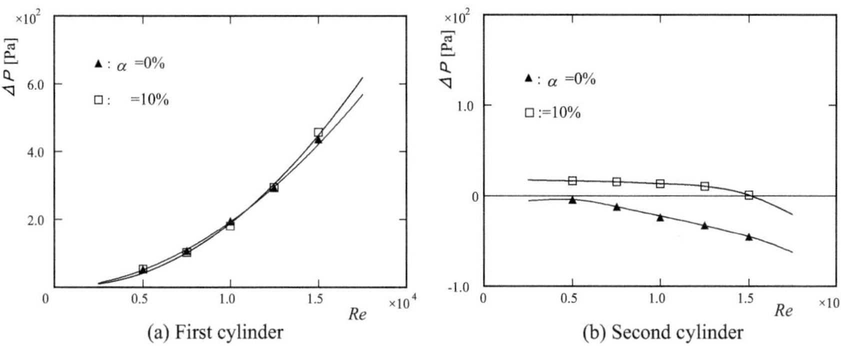

Fig. 39 Pressure difference, $\Delta P - Re$ (Two cylinders, $\alpha = 0$, 10%)

In Fig. 40, the results from the effect of α on ΔP for several Re are shown, while for the 1st cylinder a comparison is made with the case of single cylinder. As it can be seen, a slight increase caused by void fraction occurs and compared to the case of single cylinder, for every Re not much difference can be observed. On the 2nd cylinder the effect of α is also rather small.

(b) Drag on cylinder

Figure 41 is an example of pressure distributions on cylinder surface at several void fraction when $Re = 1 \times 10^4$ for both the 1st and 2nd cylinder. The pressure has the highest value at the center of the front side of 1st cylinder and decreases gradually in the direction of both sides. The pressure on rear side is almost same under certain α, and the relative pressure at the same position decreases with increasing α. On the other hands for the case of 2nd cylinder no similar form appears, while the pressure on the front side rises slightly as moving on the edges. The pressure on the rear surface is almost linear and the values increase by increasing Re. It must be noted here that only the half axis is plotted since it was confirmed that they are symmetric.

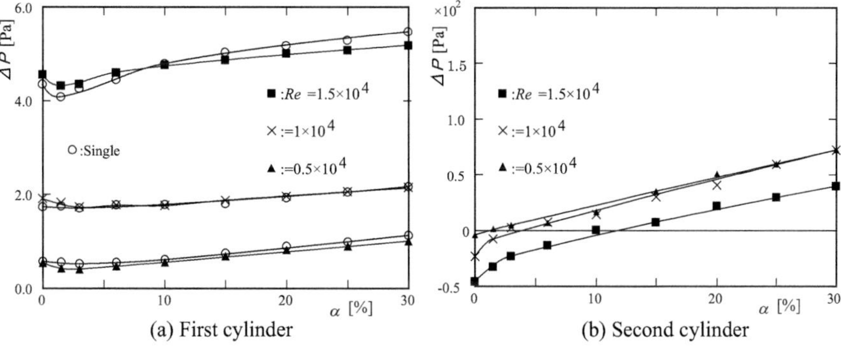

Fig. 40 Pressure difference, $\Delta P - \alpha$ [Two cylinders, $Re = (0.5 \sim 1.5) \times 10^4$]

The relation between the drag of the cylinders Dr and the pressure difference between the front and rear stagnation points is shown in Fig. 42, and it has the form of the following equation:

$$\text{1st cylinder}: Dr = 1.2 \times 10^{-3}\Delta P + 8.93 \times 10^{-3} \tag{48}$$

$$\text{2nd cylinder}: Dr = 1.36 \times 10^{-3}\Delta P + 6.68 \times 10^{-3} \tag{49}$$

In general, the form of equation has as follows:

$$\text{1st cylinder}: Dr = C_1\Delta P + C_2 \tag{50}$$

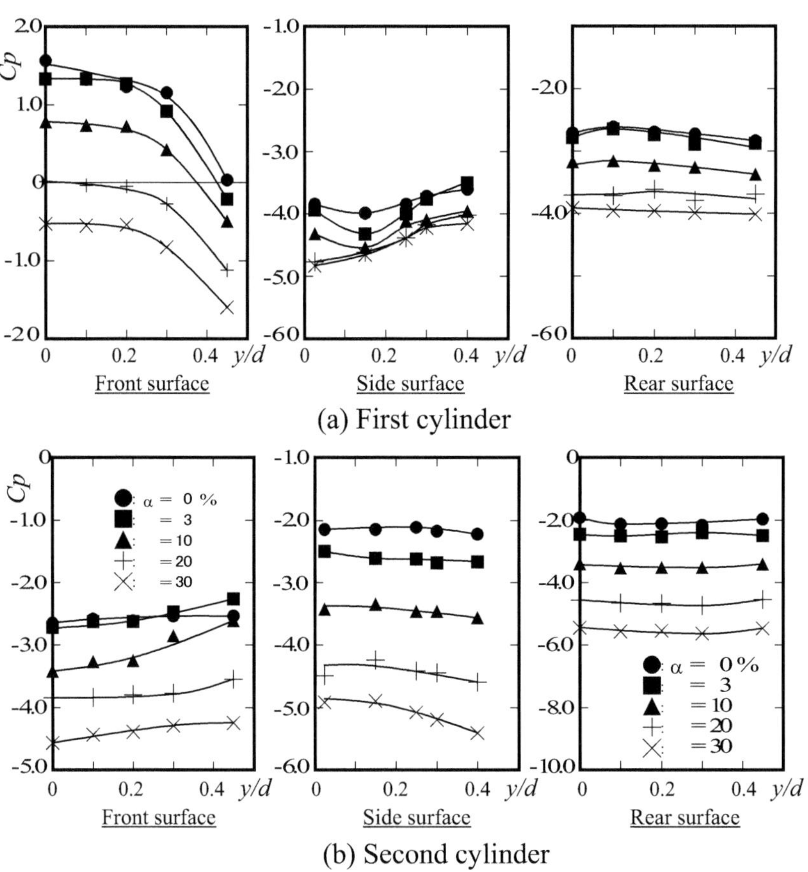

Fig. 41   Pressure distribution on cylinder surface

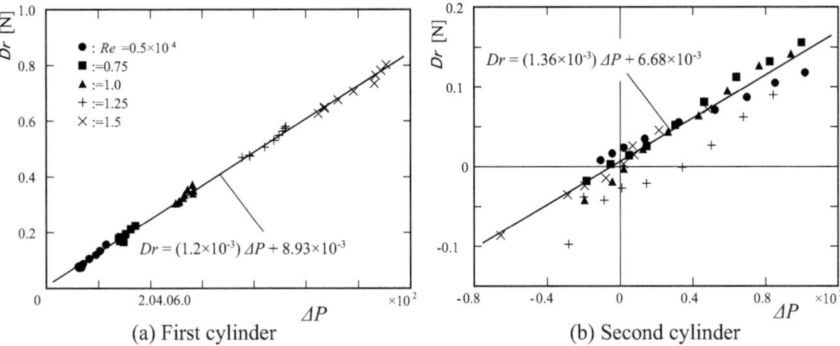

Fig. 42 Drag and pressure difference

$$\text{2nd cylinder}: Dr = C_3 \Delta P + C_4 \tag{51}$$

where, $C_1 \sim C_4$: experimental coefficients depending on s/d.

This means that the study on pressure difference between the front and rear stagnation points of cylinder(s) can characterize the drag on cylinder(s) and much work can be saved.

Here, the behavior of Karman vortex shedding from single or two-tandem rectangular cylinders in a gas–liquid, bubbly flow was examined.

***For Single Cylinder:**

(1) Karman vortex appears not only for single-phase flow but also for two-phase bubbly flow.

(2) The oscillatory frequency increases with increasing Reynolds number Re and also increases slightly with increasing void fraction.

(3) Flow resistance can be expressed by pressure difference between the front and rear stagnation points and their relation is of the form as follows:

$$Dr = B_1 \Delta P + B_2$$

and

$$\Delta P = aRe^j + c$$

where, B_1, B_2, a, j, c: experimental coefficients depending on AR.

***For Two-Tandem Cylinder:**

(4) Not only for single-phase flow but also for two-phase bubbly flow, twin vortex appears between two cylinders when $s/d = 1$ and Karman vortex appears in the wake of the second cylinder.

(5) The oscillatory frequency of Karman vortex increases with increasing Re and as also with increasing void fraction slightly.

(6) Flow resistance of the first and second cylinders can be expressed by the pressure difference between the front and rear stagnation points, and their relation can generally be expressed by the form as follows:

$$\text{1st cylinder}: Dr = C_1 \Delta P + C_2$$

$$\text{2nd cylinder}: Dr = C_3 \Delta P + C_4$$

$$\Delta P = A_1 Re^x + A_2$$

where, A_1, A_2, x, $C_1 \sim C_4$: experimental coefficients depending on *s/d* and *AR*.

4.6   Single Tube Bundle in Gas–Liquid Two-Phase Bubbly Flow

Gas–liquid two-phase cross-flow through tube bundles is found in many important industrial applications, including boilers, heat exchangers, chemical reactors, and other applications. Conventionally, these tube bundles are generally arranged in a lattice or staggered pattern, both of which are equally spaced. However, it is thought that making these arrangements unequal would change the flow conditions around the tube bundle, reducing flow resistance and improving heat transfer characteristics.

Here, we first experimentally clarify the flow (single layer water, and two-phase bubbly flow of water and air) and heat transfer characteristics of a single row of equally and unequally spaced tube banks, and then investigate the effects of tube arrangement, inflow velocity, air ratio, etc. on improving and enhancing these characteristics (Shimizu 2007; Shakouchi et al. 2010).

4.6.1   Test Section and Tube Arrangement

Figure 43 shows the test section and single row tube bank arrangements. The test flow path is width $D = 90.0$ mm, depth $H = 45.0$ mm, and length $L = 1520$ mm. As a single-row tube bank, nine tubes with a diameter $d = 15.0$ mm and a length of 45.0 mm were arranged in the flow direction at the center of the pipe (Fig. 43). The single-row tube bank was arranged in three types: equally spaced in the flow direction ($p/d = 1.5$, p: distance between the centers of the cylinders), and unequally spaced in which the positions of even-numbered tubes from the upstream side were changed in the flow direction ($p/d = 1.33, 1.17$).

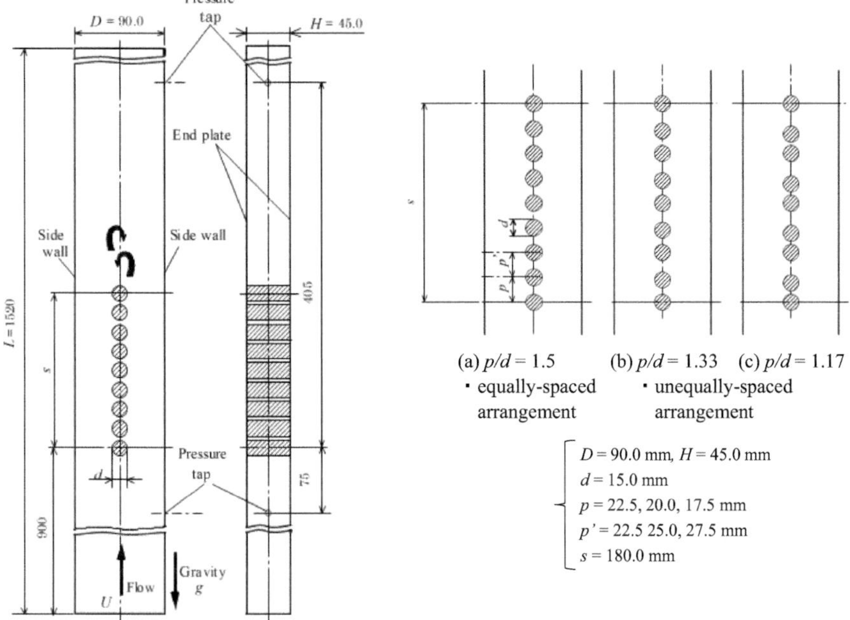

Fig. 43 Test section and tube arrangement

4.6.2 Heat Transfer Characteristics

The heat transfer characteristics of the tubes in the tube bundle were investigated as follows. A heater was installed inside a circular tube, cylinder, made of copper and it was electrically heated with a constant heat flux, and the temperature distribution was measured using a thermocouple installed from inside the tube at the center depth of the cylinder surface. The cylinder was rotatable, allowing the temperature to be measured at any circumferential position on the cylinder surface.

4.6.3 Flow Characteristics

From the visualization and observation of the flow through the single-row tube bank, and the results of the pressure distribution measurements described below, it was found that the flow becomes steady and similar from the fourth cylinder onwards.

(a) **Flow pattern**

Figure 44 shows the velocity vector diagram around the fifth and sixth cylinders at $Re = 5 \times 10^3$ and $\alpha = 0\%$ as an example of single-phase water flow. For single-phase water flow, polystyrene beads (mean diameter: 0.37 mm, specific gravity: 1.01) were mixed into the flow as a tracer, and the central cross section of the flow channel was optically cut, and the visualized flow was photographed with a high-speed camera

(PHOTORON FASTCAM-RABBIT). The image was then processed using Particle Image Velocimetry (PIV) to determine the velocity vectors of the flow field.

In the case of an equally spaced arrangement ($p/d = 1.5$), the flow along the fifth cylinder mainly flows between the cylinders, impinges with the sixth cylinder, and then flows downstream along the cylinder, then toward the center and out to the opposite side. This flow that is asymmetric with respect to the central axis irregularly becomes an axisymmetric reverse flow state, and oscillations are repeated.

(b)   Fluctuating flow characteristics and power spectrum density (PSD)

The behavior of this unsteady flow oscillation was investigated by measuring the pressure fluctuations in the flow field as follows. A small diffused semiconductor pressure transducer (PD104K, JTEKT, Japan) was installed at position A on the end plate shown in Fig. 45, and the pressure fluctuations of the flow were investigated and frequency analysis performed using FFT.

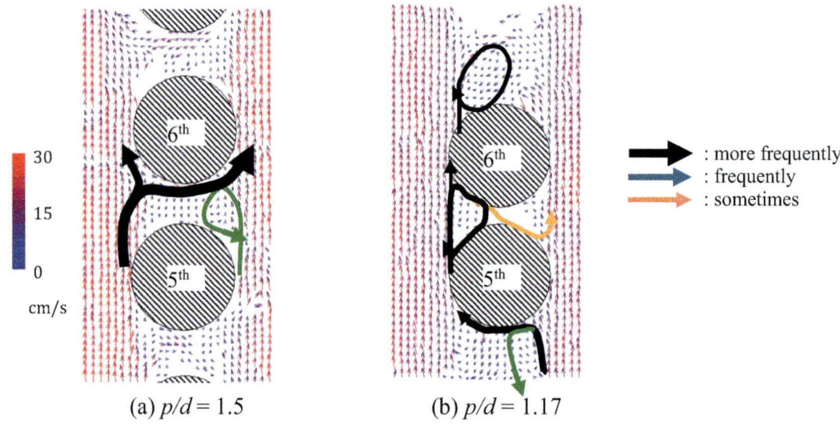

(a) $p/d = 1.5$ (b) $p/d = 1.17$

Fig. 44   Velocity vector ($Re = 5 \times 10^3$, $\alpha = 0\%$)

Fig. 45   Position of pressure transducer

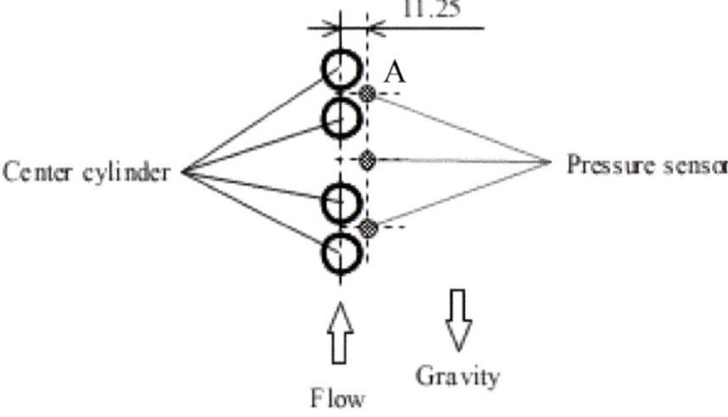

A: between 5th and 6th tubes

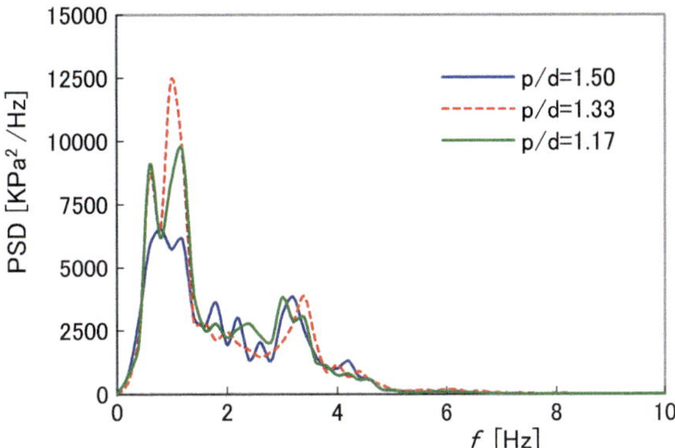

Fig. 46 PSD of pressure at position A ($Re = 5 \times 10^3$, $\alpha = 0\%$)

Figure 46 shows the results, power spectrum density (PSD) of pressure at position A, between the 5th and 6th rows of tubes, where the flow is similar and steady. The dominant frequencies were 0.8 Hz at $p/d = 1.50$, 1 Hz at $p/d = 1.33$, and 1.2 Hz at $p/d = 1.17$, which were not significantly different from those between the 1st and 2nd rows of tubes. There was also almost no difference due to the tube arrangement. The second peak of PSD is at around $f = 3.4$ Hz.

(c) Gas–liquid two-phase bubbly flow

Instead of water, a gas–liquid two-phase bubbly flow was used as the working fluid. The mean bubble diameter before passing through the single-row tube bank was $d_{m} \fallingdotseq 2.9$ mm at $Re = 5 \times 10^3$ and $\alpha = 5\%$, and was approximately the same after passing through the tube bank. The volume fraction of air, $\alpha = Q_a/(Q_w + Q_a)$, Q_w, Q_a : volumetric flow rate of water and air, respectively, is variable within the range of 0 to 15%. The air volume was calculated based on the reference pressure P_{ref} of the flow path (the pressure on the flow path side wall at a position $5d \, (= 75$ mm) upstream from the center of the first cylinder of the single-row tube bank). The Reynolds number was set to $Re = Ud/v = (2.5 \sim 7.5) \times 10^3$ (U: mean apparent velocity of water inflow into the tube bank).

As an example, Fig. 47 shows the results of FFT analysis of pressure fluctuations for each tube arrangement of $p/d = 1.5$. 1.33 and 1.17 at $Re = 5.0 \times 10^3$ and $\alpha = 5.0\%$, bubble flow.

The dominant frequency was approximately 1.0 Hz for each tube arrangement. Furthermore, for single-phase flow ($\alpha = 0\%$), the PSD value had a maximum value at approximately $f = 3.4$ Hz, but for two-phase flow ($\alpha = 5.0\%$), there was no second peak at around $f = 3.4$ Hz, and the second peak value was at around $f = 1.8$ Hz.

(d) Pressure distribution on tube surface and drag coefficient

Figure 48 shows an example of the drag force D_r acting on each cylinder when $Re = 5 \times 10^3$ and $\alpha = 5\%$, expressed as the drag coefficient $C_D \, [= 2D_r \, /(\rho U^2 A)$, A: projected area of tube, cylinder]. D_r was calculated by integrating the pressure distribution on the cylinder surface. The drag force is maximum for the first row of

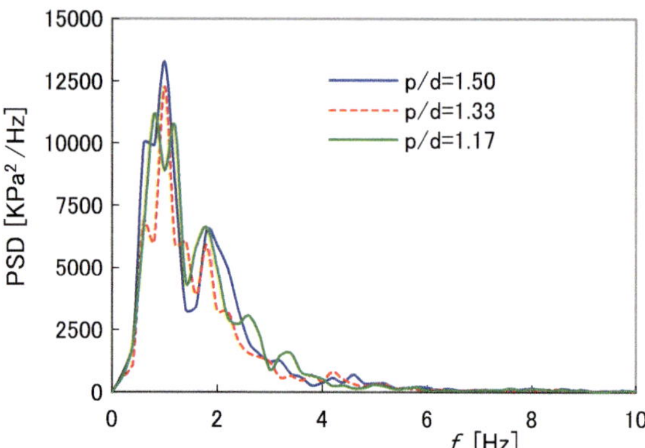

Fig. 47 PSD of pressure at position A [$Re = 5 \times 10^3$, α = 5% (bubble flow)]

cylinders, where the main flow impinges on the front surface of the cylinder, and is minimum (negative value) for the second row of cylinders, which are in the wake of the first cylinder, where the front surface of the cylinder is hardly impinges on by the flow from the front. With an evenly spaced arrangement $p/d = 1.5$, the drag force becomes almost constant from the third cylinder onwards. With an unequally spaced arrangement $p/d = 1.17$, the drag force decreases for even-numbered rows of cylinders and increases for odd-numbered rows of cylinders. Qualitatively, the results for single-phase water flow were almost the same as above.

(e) **Pressure loss**

Figure 49 shows the flow resistance, pressure loss ΔP_{tl}, in the flow passage when passing through a single row of tube bundle. ΔP_{tl} was measured between pressure holes provided in the flow passage side wall surface $5d$ upstream and $27d$ downstream from the center of the first cylinder. ΔP_{tl} increases with increasing volume fraction of air α. This is thought to be due to the fact that the presence of air bubbles increases the collision loss of the flow. Furthermore, when $\alpha = 0\%$, ΔP_{tl} increases as the Reynolds number increases, but becomes almost constant at $\alpha = 5\%$. When α increases further

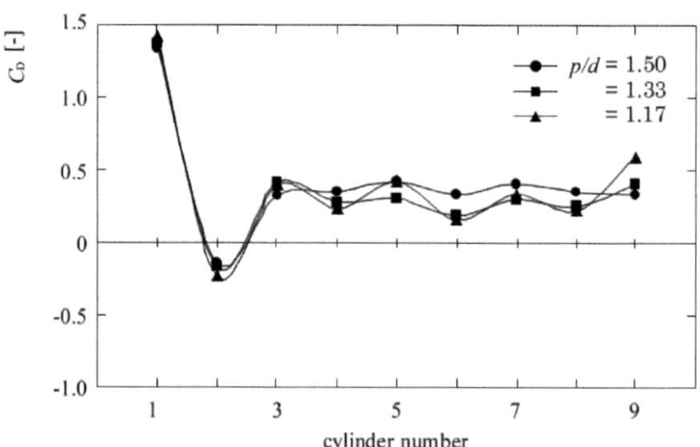

Fig. 48 Drag of cylinder [$Re = 5 \times 10^3$, $\alpha = 5\%$ (bubble flow)]

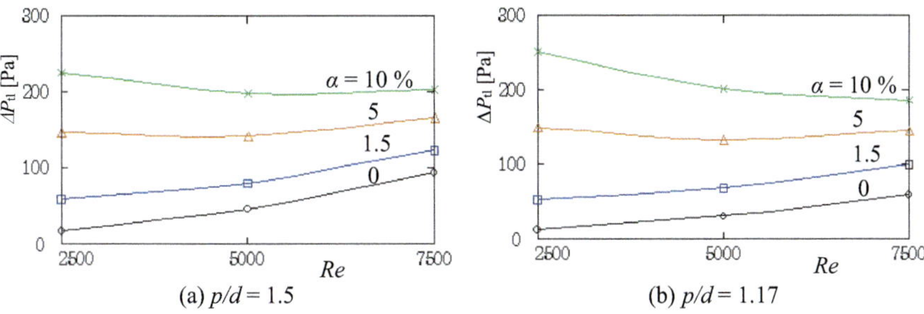

Fig. 49 Pressure loss

$(= 10\%)$, ΔP_{tl} decreases as the Reynolds number increases. This is thought to be due to a change in the flow pattern rather than an increase in collision loss due to the presence of air bubbles as mentioned above, and a decrease in friction loss.

Figure 50 shows the reduction rate $\beta[= (\Delta P_{\mathrm{tl}} - \Delta P_{\mathrm{tl},1.5})/\Delta P_{\mathrm{tl},1.5}]$ of ΔP_{tl} for an unequally spaced arrangement compared to $\Delta P_{\mathrm{tl},1.5}$ for an equally spaced arrangement ($p/d = 1.5$). β increases with increasing Reynolds number, but decreases with increasing α. The ΔP_{tl} of the unequally spaced tube arrangement is smaller than that of the equally spaced arrangement in the region of relatively large Reynolds number and small α. In other words, the operating power P_{w} ($= \Delta P_{\mathrm{tl}}\,Q$, Q: flow rate) can be reduced. For example, when $Re = 7.5 \times 10^3$ and $p/d = 1.33$, W_{d} can be reduced by approximately 35% at $\alpha = 0\%$ and approximately 9% at $\alpha = 10\%$, and when $p/d = 1.17$, P_{w} can be reduced by approximately 37% at $\alpha = 0\%$ and approximately 8% at $\alpha = 10\%$.

(f) **Heat transfer characteristics**

The heat transfer characteristics of the cylinder surface were evaluated using the Nusselt number $Nu = hd/k$. Here, k is the thermal conductivity of the flow (water), which was found using the film temperature $(T_{\mathrm{s}} + T_{\mathrm{w}})/2$. Note that T_{s} and T_{w} are

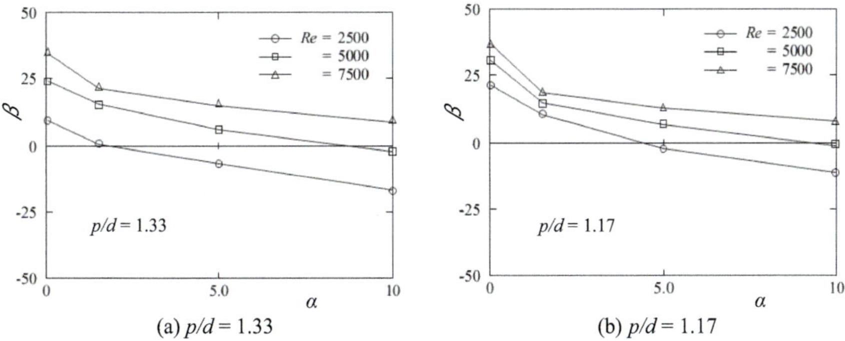

Fig. 50 Reduction rate of pressure loss

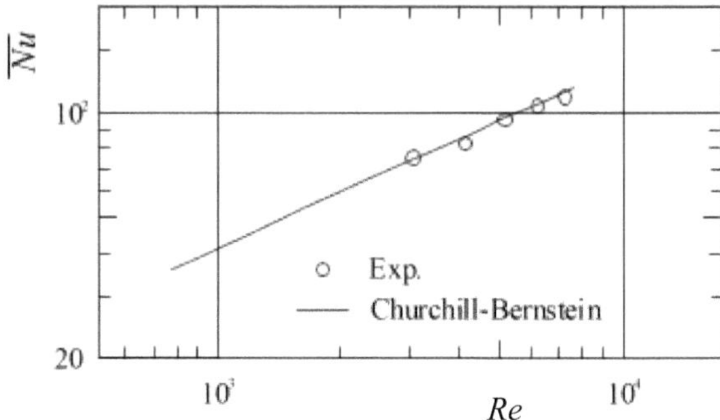

Fig. 51 Mean Nusselt number for single cylinder ($\alpha = 0\%$)

the surface temperature of the cylinder and the temperature of the flow, respectively. Furthermore, h is the heat transfer coefficient, and was found using $h = q/(T_s - T_w)$, where q is the heat flux. The mean Nusselt number $\overline{Nu}$ of the cylinder was found using the following equation.

$$\overline{Nu} = \frac{1}{\pi} \int_0^\pi Nu \, d\theta \tag{52}$$

(f.1) Single cylinder

Figure 51 shows the mean Nusselt number distribution $\overline{Nu}$ for $\alpha = 0\%$ obtained from the temperature distribution around a single cylinder. $\overline{Nu}$ increases rapidly with increasing Reynolds number, and the experimental results agree well with the Churchill-Bernstein correlation appeared by the solid line in the figure.

(f.2) Single row of cylinders, tubes

As an example, Fig. 52 shows the local Nusselt number distribution calculated from the temperature distribution around the 5th cylinder when $Re = 5 \times 10^3$ and $\alpha = 0$ and 5%. In this case, the operating power P_w for $P/d = 1.33$ and 1.17 was set to be the same as for $p/d = 1.5$ ($P_w = $ constant). Note that the distribution shape was symmetrical in the flow direction, so only half of the θ region ($\vartheta = 0° \sim 180°$) is shown. In this case, as mentioned above, with the unequally spaced tube arrangement of $p/d = 1.33$, 1.17, the flow resistance is reduced compared to the equally spaced arrangement of $p/d = 1.5$, so when $P_w = $ constant, the flow rate and velocity increase. This is also thought to contribute to improving heat transfer characteristics. When $\alpha = 0\%$, the Nusselt number for the unequally spaced tube arrangement $p/d = 1.33$ is greater than that for the equally spaced arrangement $p/d = 1.5$ over almost the entire range of θ. In other words, it has better heat transfer characteristics. Also, when $\alpha = 5\%$, the Nusselt number for $p/d = 1.33$, 1.17 is greater than $p/d = 1.5$ over almost the entire range of θ.

Figure 53a shows the mean Nusselt number of each cylinder (driving force: constant), as in one example of $Re = 5 \times 10^3$ and $\alpha = 5\%$. In the first cylinder, $\overline{Nu}$ is the maximum because the flow impinges on the front surface. In $P/d = 1.5$,

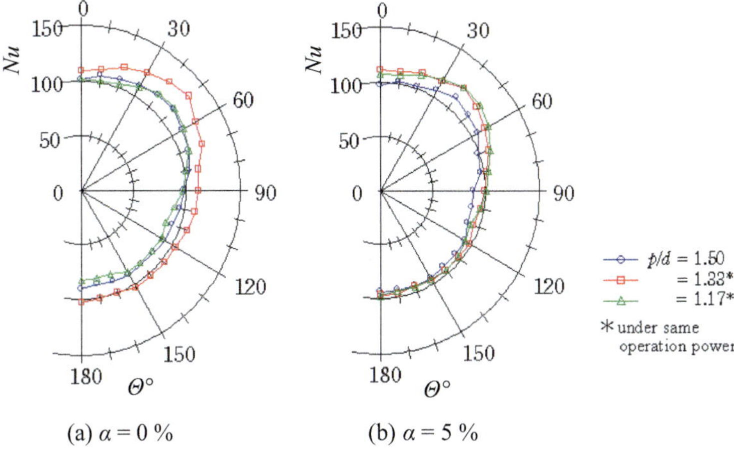

(a) $\alpha = 0\ \%$

(b) $\alpha = 5\ \%$

Fig. 52 Local Nusselt number for 5th cylinder

$\overline{Nu}$ is almost the same for the 4−8th cylinders. In comparison, in $P/d = 1.33$, the even -numbered cylinder is located in the posterior stream of the front cylinder, so the heat transfer characteristics are inferior and the $\overline{Nu}$ is reduced. $\overline{Nu}$ of the 4, 6, and 8th cylinders is almost the same ($= 104.3$), and $\overline{Nu}$ of the 5, 7 and 9th cylinders is almost the same ($= 0.111.1$).

Figure 53b shows the increase rate $\varepsilon[= (\overline{Nu} - Nu_{m1.5})/Nu_{m,1.5}]$ of $\overline{Nu}$ of each cylinder when $p/d = 1.33$, relative to $Nu_{m,1.5}$ when the cylinders are arranged at equal intervals ($p/d = 1.5$). The increase rates for the first cylinder, the odd-numbered 3rd, 5th, and 7th cylinders, and the even-numbered 4, 6, and 8th cylinders are $\varepsilon \fallingdotseq 16, 13$ and 6%, respectively. The overall increase rate is $\varepsilon \fallingdotseq 10.5\%$.

It was shown that for single-row tube bundle, uneven tube spacing can reduce flow resistance and improve heat transfer characteristics. For example, when $Re = 5 \times 10^3$, $p/d = 1.33$, the flow resistance can be reduced by approximately 35% at $\alpha = 0\%$ and approximately 9% at $\alpha = 10\%$, compared to the case of an equally spaced

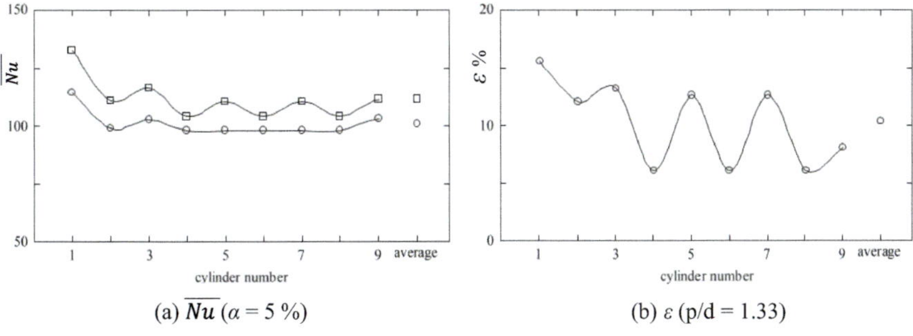

(a) $\overline{Nu}$ ($\alpha = 5\ \%$)

(b) ε (p/d = 1.33)

Fig. 53 Mean Nusselt number and increasing rate ($Re = 5 \times 10^3$) dL

arrangement $p/d = 1.5$. In addition, when $\alpha = 5\%$, the heat transfer characteristic Nu_m can be improved by approximately 10.5%.

4.7   Tube Bundle in Gas–Liquid Two-Phase Bubbly Flow

Water or air–water bubble flow passing through tube bundle can be seen in many industrial equipment such as heat exchanger and chemical equipment. In addition, tube bundle can be used as a flow-straightener, -mixer, -resistor, and -damping device.

Here, the effects of tube arrangement of equally or unequally spaced in-line and staggered tube bundle and of void fraction on the flow characteristics such as flow pattern and flow resistance of a tube or tube bundle are examined experimentally (Shakouchi et al. 2017).

4.7.1   Test Section

Figure 54 shows the schematic diagram of test section (tube bundle). The test section is made of transparent acrylic resin plate and is set vertically. The width, depth and length of the test section are $D = 90.0$, $H = 45.0$, and $L = 1520$ mm, respectively. The tube bundle is consisted of 5 rows and 9 lines of tube of diameter $d = 15.0$ mm. The pitch of tube is equal interval of $p/d = 1.5$, and unequal interval of $p/d = 1.33$, 1.17 which were realized by moving odd numbered rows. The arrangements of tube bundle, in-line and staggard, are shown in Fig. 55. Water or air–water bubble flow introduces into the test section from the bottom after passing through the inlet flow passage with an enough length. A tube has a pressure hole of diameter 0.8 mm at the half depth and which can be rotated around its own axis. The pressure loss Δp_tl of the tube bundle was measured by the pressure holes which were located on the middle depth of the side wall of the test section at the up- and down-stream of distance $5d$ and $27d$ from the first row, respectively. The visualized flow pattern by a Tracer method was photographed by a high-speed video camera (PHOTORON FASTCAM-RABBIT)). When the working fluid was a single-phase water flow, tiny polystyrene beads (mean diameter: 0.37 mm, specific gravity: 1.01) were mixed into the flow as tracers, and the flow was visualized by optical cutting at the central depth of the flow channel. In the case of gas–liquid (air–water) two-phase bubble flow air bubbles with an average diameter of about $d_\mathrm{a} \fallingdotseq 2.5$ mm served as tracers. The velocity distribution was measured by Particle Image Velocimetry, PIV, and pressure was measures by pressure hole of diameter 0.8 mm and liquid column manometer or semi-conductor type pressure transducer (JTEKT, PD104K, Japan).

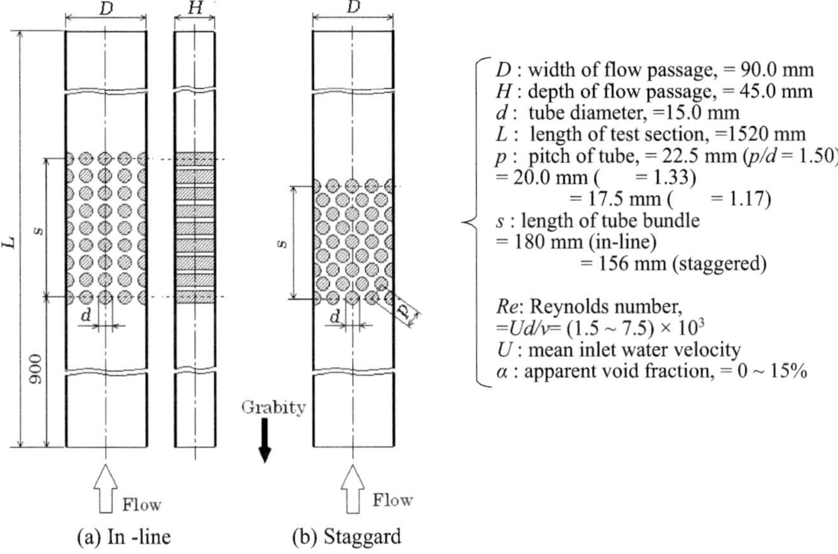

Fig. 54   Test section (tube bundle)

Fig. 55   Arrangement of
tube bundle

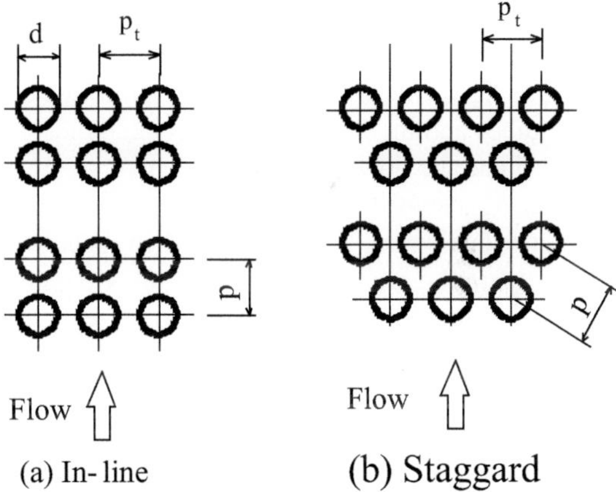

4.7.2   Flow Pattern and Velocity Distribution

Figure 56a, b show an example of the velocity vectors measured by PIV and visualized flow pattern of water flow around the 2nd to 4th rows of staggered tube bundle with $p/d = 1.5$ and 1.17, respectively, in the centre region of the test section. In the case of equal interval of $p/d = 1.5$, behind the 3rd row tube there is a wake region and the flow passing through between the 3rd tube impinge on the 4th tubes. The flow pattern for unequal interval of $p/d = 1.17$ was largely different with the $p/d = 1.5$. In the case of $p/d = 1.17$, behind the 2nd row tube there is a large wake region with

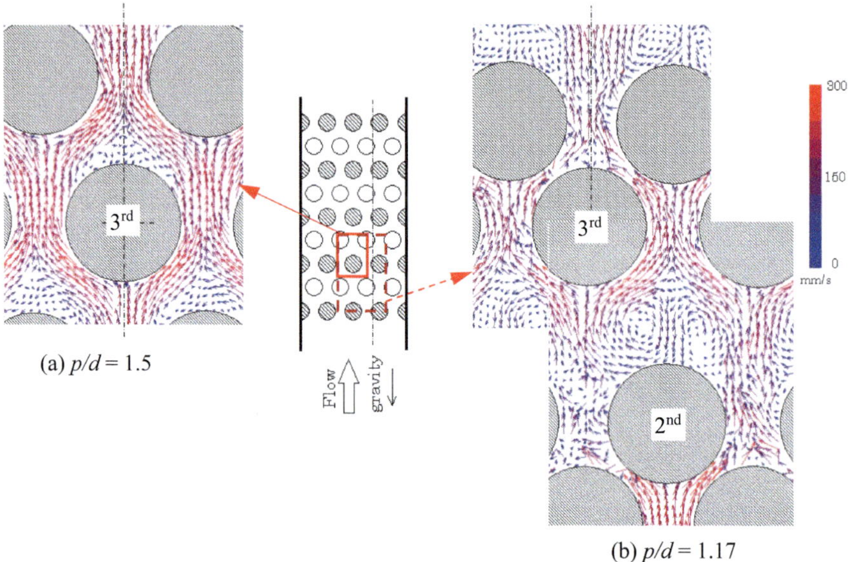

Fig. 56 Visualized flow pattern and velocity vector of staggered tube bundle, $Re = 1.5 \times 10^3$, and $\alpha = 0\%$ (water flow), (tracer: polystyrene beads with mean diameter, 0.37 mm and specific gravity, 1.01) (by PIV measurement: using 96 photos taken by 1/240 frs. with 1/1000 s.)

two opposing vortices. The flow passing through between the 3rd tubes impinges on the 4th tubes and the flow velocity between the 4th tubes is smaller than that of the $p/d = 1.5$ and the flow appears turbulent.

Figure 57 shows the results for the two-phase gas–liquid bubble flow of $\alpha = 5\%$ in the previous figure. At $p/d = 1.5$, there is almost no wake region behind the 3rd tube compared to when $\alpha = 0\%$. The flow passing through between the 3rd tubes impinges on the 4th tubes. Also, when $p/d = 1.17$, there is almost no wake region behind the 2nd tube, and the flow passing through between the 2nd tubes impinges on the 3rd tubes.

4.7.3 Flow Fluctuation

Figure 58 is the power spectral density (PSD) of the pressure fluctuation at the centre between the 2nd and 3rd tubes for the staggered tube arrangement as shown in the figure, and $Re = 5.0 \times 10^3$, $\alpha = 5.0\%$. Two dominant frequencies of $f_d = 1.0$ and 2.0 Hz are appeared. For $\alpha = 0\%$, the 2nd dominant frequency of $f_d = 2.0$ Hz did not appear. For the in-line and $\alpha = 0\%$, two dominant frequencies of $f_d = 1.0$ and 3.0 Hz, and for $\alpha = 5.0\%$ two f_d of 1.0 and 2.0 were appeared. The fluctuation characteristics are different by the arrangement of tube bundle and α.

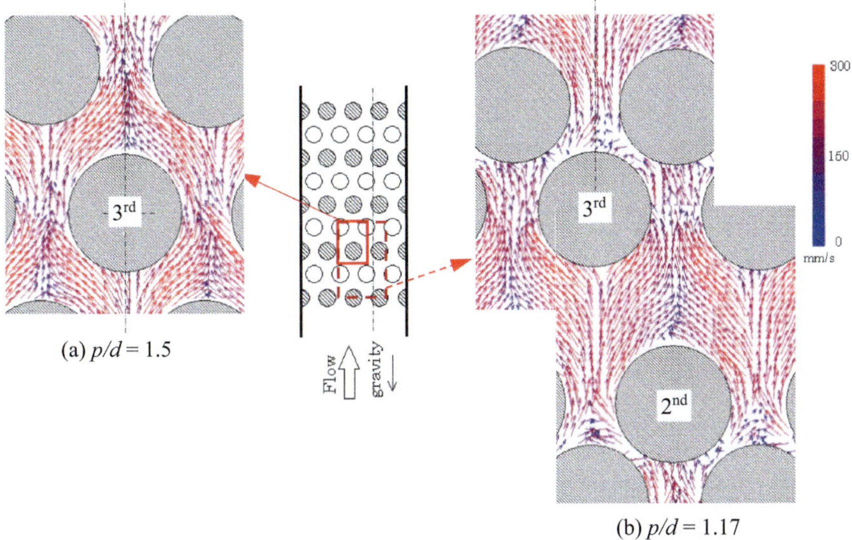

Fig. 57 Visualized flow pattern and velocity vectors of staggered tube bundle, $Re = 1.5 \times 10^3$, and $\alpha = 5\%$ (bubble flow), (tracer: air bubble with mean diameter, approximately 2.5 mm)

Fig. 58 Pressure fluctuation between 2nd and 3rd tubes at staggered tube arrangement ($Re = 5.0 \times 10^3$, $\alpha = 5.0\%$)

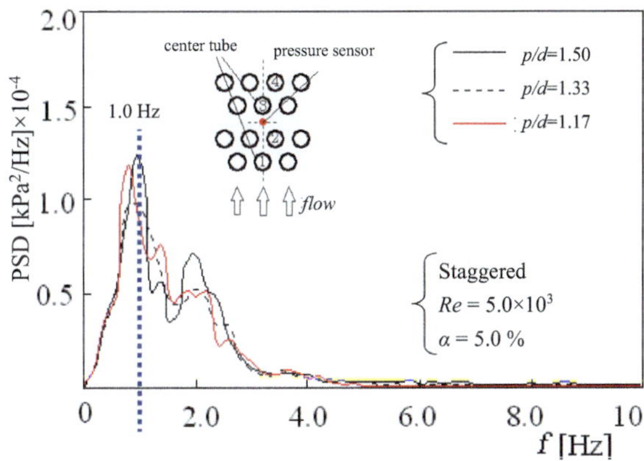

4.7.4 Drag Force of Tube

Figures 59a shows the drag force Dr of the centre tube for the in-line tube arrangement, $Re = 5.0 \times 10^3$, and $\alpha = 0\%$. Dr was obtained from the measurements of pressure distribution around the tube. For equal interval of $p/d = 1.5$ the flow drag of the 1st row of tubes is extremely large due to the impingement of the inlet flow into the flow passage and the drag of the 2nd tube is much smaller than that of the 1st tube because the 2nd tube is in the wake region of the 1st tube. After the 3rd tube Dr increases a little to the downstream. But, for unequal interval of $p/d = 1.33$, 1.17,

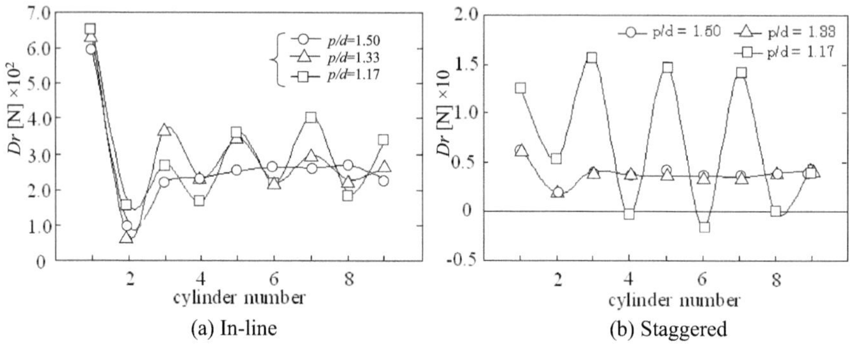

Fig. 59 Drag of center tube ($Re = 5.0 \times 10^3$, $\alpha = 0\%$)

Dr of the even numbered tube is smaller than the tube just before because it is in the wake region of the tube just before. Figure 59b is the results for staggered tube arrangement. Dr for $p/d = 1.17$ fluctuates much larger than the in-line arrangement. For bubble flow of $\alpha = 5.0\%$ of the in-line and staggered arrays, Dr for each p/d showed almost the same change although the value was a little different.

4.7.5 Pressure Loss of Tube Bundle

Figure 60a, b show the pressure loss ΔP_{tl} of the tube bundle of the in-line array for Reynolds number Re and α, respectively. For $\alpha = 0\%$, ΔP_{tl} increases with Re rapidly and is almost the same regardless of p/d. But, for the bubbly flow ΔP_{tl} of $p/d = 1.17$ at $Re = 5 \times 10^3$ (Fig. 60b) becomes a smaller than the others. For example, at $\alpha = 0$ and 10% for $p/d = 1.17$ it is smaller about 10 and 8% than the others, respectively. This means that the flow resistance of the bubbly flow through an unequally spaced tube bundle can be reduced compared to that of an equally spaced tube bank, depending on p/d, Re, bubble mean diameter d_a, and α.

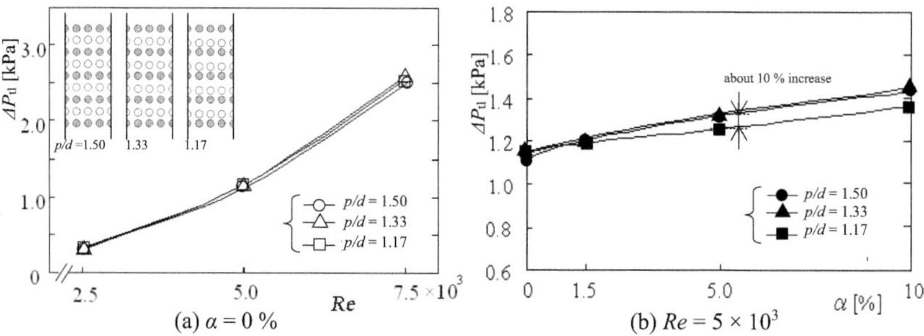

Fig. 60 Pressure loss ΔP_{tl} of tube bundle (In-line)

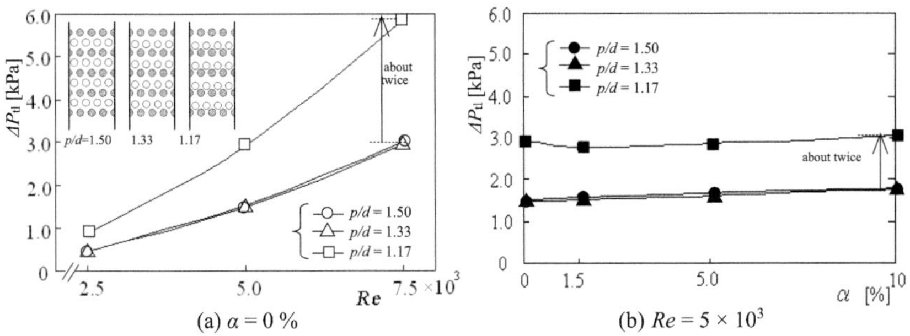

Fig. 61 Pressure loss ΔP_{tl} of tube bundle (Staggered)

Figure 61a, b show ΔP_{tl} of the staggered for Re and α, respectively. ΔP_{tl} increases with Re and the $p/d = 1.17$ is much larger than the others. For example, at $Re = (5.0 \sim 7.5) \times 10^3$ it is about twice of the others. ΔP_{tl} for bubble flow at $Re = 5.0 \times 10^3$ is almost constant for $\alpha = 0 \sim 10.0\%$, but the $p/d = 1.17$ is about twice of the others.

Here, the flow patter and resistance, drag force, of tube bundle with in-line and staggered tube arrangement in a gas–liquid bubbly flow were examined.

4.8 Flow Accelerated Corrosion on Pipe Wall in Downstream of Orifice Nozzle

The orifice is used to measure the flow rate and resistor of the fluid flowing in the pipe, but the wake flow after the orifice is disturbed, and the wall of the tube is thinning. This thinning phenomenon is known as flow-accelerated corrosion, FAC, and also occurs at the bends of the pipeline. The cause of its formation is also related to electromechanical or electrochemical corrosion and has not yet been fully clarified. Here, we examine the phenomenologically relationship between flow-accelerated corrosion and flow disturbances.

4.8.1 Effects of Pressure Fluctuation

In the piping system of power plants, chemical plants, etc. pipe wall thinning due to flow-accelerated corrosion (FAC), liquid droplet impingement (LDI) erosion and cavitation erosion (C/E) is a very serious problem because it leads to serious damage and the eventual destruction of the piping system (Yoneda et al. 2006, 2007, 2009, 2010).

Here, pipe wall thinning due to FAC in the downstream of an orifice nozzle was investigated and an experimental simulation was performed using tap water, a pipe with inner diameter $D = 40.0$ mm, and various types of orifice nozzles (Fig. 62, Shakouchi et al. 2012). The characteristics of FAC, its generation mechanism, and the estimation of the wall thinning rate and the reduction of pressure fluctuation were clarified by experimental analyses. FAC can be caused by complex factors including mechanical erosion and electromechanical or electrochemical corrosion; however, the pipe wall surface that has undergone FAC exhibits a scale-like, flaky pattern (Fig. 62), and hence, through a simulation, we can construe that one of the major reasons could be mechanical erosion. In general, it has been explained that FAC in the downstream of an orifice nozzle is mainly caused by the effects of flow fluctuation, which can be expressed by turbulent kinetic energy, near the wall; moreover, estimation of the thinning rate has also been performed (Yoneda et al. 2006). However, measuring the turbulent kinetic energy is usually slightly troublesome, and hence, we propose the measurement of pressure fluctuation p' (RMS value) instead of the turbulent kinetic energy k because of a relationship between p' and k (Hinze 1959; Kim et al. 1987). It is easy to measure p' on the pipe wall using a semi-conductor-type small pressure sensor. The corrosion pattern on the pipe wall was also examined in the experimental simulation. The simulation clarified that the occurrence of pipe wall thinning mainly depended on the amount of p'. It was also found that the thinning rate can be estimated by p', and that the suppression of p' can be realized by replacing the orifice nozzle with a tapered one having an angle α to the upstream. For example, $\alpha = 30°$ reduced p' by approximately 20% for the Reynolds number $Re = u_{\mathrm{m}}D/v = 5.0 \times 10^4$, where um and v are the mean velocity in a pipe of D and kinematic viscosity of water, respectively, and the contraction area ratio of the orifice nozzle $CR = 0.14$. This suppression of p' leads to a reduction in the wall thinning. The results of the relationship between p' and the thinning rate were also shown.

(a) Test Section and Procedure

A pump was used to send the necessary amount of water through an electromagnetic flow meter into a pipe test section; the pipe test section was made of transparent acrylic resin, had total length of $L = 2650$ mm, and was set vertically. An orifice nozzle was set at $40D$ downstream of the test section. The water passing through the test section flowed back to the water tank and circulated. A portion of the water in the tank was continuously replaced to maintain the water at a constant temperature. The pressure fluctuations in the test section due to pump operation did not appear because of a long pipe line to the test section and by-path line after the pump and inverter control of pump. The mean and fluctuating pressure distributions on the pipe wall in the up-stream and down-stream of the orifice were measured using small pressure holes of diameter 0.8 mm, a water column manometer and a semi-conductor-type small pressure transducer (JTEKT, Japan, PD104SW-100 K). The related pressure and primary resonance frequency of the transducer is 100 kPa and more than 6 kHz, respectively. The measurement time of the pressure fluctuation p' (RMS value) was 2.5 s and the average value of 10 times measurements was used. The visualized flow was achieved by a tracer method using small plastic beads with a diameter approximately 0.1 mm and density of 1.01 and a laser (Ar, 3 W) light sheet; this visualized flow was recorded using a high-speed video camera and processed by

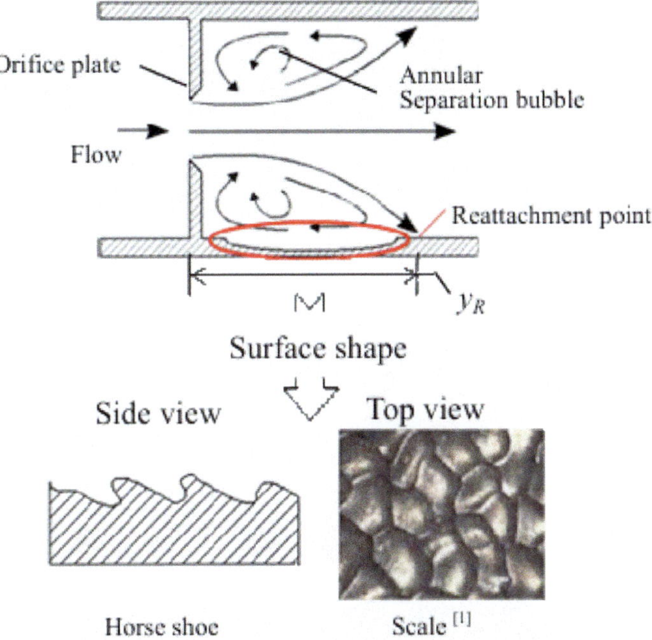

Fig. 62 Flow accelerated corrosion, FAC

particle image velocimetry (PIV) to obtain the velocity vectors of the flow field. The results were used for the detailed understanding of the flow state.

Figure 63 shows a standard orifice nozzle (hereafter referred to as Std) and the x–y coordinate system. The contraction area ratio of the orifice nozzle was $CR = 0.14$, the inner diameter of the nozzle and pipe had constant value of $d = 15.0$ mm and $D = 40.0$ mm, respectively, the plate thickness was $t = 4.0$ mm, and the clearance angle of the outflow side was constant at 45°. A special orifice nozzle used to reduce the flow fluctuation in the downstream of the orifice is shown in Fig. 64. The special orifice nozzle had a taper angle of $\alpha = 20°$ to the down-stream and up-stream, hereafter referred to as Sp20 and Sp20-rev, respectively. Special orifice nozzles having other taper angles such as $\alpha = 30$ and 45° were also used. An orifice nozzle having $CR = 0.36$ was also used.

(b)   Flow Rate Coefficient

Figure 65 shows the flow rate coefficient C of the orifice nozzle with $CR = 0.36$. The pressure loss $(p_1 - p_2)$ was obtained from a pressure difference between the pressures on the pipe wall at 10 mm up-stream and down-stream of the orifice nozzle. The flow rate coefficient is defied by

$$Q = CA_0 \sqrt{\frac{2g}{\rho}(p_1 - p_2)} \tag{53}$$

Fig. 63 Standard orifice, Std ($d = 15.0$ mm)

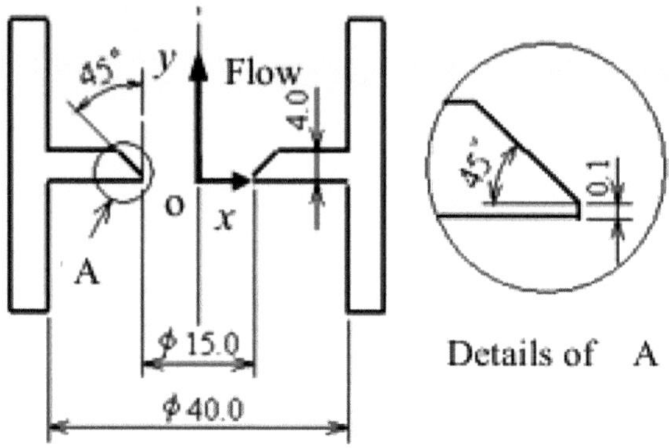

Fig. 64 Special orifice ($\alpha = 20°$)

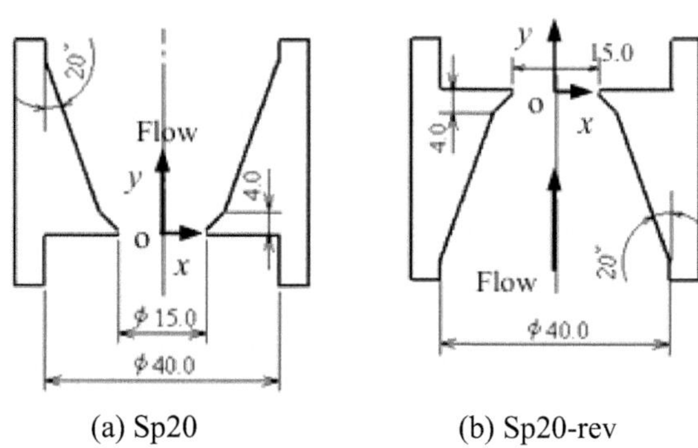

where, A_0: cross sectional area of orifice.

For Std, C had a constant value of 0.686 in the range of $1.0 \times 10^4 < Re < 6.5 \times 10^4$, this result agrees with the specification that the measurement error for the maximum

Fig. 65 Flow rate coefficient, C

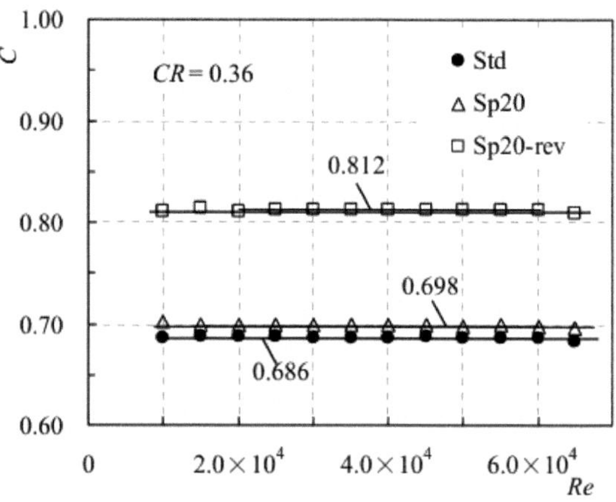

value was within $\pm 0.5\%$. Sp20 and Sp20-rev also showed constant value of C, and they can also be used as orifice flow meters. However, the value of C for Sp20-rev was much larger than that for the other orifice nozzles because its flow passage and orifice entrance had a gradual contraction. Similar results were obtained for the orifice nozzle with $CR = 0.14$.

(c) **Pressure distribution**

Figure 66 shows the pressure distribution at $Re = 5.0 \times 10^4$ by pressure coefficient C_p in the downstream of the orifice with $CR = 0.36$. Parenthetically, the pressure was corrected by the elevation head. In the downstream of the orifice, an annular vortex region with a negative pressure was formed between the pipe walls. As a result, the pressure immediately after the orifice decreased largely and then recovered at the down-stream and decreased with a constant pressure gradient. The reattachment distance was $y_R/D \fallingdotseq 3.6$ for Std and $y_R/D \fallingdotseq 2.0$ for Sp20 and Sp20-rev. The pressure distribution at a distance $y/D \fallingdotseq 4$ from the orifice nozzle agrees well with the Blasius equation.

(d) **Turbulent kinetic energy and flow accelerated corrosion (FAC)**

Yoneda and Morita (2007) performed the experimental and numerical analyses of the turbulent kinetic energy k in the downstream of an orifice nozzle, and they showed that the pipe wall thinning rate TR (Fig. 67) depends on k near the wall and is maximum where k is maximum, $(0.4\sim0.7)\,y_R$.

Figure 67 shows the results of wall thinning in A- and B-piping systems, which are condensing unit and an orifice flow meter, respectively. Although these results are interesting, in general, it is not so easy to measure the distribution of k distribution in the downstream of an orifice nozzle. Parenthetically, it is well known that there is some correlation between k near the pipe wall and the pressure fluctuation p' (Hinze 1959; Kim et al. 1987): hence, the use of p$'$ was preferred over k because it is much easier to measure p' than k.

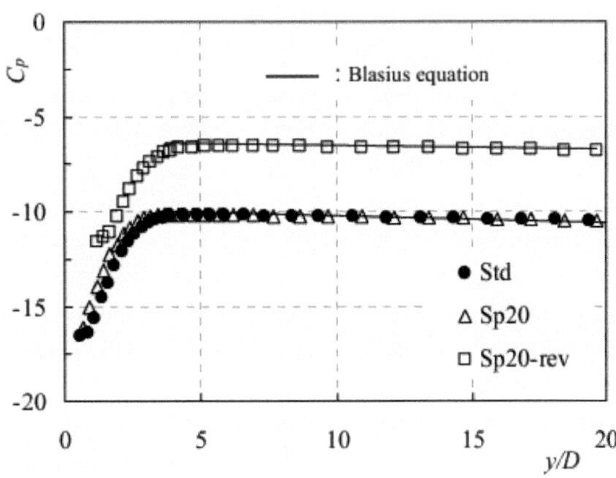

Fig. 66 Pressure distribution, $Re = 5.0 \times 10^4$

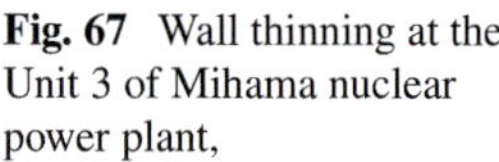

Fig. 67 Wall thinning at the Unit 3 of Mihama nuclear power plant,

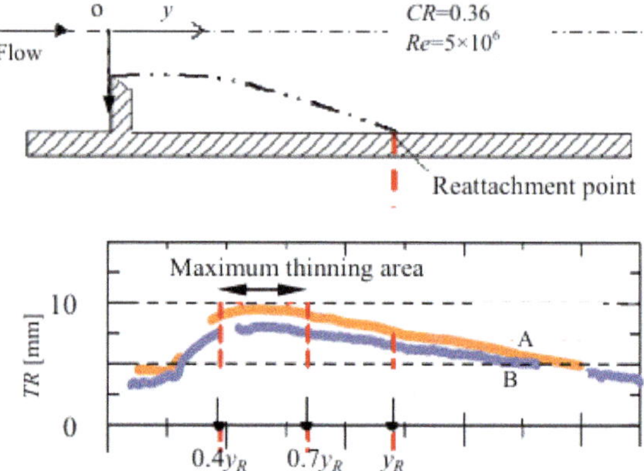

(e) **Pressure fluctuation and FAC**

Figure 68 shows typical examples of the time series of p' on the pipe wall at $y/D = 1.0 \sim 4.0$. The reattachment points for Std and Sp20-rev were at $y/D \fallingdotseq 2.4$ and that for Spd20 was at $y/D \fallingdotseq 2.0$. A larger pressure fluctuation appeared at $y/D = 2.0$, and the pressure fluctuation for Sp20 and Sp20-rev were larger and smaller than that for Std, respectively.

Figure 69 shows a result of the fast Fourier transform (FFT) analysis of the example shown in Fig. 68a. The dominant frequencies of the fluctuation were in approximately between 2 and 10 Hz. The results of the analysis of the examples in Fig. 68b, c were almost same as the result shown in Fig. 69.

Figure 70 shows the pressure fluctuation for Sp20 and Sp20-rev. The maxima $(p'_{RMS})_{max}$ of the pressure fluctuation p'_{RMS} for Sp20 and Sp20-rev was approximately 28% larger and 21% smaller than that for Std, respectively. This implies that the turbulent kinetic energy near the wall around (p'_{RMS}) max for Sp20-rev was less than that for Std. The results for $CR = 0.36$ were almost the same as the above-mentioned results for $CR = 0.14$.

The lower part of Fig. 70 shows the peeling of the grease applied on the inner wall surface of the tube (details are shown in Fig. 74 later). In the surface shape of the grease, small dimples in the form of scales separated from the traces of grease drag in the area where p' is large are observed. However, there is no change in the surface shape of the Sp-rev with a small p'. This scale-like pattern is similar to the surface shape of the piping seen in the thinning phenomenon caused by FAC in the actual machine, and is thought to be caused by the flow of pipe thinning downstream of the orifice. In other words, it can be seen that the thinning phenomenon of the actual piping is strongly affected by the flow (fluctuating pressure, etc.).

Figure 71 shows the relationship between (p', RMS) max and the mean velocity um. The results for the orifice nozzle with $\alpha = 30$ and $45°$ are also shown. (p'_{RMS}) max increases quadratically or cubically with um and it increases with α. Moreover, $(p'_{RMS})_{max}$ for "Sp20–rev" is much smaller than that of the other orifice nozzles at large values of u_m.

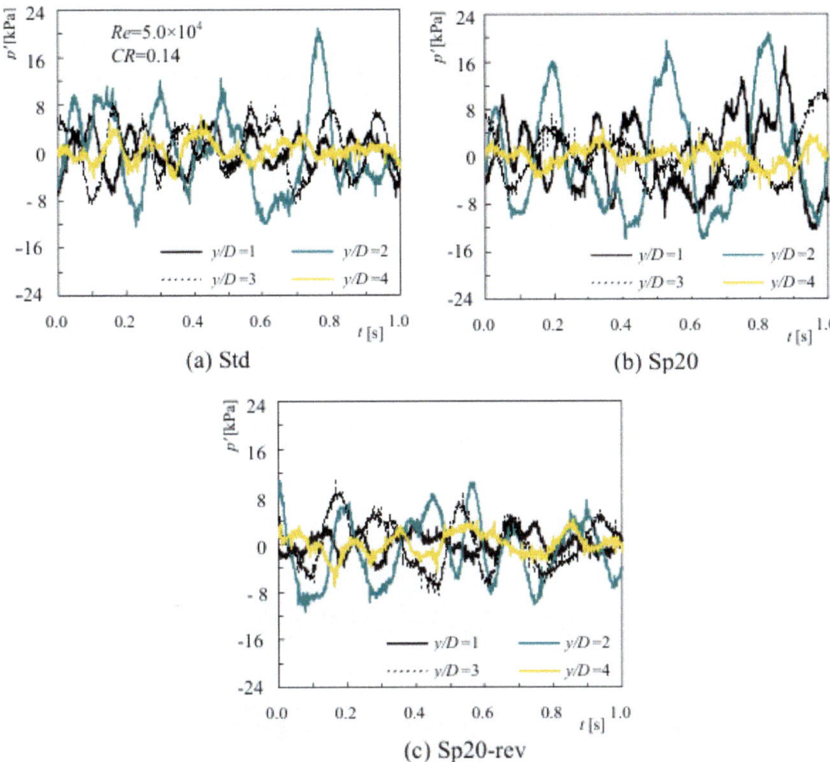

(a) Std

(b) Sp20

(c) Sp20-rev

Fig. 68   Pressure fluctuation ($CR = 0.14$, $Re = 5.0 \times 10^4$)

Fig. 69   FFT analysis (Std, $CR = 0.14$, $Re = 5.0 \times 10^4$)

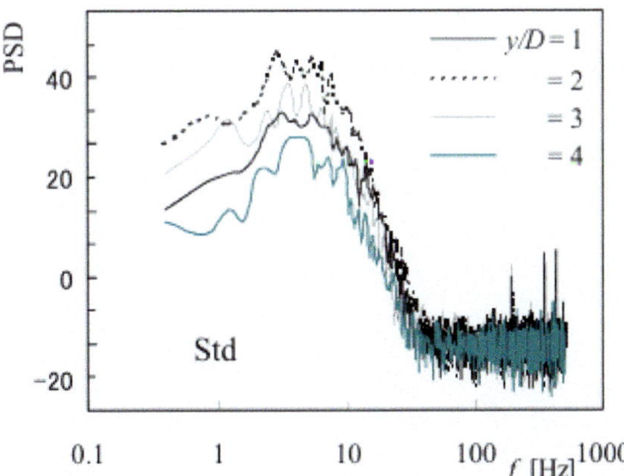

Fig. 70 Pressure fluctuation, p'(RMS) ($CR = 0.14$, $Re = 5.0 \times 10^4$)

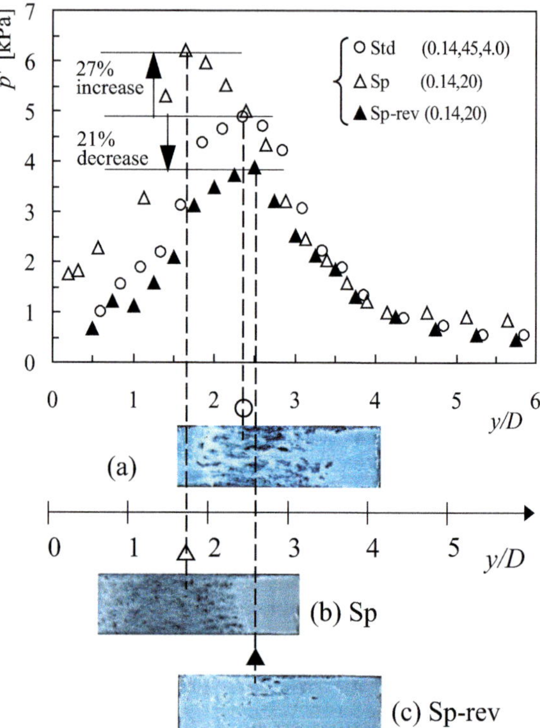

Fig. 71 Maximum pressure fluctuation and u_m ($CR = 0.14$, $Re = 5.0 \times 10^4$)

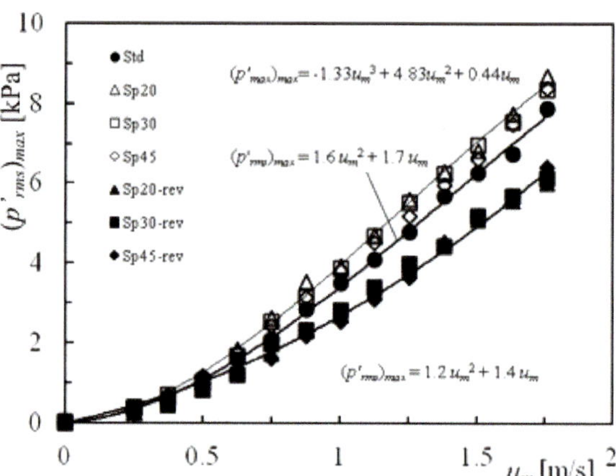

The relation between (p'_RMS) max and um can be well expressed by a quadratic or cubic function obtained from least squares: some of these functions are as follows.

$$\text{For Std:} \quad \left(p'_\mathrm{RMS}\right)_\mathrm{max} = 1.6u_\mathrm{m}^2 + 1.7u_\mathrm{m} \tag{54}$$

$$\text{Sp20}: \quad \left(p'_{\text{RMS}}\right)_{\text{max}} = -1.33u_{\text{m}}^3 + 4.83u_{\text{m}}^2 + 0.44u_{\text{m}} \tag{55}$$

$$\text{Sp20–rev}: \quad \left(p'_{\text{RMS}}\right)_{\text{max}} = 1.2u_{\text{m}}^2 + 1.4u_{\text{m}} \tag{56}$$

(f) Maximum pressure fluctuation $\left[\left(p'_{\text{RMS}}\right)_{\text{max}}\right]$ and FAC

As mentioned before, $(p'_{\text{RMS}})_{\text{max}}$ for Sp20-rev is the smaller than those for Std and Sp20.

Figure 72 shows the mean and fluctuating velocities, i.e., u/um and u'/um, respectively, at $y/D = 0$ for $CR = 0.14$. The working fluid was air, and Re was 1.0×10^4; the velocities were measured by a hot-wire anemometry and single probe. Because the profiles were symmetric, only half of the region along the radius direction x is shown. The mean velocities for Std and Sp20 were almost the same, and Sp20-rev had the lowest mean velocity. This means that the effect of the contraction rate of the flow at the orifice exit was larger for Std and Sp20 than that for Sp20-rev. The fluctuation velocity u'/u_{m} at the shear layer for Sp20-rev was much lower than that for the others. This is the reason why Sp20-rev had the smallest $(p'_{\text{RMS}})_{\text{max}}$.

(g) Wall thinning rate and its estimation by $\left(p'_{\text{RMS}}\right)_{\text{max}}$

Figure 73 shows the relationship between the maximum thinning rate TR_{max} and the mean velocity u_{m} measured by Bignold et al. (1981) and Yoneda and Morita (2006). TR_{max} increases with u_{m}, and their relationship can be well expressed by a quadratic function. It is known that the piping system is different between them; however, TR_{max} and um have some correlation. As mentioned above and from, $(p'_{\text{RMS}})_{\text{max}}$ can be well expressed as a quadratic or cubic function of u_{m}. Then, from this function

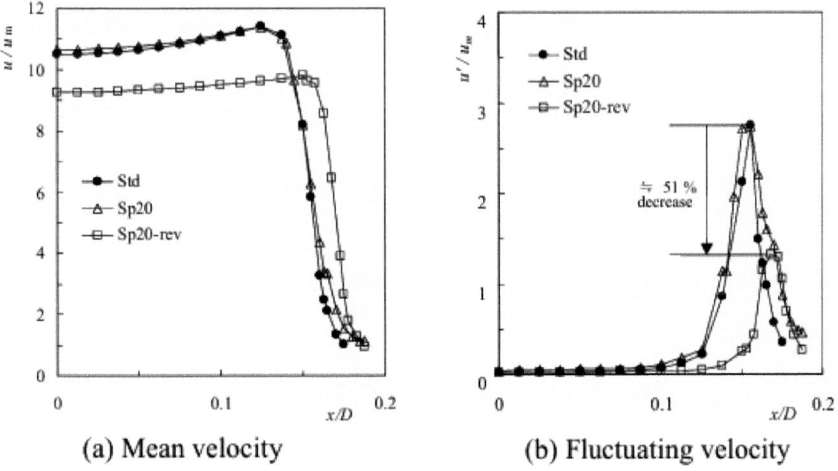

(a) Mean velocity   (b) Fluctuating velocity

Fig. 72 Mean and fluctuating velocities ($Re = 1.0 \times 10^4$)

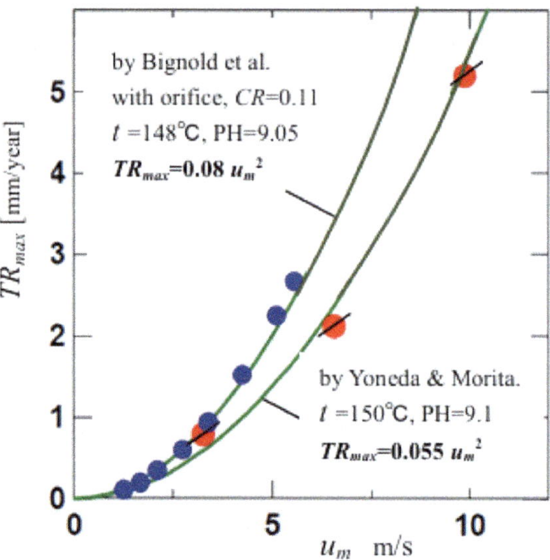

Fig. 73 Wall thinning, maximum thinning rate, $TR_{\max} - u_m$ (Yoneda et al. 2006; Bignold et al. 1981)

and the results shown in Fig. 73, the thinning rate $TR_{\max}$ can be estimated using $(p'_{\text{RMS}})_{\max}$.

$$TR_{\max} \propto \left(p'_{\text{RMS}}\right)_{\max} \tag{57}$$

TR for Sp20-rev will become much smaller than that for Std because p'_{RMS} for Sp20-rev is much smaller than that.

(h) Mechanism of wall thinning and the practical test

It is very difficult to perform an actual test of pipe wall thinning because such a test requires a very huge test plant, long periods of time, and high cost; hence the following simulation test was performed. First, the grease (Shin-Etsu Silicones, Japan, HIVAC-G) was applied with a constant thickness of 0.5 mm on the pipe wall in the downstream of the orifice nozzle. Then, after allowing the water to flow for 15 min at $Re = 5.0 \times 10^4$, the separation of grease was observed and measured using a laser displacement meter (KEYENCE, Japan, LT-8010).

Figure 74a shows some photographs of the change in the surface shape of the grease after the 15 min simulation test for Sp20. In the left-hand side of Fig. 69, the surface separation of the grease to the downstream is shown and the separation range is $y/D \doteqdot 0.65-2.5$. In the case of Std and Sp20-rev, the separation range was $y/D \doteqdot 1.7 \sim 3.6$ and $1.8 \sim 3.4$, respectively. Figure 74a also shows detailed photographs of the change in surface shape at $y/D \doteqdot 1.2$ and 2.2. It is very interesting to note that a flake-like pattern with a size of a few millimeters. This pattern is considered to be caused by a mechanical force due to flow fluctuation, because this pattern was observed after a very short time, (15 min). Figure 74b shows the results for the actual equipment at Mihama nuclear power plant, Japan, Unit 3. The white-colored area is concave in shape. In this case as well, the flake-like pattern with a size of a few millimeters is observed. Both the results are very interesting because they suggest that one of the major reasons for FAC could be the mechanical force caused by flow fluctuation. However, this result certainly needs a more detailed examination.

Here, pipe wall thinning rate TR in the downstream of an orifice nozzle was experimentally analyzed; and it was found that TR_{max} can be estimated using the maximum pressure fluctuation $(p'_{RMS})_{max}$ and that Sp20-rev had the smallest $(p'_{RMS})_{max}$ and TR_{max} as compared to Std. The major results are as follows.

(1) The position of the maximum pressure fluctuation $(p'_{RMS})_{max}$ is almost midway between the orifice and the reattachment point. This position is almost the same as that where pipe wall thinning has maximum TR_{max}.
(2) $(p'_{RMS})_{max}$ for Sp20-rev and Sp20 is approximately 21% smaller and 27% larger than that for Std.
(3) The maximum pipe wall thinning TR_{max} can be estimated using $(p'_{RMS})_{max}$ and the relationship between them is given by $TR_{max} \propto \left(p'_{RMS}\right)_{max}$
(4) Sp20-rev reduces TR to a greater extent as compared to Std.
(5) One of the major reasons for FAC could be mechanical force due to flow fluctuation.

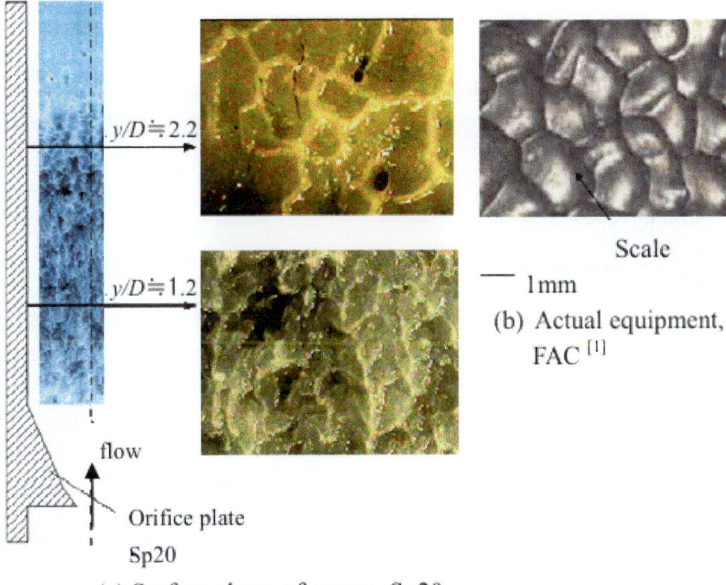

(a) Surface shape of grease, Sp20

Fig. 74 Surface shape of FAC for Sp20 and actual equipment ($CR = 0.14$, $Re = 5.0 \times 10^4$)

4.8.2 Effects of Geometries of Flow Passage on FAC

Here, the pipe wall thinning (Fig. 75) by FAC in the downstream of an orifice, flow meter, was investigated by experimental and numerical analyses. The experimental analysis was performed by using tap water, pipe with inner diameter of $D = 40.0$ mm, and a few types of nozzles (Fig. 76). The characteristics of FAC, the generation mechanism, the estimation of the wall thinning rate and a suppression method were examined by experimental and numerical analyses. FAC is caused by some complex factors including mechanical erosion and electro-mechanical or electro-chemical corrosion, and then it has to be examined from the point of view of material science, electro-chemical and hydrodynamics. The pipe wall surface corroded by FAC exhibits a scale-like or flaky pattern (Figs. 62 and 74), and through the present experimental analysis we can construe that one of the major reasons could be mechanical erosion by flow fluctuation. In general, it has been explained that FAC in the downstream of an orifice is mainly caused by the effects of flow fluctuation, which can be expressed by turbulent kinetic energy k or fluctuated wall shear stress τ_{wx} near the wall; moreover, estimation of the thinning rate has also been performed as shown in Fig. 75 (Yoneda and Morita 2006, 2007a). However, measuring k or τ_{wx} is usually slightly troublesome, and hence, the author proposed the pressure fluctuation p' as a governing parameter instead of k or τ_{wx} with the aid of a relationship between p' and k or τ_{wx} (Shakouchi et al. 2012; Utanohara, et al. 2012; Kiwata et al. 2013; Bignold et al. 1981; Hinze 1959; Kim et al. 1987). It is easy to measure p' on the pipe wall using a semi-conductor-type small pressure sensor. It was also found that the thinning rate TR can be estimated by p', and that the suppression of p' could be realized by replacing the orifice nozzle to that with a tapered angle to the upstream or by replacing the downstream pipe of orifice nozzle to a smaller one. The suppression of p' leads to a reduction of the wall thinning. The results of the relation between p' and the thinning rate were also examined.

(a) Test section and procedure

The pipe test section was made of transparent acrylic resin whose total length of $L = 2,650$ mm and was set vertically. An orifice nozzle was set at $40D$ downstream of the inlet of the test section. The water passing the test section flowed back to the water tank and was circulated. A portion of the water in the tank was continuously replaced in order to maintain the water at a constant temperature. The pressure fluctuation in the test section due to pump operation did not appear because of a pipe line long enough to the test section, by-path line downstream of the pump and inverter control of pump. The mean and fluctuating pressure distributions on the pipe wall in the up-stream and down-stream of the orifice were measured by small pressure holes of a diameter 0.8 mm, a water column manometer and a semi-conductor type small pressure transducer (JTEKT, Japan, PD104SW-100 K). The related pressure and primary resonance frequency of the transducer is 100 kPa and more than 6 kHz, respectively. The measurement time of the pressure fluctuation p' was 2.5 s and

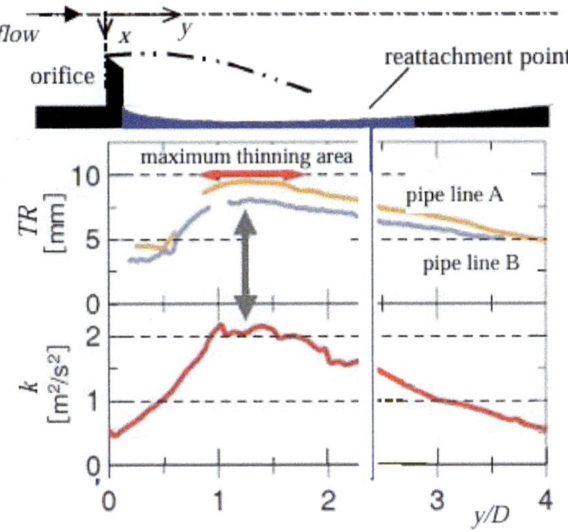

Fig. 75 Wall thinning, *TR*, and turbulent kinetic energy, *k* (Yoneda et al. 2006)

sampling frequency was 1.0 kHz, and the mean value of 10 times measurements was used. The dominant frequency of p' was in about 6 ~ 8 Hz in the present study.

Figure 76a shows a standard orifice (made according to JIS Z 8762, hereafter referred to as Std) and the *x*–*y* coordinate system. The contraction area ratio of the orifice was $CR = 0.36$, the inner diameter of the nozzle and pipe had constant value of d = 24.0 mm and D = 40.0 mm, respectively, the plate thickness was $t = 4.0$ mm, and the clearance angle of the outflow side was constant at 45°. The Std-rev orifice which reversed the direction of the orifice was also used to reduce the flow fluctuation in the downstream of the orifice. Furthermore, the nozzle called Pipe nozzle and the nozzle with a taper called Tapered pipe nozzle shown in Fig. 76b and Std nozzle with the downstream pipe of $D_p < D = 40.0$ mm called Std-D_p shown in Fig. 76c were also used.

(b) Flow rate coefficient

Figure 77 shows the flow rate coefficient *C* of the orifice with $CR = 0.36$. The pressure loss ($p_1 - p_2$) was obtained as an estimated pressure difference just before and after the orifice. p_1 and p_2 was estimated by the experimental result of linear pressure distribution following the Blasius relation in the upstream and downstream of orifice, respectively (Kinoshita 2014). For Std, C had a constant value of 0.687 in the range of $1.0 \times 10^4 < Re < 6.5 \times 10^4$, this result agrees well with the specification of the JIS with less than ± 0.5% of errors for the maximum. Std-rev, Pipe-24.0 and Std-$D_p = 35.2$ mm are also shown a constant value of *C*, and they can use as orifice flow meter. However, the value of *C* of Std-rev and Pipe-24.0 were much larger than that of Std because the outflow from the orifice is more stable. A large *C* means a

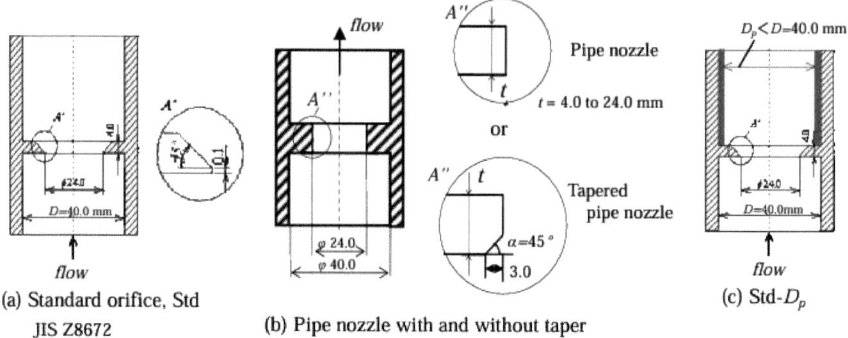

Fig. 76 Configurations of Orifice and nozzle

small pressure loss under a same flow rate, and then a more highly precise pressure measurement is necessary.

(c) **Turbulent kinetic energy, shear stress and FAC**

Yoneda and Morita (2006, 2007a) performed the experimental and numerical analyses for the turbulent kinetic energy k in the downstream flow of an orifice nozzle, and they showed that the pipe wall thinning rate TR (Fig. 75) depends on k near the wall and is maximum at where k is maximum, $y = (0.4 \sim 0.7)\, y_R$, where y_R: reattachment distance. Recently, Utanohara, et al. (2012) and Kiwata, et al. (2013) showed that wall shear stress τ also has some relation with TR.

Figure 75 shows the results of wall thinning in A- and B-piping systems of Mihama nuclear power plant, Japan, plant 3, which are condensing unit and an orifice flow meter, respectively. Although their results are interesting, in general, it is not so easy to measure the distribution of k or τ in the downstream of an orifice. It is well known that there is some correlation between k or τ near the pipe wall and the pressure

Fig. 77 Flow rate coefficient, C ($CR = 0.36$)

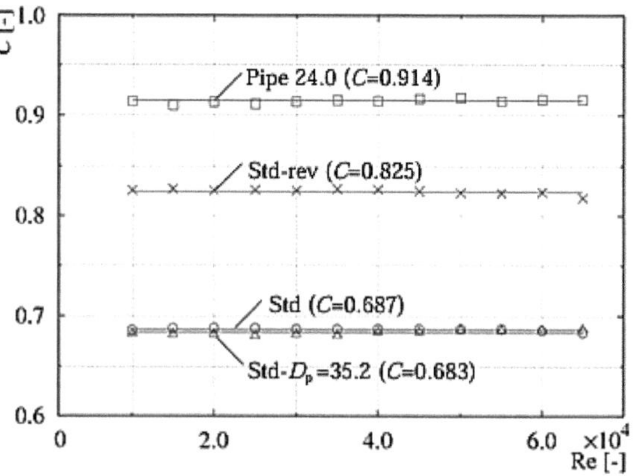

fluctuation p' (Shakouchi et al. 2012; Utanohara et al. 2012; Kiwata et al. 2013; Bignold et al. 1981; Hinze 1959; Kim et al. 1987), hence, the use of p' was preferred over k or τ because it is much easier to measure p' than k or τ.

(d) Pressure fluctuation and wall thinning

Bignold et al. (1981) and Yoneda et al. (2006, 2007a) showed the relation between the maximum wall thinning rate $TR_{\max}$ and the mean velocity u_{m} in the pipe (Fig. 73) by

$$TR_{\max} \propto u_{\mathrm{m}}^2 \tag{58}$$

where, $TR_{\max}$: [mm/year], u_{m}: [m/s].

The author showed the relation between the maximum pressure fluctuation $p'_{\max}$ and u_{m} (Fig. 78) by

$$p'_{\max} \propto u_{\mathrm{m}}^2 \tag{59}$$

From these equations, it is known that $TR_{\max}$ can be expressed by $p'_{\max}$.

Figure 79 shows the pressure fluctuation p' (RMS value) of Pipe nozzle (Fig. 76b). The distribution of p' in the downstream of Pipe nozzle almost corresponds to the wall thinning (Fig. 75). p' and the maximum pressure fluctuation $p'_{\max}$ of Pipe-24.0, Pipe with $t = 24.0$ mm, takes much smaller values than those of Std and Pipe-8.0.

Figure 80 shows $p'_{\max}$ of Pipe nozzle. $p'_{\max}$ decreases rapidly with decreasing the nozzle thickness t, and it can be expressed by

$$p'_{\max} = -5.22(t/D)^3 + 9.32(t/D)^2 - 5.53(t/D) + 1.82 \tag{60}$$

where, $p'_{\max}$: [kPa].

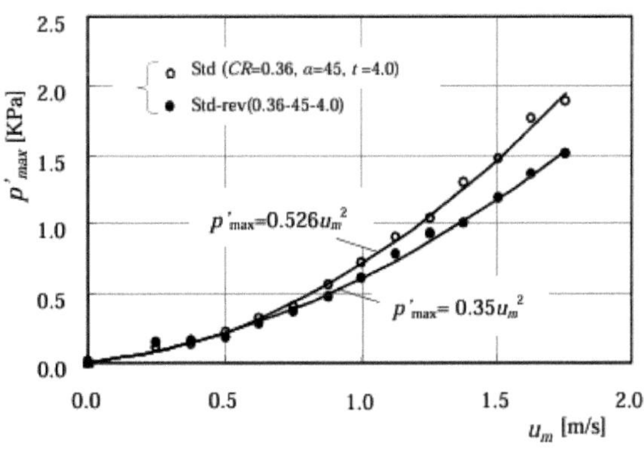

Fig. 78 $p'_{\max}$ and u_{m} ($CR = 0.36$).

Fig. 79 Pressure fluctuation, p' (Pipe nozzle, $CR = 0.36$, $t = 8.0 \sim 24.0$ mm)

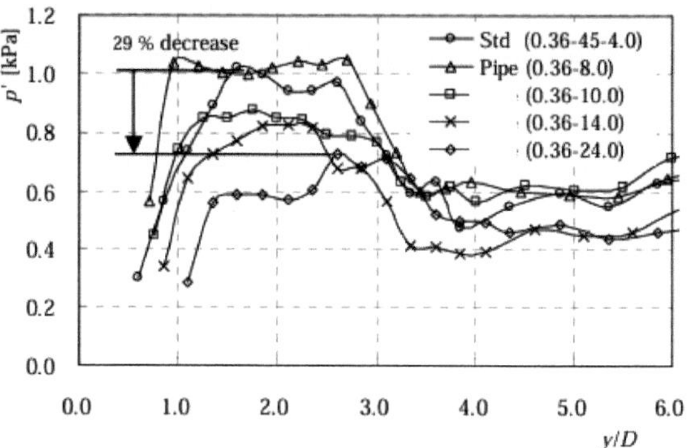

Fig. 80 Maximum pressure fluctuation, p'_{max} (Pipe nozzle)

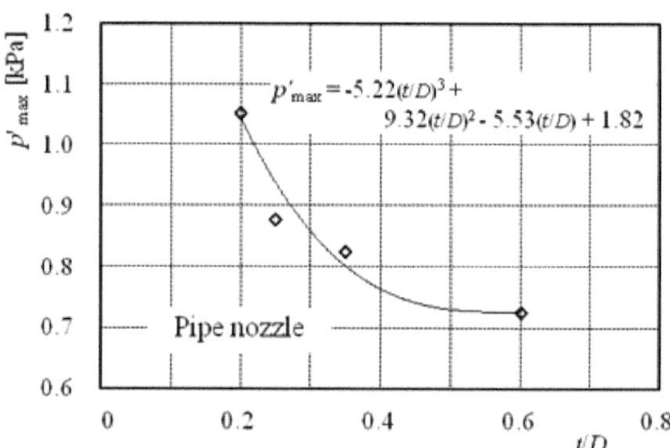

For example, p'_{max} of Pipe-24.0 is approximately 29% smaller than that of Std. This depends on that the flow separated from leading edge of the Pipe nozzle reattaches on the inner wall of the Pipe nozzle and then flows out smoothly from the rear edge of the nozzle.

Figure 81 shows the pressure fluctuation p' in the downstream of orifice of Std-D_p. p' of Std-D_p takes a smaller value than that of Std and decreases approximately with decreasing the pipe diameter, D_p.

Figure 82 shows p'_{max} of Std-D_p. p'_{max} decreases with decreasing D_p, and it can be expressed by

$$p'_{max} = 0.003D_p^3 - 0.369D_p^2 + 13.48D_p - 1.63 \tag{61}$$

where, p'_{max}: [kPa], D_p: [mm].

For example, p'_{max} of Std-$D_p = 35.2$ mm is approximately 16% smaller than that of Std. Because, as decreasing D_p the size of the vortex region in the down-stream of the orifice decreases and the restriction of the flow increases.

(e) Numerical Analysis

Fig. 81 Pressure fluctuation, p' (Std-D_p)

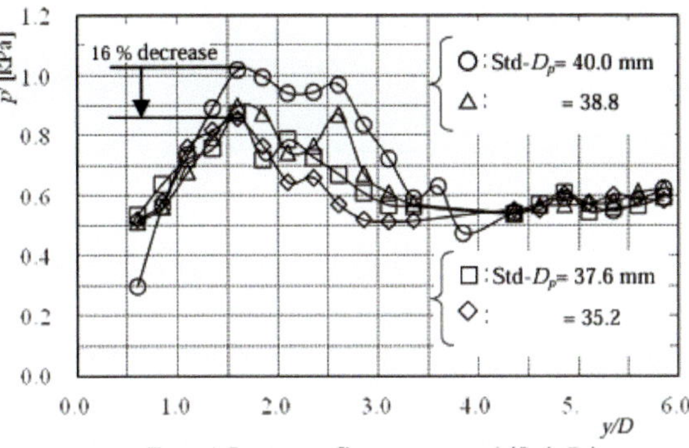

Fig. 82 Maximum pressure fluctuation, p'_{max} (Std-Dp)

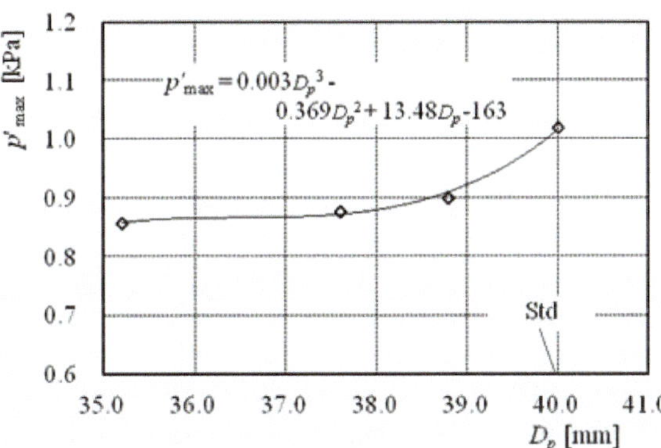

In order to estimate the flow pattern and the distribution of turbulent kinetic energy for various flow passages, numerical analysis by CFD-2000 using k–ε turbulent model was conducted.

Figure 83a, b shows the velocity vector and the contour map of turbulent kinetic energy $k\left[= \left(u'^2 + v'^2 + w'^2\right)/2\right]$ obtained by numerical analysis for Std of $CR = 0.36$, $Re = 5.8 \times 10^4$, respectively. The flow passing through the orifice generates a shear layer with a large velocity gradient and a large turbulent kinetic energy.

Figure 84a shows the velocity vector of Std-rev. Since Std-rev has a tapered portion on the upstream side of the orifice edge the stream flows in smoothly and the turbulence level appeared in Fig. 84b decreases comparing with Std [Fig. 83b].

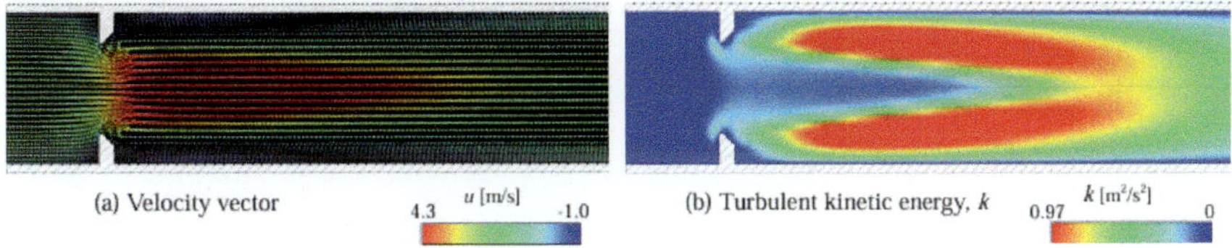

Fig. 83 Std (Numerical analysis, $CR = 0.36$, $Re = 5.8 \times 10^4$)

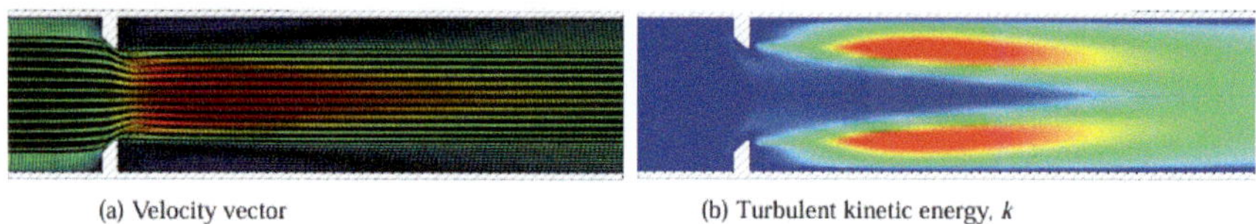

(a) Velocity vector   (b) Turbulent kinetic energy, k

Fig. 84 Std-rev

Figure 85 shows the k of Pipe nozzle with the thickness $t = 8.0, 24.0$ mm. k of Pipe-8 is smaller.

than that of Std, and as increasing t ($= 24$ mm) k decreases more.

These agreed well with experimental results (Kinoshita 2014). In order to decrease k more, Tapered pipe was tried which is shown in Fig. 86. k of Tapered pipe is smaller than that of Pipe nozzle or Std, and increasing t ($= 24.0$ mm) it becomes much smaller.

Figure 87a, b shows the contour map of k of Std-$D_p = 37.6, 35.2$ mm, respectively. The pipe diameter D_p of orifice downstream is smaller than that $D = 40.0$ mm of upstream. Turbulent kinetic energy k of Std-D_p decrease with decreasing inner diameter, D_p.

Figure 88 shows k and C of Std, Std-rev, Pipe and Tapered pipe. Std takes a maximum k of $k_{max} = 1.373$ [m^2/s^2], and k becomes smaller in order of Std-rev, Pipe, Tapered pipe, but C of Std is a minimum and becomes larger in the same order. Tapered pipe-16.0 and -24.0 take a minimum k of 0.78 [m^2/s^2], which is smaller approximately 43% than that of Std, and a maximum C of 0.977. Tapered pipe is effective to reduce k, but to measure a same flow rate an accurate pressure loss ($p_1 - p_2$) measurement is needed because C is larger than that of Std. In order to use an orifice as a flow meter and to reduce wall thinning, it is desired to have an almost same value of C as that for Std and a smaller k than that of Std.

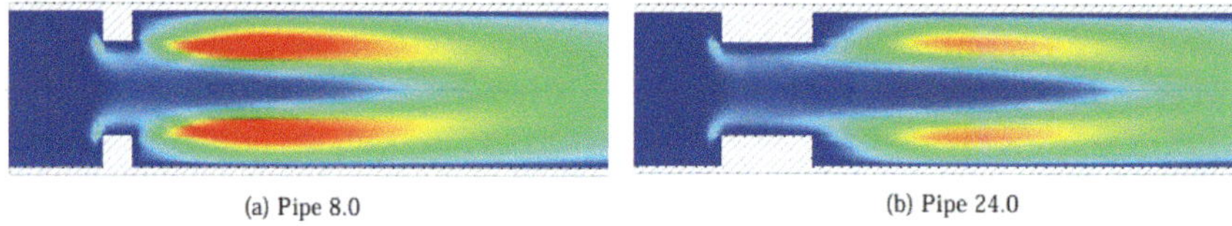

(a) Pipe 8.0   (b) Pipe 24.0

Fig. 85 k of Pipe-8.0, −24.0

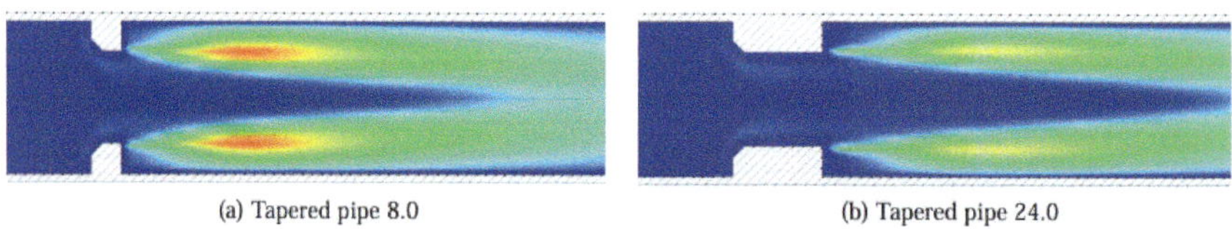

(a) Tapered pipe 8.0   (b) Tapered pipe 24.0

Fig. 86 k of Tapered pipe-8.0, −24.0

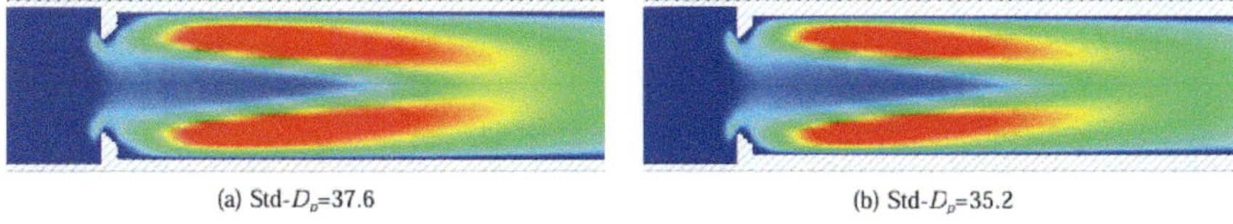

Fig. 87 k of Std-D_p = 37.6, 35.2 mm ($< D$).

Fig. 88 C and k of various orifice and nozzle (Numerical analysis, $CR = 0.36$, $Re = 5.8 \times 10^4$)

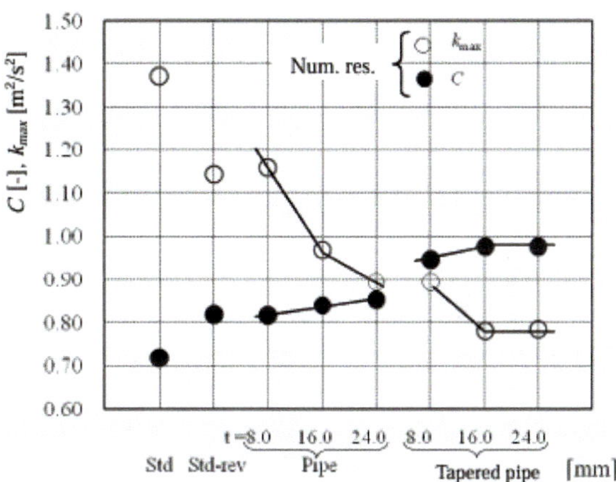

Figure 89 shows C and k of Std-D_p. In the figure, the experimental results of C are also shown. C is almost constant and numerical results take a little larger value than those of the experimental results, but the numerical analysis of C can express the experimental results well. k decreases with decreasing D_p, where $t' = (D-D_\mathrm{p})/2$. For example, k of Std-$D_\mathrm{p} = 35.2$ mm decreases approximately 9.5% than that of Std. This is approximately same as the experimental results of p'. In consequence, in this study it is thought that Std-D_p, for example $D_\mathrm{p} = 35.2$ mm, is the most suitable orifice in order to use as a flow meter and to reduce wall thinning because it has almost same flow rate coefficient C of Std and a smaller pressure fluctuation p' or turbulent kinetic energy k than that of Std.

Here, the pipe wall thinning in the downstream of an orifice nozzle, FAC, was examined.

(1) The wall thinning rate TR in the downstream of an orifice nozzle can be estimated by the pressure fluctuation p' on the pipe wall. Where, the decrease of p' or turbulent kinetic energy k leads to the decrease of TR.

(2) Pipe- and Tapered pipe-nozzles can reduce p', for example Tapered pipe with $t = 16.0$ mm can decrease it approximately 43% than that of Std. However, the flow rate coefficient C is larger than that of Std.

(3) Std-D_p has an almost same C with Std and p' or p'_{max} is smaller than that of Std, for example p'_{max} of Std-$D_\mathrm{p} = 35.2$ mm is approximately 16% smaller than that of Std.

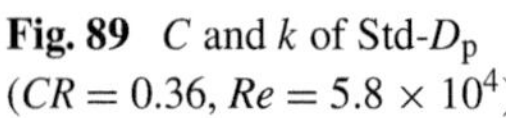

Fig. 89 C and k of Std-D_p
($CR = 0.36$, $Re = 5.8 \times 10^4$)

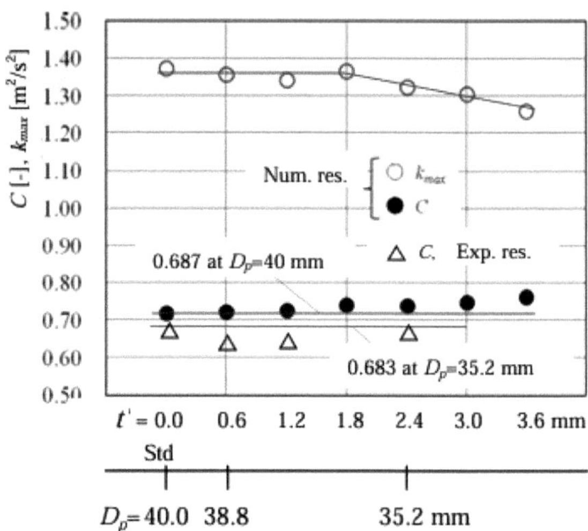

References

Artana G, Sosa R, Moreau E, Touchard G (2003) Control of the near-wake flow around a circular cylinder with electrohydrodynamic actuators. Exp Fluids 35:580–588 (2003). https://doi.org/10.1007/s00348-003-0704-z

Bignold GJ, Whalley CH, Garbett K (1981) Erosion corrosion of mild steel in 138 J Fluid Sci Technol 7(1) (2012); Ammoniated water. In: Proceedings of 8th international congress metallic corrosion, pp 1548–1554

Blevins RD (1984) Applied fluid dynamics handbook. Van Nostrand Reinhold Co. ISBN 0-442-21296-8

Buyruk E (1999) Heat transfer and flow structures around circular cylinders in cross flow. J Eng Environ Sci 23:299–315

Cheong KH, Iwaki C, Monji H, Matsui G (2003) Flow characteristics of bubbly flow crossing tube bundle. Trans JSMF-2003 185–186

Churchill SW, Bernstein M (1977) A correlating equation for forced convection from gases and liquids to a circular cylinder in cross flow. J Heat Transfer Trans ASME 94:300–3006. https://doi.org/10.1115/1.3450685

Hara F, Ohtani L (1982a) Vibration of circular cylinder in cross two-phase flow—1st Rep. Karman vortex shedding and pressure fluctuations. Trans JSME 48(431) C:962–971

Hara F, Ohtani L (1982b) Two-phase cross-flow induced vibrations in a cylindrical systems—2nd Rep. Characteristics of unsteady lift and drag forces. Trans JSME 48(433) C:1371–1379

Hinze JO (1959) Turbulence. Mc-Graw-Hill

Homann F (1936) Einfluß großer Zähigkeit bei Strömung um Zylinder. Forschg Ing-Wes 7:1–10

Hulin J-P, Fierfort C, Coudol R (1982) Experimental study of vortex emission behind bluff obstacles in a gas liquid vertical two-phase flow. Int J Multiphase Flow 8(5):475–490. https://doi.org/10.1016/0301-9322(82)90019-2

Inoue A, Kozawa Y, Yokosawa M, Aoki S (1986) Studies on two-phase cross flow, Part 1; flow characteristics around a cylinder. Int J Multiphase Flow 12(2):149–167

Joo Y, Dhir VK (1994) An experimental study on a single tube and on a tube in an array under two-phase cross flow. Int J Multi-Phase Flow 20(6):1009–1019

Kim J, Moin P, Moser R (1987) Turbulence statistics in fully developed channel flow at low Reynolds number. J Fluid Mech 177:133–166

Kinoshita K (2014) Flow accelerated corrosion on the pipe wall in the downstream of orifice, and its prediction and suppression, Master's thesis of Graduate School of Eng., Mie University, Japan

Kiwata T, Utanohara Y, Nakamura A, Kimura S, Komatsu N, Yamada K, Sugawara Y, Nakamichi J (n.d.) Time-averaged and fluctuated wall shear stresses downstream from an orifice in a circular pipe. Trans JSME 79(799 B):258–262

Nakayama Y (1989) Fluid mechanics. Yokendo Ltd. Publisher, Japan. ISBN-13 978-4-8425-0478-0 C3053, ISBN-104-8425-0478-1

Noghreehkar GR, Kawaji M, Chen AMC (1999) Investigation of two-phase flow regimes in tube bundles under cross-flow conditions. Int J Multiphase Flow 25(5):857–874. https://doi.org/10.1016/S0301-9322(98)00075-5

Pettigrew MJ (1994) Two-phase flow-induced vibration: an overview. J Pressure Vessel Technol ASME 116(3):233–253. https://doi.org/10.1115/1.2929583

Pettigrew M, Taylor CE, Kim BS (1989) Vibration of tube bundle in Two-phase cross-flow, Part 1—hydrodynamic mass and damping. Transf ASME J Pressure Vessel Technol 111(4):466–477. https://doi.org/10.1115/1.3265705

Roshko A (1955) The wake and drag of bluff bodies. J Aeronaut Sci 22:124–132

Schlichting H (1968) Boundary layer theory, 6th edn. McGraw Hill

Shakouchi T, Tian D, Ida T (2001) Behavior of wake flow behind obstacle in a gas-liquid vertical two-phase flow—effects of blockage ratio. 67(654):377–383. https://doi.org/10.1299/kikaib.67.377

Shakouchi T, Tian D, Ida T (2002) Behavior of vertical upward gas-liquid two-phase flow past obstacle in rectangular channel–effect of blockage ratio. JSME Int J Ser B 45(3):686–693. https://doi.org/10.1299/jsmeb.45.686

Shakouchi T, Shimizu T, Tsujimoto K, Ando T (2010) Flow and heat transfer of gas-liquid two-phase flow through a single raw tube bundle—effects of unequal space pipe pitch arrangement. J Multiphase Flow JSMF 23(5):555–561. https://doi.org/10.3811/jjmf.23.555

Shakouchi T, Suzuki T, Kugimoto M, Tsujimoto K, Ando T (2012) Effects of pressure fluctuation on flow accelerated corrosion in the downstream of orifice nozzle. J Fluid Sci Technol JSME 7(1):129–138

Shakouchi T, Kitamura T, Tsujimoto K, Ando T (2017) Interaction between water or air-water bubble flow and tube bundle—effects of arrangement of tube bundle and void fraction. In: Fluid-structure-sound interactions and control. Springer, pp 111–116. ISBN 9789811075414

Shimizu T (2007) Flow and heat transfer characteristics of gas-liquid two-phase flow passing through single raw tube bundle with non-equal spaces, Thesis of Master Course. Mie University, Japan

Utanohara Y, Nagaya Y, Nakamura A, Murse M (2012) Influence of local flow field on flow accelerated corrosion downstream from an orifice. J Power Energy Syst JSME 6(1):18–33

Willmarth WW (1957) Pressure fluctuations beneath turbulent boundary layers. Ann Rev Fluid Mech 5:13–23

Yokosawa M, Kozawa Y, Inoue A, Aoki S (1986) Studies on two-phase cross flow, Part III; Characteristics on unstable flow behaviour. Int J Multiphase Flow 12(2):185–202

Quantitative evaluation of effective factors on flow accelerated corrosion—Part 1, Corrosion of hydraulic factors and thinning rate. Nuclear Res. Lab., CRIEPI, Rep., L06007, pp 1–22 (in Japanese)

Yoneda K, Morita R (2006) Investigation of flow characteristics affecting on pipe wall thinning—Part 1, Turbulent properties at orifice downstream in single-phase flow. Nuclear Research Lab., Central Res. Inst. of Electric Power Industry, CRIEPI, Rep. L05007, pp 1–23

Yoneda K, Morita R, Satake M, Fujikawa K (2009) Quantitative evaluation of effective factors on flow accelerated corrosion (Part 3)—proposition of wall thinning prediction model. Nuclear Res. Lab., CRIEPI, Rep. L08016, pp 1–33 (in Japanese)

Yoneda K, Morita R, Fujikawa K (2010) Quantitative evaluation of effective factors on flow accelerated corrosion—Part 4, evaluation of wall thinning profile in piping elements. Nuclear Res. Lab., CRIEPI, Rep. L09006, pp 1–50 (in Japanese)

Zdravkovich MM (1997) Flow around circular cylinder, vol 1. Oxford University Press